MOSQUITOES
AND THEIR CONTROL

MOSQUITOES
AND THEIR CONTROL

Norbert Becker

German Mosquito Control Association (KABS)
Waldsee, Germany

and

Dušan Petrić

University of Novi Sad
Novi Sad, Yugoslavia

Marija Zgomba

University of Novi Sad
Novi Sad, Yugoslavia

Clive Boase

The Pest Management Consultancy
Haverhill, UK

Christine Dahl

Uppsala University
Uppsala, Sweden

John Lane

London School of Hygiene and Tropical Medicine (retired)
London, England

Achim Kaiser

German Mosquito Control Association (KABS)
Waldsee, Germany

Kluwer Academic / Plenum Publishers
New York, Boston, Dordrecht, London, Moscow

ISBN 0-306-47360-7

©2003 Kluwer Academic / Plenum Publishers, New York
233 Spring Street, New York, New York 10013

http://www.wkap.nl/

10 9 8 7 6 5 4 3 2 1

A C.I.P. record for this book is available from the Library of Congress

Printed in the United States of America

Foreword

Mankind has been plagued by mosquitoes as nuisances and as vectors of mosquito-borne diseases for centuries, resulting in inestimable economic losses and indeterminable human suffering. Mosquitoes transmit some of the deadliest diseases known to man—malaria and yellow fever—as well as dengue, encephalitis, filariasis and a hundred or so other maladies. In spite of decades of mosquito control efforts throughout affected regions worldwide, this scourge has not left us and our present-day overpopulated, jet-linked world remains on the edge of resurgence and outbreaks of old and new mosquito-borne disease epidemics.

Ninety-two mosquito species of more than 3200 recorded worldwide, traverse the European continent. In Europe malaria was eradicated ca. 50 years ago. Current trends in re-emerging mosquito-borne infectious diseases, exemplified by increasing numbers of imported malaria cases and recent outbreaks of West Nile Fever (WNF) virus, however, have given rise to growing public concern. Seasonal outbreaks of nuisance mosquito populations, which plague ecologically sensitive tourist and urban areas cause significant economic damage and constitute the major problem in Europe.

This book is the product of a monumental task of collecting, processing and organising vital information on the mosquito populations of Europe. It presents a multitude of information on the bionomics, systematics, ecology and control of both pestiferous (nuisance) and disease vectors in an easily readable style providing practical guidance and important information to both professional and layman alike. It is conveniently divided into four parts containing sixteen chapters. Part one deals with general information on systematics, morphology and biology of mosquitoes, their medical significance and a very useful subchapter on mosquito research that includes important techniques and technologies utilised in mosquito surveys for sampling eggs, larvae and adult mosquito populations. Part two contains keys to identification of larval and adult mosquito genera. Part three gives a very detailed and instructive account of the morphology, ecology and distribution of all 92 European species included in 8 genera: *Anopheles, Aedes, Ochlerotatus, Culex, Culiseta, Coquillettidia, Orthopodomyia* and *Uranotaenia*. This part is extremely important for species identification in any successful mosquito control or mosquito-borne disease surveillance programme.

Part four dealing with control of mosquitoes is well ordered in seven sections: biological, chemical and physical control as well as personal protection, integrated pest management, implementation of survey methodologies and management. This information is particularly necessary for a comprehensive approach to both studying and controlling mosquitoes. This part distinctly demonstrates the tremendous experience that the authors possess over decades in mosquito biology and control. The section on biological control contains important information on predators, parasites and pathogens is of special significance in light of the increasing interest in replacing detrimental chemical pesticides with environmentally friendly integrated biological control (IBC) technologies.

The authors of this book collectively possess a vast amount of knowledge and experience in the relevant topics and they have meticulously gathered and incorporated the experience and knowledge of others to compose the first and only comprehensive treatise on the subject of European mosquitoes.

This book is a valuable tool for vector ecologists, entomologists, and all those involved with mosquito control, biology, ecology and systematics in Europe. It's primary beneficiaries will be the students, scientists and professionals dealing on a day-to-day basis with mosquitoes and their control. Society as a whole stands to gain from improved, environmentally responsible mosquito management programmes designed on the basis of the broader understanding of mosquitoes and their control provided in this enlightening book.

Yoel Margalith, Ph.D
Tyler Prize Laureate
Center for Biological Control
Ben Gurion University of the Negev, Israel

Preface

Throughout the world, mosquitoes interact with man in many different ways. However, despite the very extensive research on these interactions, there remains much that is not yet fully understood. The attempt by the authors of this volume, is to highlight the importance of a basic knowledge of mosquito biology as a basis for successful control operations.

When compared to the Curculionidae, the largest family in the animal kingdom, with 35 000 known species, the Culicidae, numbering more than 3200 species, could be ranked as a family of only small to moderate size. Even though yield losses caused by weevils could be estimated in billions of dollars, mosquitoes are able to carry many lethal diseases in their bodies. By the time you have read this Preface (5 min.) ten human lives will have been lost due to plasmodian infections. Apart from being the well known vectors of life-threatening diseases, in some parts of the world mosquitoes may also occur in enormous numbers thus causing a significant reduction in human life quality and serious economic damage for instance in livestock.

Ramsdale and Snow (1999) published a list of currently recognised European mosquito taxa with synonyms. They included 98 species in 7 genera and 17 subgenera. The authors of this volume, in deciding upon the species which should be covered, have come to the conclusion that the pertinent species are those which were recorded on more than one occasion at least, where type specimens have been deposited, identification material has been available, and where the position regarding the validity of the species is satisfactory. In addition to these requirements, information about geographical distribution with substantiating references was considered. Accordingly to this, 92 species and subspecies belonging to 8 genera and 18 subgenera are described and included in the keys.

The following species or subspecies previously reported from Europe are not included in the list: *Ochlerotatus gilcolladoi* (Villa, Rodrigez, and Llera, 1985), was named from a form from central Spain differing in certain features of larval chaetotaxy from both *Oc. echinus* and *Oc. geniculatus*. The position regarding the validity of this species is unsatisfactory. Type specimens were not deposited, and material is not available for examination (Ramsdale and Snow, 1999). *Oc. thibaulti* (Dyar and Knab, 1909) is a Nearctic species, in Europe recorded from the river Dnieper, Ukraine (Gutsevitch and Goritskaya, 1970), but no longer found there according to Gutsevitch and Dubitsky (1987). *Oc. atropalpus* (Coquillett, 1902) is not considered here because it was only once reported from Italy and it is not regarded as a permanent species in the European area.

For the species *Oc. krymmontanus* Alekseev 1989, *Oc. coluzzii* Rioux, Guilvard and Pasteur 1998, *Oc. duplex* Martini 1926, and *Culex deserticola* short notes are given in the description of the species.

The principal objective, regarding the taxonomy and morphology chapters, is to provide an identification key to both adult females and males as well as to fourth-instar larvae, which incorporate all 92 species. At the beginning of chapters 6–8, keys for the identification of genera are given. Recently, Reinert (2000c) divided the composite genus *Aedes* Meigen into 2 genera, *Ochlerotatus* Lynch Arribalzaga and *Aedes*. Adults of these genera are distinguished primarily on the characters of the genitalia that require dissection. For this reason, the species belonging to these genera are treated in a single key. The illustrated keys are followed by a detailed description of the morphology, biology and distribution of each species. The morphological terms engaged in this volume are somewhat changed from those used by Harbach and Knight (1980, 1981), taking into consideration homology, phylogeny and their general use among the dipterous insects.

When a specimen has been keyed out, it should be compared with the description of that species to use the additional information as a cross-check. The description of the species is given according to the examination of the Peus collection, as well as a considerable number of species sampled in different European regions and various literature sources.

Characters of fourth-instar larvae are often clearer than those of the adults, and many taxonomists prefer to identify mosquitoes in this stage. Since there is a certain degree of variation among larvae of a species, it is best to identify, if possible, from a series of specimens. Rare exceptions to key characters can almost always be found. More important variations and their relative frequency are indicated in the systematic section of the species.

In the keys the larval chaetotaxy of the thorax and abdomen is not used to such an extent as by other authors. This seems to be a little bit "old fashioned", but on the other hand, quite often setae may be broken off, lying in a misleading or barely visible position in slide preparations or, in the worst case, are totally missing. If the latter occurs, an experienced eye is needed to see the alveola or tubercle of the missing seta.

Although eggs, early instar larvae and pupae of most European species are known, they are more difficult to identify than adults and fourth-instar larvae. Rearing eggs to fourth-instar larvae and pupae to adults is easier and less time consuming than to identify them in the early stages.

The authors suggest that the user should study the sections on general morphology before starting to identify specimens. The user should also be familiar with the proper sampling and mounting techniques of adults and larvae, because the presence of a full set of scales, setae and other features is essential for identification. Updated mosquito distribution throughout Europe, together with the bioecological conditions to be met for each species, should also help in the species identification. The territory of Europe, despite not being a distinct zoogeographical region, is chosen in an attempt to provide for the first time a unique key for the whole European region.

The authors' intention is to encourage and give support to every person who intends to start, or already has some experience in, mosquito control. The concept of the book is also based on several fields of knowledge which are important for everyone who deals with mosquitoes. Overviews on mosquito taxonomy, morphology, biology, biological, physical and chemical control measures are given, to complete the information needed for a comprehensive approach to both studying and controlling mosquitoes.

Mosquito control measures are dependent on many complex and interacting factors ranging from biological (species dependent), abiotic and physical factors influencing the phenology and abundance of mosquitoes (terrain features, climate, types of breeding sites etc.) to administrative, organisational and most certainly economical conditions. A decision to use one or another mosquito control measure is highly dependent on a basic knowledge of all aspects of the target species, and the impact of a chosen control method on the target and non-target organisms, as well as on the environment. A professional control programme should combine cost effectiveness, acceptable level of mosquito population suppression, and an environmentally sensitive approach. In some situations it is possible to rely on inexpensive and simple methods such as applying fragments of copper wire into flower pots for the control of *Aedes albopictus* larvae (Bellini, pers. comm.), while in others only pure biological, highly selective control measures are allowed, such as in the Upper Rhine Valley where almost the entire mosquito control programme in the river flood-plains has relied solely on *Bacillus thuringiensis ssp. israelensis* larviciding for several decades. Sections on different approaches in mosquito control (chemical, biological, physical, integrated or personal protection) provide basic information about different methods of using products with different formulation and toxicological features, effectiveness on target species, as well as their impact on non-target organisms. Information from this part of the volume attempts to serve as a basis for an appropriate mosquito control operation, allowing the user to live in relative safety from some vector-borne diseases, to alleviate the effects of abundant nuisance populations, to re-establish wetlands, and to share and enjoy nature by conserving the biodiversity, by using environmentally friendly control tools.

Up to now a comprehensive book in which the taxonomy, biology and distribution of all currently known European mosquito species are described, as well as the options for their control, has been missing. This volume should fill the gap and be a valuable help to scientists and indeed all those interested in, or working in any of the fields related to the Culicidae. It should provide guidance to field workers concerned with mosquito control and who wish, for example, to learn more about the behaviour of the species in their region, about mosquito breeding sites, or about the mosquito control techniques and options that may be suitable for each specific environment. Since there is still much information which for some, despite greatly increased access to the Internet, may be difficult to acquire, we have tried to include and summarise all available information, so that entomologists can apply it to their own situations.

The authors

Acknowledgements

We are most indebted to Paul Schädler, President of the German Mosquito Control Association (KABS, Waldsee) and the members of the KABS, Herbert W. Ludwig (University of Heidelberg, Germany) and Wolfgang Schnetter (University of Heidelberg, Germany), who encouraged the preparation of this work.

This book developed thanks to valuable information provided by our colleagues and friends worldwide. Herewith we would like to thank the following contributors who supplied a variety of information: Carlos Aranda (Consell Comarcal del Baix Llobregat, Spain), Andreas Arnold (KABS, Germany), Romeo Bellini (Centro Agricoltura Ambiente, Italy), Roger Eritja (SCM, Consell Comarcal del Baix Llobregat, Spain), Raul Escosa (CODE, Ebro Delta, Spain), Remi Foussadier (E.I.D., Rhone-Alpes, France), José Carlos Galvez (SCM, Huelva County Council, Spain), Aleksandra Gliniewicz (National Institute of Hygiene, Poland), Raymond Gruffaz (E.I.D., Rhone-Alpes, France), Klaus Hoffmann (KABS, Germany), Karch Saïd (SIIAP Paris, France), Beata Kubica-Biernat (University of Gdynia, Poland), Oszkar Kufcsak (Bay Zoltan Foundation, Szeged, Hungary), Christophe Lagneau (E.I.D. Mediterranee, France), Peter Lüthy (University of Zürich, Switzerland), Jan Lundström (University of Uppsala, Sweden), Minoo B. Madon (Greater Los Angeles County Vector Control District, USA), Eduard Marquès i Mora (SCM, Roses Bay and Lower Ter, Spain), Jean-Pierre Mas (E.I.D. Atlantique, France), Enrih Merdić (University of Osijek, Croatia), Spiros Mourelatos ((Environment, Public Health and Ecodevelopment, Greece), Odile Moussiegt (E.I.D. Mediterranee, France), Francoise Pfirsch (Mosquito Control Organization, Bas-Rhin, France), Dirk Reichle (KABS, Germany), František Rettich (National Institute of Public Health, Czech Republic), Martina Schäfer (University of Uppsala, Sweden), Francis Schaffner (E.I.D. Mediterranee, France), Elzibeta Wegner (Institute of Zoology, Warszawa, Poland), and Thomas Weitzel (KABS, Germany).

The authors express appreciation for the skillful and diligent preparations of the illustrations to Miguel Neri (University of San Carlos, Cebu, Philippines), Michael Gottwald (University of Heidelberg, Germany), Vlada Vojnić-Hajduk (University of Novi Sad, Serbia and Monte Negro) and Aung Moe (University of Heidelberg, Germany).

We are grateful to numerous individuals and organizations who assisted in various ways towards the final form of this book, especially to Matthias Beck (Waldangelloch, Germany), Geoffrey L. Brown (Heidelberg, Germany), Djamal Chakhmaliev (Heidelberg, Germany), Jeff Charles (Institute Pasteur, Paris, France), A.D. Ciklonizacija (Novi Sad, Serbia and Monte Negro), Culinex GmbH (Ludwigshafen, Germany), Major Dhillon (Northwest MVCD, Corona, USA), Elisabeth and Albert Gasparich (Heidelberg, Germany), Gesellschaft zur Förderung der Stechmückenbekämpfung (GFS, Waldsee, Germany), Thomas Heeger (Böhl-Iggelheim, Germany), Mikael Henriksson (Borgå, Finland), Icybac GmbH (Speyer, Germany), Thomas Imhof

(Heidelberg, Germany), Michael Kinzig (Heidelberg, Germany), Roland Kuhn (University of Mainz, Germany), Milagros Mahilum (University of Heidelberg, Germany), Jessica Munns (Pioneer, USA), Charles M. Myers (Glendora, USA), Beate Ruch-Heeger (Böhl-Iggelheim, Germany), Volker Storch (University of Heidelberg, Germany), Alexander Tahori (Ben Gurion University, Israel), Thin Thin Oo (University of Heidelberg, Germany), and Ute Timmermann (Weinheim, Germany).

Contents

II. THE IDENTIFICATION KEYS

IV. CONTROL OF MOSQUITOES

Introduction

The mosquitoes (family Culicidae) are at the centre of worldwide entomological research primarily because of their medical importance as vectors of dangerous diseases, such as malaria, yellow and dengue fever, encephalitis and lymphatic filariasis. More than half of the world's population lives under the risk of becoming infected by the causative agents of these diseases. Estimates made by the World Health Organisation (WHO), show that many hundreds of millions of people become ill, and some million people die annually (WHO, 1993). Although approximately three quarters of all mosquito species occur in the humid tropics and subtropics, mosquitoes are not just a problem of these regions. They may also cause a considerable nuisance in temperate latitudes. The mosquito species most commonly involved are the so-called floodwater mosquitoes, such as *Aedes vexans* and *Ochlerotatus sticticus* in river valleys that regularly flood, the snow-melt mosquitoes, *e.g. Oc. communis, Oc. punctor, Oc. hexodontus* in swampy woodlands and tundra areas, the halophilous species *Oc. caspius* and *Oc. detritus*, which breed particularly in the shallow lagoons found along the coasts of southern Europe, Asia Minor and north Africa, or the rock-pool mosquito *Oc. mariae* found along parts of the mediterranian rocky coasts, where mass occurrences can become a great nuisance. *Culex pipiens pipiens* biotype *molestus*, which is known as the "house mosquito" because of its presence close to human settlements can likewise make itself noticed in temperate zones as a nuisance.

Mosquitoes are extremely successful organisms due to their ability to adapt to a wide range of habitats. They are found throughout the world, except in deserts and permanently frozen areas. Mosquito larvae colonise a wide range of water-bodies, temporary and permanent, highly polluted as well as clean, large or small, stagnant or flowing, even in the smallest accumulations such as water-filled buckets, flower vases, old tyres, hoof prints or leaf axils. Adult mosquitoes vary greatly in their bionomics, *e.g.* concerning the host seeking, biting and migration behaviour and strategy for reproduction. It is the medical importance and the troublesome behaviour of mosquitoes that has aroused the interest of scientists. Their importance as vectors of malaria and yellow fever were suspected by Joseph Nott in 1848. In 1878 Sir Patrick Manson showed that the roundworm *Wuchereria bancrofti* is transmitted by *Culex p. quinquefasciatus*. Only three years later Carlos Finley postulated that yellow fever was transmitted by mosquitoes, which was proven by Walter Reed and his co-workers in 1901. Sir Ronald Ross made a further pioneering discovery in 1897, when he recognised the importance of the anophelines as vectors of malaria. The discovery of the transmission cycle of most vector-borne diseases gave hope to being able to successfully control this scourge of humanity. The foundations for the control of the mosquitoes were established at the beginning of the 20th century, which included source reduction as a means to reduce human vector contact. The development and use of DDT as a residual insecticide, achieved excellent results in the control of mosquitoes. In the 1950s it was believed that malaria would be exterminated by the use of DDT and chloroquine, but disillusionment quickly followed, because mosquitoes became resistant to the insecticide in many areas. In addition the control efforts against the vectors were not consistent. Toxicological and ecotoxicological problems were undesirable disadvantages caused by the use of unspecific and highly persistent insecticides. Despite considerable efforts of national and international organisations like the WHO, prevention of a more dramatic increase in vector-borne diseases is mainly what has been achieved up to date. Not only do the vectors and pathogens have tremendous adaptability, but also new types of diseases appear, such as dengue haemorrhagic fever that was observed for the first time in southeast Asia in 1954. Altogether the greater mobility of people by modern means of transport, the intensified international trade, as well as fluctuations in climate, have resulted in a wider distribution of vector mosquitoes and disease-causing agents. The risk of becoming infected with a vector-borne disease has increased again not only in the tropics but also in Europe. The essential foundation for successful action against the vectors requires not only an integrated control programme, in which all appropriate methods for control are used, but also knowledge of the biology and ecology of the target organism. The importance of a vector or nuisance species is determined above all by its physiological characteristics, such as reproduction, migration, host-seeking and biting behaviour. Accurate identification is a basic prerequisite to a study of the autecology of a species as well as its biocoenotic relationship in the ecosystem.

I

General Aspects

Systematics

The insect order Diptera traditionally has been divided into two suborders, the Nematocera with about 26 families, and the Brachycera with about 110 families. According to the phylogenetic principle of Hennig (1966), the Brachycera (including the Cyclorrhapha) are monophyletic, whereas the Nematocera were found to be a polyphyletic clade. Within the nematoceran clade, the Bibionomorpha were pointed out as the sister-group of the Brachycera (Hennig, 1973).

Taxa such as families, genera, subgenera, species groups and complexes are all interpretations of definitions of morphological similarity or dissimilarity, reached by consensus between specialists. Cladistics with the concept of monophyly, that is a shared common ancestor, has helped to give these taxa a more theoretical basis.

The clade Nematocera is defined by filamentous antennae, more or less slender legs, and an abdomen with ten identifiable segments in the larvae and males. The male genitalia, the so called hypopygium, is derived from the ninth and tenth segments. Females have cerci and/or ovipositors, which between families can vary as to segmental contributions. Depending on different results of various phylogenetic analyses, the Nematocera were assigned between four and seven further divisions or infraorders (Hennig, 1973; Wood and Borkent, 1989; Sinclair, 1992). Oosterbrook and Cortney (1995) postulated that within a group of higher Nematocera including the Anisopodidae, the Tipulidae and Trichoceridae the stem group for the Brachycera is to be found. These ongoing discussions do not affect the general consensus about the monophyly of Culicomorpha. However, different taxa were included by some authors.

The Culicomorpha with the Culicidae, Chaoboridae, Dixidae and the Chironomidae was regarded a monophyletic group by Hennig (1973). This has been accepted by Knight and Stone (1977). In a more recent analysis of Culicopmorpha also the Corethrellidae, Simuliidae and Ceratopogonidae were included in Culicomorpha, as well as the Thaumaleidae and the Nymphomyiidae as a sistergroup (Saether, 2000) Fig. 1.1.

The monophyly of Culicoidea is confirmed by the 100* value. The Culicomorpha include a sister group to Culicoidea. The Neodiptera*, according to Michelsen (1996), comprise the bibionomorph Nematocera and the Brachycera.

On the family level, the monophyly of the Culicidae with three subfamilies Anophelinae, Culicinae and Toxorhynchitinae was established by Edwards (1932), and subsequently accepted by Belkin (1962) and Knight and Stone (1977). The first attempt to elucidate higher phylogeny within subfamilies by creating tribes, was made by Belkin (1962). He discussed the general assumption that the Anophelinae are closer to the stem of the superfamily, but did not accept it because of lack of evidence. A cladistic phylogentic study of the family by Harbach and Kitching (1998) has now confirmed the basic position of the Anophelinae as a subfamily. They also partly affirmed the currently used systematics of the family, and partly gave other results. The Anophelinae and Culicinae are

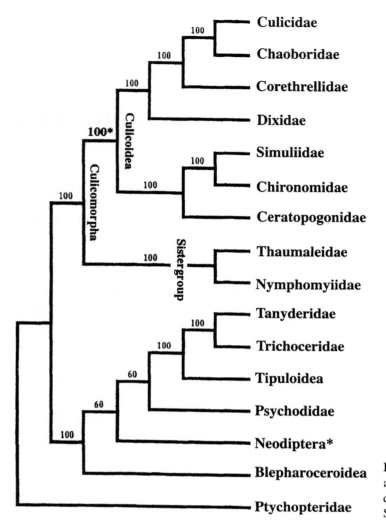

Figure 1.1. The phylogeny of Culicomorpha and other groups of Nematocera after a strict consensus and majority rule cladogram by Saether (2000).

regarded as valid monophyletic subfamilies, whereas the Toxorhynchitinae were lowered to tribal rank under the Culicinae. Their phylogeny of 11 tribes broadly corresponds to the classic divisions. They discussed the relationship between lower categories as genera and brought the tribes Aedini and Mansoniini into one sister group, and the tribes Culicini and Sabethini into another. Relationships within the Aedini are not yet fully resolved. However, recently Reinert (2000c) raised the subgenus *Ochlerotatus* Lynch Arribalzaga of the genus *Aedes* Meigen to generic rank, placed other former subgenera of *Aedes* in the new genus and divided it into two sections of species based on

larval and pupal characters. This new systematics is included into our citation of taxa, as well as other changes, which followed after the catalog of Knight and Stone (1977) and its supplements (Fig. 1.2.) (Reinert, 2001).

At the level of genera, species groups and complexes, many problems are still unsolved. However, many outstanding species revisions of genera or subgenera are available, especially on tropical material. Further systematic evolutionary and cladistic studies of relationships between and within different categories of the family, will contribute to the understanding of the events which have made the Culicidae such a successful group.

Subfamily Anophelinae
    Genus *Anopheles*
        Subgenera *Anopheles, Cellia, Kerteszia, Nyssorhynchus*
    Genus *Chagasia*

Subfamily Culicinae
Tribus Aedeomyiini
    Genus *Aedeomyia*
        Subgenera *Aedeomyia, Lepiothauma*
Tribus Aedini
    Genus *Aedes*
        Subgenera *Aedes, Aedimorphus, Alanstonea, Albuginosus, Belkinius, Bothaella, Cancraedes, Christophersiomyia, Diceromyia, Edwardsaedes, Fredwardsius, Huaedes, Indusius, Isoaedes, Leptosomatomyia, Lorrainea, Neomelaniconion, Paraedes, Pseudarmigeres, Scutomyia, Skusea, Stegomyia*
    Genus *Armigeres*
        Subgenera *Armigeres, Leicesteria*
    Genus *Ayurakitia*
    Genus *Eretmapodites*
    Genus *Haemagogus*
        Subgenera *Conopostegus, Haemagogus*
    Genus *Heizmannia*
        Subgenera *Heizmannia, Mattinglyia*
    Genus *Ochlerotatus*
        Subgenera section I*: *Chaetocruiomyia, Finlaya, Geoskusea, Halaedes, Kenknightia, Levua, Macleaya, Molpemyia, Mucidus, Nothoskusea, Ochlerotatus, Protomacleaya, Pseudoskusea, Rhinoskusea, Rusticoidus, Zavortinkius*
        Subgenera section II*: *Abraedes, Aztecaedes, Gymnometopa, Howardina, Kompia*
    Genus *Opifex*
    Genus *Psorophora*
        Subgenera *Grabhamia, Janthinosoma, Psorophora*
    Genus *Udaya*
    Genus *Verrallina*
        Subgenera *Harbachius, Neomacleaya, Verrallina*
    Genus *Zeugnomyia*
Tribus Culicini
    Genus *Culex*
        Subgenera *Acalleomyia, Acallyntrum, Aedinus, Afroculex, Allimanta, Anoedioporpa, Barraudius, Belkinomyia, Carrollia, Culex, Culiciomyia, Deinocerites, Eumelanomyia, Kitzmilleria, Lasiosiphon, Lophoceraomyia, Lutzia, Maillotia, Melanoconion, Micraedes, Microculex, Neoculex, Phenacomyia, Thaiomyia, Tinolestes*
    Genus *Galindomyia*
Tribus Culisetini
    Genus *Culiseta*
        Subgenera *Allotheobaldia, Austrotheobaldia, Climacura, Culicella, Culiseta, Neotheobaldia, Theomyia*
Tribus Ficalbiini
    Genus *Ficalbia*
    Genus *Mimomyia*
        Subgenera *Etorleptiomyia, Ingramia, Mimomyia*
Tribus Hodgesiini
    Genus *Hodgesia*
Tribus Mansoniini
    Genus *Coquillettidia*
        Subgenera *Austromansonia, Coquillettidia, Rhynchotaenia*
    Genus *Mansonia*
        Subgenera *Mansonia, Mansonioides*
Tribus Orthopodomyiini
    Genus *Orthopodomyia*
Tribus Sabethini
    Genus *Isostomyia*
    Genus *Johnbelkinia*
    Genus *Limatus*
    Genus *Malaya*
    Genus *Maorigoeldia*
    Genus *Onirion*
    Genus *Runchomyia*
        Subgenera *Ctenogoeldia, Runchomyia*
    Genus *Sabethes*
        Subgenera *Davismyia, Peytonulus, Sabethes, Sabethinus, Sabethoides*
    Genus *Shannoniana*
    Genus *Topomyia*
        Subgenera *Suaymyia, Topomyia*
    Genus *Trichoprosopon*
    Genus *Tripteroides*
        Subgenera *Polylepidomyia, Rachionotomyia, Rachisoura, Tricholeptomyia, Tripteroides*
    Genus *Wyeomyia*
        Subgenera *Antunesmyia, Caenomyiella, Cruzmyia, Decamyia, Dendromyia, Dodecamyia, Exallomyia, Menolepis, Nunezia, Phoniomyia, Prosopolepis, Wyeomyia, Zinzala*
Tribus Uranotaeniini
    Genus *Uranotaenia*
        Subgenera *Pseudoficalbia, Uranotaenia*
Tribus Toxorhynchitini
    Genus *Toxorhynchites*
        Subgenera *Afrorhynchus, Ankylorhynchus, Lynchiella, Toxorhynchites*

---

* formerly in genus *Aedes*

Figure 1.2.   Systematic order of the Family Culicidae.

Up to 1992 there were 3209 valid species recognised worldwide (Ward, 1992), with 38 genera (Harbach and Kitching, 1998). Since then, more species, subgenera and genera have been described, synonymised or resurrected, and currently there are 39 genera and 135 subgenera recognised (Reinert, 2001). In the present publication, 92 out of 98 listed European species (Ramsdale and Snow, 1999) are considered. The difficulties in delineating species are shown by the numerous synonyms available for some species such as *Cx. pipiens*. This indicates the amount of variability in external morphological characters as body size, scaling, colour, and position of setae in different life stages. Sibling species and species group identifications on non-morphological criteria are of great value. Behaviour and physiology studies add much to our knowledge and to the recognition of species in the field. The introduction of new biochemical and recently developed molecular and cytogenetic methods, has opened up new fields of research. However, such studies should be complemented by renewed studies on external morphology. All taxonomic entities below subgenus level such as species complexes, species, subspecies and genetically, behaviourally or physiologically different strains, or populations, are important in analyses of vector capacity and approach to their control. A combination of different methods may enhance chances to identify these vector species more efficiently. Therefore more studies of inter- and intraspecific variability on several morphological levels, are much needed. Cross mating studies can also add to further understanding. Munsterman (1995) summarised advances in molecular systematics. He also gave several examples of molecular and chromosomal evolutionary studies which have contributed substantially to the understanding of evolutionary trends between species or within species complexes.

Biology of Mosquitoes

In view of their special adaptation mechanisms, mosquitoes are capable of thriving in almost all kinds of environments. There is hardly any aquatic habitat that does not lend itself as a breeding site for mosquitoes.

In temporary flooded areas along rivers or lakes with water fluctuations, floodwater mosquitoes such as *Ae. vexans* or *Oc. sticticus* develop in large numbers, becoming a tremendous nuisance, even in places located far away from their breeding sites (Mohrig, 1969; Becker and Ludwig, 1981; Schäfer *et al.*, 1997).

In swampy woodlands, snow-melt mosquitoes such as *Oc. cantans, Oc. rusticus, Oc. communis* or *Oc. punctor* encounter ideal conditions for development in pools that are formed after the snow thaws, or during heavy rainfalls.

In floodplains along coastal areas of the Mediterranean, North and Baltic Seas, halophilous species such as *Oc. caspius* and *Oc. detritus*, which prefer salt water habitats, develop in huge numbers.

Tree-holes are the habitat of arboreal species such as *Oc. geniculatus, Ae. cretinus, Anopheles plumbeus* and *Orthopodomyia pulcripalpis*.

Species like *Cx. p. pipiens* can even breed in small water recipients, *e.g.* in rain water drums, old tyres, or in small clay pots. In the early 1990s *Ae. albopictus* was introduced into Europe due to the uncontrolled trade in used tyres. Apart from its potential as a vector of serious pathogens in some Mediterranean countries, this mosquito species is causing a serious nuisance problem where artificial and natural water containers are abundant as breeding sites. These few examples illustrate the tremendous ecological flexibility mosquitoes have developed.

Like all Diptera, mosquitoes exhibit complete metamorphosis. All mosquitoes need aquatic habitats for their development. After hatching they pass through four larval instars, and a pupal stage where the transformation to the adult takes place. Most species are unautogenous, it means after copulation the females have to take a blood-meal to complete egg development. Only a few species have populations that are autogenous. They develop first egg batches without a blood-meal (*e.g. Cx. p. pipiens* biotype *molestus*).

2.1. OVIPOSITION

Female mosquitoes usually lay between 50 and almost 500 eggs, two to four days (or even more in cool temperate climates) after the blood-meal. In general, the mosquitoes can be divided into two groups depending on the egg laying behaviour (Barr, 1958) and whether or not the embryos enter into a period of dormancy (externally triggered resting period) or diapause (genetically determined resting period).

In the first group females deposit the eggs onto the water surface either singly (*Anopheles*) or in egg batches (*e.g. Culex, Uranotaenia, Coquillettidia, Orthopodomyia* and subgenus *Culiseta* Fig. 2.1).

Culex females lay the eggs in rafts comprised of several hundred eggs locked together

9

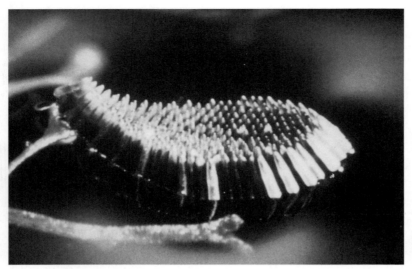

Figure 2.1. Egg raft of *Culiseta annulata* (size approx. 5 mm).

in a boat-shaped structure. During oviposition the females stand on the water surface with the hind-legs in a V-shaped position. The eggs are released through the genital opening and grouped together between the hind-legs, forming a raft where the eggs stand vertically on their anterior poles attached together by chorionic protrusions (Clements, 1992). The anterior pole of each egg has a cup-shaped corolla with a hydrophilic inner surface which lies on the water surface. The outer surface is hydrophobic. The resulting surface tension forces keep the egg raft in position and when rafts drift to boundaries they tend to remain there. Immediately after oviposition the eggs are soft and white, but they sclerotise and darken within 1–2 hours.

Anophelines lay single eggs while standing on the water or hovering over it. The eggs of this subfamily are adapted for floating and can easily be killed by dessication.

The embryos of the first group do not enter a dormancy or diapause and hatch when the embryonic development is completed. Species having these non-dormant eggs usually produce several generations each year. Their developing stages are found for the most part in more permanent waters where one generation succeeds another during the breeding season. The number of generations depends on the length of the breeding season, as well as abiotic and biotic conditions, most important of which is the temperature which influences the speed of development.

The parameters that determine the choice of a breeding site by the females laying their eggs onto the water surface, are still unknown for many species. Factors such as water quality, incidence of light, existing eggs, available food, and local vegetation, are decisive factors in the choice of a breeding site.

For *Cx. p. pipiens* it is known that the content of organic material in the water plays an important role in attracting the females when they are about to lay eggs. Apparently, gaseous substances such as ammonia, methane, or carbon dioxide, which are released when organic material decomposes, have the effect of attracting the females of *Cx. p. pipiens* (Becker, 1989). They recognise that such a site has adequate food and that favourable conditions prevail for the development of their brood.

A few other examples show that the egg-laying behaviour reflects the ecological conditions in the breeding site. The submerged larvae and pupae of *Coquillettidia* obtain the oxygen they need by inserting their breathing apparatus into the aerenchyma (air filled tissue) of certain plants under water. Therefore, in order to ensure the development of the larvae the females must recognise the appropriate water plants at the time they lay their eggs, in order to ensure the development of the larvae.

The second group lays eggs which do not hatch immediately after oviposition (Fig. 2.2). Most interesting is the egg-laying behaviour of the floodwater mosquitoes (*e.g. Ae. vexans*), and the subgenus *Culicella* of genus *Culiseta*, which lay their eggs not on water but as single eggs into the moist soil of their breeding places, which is subsequently flooded when the water level rises. The eggs are laid into small depressions or between particles of moss with a high degree of soil moisture, to protect the sensitive eggs from drying out during the embryogenesis (Barr and Azawi, 1958; Horsfall *et al.*, 1973).

In the case of *Ae. vexans* and *Oc. caspius*, which breed in flooded areas where the level fluctuates frequently, the appropriate egg-laying behaviour is crucial to ensure successful development of the immature stages. A suitable egg-laying site for floodwater species must meet the following prerequisites:

(a) it must be wet enough at the time the eggs are laid in order to ensure that the freshly laid eggs, which are very sensitive to any water loss, do not dry out before their impermeable endochorion has been tanned and the wax layer of the serosal cuticle is formed (Horsfall *et al.*, 1973; Clements, 1992);

(b) there must be a subsequent and sufficient flooding of the soil where the eggs have been laid, so that the complete process can take place from hatching all the way to imago emergence;

(c) the water body for subsequent breeding should have as few mosquito predators as possible, to ensure that the larvae are not preyed upon by natural enemies when they hatch.

The ability of a floodwater mosquito female to find appropriate places for egg-laying which guarantees maximum breeding success is yet not fully understood. However, respect is due to

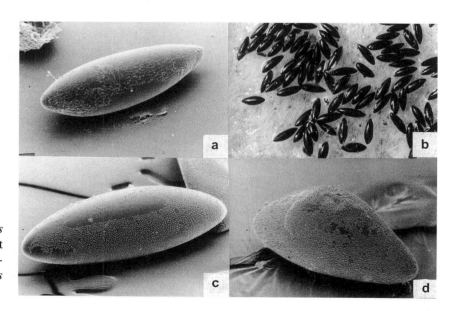

Figure 2.2. Eggs of *Ae. vexans* (a: SEM-photo, 50×; b: light microscopical photo 8×), *Oc. cantans* (c, SEM, 50×), *Oc. rusticus* (d, SEM, 50×).

these tiny insects which have adapted their behaviour to overcome the hostile conditions in their breeding sites. If the females chose to lay their eggs in low-lying areas with almost permanent water flow this would have crucial disadvantages: low-lying areas are flooded for long periods of time and have, therefore, a very unfavourable alternating sequence of dry and flood spells, so allowing only very few populations of floodwater mosquitoes to develop. Areas with almost permanent water flow generally have a high concentration of natural enemies, such as fish, so that the risk of being eaten would be very high for the mosquito larvae.

However, flooded areas with a very short period of water flow are also unfavourable egg-laying sites, because they entail the risk of premature drying out. This kind of terrain becomes flooded for a short period of time, and only in the years when water is abundant. Thus, the wet and dry sequence is not favourable for the development of several consecutive generations. These areas dry very rapidly after a flood so that *Aedes* or *Ochlerotatus* eggs run the risk of dessication since the developing embryos are very sensitive to water loss.

In Fig. 2.3 the preferred egg-laying sites of the floodwater mosquitoes (*e.g. Ae. vexans*) are shown. Usually these are areas of dense vegetation and silty soil, with the reed (*e.g. Phragmites australis*) being highly attractive to female mosquitoes that are about to lay eggs.

Usually the reed zone precisely demarcates the mid-water level in river systems, since reeds need a great deal of water.

The fluctuating curve of the Rhine river, as an example, makes it quite clear how important such egg-laying behaviour is for the development of the floodwater mosquitoes (Fig. 2.4). In the zone between 4 and 5 metres, there is the ideal timing of dry and wet periods, with optimum water flow to guarantee the development of a large number of populations. Furthermore predators, especially fish, are usually low in numbers or absent.

But how do these female floodwater mosquitoes find the right oviposition site? Gravid females obviously recognise the wet, silty, riverside clay soil as an appropriate egg laying substrate and are attracted to it. However, these criteria alone are not enough, given that in rainy periods there would be many places which would seem adequate due to their high degree of moisture.

It is likely that the floodwater mosquitoes are able to differentiate between various soil types. The soil in most floodplains consists of a high percentage of clay and low percentage of humus or organic materials, in contrast to many other soils (Ikeshoji and Mulla, 1970; Strickman, 1980a,b; Becker, 1989).

It is also possible that flooded areas along the riverbanks produce pheromone-like odours, which may be recognised by the female

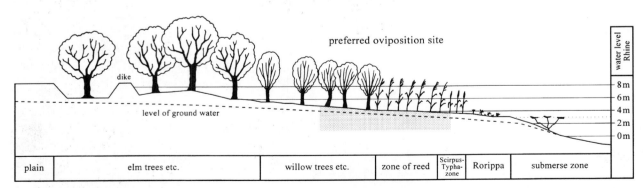

Figure 2.3. Transsection through the Upper Rhine Valley with zones of vegetation and preferred oviposition sites of floodwater mosquitoes (*e.g. Aedes vexans*).

River water level

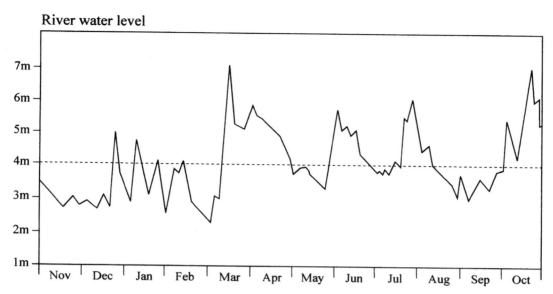

Figure 2.4. Water fluctuation of the Rhine river which is favourable for the development of floodwater mosquitoes (--- indicates the beginning of flooding of the most productive breeding sites).

mosquitoes and which induce them to lay their eggs. These odours could come from eggs, which have already been laid in the soil, or from particular associations of plants which are indicators of a specific moisture level in the soil, and the occurrence of regular floods. As a whole, the oviposition behaviour of the flood-water mosquitoes shows an astonishing degree of adaptation to their habitat, an adaptation which has developed throughout the course of evolution.

2.2. EMBRYONIC DEVELOPMENT

Mosquito embryogenesis has been described in detail by Clements (1992). The embryonic development starts almost immediately after the eggs have been laid. Depending on the temperature it takes about two days to a week or more until the embryos are fully developed.

The course of embryonic development reflects also a special adaptation to various abiotic conditions in the larval habitat (Becker, 1989). The non-dormant eggs of *Culex, Coquillettidia, Uranotaenia, Orthopodomyia*

and the subgenus *Culiseta* usually hatch after a short time when the embryonic development is completed. The length of time required is dependent almost entirely on temperature. At a temperature of 30°C, *Cx. p. pipiens* larvae hatch one day after the eggs have been laid, at 20°C and 10°C it takes three and 10 days respectively, and at 4°C, the embryonic development of *Cx. p. pipiens* cannot be completed (Fig. 2.5).

The embryonic development of *Aedes/ Ochlerotatus* species usually takes significantly longer. For instance, larvae of *Ae. vexans* are ready to hatch four or eight days after oviposition, when the eggs are kept at 25°C and 20°C, respectively (Horsfall *et al.*, 1973; Becker, 1989). Hatching experiments with freshly laid *Ae. vexans* eggs kept at 20°C have shown that eight days after the eggs had been laid, almost 50% were ready to hatch. This means that the embryonic development of *Cx. p. pipiens* usually takes only half as long as that of *Ae. vexans*, which assures a quick generation renewal of the former. The relatively slow embryonic development of floodwater *Aedes/Ochlerotatus* species can be explained by the fact that these mosquitoes lay their eggs in flooded areas where

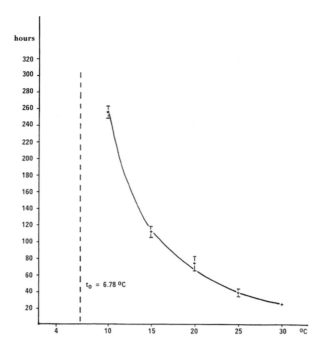

Figure 2.5. Duration of the embryonic development of *Culex. p. pipiens* at various temperatures (t_0 = no development possible according to Tischler, 1984).

there are few ecological factors requiring rapid embryonic development. It usually takes more than one week until the next flooding occurs. Therefore, there is little ecological advantage in the fast sequence of generations that would result from rapid development of the embryos.

2.3. HATCHING

Aedes/Ochlerotatus mosquitoes have developed a highly sophisticated mechanism which regulates the hatching process, as a direct adaptation to the greatly fluctuating abiotic conditions existing in the temporary waters where these mosquitoes breed (Gillett, 1955; Telford, 1963; Horsfall *et al.*, 1973; Beach, 1978; Becker, 1989). The timing of larval hatching to coincide with the presence of ideal developmental conditions, is a prerequisite for successful development in temporary water bodies.

The difference in hatching behaviour between the snow-melt mosquitoes (*e.g. Oc. cantans*, *Oc. communis*, and *Oc. rusticus*) and the floodwater mosquitoes (*e.g. Ae. vexans*) clearly illustrates the extent to which the hatching behaviour of each *Aedes/Ochlerotatus* species is adapted to the abiotic conditions in their breeding waters.

The breeding waters of snow-melt mosquitoes, for example depressions and ditches in marshy regions covered frequently with alder trees in central Europe, are usually flooded for long periods of time with relatively cold water.

In Fig. 2.6 the phases of development and diapause of the univoltine (monocyclic, one generation per year) snow-melt mosquitoes, are shown as a function of the water level variation of a pool in the swampy woodlands in central Europe.

The breeding places of the snow-melt mosquitoes are usually flooded in the late fall and after the snow has melted. The water level usually reaches its peak in early spring. Under normal conditions, it then recedes slowly but steadily throughout the summer until the pools are dry again.

Snow-melt mosquitoes have adapted perfectly to these conditions in their breeding sites, by their diapause patterns and appropriate reaction to hatching stimuli. After the eggs have been laid, usually in early summer, the embryos of the majority of the snow-melt mosquitoes automatically enter diapause. They are unable to hatch during the summer months so avoiding the risk of prematurely emerging during the dry spells of summer.

In central Europe some species *e.g.* the snow-melt mosquito *Oc. rusticus* and *Cs. morsitans* are ready to hatch at the beginning of winter after sensing the continuous temperature decrease during autumn. The diapause of most snow-melt mosquitoes (*e.g. Oc. cantans* and *Oc. communis*) is interrupted when the temperature has dropped in autumn and the cold winter period has set in. Consequently,

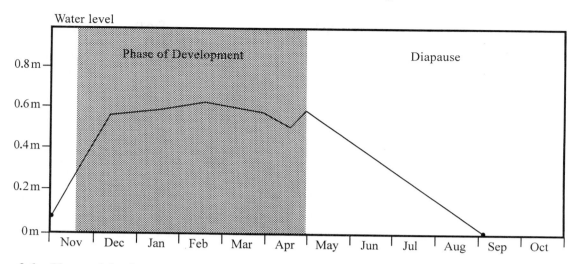

Figure 2.6.   Phases of development and diapause of univoltine snow-melt mosquitoes in a pond in central Europe.

these mosquitoes are ready to hatch during the snow-melt in the next spring and shortly afterwards. This factor, along with their ability to hatch in very cold and oxygen-rich waters, enables these mosquitoes to be ready to hatch at a time when favourable water level conditions are present. After hatching, the semi-permanent water in forest pools provides ideal conditions for slow development. In central Europe, this usually takes place between the end of April and the beginning of May.

Even more sophisticated is the hatching behaviour of the floodwater mosquitoes. In Fig. 2.7 the phases of development and diapause of *Ae. vexans* are shown as a function of the water level variation (here the Rhine river is taken as an example).

Unlike the semi-permanent status of the water in the breeding sites of snow-melt mosquitoes, the breeding sites of the floodwater mosquitoes are characterised by temporary water flow caused by rapid, substantial fluctations in the water level of the rivers following heavy rains in early and mid-summer. By contrast, late summer and winter are usually marked by extensive periods of low water.

As a consequence of this, the best developmental conditions for larvae of floodwater mosquitoes, occur between April and September.

Thus, they diapause during autumn, winter, and early spring (Telford, 1963). Due to the extremely variable nature of the water flow, floodwater mosquitoes have to hatch when the high summer water temperature enables rapid development to take place. Moreover, their being multivoltine (polycyclic, several generations per year) makes sense from an ecological viewpoint, since they can go through several generations coinciding with the fluctuations in the water level. This factor is mainly responsible for the huge reproduction rate of these species, often creating a tremendous nuisance.

If the hatching behaviour of the floodwater mosquito is analysed in detail, a well-adjusted control mechanism can be seen, which is mainly influenced by the following factors:

(a) Dissolved oxygen. During periods of high water level, most of the flooded areas along rivers are covered with flowing and oxygen-rich water. Hatching at that time would create the risk of the larvae being swept away. Moreover, fish usually invade deep flooded areas in search for food. In order to avoid these dangers, the floodwater mosquitoes have developed a specific hatching behaviour. The decline in dissolved oxygen usually

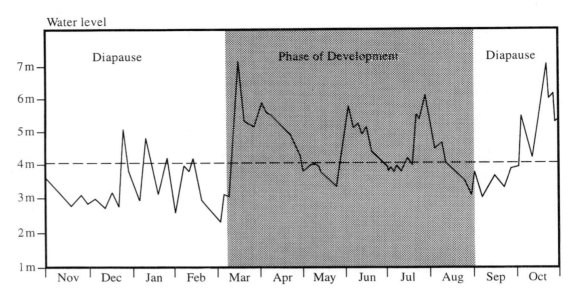

Figure 2.7.   Phases of development and diapause of the floodwater mosquito *Aedes vexans* in central Europe.

triggers hatching of the larvae of the floodwater mosquitoes (Hearle, 1926; Gjullin *et al.*, 1941; Borg and Horsfall, 1953; Travis, 1953; Judson, 1960; Horsfall *et al.*, 1958, 1973; Burgess, 1959; Becker, 1989). They only hatch once the flooded pools become stagnant and the oxygen content starts to decrease (for example, due to bacterial degradation processes) very rapidly. The decreasing level of oxygen in the water signals to the unhatched larvae that the water will be stagnant at the time of their hatching, thus, ensuring that the risk of being washed away has passed. Moreover, the bacterial action causing decomposition of organic material ensures that the larvae will have an adequate food supply. In addition, the shallow, stagnant waters are not a suitable environment for the larvae's main predator, the fish.

(b) Water temperature plays a fundamental role in the hatching process of the flood-water mosquito. Premature hatching during cold weather would greatly delay the development of the larvae, since the process is very temperature dependent.

Apart from the diapause of floodwater mosquitoes from autumn until springtime, the temperature-dependent hatching behaviour of, for example *Ae. vexans* also ensures that the larvae will not hatch before the temperature of the water is warm enough to permit rapid development. This ensures that the brood will not dry out due to rapid changes in the water levels. For instance, in the Upper Rhine Valley *Ae. vexans* larvae hatch in springtime when the water temperature reaches 10°C or more. An interesting phenomenon is the seasonal-related hatching process. After the winter cold phase and after the subsequent increase in temperature, a small percentage of *Ae. vexans* larvae are ready to hatch even at 4°C. The adaptation of the temperature-dependent hatching behaviour to the climatic conditions and water flow in an area in central Europe, can be demonstrated with the example of the hatching response to water of 15°C. In springtime the hatching rate reaches its peak at 15°C. In the months of March and April, apart from the gradually increasing water temperatures in the spring, the water level usually rises due to spring rainfall. This creates favourable conditions for the development of the floodwater mosquitoes. On the other hand, water of the same temperature induces a reduced hatching response during late summer and autumn. Larvae would not be able to complete their development if they were to hatch at a temperature of 15°C in October or November, since at this time of year the falling temperature prolongs the larval development, and the water level is receding.

It is remarkable that there are also differences in the hatching behaviour of *Ae. vexans* populations in different river systems. It appears that *Aedes/Ochlerotatus* species are adapted to the hydrological characteristics of each river system. In river systems with a lower discharge of water the periods of inundation are usually shorter, which means that the development of the mosquito has to be rapid. Therefore, *Aedes* and *Ochlerotatus* mosquitoes breeding in these areas have an extended diapause until summer, to allow a faster development at high temperatures.

Let us have a closer look at the diapause of the ready to hatch larvae inside the eggs of the floodwater mosquito *Ae. vexans*, especially concerning the change from a state of hatching inhibition to one of the hatching readiness. This process is called "conditioning" whereas, the onset of hatching inhibition is called "deconditioning" (Horsfall, 1956; Horsfall and Fowler, 1961; Clements, 1963; Horsfall *et al.*, 1973).

The factors that are most likely to have an influence on diapause, or on hatching inhibition and readiness, are temperature fluctuations, varying degrees of moisture in the air and soil as well as changes in the day-length (Brust and Costello, 1969). Larvae of *Ae. vexans* are able to hatch to some extent during a flood in the same year as that in which the embryogenesis was completed, providing the temperature remains above 20°C. Decreasing temperatures

below 15°C lead to a hatching inhibition in autumn. After a cold (winter) phase, temperatures of 10°C and above have a conditioning effect, and interrupt the diapause in the next springtime. It is worth mentioning that after a cold spell, the readiness to hatch is positively correlated with the rise in temperature. The higher the temperature during egg laying and the lower the temperature in winter, the higher is the hatching response in the following summer. The complex diapause behaviour allows the larvae of *Ae. vexans* to distinguish between favourable developmental conditions in the springtime, and unfavourable conditions in late summer. It is remarkable that in winter even extreme temperatures well below the freezing point will not kill the diapausing larvae.

Another behaviour that represents a sophisticated adaptation to the very variable water flow in the breeding sites, is the so-called "hatching in installments". Even within a batch of eggs deriving from one female subjected to the same microclimatical conditions, not all of the larvae hatch uniformly. Without a cold phase, only a few individuals from a freshly laid egg batch of one female are ready to hatch, whereas, after a cold phase, the readiness to hatch is far greater. Apart from their inherited variability, the conditions that each egg had experienced (for instance, the location of the egg in the ovariole during maturation, the timing of the oviposition, as well as differing microclimatic factors at the egg-laying site) determine whether a larva will hatch under certain conditions or not. Thus, the larvae hatch "in installments" (Wilson and Horsfall, 1970; Becker, 1989). For instance, soil samples containing eggs of *Ae. vexans* and kept at 25°C were flooded several times, with dry phases of four weeks between each flooding step. After the first flooding 57% of the total number of larvae hatched, 10% after the second, 25% after the third, and 8% hatched the fourth time (Becker, 1989). This behaviour pattern assures long-term survival for mosquito species that develop in temporary bodies of water. If, for example, all of the larvae were to hatch at the same time under ideal hatching conditions and if, as a consequence of a sudden dry spell, all of the breeding waters were to dry out before the brood could complete its development cycle, one single natural event could virtually wipe out the entire mosquito population. By "hatching in installments", the floodwater mosquito population can survive such potentially catastrophic events. There is still a large contingent of unhatched larvae remaining in the breeding area. At the time of the next flooding these larvae will still have an opportunity to successfully produce a new generation without new eggs having to be laid. Incidentally, this is also what happens after a treatment with larvicides. It is remarkable that the unhatched larvae are able to persist for at least 4 years without losing their ability to hatch (Horsfall *et al.*, 1973).

After the content of oxygen decreases in the water of the breeding site, the larvae initiates the shell rupture by pressing the so-called "egg tooth", an egg burster located posterodorsally on the head capsules of the larva, onto the egg shell. As a result the shell splits along a particular line at the anterior end of the egg. A cap (anterior part of the egg shell) comes away and the larvae escape by swallowing water into the gut which forces the body from the shell (Clements, 1992). The whole process of hatching takes only a few minutes.

2.4. LARVAE

The legless (apodous) larval body is divided into three distinct parts: (a) the head with mouthparts, eyes and antennae; (b) the broader thorax and (c) the abdomen which is composed of seven almost identical segments and three modified posterior segments. These posterior segments bear 4 anal papillae to regulate electrolyte levels. At the abdominal segment VIII a siphon in culicines, or only spiracular lobes in anophelines, are developed where the tracheal trunks open at spiracles for the intake of

oxygen. Usually the culicine larvae hang head downwards from the water surface (Fig. 2.8a). Anopheline larvae lie horizontally at the water surface. Their body is held horizontally by specialised setae (palmate setae), the notched organ located at the anterior margin of the prothorax and the spiracular lobes which are flush with the dorsal surface of the larval body and have direct contact with the air (Fig. 2.8b).

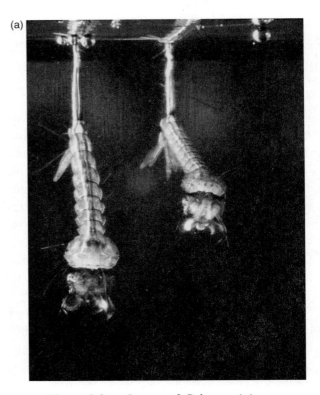

(a)

Figure 2.8a.   Larvae of *Culex p. pipiens.*

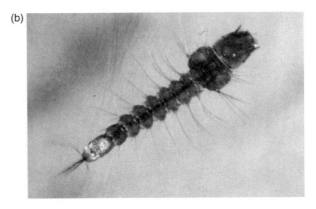

(b)

Figure 2.8b. Larva of *Anopheles plumbeus.*

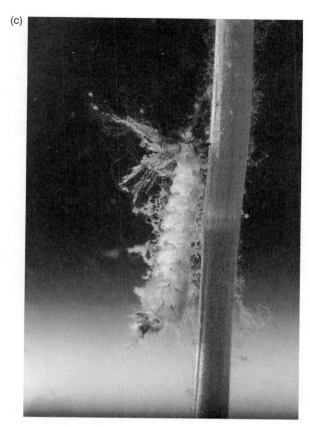

(c)

Figure 2.8c. Larva of *Coquillettidia richiardii* attached to plant tissue (photo, Hollatz).

When the larvae leave the water surface the lobes are retracted and the spiracles closed. When they reach the water surface the flaplike 5 or 4 spiracular lobes are pulled into their extended position by the surface tension forces. A gland adjacent to the spiracles of larvae secrete hydrophobic substances to avoid the influx of water into the respiratory system.

The larvae of *Coquillettidia* and *Mansonia* live submerged. They therefore possess a siphon which is modified for piercing submerged parts of aquatic plants to take the oxygen from the aerenchyma. The spiracular apparatus at the distal end of the siphon contains hooks and a saw-like blade with teeth to pierce the plant tissue. These larvae have a more sessile habit, hanging head downwards whilst attached to the plant tissue and filtering the water column for food

(Fig. 2.8c). They are therefore not easily recognised by predators such as fish.

The larval food consists of microorganisms, algae, protozoa, invertebrates and detritus. On the basis of their feeding behaviour they may be classified into filter or suspension feeders, browsers or predators. The filter feeders collect food particles suspended in the water column (especially larvae of *Culex*, *Coquillettidia*, subgenus *Culiseta* or to some extent *Aedes/Ochlerotatus* larvae). The larvae of the filter feeders typically hang on the water surface, filtering the water column beneath the surface by beating their head brushes (lateral palatal brushes) towards the preoral cavity. This generates water currents which carry food particles towards the mouth (Dahl *et al.*, 1988). Mosquito larvae are usually not discriminatory in what they ingest. However, the size of the particles is usually less than 50 μm. Larvae can also move slowly in the water column while filter feeding. The browsers (*e.g.* most *Aedes* and *Ochlerotatus* species) collect food by resuspension, scraping or shredding particles, microorganisms, algae, and protozoa from the surface of submerged substrates or the microbial film at the air–water interface (*Anopheles* larvae). Even small parts of dead invertebrates and plants can be bitten off with the mouth parts.

The anopheline larva hangs horizontally under the water surface with its dorsal side uppermost and the mouthparts directed downwards. When feeding, the larva rotates its head through 180° and creates a water current by beating its head brushes to collect the food organisms in the surface film.

Predacious larvae of the genera *Toxorhynchites*, *Aedes*, *Psorophora* and *Culex* which feed upon insects (often other mosquito larvae) do not occur in Europe.

Disturbances of the water surface cause the larva to dive for a short period of time. They dive by flexing the abdomen and moving backwards. When the larva returns to the water surface, it swims backwards until the abdomen comes into contact with the surface.

Larvae moult four times at intervals before reaching the pupal stage. At each moult the head capsule is increased to the full size characteristic of the next instar, whereas the body grows continuously. Thus the size of the head capsule is a fairly good morphometric indicator for the larval instar. Each moult is coordinated by the relative concentrations and interactions of juvenile hormone and ecdysone, a moulting hormone.

The development of larvae is temperature dependent. There are great differences in the optimum temperature for the development of different mosquito species (Figs 2.9a–c). For instance, the snow-melt mosquitoes can complete their development at temperatures as low as 10°C, whereas they are incapable of developing successfully at temperatures above 25°C (Fig. 2.9a). Usually they hatch in southern and central Europe during February, or later in the northern parts, and the adults emerge two to three months later. Larvae of those species which overwinter in the larval stage such as

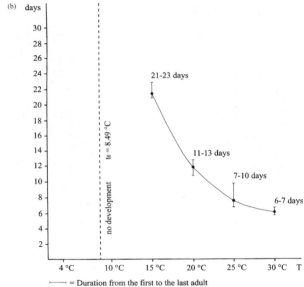

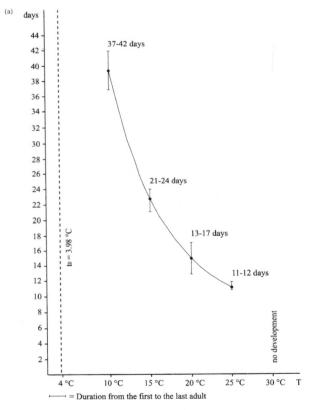

Figure 2.9a. Larval and pupal development of *Ochlerotatus cantans* in relation to water temperature. (t_0 = theoretical temperature where no development is possible.)

Figure 2.9b. Larval and pupal development of *Aedes vexans* in relation to water temperature.

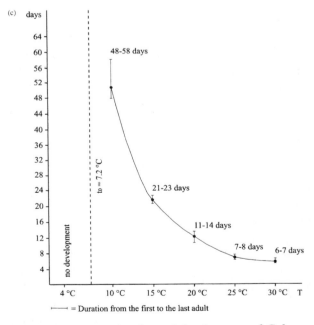

Figure 2.9c. Larval and pupal development of *Culex p. pipiens* in relation to water temperature.

Oc. rusticus or *Cs. morsitans* will survive in water close to freezing, or even in waters which become regularly coated with ice.

In contrast, the floodwater mosquitoes, *e.g. Ae. vexans*, develop successfully at high temperatures within a short period of time (6–7 days from hatching to emerging at 30°C) (Fig. 2.9b). Larvae of *Cx. p. pipiens* can successfully develop in a wide range of temperatures (10–30°C) (Fig. 2.9c). The development of the aquatic stages represents also an adaptation to the ecological environment in the breeding water.

Mosquito larvae (*e.g. Ae. vexans*) sometimes aggregate at particular places in the breeding site. This crowding effect may be a mechanism to reduce the chance of predation of any single larva.

2.5. PUPAE

The pupae are also aquatic. The pupal stage usually lasts about two days. However, this period may be reduced or extended, at higher or lower temperatures repectively. During the pupal stage the process of metamorphosis takes place. Some larval organs are histolysed, whilst the body of the adult is formed through the development of imaginal discs (cells or groups of cells that stayed quiescent in the larval body until the pupal stage). In particular the fat body of the larvae is transferred to the adult stage, and used as a source of vitellogenines for autogenous egg formation or as a reserve for hibernation.

Characteristically, the head and thorax of the pupae are fused into a prominent cephalothorax which carries anterolaterally two respiratory trumpets (see Chapter 5, Fig. 5.18), which are connected with the mesothoracic spiracles of the developing adult to provide the organism with oxygen. The abdomen terminates with two paddles and is kept flexed under the cephalothorax. When at rest the pupae float motionless at the water surface, with the tip of the trumpets and the palmate setae of the first abdominal segment in contact with the water surface. The hydrophobic rim of the trumpets protrude through the water surface for respiration. An air bubble between the appendages of the cephalothorax makes the pupa positively buoyant. Mosquito pupae, unlike the pupae of most other insects, are quite mobile. When the pupa is disturbed it dives by straightening the abdomen and spreading the paddles, then it rapidly flexes the abdomen which has retained the larval musculature. In contrast to larvae which have to swim actively to the water surface, the pupa floats passively back to the surface after diving. Pupae of most mosquitoes are relatively tolerant of dessication, and adults can emerge successfully even if pupae have been stranded, or when the breeding sites have almost dried out. Unlike larvae, pupae do not feed.

In *Mansonia* and *Coquillettidia* the trumpets are modified for penetrating plant tissues. The pupae, like the larvae, take the oxygen from the aerenchyma of the submerged parts of the plants.

2.6. ADULTS

2.6.1. Emergence

When the final stage of metamorphosis is complete, gas is forced between the pupal and the pharate adult cuticle, and into its midgut. The pupa straightens the abdomen into a horizontal position, and by swallowing air it further increases the internal pressure. The cephalothoracic cuticle of the pupa then splits along the ecdysial line and the adult slowly emerges from the pupal skin (Fig. 2.10). The emerging adult moves particularly carefully to avoid falling onto the water surface, whilst its appendages are still partly in the exuvia. In this phase the emerging individual is highly susceptible to strong winds.

The pupae of the genus *Coquillettidia* are fixed to the plant tissue in the water body. At the end of the pupal development they have to float to the water surface. Therefore, the tips of the pupal trumpets break to release the pupa from the plant before emergence (Mohrig, 1969).

Figure 2.10.   Emerging mosquito.

After emergence the adult increases the haemolymph pressure which causes the legs and wings to stretch. It then immediately ejects droplets of fluid to empty the gut, while air disappears from the gut some hours later. Within a few minutes, when the soft cuticle has sclerotised, it is able to fly. However a further 1 or 1.5 days are required for males and females respectively to adjust their metabolism (Gillett, 1983).

There is also a difference between male and female sexual maturity at the time of emergence. Male mosquitoes are not sexually mature at emergence. They have to rotate their hypopygium through 180° before they are ready to mate, which takes about one day. The males of one population emerge therefore 1–2 days before the females, in order to achieve sexual maturity at the same time as the females. Since the pupal stage of the two sexes appears to be about the same length, the shortening in development of males takes place primarily in the larval stage. Consequently, the male pupae and adults of a population, are smaller in size than the corresponding females.

Following emergence the adults are ready to fulfil their designation of mating, feeding and oviposition.

2.6.2. Mating

Mating of most mosquitoes in the palaearctic region takes place when females enter swarms of flying males. Such a swarm can consist of only a few or up to several thousands of male mosquitoes. Usually males form swarms over a marker at low light intensities especially in the evening and morning. Markers are projecting objects which contrast with the surroundings such as bushes. When swarming, males face into the wind and fly forward and backwards, up and down over the marker. This oscillating flight pattern is often called "dancing flight". The sound produced by the male wing beat has a frequency of about $600 \, cs^{-1}$. The frequency of the wing beat of the females is lower than that of the males at $500–550 \, cs^{-1}$ and even less when engorged.

The plumose antennae of the males are especially receptive to the sound generated by the female. The flagellum of the antenna starts to vibrate and stimulate the Johnston's organ which is located in the swollen second segment of the antenna, the pedicel (Clements, 1963; McIver, 1982). Contact pheromones may also be involved in the mating behaviour.

When a female enters a swarm it will be seized immediately by a male. Usually the male and female copulate face to face when flying outside the swarm (Fig. 2.11). The copulation requires a complex merging of the male and female reproductive structures. Usually it takes less than half a minute for the male to deposit the spermatozoa in the bursa copulatrix of the female (Clements, 1963), the sperm then moves to the spermathecae. The male accessory glands secretions contain a substance, known as matronae, which after copulation makes the female unreceptive for the rest of her life. The females store sufficient sperm in their spermathecae to fertilise several egg batches without further copulation. In contrast to females, male mosquitoes may mate many times. The time and preferred location of swarming is species specific. Swarming (eurygamy) is not necessary for all species, and

Figure 2.11. Mating mosquitoes. (Photo courtesy of Roland Kuhn, University of Mainz, Germany).

some species may mate without it (stenogamy). Usually mating takes place soon after emergence, since biting females are almost invariably inseminated. After insemination the search for a host for the blood-meal is the next important phase in the reproductive life of the female.

2.6.3. Dispersal and Host-seeking Behaviour

In most mosquito species oogenesis can only be completed when the females take a blood-meal. Therefore, they have developed a complex host-seeking behaviour to locate a potential host. Primarily the location of the host is based on olfactory, visual and thermal stimuli. Females possess numerous antennal receptors which respond to host odours. The main olfactory stimuli are carbon dioxide, lactic acid, octenol, acetone, butanone and phenolic compounds. The host-seeking process may differ within species depending on the season and the availability of certain hosts. However, it can usually

be divided into three phases (Sutcliffe, 1987):

(a) non-oriented dispersal behaviour which enhances the likelihood of the female coming into contact with stimuli derived from a potential host.
(b) oriented host location behaviour resulting from contact with host stimuli. The strength of the stimuli are increased as the mosquito and host come closer together;
(c) attraction to a suitable candidate host, once the female has identified it in her immediate vicinity.

The extent of the non-oriented dispersal behaviour differs from species to species. In general it can be separated into (a) species which usually breed and rest close to the habitat of their hosts and therefore do not fly long distances (most container breeders *e.g. Cx. p. pipiens*); (b) species which disperse moderate distances from their breeding or resting places to the hosts' habitats (some snow-melt mosquitoes *e.g. Oc. rusticus*); (c) species, which migrate considerable long distances to invade new habitats for biting, and/or egg-laying when suitable habitat is available (some floodwater mosquitoes *e.g. Ae. vexans*).

The flight behaviour is influenced by temperature, humidity, illumination levels, wind velocity and the physiological stage of a female. For instance, most *Aedes/Ochlerotatus* species migrate during the twilight when the temperature is dropping and the humidity is increasing. They are usually more active on nights with moonlight (Bidlingmayer, 1964).

Species with a tendency for extensive flight activities usually show two different non-oriented dispersal behaviours (Provost, 1953), a drift with the wind (so-called passive migration) and an active dispersal (so-called appetitive flight).

During passive migration, the mosquitoes ascend in swarms and use the wind to drift long-distances, and may occur suddenly in large numbers far away from their breeding places. This non-oriented flight activity is especially influenced by the speed and direction of the wind and guiding landmarks. The passive migration in swarms occurs only a short time after emergence (Bidlingmayer, 1985).

During the appetitive flight female mosquitoes, usually at least 24 hours after emergence, disperse actively. They fly upwind when the wind velocities are below mosquito flight speed, which is approximately 1 m/sec. (Bidlingmayer and Evans, 1987). The flight against the wind increases the likelihood of encountering stimuli deriving from a host. However, strong wind prevents active dispersal. This behaviour is species specific and depends on various features of the terrain and meteorological factors. The microclimate influenced by the vegetation type, which causes *e.g.* increased humidity and reduced wind, strongly affects the dispersal behaviour. Therefore, females usually fly close to the ground or slightly above the top of the vegetation. According to the preferred microclimate requirements, some species occur in greatest numbers in open areas (mostly strong flyers), others in woodlands (woodland species are moderate flyers), a third group prefers edges of fields and forests and finally the fourth group comprises the urban domestic species, which are usually weak flyers (Gillies, 1972; Bidlingmayer, 1975). Experiments show that *Ae. vexans* migrates approximately 1 km per night during warm and humid weather periods with moderate wind speeds. Increasing numbers of *Ae. vexans* females have been caught in CDC-traps at a distance of about 5 km from their breeding place eight days after emergence, and within two weeks at a distance of 10 km and more. Clarke (1943) recorded migration distances of marked *Ae. vexans* females of 22 km and Gjullin and Stage (1950) and Mohrig (1969) up to 48 km.

In contrast, snow-melt mosquito species stay near their breeding sites and do not regularly migrate long distances (Schäfer *et al.*, 1997). In mark-release-recapture experiments Joslyn and Fish (1986) collected *Oc. communis* females at distances up to 1600 m from their breeding sites. Nielsen (1957) reported a maximum flight range of about 1600 m for *Oc. communis* and *Ae. cinereus*, with an average dispersal range of less than half this distance.

In Germany, *Oc. rusticus* females have been found resting in the forest during the daytime, and flying to the forest edge and the adjoining fields with approaching dusk. Females preferred to disperse along rows of trees in open areas. It can be assumed that in the absence of other attractants or adverse meteorological conditions, the flight of these mosquitoes was guided by the reduced level of illumination beneath the forest canopy, the contrast in illumination with the adjoining forest edge as well as the visual image of the rows of the trees. The mosquitoes obviously follow their hosts, mostly red deer, when these animals browse on the meadows next to the forest. The flight distance observed was only a few hundred metres (Schäfer *et al.*, 1997).

Dispersal serves mostly to bring the blood-sucking insects into contact with a suitable signal from a potential host animal. It is likely therefore that species which breed in areas where few hosts are available, develop a stronger tendency for migration than those which breed in the vicinity of their hosts. For instance, *Cx. p. pipiens* which breeds in human settlements, migrates usually less than 500 m. It is likely that females searching for a host will find stimuli from a suitable host within a few hundred metres.

In field studies it was shown that both a horizontal and a vertical dispersal behaviour assists the host-finding process. Females of *Aedes* spp. and *Ochlerotatus* spp. (*Ae. vexans*, *Ae. rossicus*, *Ae. cinereus* and *Oc. sticticus*) were most frequently captured in traps between ground level and 4 m, whereas in a height of 10 m *Cx. p. pipiens* was, with 99.2%, by far the most abundant species. There is an interaction between the availability of suitable hosts and the distribution of mosquito species. For blood-seeking females of ornithophilic species *e.g. Cx. p. pipiens* and *Cs. morsitans* it is an advantage to search for birds in the canopy. In contrast, *Aedes* and *Ochlerotatus* species prefer mammals as hosts, which explains the dominance of these species at ground level.

After encountering the host stimuli, the female mosquito changes its behaviour from the non-oriented flight pattern to an oriented host location. Initially the mosquito is activated by host odour and then it uses this odour to track the host from a distance of more than 20 m. It is the release of carbon dioxide by the host, and the change in concentration of carbon dioxide in combination with other stimuli, which elicits behavioural responses. Mosquitoes are sensitive to very small changes in carbon dioxide levels. The receptors on mosquito palps show responses to changes as small as 0.01% (Kellogg, 1970). There are many other components of host breath and odour which stimulate the antennal receptors of female mosquitoes when mixed with carbon dioxide. For instance, lactic acid is an activating and orientating stimulus for mosquitoes, but only if carbon dioxide is also present in the air stream (Smith *et al.*, 1970; Price *et al.*, 1979). Interaction or synergism of the components of the host odour in attracting a given species, is a very complex process which developed in the course of evolution between the insect and the target organisms. The components of the host odour only stimulate the mosquito female when they occur in a distinctive mixture typical of the host. This enables it to distinguish between different hosts and to trace the plume as a series of packets, lamellae and filaments of odour mixed and dispersed by wind (Murlis, 1986). The female mosquito flies in an upwind, zigzagging pattern which holds the mosquito within the plume and brings it closer to the odour source. In the final stages of orientation mosquitoes, especially those which bite during daytime or in twilight, use visual contact to locate the host. The compound eyes serve to discriminate between form, movement, light intensity, contrast and colour. Mosquitoes respond particularly to blue, black and red colour, whereas least attraction is caused by white and yellow (Lehane, 1991). It is unlikely that the utilisation of colour information is well developed in mosquitoes active at night, but they may be particularly sensitive to intensity contrast between the background and the target. When the mosquito is close to the host it may also distinguish between three-dimensional targets, and infrared radiation may also be involved in host location.

In the immediate vicinity of the host, odour is important once again as well as its body heat. Mosquitoes can easily detect temperature differences of 0.2°C. Water vapour in short-range orientation-attraction may also play a role (Lehane, 1991).

2.6.4. Feeding

Mosquitoes developed piercing and sucking mouthparts for the uptake of plant juices as a source of carbohydrates, but only those of the females can also be used to pierce the skin to obtain blood for egg maturation (Magnarelli, 1979; Clements, 1992). The mouthparts are extended into a proboscis which consists of the stylet bundle (fascicle) and the labium which encloses it when the mosquito is not feeding. The stylet bundle is formed from six long thin stylets: the labroepipharynx, hypopharynx, and two of both maxillae and mandibles. These stylets are held tightly together. The mandibles are sharply pointed and used to rupture the skin for passage of the other stylets. The maxillae have a pointed tip and recurved teeth at their distal ends. They are the main penetrative elements of the mouthparts, which are thrust alternately by using the teeth to anchor themselves in the tissues. Thus the labroepipharynx (food channel) and hypopharynx (salivary channel) are pulled with them into the tissue (Robinson, 1939). The penetration of the stylets into the tissue is caused by an alternating rotary movement of the head. Four muscles are associated with each maxilla, to protract and retract the maxillary stylets (Clements, 1992).

After the females land on the host, they may probe the skin a few times with the labellae while they are searching for a capillary for the intake of blood. Thickness and temperature of the skin are probably important probing stimuli for mosquitoes, since the surface temperature of the skin is related to the number of blood vessels in the skin (Davis and Sokolove, 1975). Sensilla on the ventral side of the pair of labellae, and on the distal part of the labium, contain receptors for stimuli that may indicate a suitable site for piercing the skin.

Having successfully punctured the skin, the female starts ingesting blood. When feeding, only the stylet bundle passes into the skin, while the labium becomes progressively bent backwards as the mosquito probes deeper into the tissue. Two pumps, the cibarial pump below the clypeus and the pharyngeal pump, are the sucking organs which pump the blood or plant juices into the gut.

It is important for the female mosquito that the blood remains in a liquid form and does not coagulate, as the mosquito would not then be able to complete the blood-meal. Therefore, the mosquito injects saliva into the wound. This usually contains anticoagulants which are similar to hirudin produced by blood-sucking leeches (Parker and Mant, 1979). The introduction of saliva into the host tissue usually stimulates an immune response which can cause an inflammatory reaction of the host at the site of piercing. These sites often itch, and when the host is scratching, the small wound can be infected with bacteria. When the stylets reach a blood vessel, blood-associated components (*e.g.* ADP and ATP) function as phagostimulants, and the mosquito starts to take in blood through the food channel.

The female mosquito can ingest more than three times its mean body weight (Nayar

and Sauerman, 1975). Larger species such as *Oc. cantans* can ingest more than 6 μl and smaller species such as *Ae. cinereus* only 3.7 μl. The blood and especially its protein ingredients are essential for egg-production by anautogenous females. Only a few autogenous species such as *Cx. p. pipiens* biotype *molestus* are able to produce their first egg batch without a bloodmeal. When they originate from eutrophic breeding sites (*e.g.* septic tanks) where the larvae can develop a prominent fat body due to the high level of nutrition, the fecundity is quite high, but still lower than after a blood-meal by the female. The fat body is obviously enough to complete ovarial development without a further blood-meal. The blood is used more for egg production and less as a source of energy. Both sexes of mosquitoes require plant juices as an energy source, mostly for flight. Plant sugars such as floral nectar, damaged fruits and honeydew are the main energy source during the adult life of both sexes (Briegel and Kaiser, 1973).

Mosquitoes differ in their feeding and resting behaviour. Species which preferably feed indoors are called endophagic (endophagy), and those which feed mainly outdoors, are called exophagic (exophagy). The females which rest after feeding or during the day outdoors are called exophilic (exophily), and those which rest indoors are called endophilic (endophily). Ornithophily is expressed when females prefer to feed on birds (ornithophilic species), zoophily when they feed on other animals (zoophilic species) and the term anthropophily is used when they prefer to feed on humans (anthropophilic species).

2.7. HIBERNATION

Mosquitoes of temperate zones have developed efficient overwintering mechanisms in the egg, larval or adult stages. Some species such as *Oc. rusticus* and *Cs. morsitans* can overwinter in more than one stage, *e.g.* in the larval and the egg-stage. Several factors, especially the latitude (cold) and hydrological conditions (droughts) determine the duration of hibernation and can differ within one species according to the latitude.

2.7.1. Egg Stage

Hibernation in the egg-stage is practised by most of the *Aedes/Ochlerotatus* species in the temperate zone. Their diapause is induced in such a way that they do not hatch when unsuitable climatic and hydrological conditions prohibit successful development to the adult stage (see Chapter 2.3). The occurrence of larvae of these species overwintering in the egg stage can vary greatly within the species of *Aedes* and *Ochlerotatus*. Within some species the hatching time is closely related to snow-melt, others hatch in late spring or summer.

2.7.2. Larval Stage

Some mosquitoes are known to overwinter in the larval stage and can even survive for days in breeding sites with a frozen surface. During the cold season their metabolism is reduced and the larval development delayed. For instance, larvae of *Oc. rusticus* and *Cs. morsitans*, which hatched in autumn, hibernate in the second and third larval instar. The high content of dissolved oxygen in cold water or bubbles of oxygen under the ice enable the larvae to cover their demand of oxygen and to survive. However, during a severe winter the mortality rate can be very high. Some anopheline species such as *An. claviger* and *An. plumbeus* hibernate as larvae in pools or tree holes, respectively. Usually, hibernation takes place in the 3rd or 4th larval stage in water bodies that do not entirely freeze or only for a short time. This is also true for the hibernating larvae of *Orthopodomyia pulcripalpis*. In contrary to the former mentioned species, the larvae of *Cq. richiardii*, which usually hibernate in the 3rd or 4th-instar are not sensitive to long frost periods, because they live submerged in permanent water bodies.

Larvae of *Cx. p. pipiens* can frequently be found during winter. Whereas the females of anautogenous, ornithophilic, eurygamous *Cx. p. pipiens* are overwintering in diapause, its biotype *molestus* reproduces during the winter. Therefore, all developing stages of this biotype can be found in the breeding habitat (usually underground breeding sites) within the temperate zones during winter.

2.7.3. Adult Stage

Most of the mosquitoes, which overwinter as adult females belong to the genera *Culex, Culiseta, Uranotaenia* and *Anopheles*. They seek for hibernating shelters (locations free of frost such as caves, stables, cellars, canalisations and earth burrows) during autumn and leave these places in spring when the temperature increases. Usually, females of these species use the larvel fed body and feed intensively on plant juices during autumn to synthesise huge lipid reserves for diapausing. Females of some species within the Anopheles Maculipennis Complex can take occasional blood-meals during winter to withstand the long period of starvation (Clements, 1992).

3

Medical Importance of Mosquitoes

Mosquitoes are responsible for the transmission of many medically important pathogens and parasites such as viruses, bacteria, protozoans, and nematodes which cause serious diseases like malaria, dengue, yellow fever, encephalitis or filariasis (Kettle, 1995; Beaty and Marquardt, 1996; Lehane, 1991). Transmission can be mechanical (*e.g.* myxoma virus causing myxomatosis in rabbits) or biological. The latter is more complex because it involves an obligatory period of replication and/or development of the pathogen or parasite in the vector insect. Due to their blood-sucking behaviour, mosquitoes are able to acquire the pathogens or parasites from one vertebrate host and pass them to another, if the mosquito's ecology and physiology is appropriate for transmission. Highly efficient vectors have to be closely associated with the hosts, and their longevity has to be sufficient to enable the pathogens/parasites to proliferate and/or to develop to the infective stages in the vector. For successful transmission, usually multiple blood-meals are necessary.

In terms of morbidity and mortality caused by vector-borne diseases, mosquitoes are the most dangerous animals confronting mankind. They threaten more than two billion people in tropical and subtropical regions, and have substantially influenced the development of mankind, not only socio-economically but also politically. Undoubtedly insect-transmitted pathogens leading to epidemics and pandemics have been instrumental in the development, decline and fall of empires *e.g.* in Greece and Rome. Malaria was the dominant health problem in the latter days of the Roman Empire (Bruce-Chwatt and de Zulueta, 1980). The Roman marshes were notorious for "*mala aria*" (bad air). This disease killed also Alexander the Great and prevented the conqueror from extending his empire, to mention but a few examples.

3.1. MALARIA

Human malaria caused by the protozoans *Plasmodium* spp. continues to be the most important vector-borne disease. It affects more than 100 tropical countries, placing more than 40% of the world population at risk. Some 300 million people are believed to be infected with malaria parasites, with 90% of them living in tropical Africa (WHO, 1993; WHO, 1997a,b). In Africa, the disease is probably responsible for no less than 500 000 to 1.2 million deaths annually, mainly among children below the age of five. The enormous total of lives and days of labour lost, the costs of treatment of patients, and the negative impact of the disease on development make malaria a major social and economic burden. The annual costs of malaria in Africa alone were estimated to be almost 2 billion US$ (WHO, 1993).

Four species of the genus *Plasmodium* (*P. falciparum, P. vivax, P. ovale* and *P. malariae*) cause human malaria and are transmitted solely by anopheline mosquitoes. About 20 *Plasmodium* species occur in other primates, a similar number in other mammals, and about 40 each in birds and reptiles (Garnham, 1980, 1988).

The *Plasmodium* species have a complex replication and transmission cycle with the sexual replication in mosquitoes and the asexual replication in vertebrates (Fig. 3.1). Shortly after the ingestion of blood from infected vertebrates containing sexual forms of the parasite, the gametes fuse in the mosquito gut to form a zygote which elongates and develops into a motile ookinete (Fig. 3.2A). It penetrates to the outside of the midgut epithelium, settles there and forms an oocyst (Fig. 3.2B). Meiotic and subsequent mitotic divisions (sporogony) within the oocyst

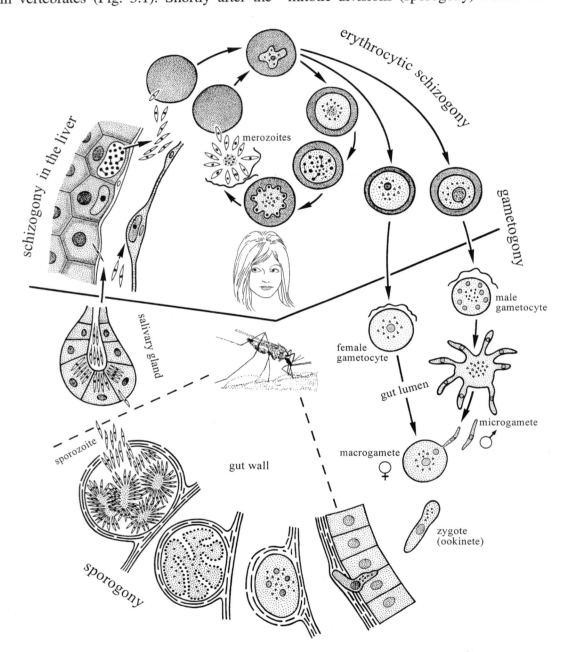

Figure 3.1.   Life cycle of malaria parasites in *Anopheles* and the human host.

result in the formation of many haploid, spindle-shaped sporozoites which burst the wall of the oocyst and migrate through the haemocoel and accumulate in the salivary glands (Fig. 3.2C-F). The infected mosquito is now able to inject the sporozoites with saliva into the next host. Once inside a vertebrate, the sporozoites infect the parenchyma of the liver where they immediately undergo a cycle of exoerythrocytic schizogony, or develop to latent hypnozoites which sometimes later undergo schizogony to cause relapse. In each large schizont several thousand merozoites are formed and released into the circulating blood (end of the prepatent period), where they invade the erythrocytes to commence the erythrocytic schizogony. In the infected erythrocyte the merozoite becomes a feeding trophozoite and then, when a fully grown schizont, produces a small number of new merozoites (Garnham, 1966). The merozoites burst the erythrocytes, are released and then invade other erythrocytes to repeat the schizogonic cycle. Each release of the merozoites from the erythrocytes causes an attack of malaria with fever and other clinical signs. The length of the schizogonic cycle determines the interval between the fever attacks.

P. falciparum causes malignant tertian, *P. vivax* and *P. ovale* benign tertian malaria recurring at 48 h intervals (fever attack at the third day); and *P. malariae* quartan malaria recurring at 72 h intervals (Kettle, 1995). After several cycles of schizogony, some trophozoites do not produce merozoites but become gametocytes which have to be ingested by anopheline mosquitoes to conclude the cycle of development.

In humans malignant malaria caused by *P. falciparum* is the most severe form resulting in life-threatening complications such as anaemia and cerebral malaria. This is a frequent cause of mortality in children and can kill up to 25% of non-immune adults within two weeks. This form of malaria is called malaria tropica and occurs mostly in tropical and subtropical areas, being limited by a summer isotherm of 20°C which is necessary to complete sporogony of the parasite

in the mosquito. In contrast, *P. vivax* can complete sporogony in mosquitoes in areas with a summer isotherm of 16°C (Wernsdorfer, 1980).

In Europe, malaria threatened human life until the first half of the 20th century. Although the impact of the disease was more severe in southern Europe, it is well documented that even in northern Europe malaria was a well-known hazard of life (Marchant *et al.*, 1998). It is known that for instance, Napoleon lost large numbers of soldiers due to malaria when he invaded the Upper Rhine Valley in Germany. The two main *Plasmodium* species found in Europe were *P. vivax* and *P. falciparum*. Whereas the former occurred throughout the continent, the latter was restricted to southern Europe (Jetten and Takken, 1994). In northern Europe the parasite must have been *P. vivax* because of its adaptation to the moderate climate. Furthermore, it is likely that the parasite could survive as hypnozoites in the human liver during phases too cold for mosquito transmission (Marchant *et al.*, 1998). Nowadays, *P. vivax* seldom causes a lethal disease, which suggests that *P. vivax* has evolved a reduced virulence over the last century (Kettle, 1995).

Before World War II, endemic malaria was spread throughout Europe (Bruce-Chwatt and de Zulueta, 1980). The most endemic malaria areas were found in the south, where a continuous transmission occurred from spring to autumn. Greece was considered to be the country with the highest incidence of malaria. In the early 1930s the annual number of people in Greece infected with malaria ranged from one to two million. Other severe epidemics were reported from the Dalmatian coast in Croatia, in coastal areas in southern Spain and in the south of Italy and the island of Sardinia. In central Europe the malaria incidence was much lower than in the south (Jetten and Takken, 1994). In northern and western Europe the transmission of the disease was discontinuous with annual maxima. Malaria epidemics were mainly restricted to coastal areas in southern Sweden and southern Finland,

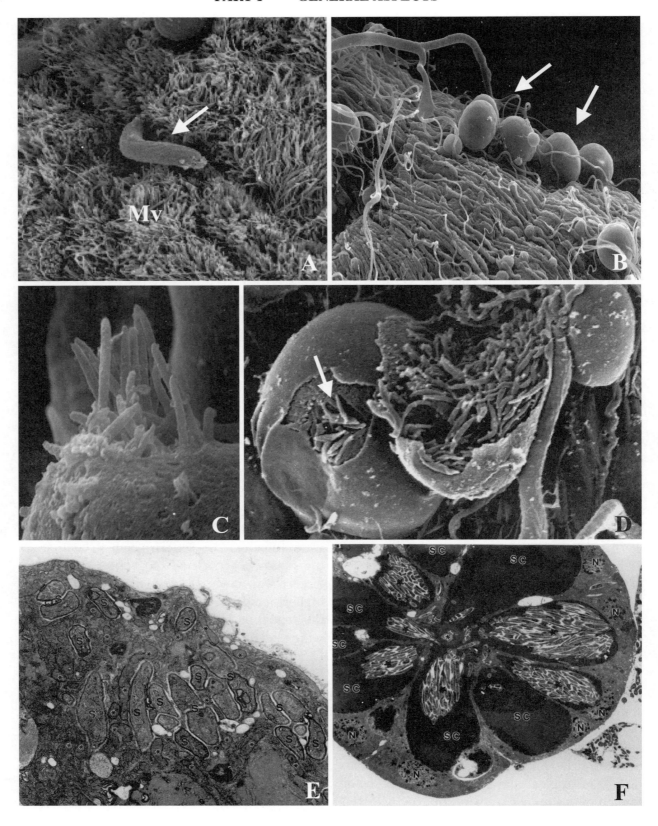

Denmark, the Netherlands, Belgium, Germany and northern France. In the eastern parts of Europe malaria was mainly recorded from the southern Ukraine and along the lower Volga River. After World War II, malaria slowly disappeared from the continent. This was mainly due to the reduction in natural breeding habitats through improved agricultural techniques, improved social and economic conditions and better sanitation. An important role was also played by the malaria eradication campaigns with the application of residual insecticides and the availability of new drugs. The last reported focus of indigenous malaria in continental Europe disappeared in Greek Macedonia in 1975 (Bruce-Chwatt *et al.*, 1975).

There are indications that malaria was mainly transmitted in Europe by species of the Anopheles Maculipennis Complex, which are widely distributed in the Palaearctic region. However, the distribution of *An. maculipennis s.l.* was not directly related to the distribution of malaria (Jetten and Takken, 1994). What was the reason for the "Anophelism without malaria"? Following intensive research it was established that the former described species *An. maculipennis* is not a single species but a species complex consisting of more than a dozen separate species, of which eight occur in Europe (White, 1978). The knowledge of species complexes, that contain species that are morphologically very similar but differ greatly in their vector competence, has generated interest in the control of malaria by genetic manipulation (Crampton *et al.*, 1990; Crampton, 1992; Crampton and Eggleston, 1992; Kidwell and Ribeiro, 1992; Carlson, 1995; Rai, 1995).

In addition to the species of the Anopheles Maculipennis Complex, some other European anophelines are known to be potential malaria vectors, such as *An. claviger, An. sergentii, An. cinereus hispaniola, An. algeriensis, An. superpictus* and *An. plumbeus*. The latter is increasingly becoming of interest to malariologists, because in recent decades it has proliferated in huge numbers as a result of its adaptation from natural (tree holes) to artificial breeding sites (water catch basins and septic tanks with water contaminated with organic waste). In recent studies it has been demonstrated that in contrast to *An. atroparvus* which is more or less refractory to *P. falciparum, An. plumbeus* is able to develop oocysts of *P. falciparum* when fed with blood containing gametocytes (Marchant *et al.*, 1998). Although *An. plumbeus* lives in close proximity to man, the risk of malaria epidemics is very low due to good malaria notification and the absence of cases of indigenous transmission. However, the effect of global warming which favours the completion of the sporogonic cycle of *Plasmodium* in anophelines, and the hundreds of malaria cases mostly caused by *P. falciparum* acquired in the tropics and imported to Europe, could increase the risk of some indigenous malaria transmission, even though malaria epidemics can be excluded in Europe (Fig. 3.3). Nevertheless imported malaria cases regularly cause a few deaths if diagnosis is delayed.

Figure 3.2. A–D (from a scanning electron microscope): A – *Plasmodium* ookinete (arrow) attached to the mosquito midgut microvillar epithelium (Mv). The anterior part of the parasite is already invading between the microvilli. B – View of an infected midgut (6 days after the infective bloodmeal). Several immature oocysts (arrows) are attached to the outer midgut wall. C – Mature oocyst releasing the sporozoites to the hemocele (10 days after infection). D – Two broken oocysts in different phases (12 days of infection). One oocyst at the beginning of sporozoite release (arrow) and the other opened showing tens of sporozoites ready to escape. E, F (from a transmission electron microscope): E – Initial phase of salivary gland invasion by sporozoites (10 days after infection). The whole cytoplasm of the secretory cell is full of sporozoites (s). F – Late phase of salivary gland invasion (15 days after infection). All sporozoites are inside the secretory cavities (sc) forming longitudinal arrangements (*). Some parasites are in the salivary duct (N = nucleus). (Micrographs courtesy of Paulo Pimenta, Fundacao Oswaldo Cruz, Brazil).

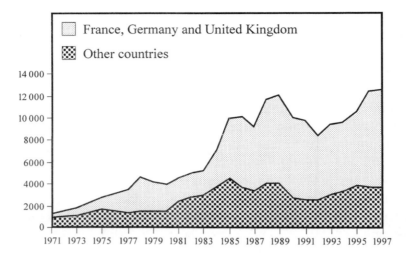

Figure 3.3. Malaria cases imported in Europe in 1971–1997.

Not only infected people import the parasite to Europe; infected anophelines can also be carried by aircraft into Europe. These insects are a threat for people working in or living close to international airports. Since 1963, more than 60 cases of airport malaria have been reported. In 1994, in the vicinity of the airport Charles de Gaulle in Paris alone, 7 people were infected by *P. falciparum* although they had never visited the tropics. Similar cases are reported from Gatwick Airport in London and Frankfurt International Airport.

3.2. VIROSIS

Arboviruses (**ar**thropod-**bo**rne-viruses) are defined as viruses that replicate in arthropods and are transmitted by arthropods to vertebrates. The arthropod becomes infected by feeding on blood from an infected vertebrate during viraemia (virus circulation in the peripheral blood vessels), and after proliferation in the vector the virus can be transmitted to another vertebrate-host (horizontal transmission). Arboviruses can also be passed from one arthropod generation to another by transovarian transmission (vertical transmission). Thus, some of these viruses are known to be capable of overwintering in the egg stage of the vector (*e.g.* a few *Aedes/Ochlerotatus* species).

More than 300 arboviruses are listed by Francki *et al.* (1991) and more than 500 by Karabatsos (1985). Approximately 100 viruses infect humans and 40 infect livestock (Monath, 1988). The most important viruses transmitted by mosquitoes to humans or other vertebrates are found in three families: the Togaviridae with the genus *Alphavirus* (*e.g.* Sindbis virus, equine encephalitis viruses), Flaviviridae with the genus *Flavivirus* (*e.g.* yellow fever virus, dengue 1–4 viruses, West Nile virus, Japanese and St. Louis encephalitis viruses) and the Bunyaviridae with the genera *Bunyavirus* (*e.g.* the California group), and *Phlebovirus* (Rift Valley fever) (Murphy *et al.*, 1995). Human arboviral diseases are classified by the major clinical symptoms they cause such as encephalitis, febrile illness accompanied by rash and arthritis as well as haemorrhagic fever. Infections can cause a wide range of mild or severe symptoms with significant morbidity and mortality, especially in tropical countries.

Dengue and dengue haemorrhagic fever, primarily transmitted by the mosquito *Ae. aegypti*, constitute an important burden to mankind in terms of morbidity and mortality. About 1.5 billion people in the tropics, mainly in Asia, the Western Pacific region, the Caribbean, as well as Central and South America, live under the risk of dengue virus infection (Halstead, 1980, 1982, 1992; Becker *et al.*, 1991; Gratz, 1999). In Europe, one of the most recent outbreaks of dengue occurred in Athens during 1927 and 1928, when the dengue viruses caused a devastating epidemic of fever and polyarthritis. Approximately 1 million people were infected (80% of the inhabitants at that time) with more than 1500 fatalities (Papaevangelou and Halstead, 1977). It is most likely that the vector of this flavivirus was *Ae. aegypti*, a species no longer present in the region (Mitchell, 1995). Dengue cases also occurred in Spain, Italy, Austria (Vienna had a large epidemic in 1941) and Yugoslavia. In Europe, especially in countries bordering the Mediterranean, people may still suffer from this disease.

In Europe, arbovirus diseases were considered a minor threat for humans after the disappearance of indigenous dengue cases. This changed in the 1980s when Ockelbo virus (*Alphavirus*, subtype of Sindbis) caused outbreaks of rash and polyarthritis in northern Europe. More recent severe outbreaks of West Nile fever in Romania in 1996 and 1997 as well as in southern Russia ocurred (Rodhain and Hannoun, 1980; Lundström *et al.*, 1991; Lundström, 1999; Tsai *et al.*, 1998).

Since the first isolation of an arbovirus from mosquitoes in central Europe (Bardos and Danielova, 1959), the Tahyna virus (*Bunyavirus*) (named after a village in Slovakia where the mosquitoes were caught), numerous reports on arboviruses in Europe have been published (Spieckermann and Ackermann, 1972, 1974; Filipe, 1980; Lozano, 1980; Papadopoulos, 1980; Verani, 1980; Vesenjak-Hirjan, 1980; Malkova *et al.*, 1986; Pilaski, 1987; Mitchell, 1995; Danielova, 1992; Lundström, 1994; Aspöck, 1996; Lundström, 1999) (Table 3.1).

Besides the clinical symptoms, records of arboviruses are based on virus isolation from mosquitoes and vertebrates including the sentinel-method (serologicaly negative vertebrates, *e.g.* rabbits or chickens, are exposed to mosquito bites in the field; after a while the blood is tested for virus antibodies) as well as serological investigations to detect virus antibody in vertebrates including humans. More recently molecular techniques *e.g.* polymerase chain reaction (PCR) have been applied (Puri *et al.*, 1994).

Table 3.1. Main vectors and hosts of Arboviruses transmitted by mosquitoes in Europe

Family Genus Virus	Vectors in the enzootic cycle	Vectors for transmission to humans	Vertebrate hosts
Togaviridae			
Alphavirus			
Sindbis	*Culex torrentium* (V)	*Aedes cinereus* (V)	*Turdus* spp. (V)
	Culiseta morsitans (P)		other Passeriformes (P)
Flaviviridae			
Flavivirus			
West Nile	*Culex p. pipiens* (P)	*Culex p. pipiens* (P)	Birds (P)
Bunyaviridae			
Bunyavirus			
Tahyna	*Aedes vexans* (P)	*Aedes vexans* (P)	Mammals (P)
	Ochlerotatus caspius/dorsalis (P)	*Ochlerotatus caspius/dorsalis* (P)	
	Ochlerotatus cantans/annulipes (P)	*Ochlerotatus cantans/annulipes* (P)	
Inkoo	*Ochlerotatus communis* (P)	*Ochlerotatus communis* (P)	Mammals (P)
	Ochlerotatus punctor (P)	*Ochlerotatus punctor* (P)	
Batai	*Anopheles maculipennis s.l.* (P)	*Anopheles maculipennis s.l.* (P)	Mammals (P)
Lednice	*Culex modestus* (P)		Birds (P)

Letters in parenthesis V = Verified vector or vertebrate host, P = Potential vector or vertebrate host.

At present 6 arboviruses carried by mosquitoes are reported in Europe: one *Alphavirus* (Sindbis virus), one *Flavivirus* (West Nile virus), and four *Bunyaviruses* (Tahyna, Inkoo, Batai and Lednice virus). Additionally, antibodies of the Semliki Forest complex (*Alphavirus*) have been detected in vertebrates in the former Czechoslovakia and Romania (Lundström, 1999).

3.2.1. Togaviridae (*Alphavirus*)

(a) The Sindbis Virus Complex

Four viruses of the Sindbis complex (Sindbis/subtype Ockelbo, Whataroa, Babanki, and Kyzylagach viruses) are widely spread in the Old World (Strauss and Strauss, 1994). In terms of nucleotide sequences, European strains are more or less identical with the strains in Africa and it can be assumed that they were transported by migrating birds between the two continents (Shirako *et al.*, 1991; Norder *et al.*, 1996). Antibodies of Sindbis virus have been detected in birds of several orders, *e.g.* Passeriformes, Galliformes, Anseriformes and Ciconiiformes and also in domestic and wild mammals, *e.g.* in bovids, in several European countries (Lundström, 1999). Additionally, Sindbis viruses were isolated from hamsters and frogs as well as from sentinel chickens and rabbits (Gresikova *et al.*, 1973; Kozuch *et al.*, 1978; Aspöck, 1996).

In humans, Sindbis virus infections have resulted in hundreds of clinical cases in northern Europe, especially in Sweden, Finland, Russia and Norway (Lundström *et al.*, 1991). The disease results in a rash, as well as pain in muscles and joints with joint swelling and fever (Espmark and Niklasson, 1984).

Sindbis virus was isolated for the first time in Sweden in 1984 from *Culiseta* spp. (Niklasson *et al.*, 1984) and later from *Cx. p. pipiens* and/or *Cx. torrentium, Cs. morsitans* and *Ae. cinereus* (Francy *et al.*, 1989) as well as from mosquitoes in Russia and Norway (Lvov *et al.*, 1984; Norder *et al.*, 1996). The virus circulates in bird populations during the summer and is transmitted by the ornithophilic mosquitoes *Cx. p. pipiens* and/or *Cx. torrentium* (Lundström, 1994). In contrast to these *Culex* species, *Aedes/Ochlerotatus* species such as *Ae. cinereus* are less specific concerning host choice. They bite birds as well as humans and are thus able to transmit the virus infection into the human population (Aspöck, 1996; Lundström, 1999). Outbreaks of Sindbis diseases occur usually in July and August and disappear at the end of autumn. The initiation of Sindbis virus transmission in summer is not fully understood, the virus may either overwinter in the mosquito vectors or in the bird hosts. Re-infections can be caused each year by migrating birds from Africa carrying the viruses.

(b) Viruses of the Semliki Forest Complex

Nine viruses of this complex occur in the Old World (Strauss and Strauss, 1994). In western Europe, the only record of human disease possibly caused by a virus of the Semliki Forest complex is from Albania (Eltari *et al.*, 1987; Lundström, 1999). The viruses could occur in humans and other mammals (*e.g.* bovids and ovids) at low frequency. Clinical cases have not been reported from western Europe. Although antibodies have been detected in a number of birds (*e.g. Ardea cinerea, Acrocephalus* spp., *Remiz pendulinus*) and viruses have been isolated from *Oc. euedes* in central Russia (Mitchell *et al.*, 1993), these viruses are of no or minor epidemiological importance.

3.2.2. Flaviviridae (*Flavivirus*)

Since dengue is no longer recorded as an endemic disease in Europe, West Nile virus is the only *Flavivirus* transmitted by mosquitoes in Europe (Kunz, 1969; Aspöck, 1979; Filipe, 1990; Vesenjak *et al.*, 1991). Outside Europe, this virus occurs in the Middle East, Africa, India, and recently in the United States. Following infection it takes 3–6 days until the symptoms of the disease occur, *e.g.* malaise with fever, headache and muscle pain, sore throat, rash and swollen lymph nodes (Lundström, 1999). The disease usually lasts for about one week. In most endemic areas it is a childhood disease, as adults have acquired immunity (Manson-Bahr and Bell, 1987; Tesh, 1990). However, severe cases of meningoencephalitis, myocarditis, acute pancreatitis and hepatitis are reported, which can be fatal. In 1935, 1942 and the early 1960s outbreaks occurred in the Camargue (southern France) with benign to fatal encephalitis symptoms (Panthier *et al.*, 1968). More recently, West Nile fever epidemics were reported from southeastern Romania (the Danube plains and the city of Bucharest) with 393 acute human cases and a case fatality rate of 4.3% in 1996 (17 deaths). In 1997, 14 cases with two deaths occurred between July and the end of September (Ceianu, pers. comm.). Virus isolates are known from France, Portugal, former Czechoslovakia and Romania. Serological surveys have revealed antibodies in human sera from Albania, Austria, Germany, Greece, Italy, Poland, Spain and former Yugoslavia (Hubalek and Halouzka, 1999; Lundström, 1999). Besides humans, West Nile virus has also been isolated from horses with encephalomyelitis in France, Italy and Portugal (Jourbert *et al.*, 1970; Filipe, 1972; Cantile *et al.*, 2000). Antibodies have been detected in other domestic mammals *e.g.* bovids as well as in wild mammals, *e.g.* mice, bats and even in the water snake, *Natrix natrix* (Aspöck, 1996).

West Nile virus circulates usually in birds with *Cx. p. pipiens* being the most likely enzootic vector (Mouchet *et al.*, 1970). In the city of Bucharest the infection rate was 0.20/1000 in *Cx. p. pipiens* (Savage, pers. comm.). Virus isolates are also known from *Cx. modestus* in southern France (Hannoun *et al.*, 1964), from *An. maculipennis s.l.* in Portugal (Filipe, 1972) and from *Oc. cantans* in former Czechoslovakia (Labuda *et al.*, 1974). Obviously virus amplification can occur in a wide range of birds (*e.g.* Passeriformes, chickens, ducks, geese and pigeons). There is strong evidence that the West Nile virus is sporadically introduced from Africa to Europe by migrating birds. The virus is then transmitted in a bird–mosquito cycle in endemic areas in Europe. Humans are infected in summer when suitable mosquito vectors are abundant. However the possibility of transovarial transmission, with the virus surviving the winter in hibernating *Culex* females in Europe cannot be excluded, as transovarial transmission has been demonstrated in the laboratory by Baqar *et al.* (1993).

In Israel, the Israel Turkey Meningoencephalitis virus (Flaviviridae) causes economic losses among turkey growers. The virus has been isolated from biting midges and blood-engorged *Cx. p. pipiens*. However, so far the virus is not known to infect humans (Mitchell, 1995).

3.2.3. Bunyaviridae (*Bunyavirus*)

In Europe, three groups of *Bunyavirus*es have been recorded, namely the California group including Tahyna and Inkoo viruses, the Bunyamwera complex including the Batai (Calovo) virus, and the Turlock group including the Lednice virus (Lundström, 1999).

In contrast to the Togaviridae and Flaviviridae which are most likely sporadically introduced to Europe by migrating birds with viraemiae, the Bunyaviridae are endemic in Europe.

The Tahyna virus is widespread in Europe (Traavik *et al.*, 1978; Aspöck, 1979; Pilaski and Mackenstein, 1985; Danielova, 1992; Lundström, 1994). It has frequently been isolated from *Ae. vexans*, but also from other species such as *Ae. cinereus, Oc. sticticus, Oc. cantans, Oc. flavescens, Oc. caspius, Cs. annulata, Cx. modestus, Cq. richiardii* (Lundström, 1994; Aspöck, 1996) and even in biting midges *Culicoides* spp. (Ceratopogonidae) (Halouzka *et al.*, 1991). Danielova and Ryba (1979) proved in experiments that the virus can be transmitted transovarially in *Ae. vexans*. Therefore, it can be assumed that the virus overwinters within the eggs and with the mass development of *Ae. vexans* after spring floods, the virus circulation starts again. Virus overwintering in females of *Culiseta* spp. and *Culex* spp. seems to be of minor importance (Chippaux *et al.*, 1970).

In contrast to the Sindbis and West Nile viruses which need birds for their circulation, viruses of the California group need mammals as amplification hosts (Lundström, 1994, 1999; Aspöck, 1996). Antibodies have been detected in wild and domestic mammals such as hares, rabbits, bovids, cervids (reindeer), and in carnivores as well as in hedgehogs (*Erinaceus europaeus*). Humans are also susceptible to Tahyna and Inkoo viruses. Infected people develop influenza-like symptoms which can lead to morbidity. However, in Europe the California-group is still not well characterised in terms of the occurrence of various virus strains, their specific vectors and medical importance (Lundström, 1999).

The Batai virus (strain: Calovo virus) is mainly transmitted by *An. maculipennis s.l.* (Bardos and Cupkova, 1962). However, some isolates have also been obtained from *An. claviger, Cq. richiardii* and *Oc. communis* (Francy *et al.*, 1989; Traavik *et al.*, 1985). It can be assumed that the virus overwinters in hibernating anopheline females. Following infection humans can develop febrile disease with bronchopneumonia, catarrh and gastritis (Sluka, 1969). However, in Europe Batai virus antibodies are usually found in less than 1% of the human population (Lundström, 1999). Bovids are often infected with Batai virus. Antibodies have also been detected in pigs and deer; birds are not important for the virus circulation. The virus has been reported from several European countries *e.g.* Finland, Sweden, Germany, Austria, former Czechoslovakia and Yugoslavia (Lundström, 1999).

The Lednice-Virus has been isolated from *Cx. modestus*. It is likely that the virus is transmitted vertically in the *Culex* population. Only rarely are vertebrates, mainly birds, infected (Aspöck, 1996; Lundström, 1999). Antibodies have never been found in humans or wild mammals, except in two hares (Wojta and Aspöck, 1982), so the Lednice virus seems to be of no epidemiological importance.

3.3. FILARIASIS

In the tropics lymphatic filarial diseases caused by the nematodes *Wuchereria bancrofti*, *Brugia malayi* and *Brugia timori* affect an estimated 120 million people in Asia, Africa and South America; 90% of the infections are caused by *W. bancrofti*. An estimated 905 million people are directly exposed to the infection by transmission by various genera of mosquitoes, the most important being *Cx. p. quinquefasciatus* and *Mansonia* spp. (Eldrige and Edman, 2000).

In Europe, filariasis is of no medical importance. The dog heartworm, *Dirofilaria immitis* (Onchocercidae) causes canine cardiovascular dirofilariasis in dogs and other canids, but rarely cats (Boreham and Atwell, 1988). Infection in humans is asymptomatic because the worm does not mature in humans. The adult worms are 12–31 cm in length. Usually they are found in the heart (hence heartworm) and the pulmonary artery of the host. Infected animals can suffer from cardiac insufficiency and heart failure. Mature female worms release microfilariae into the blood. They exhibit a nocturnal periodicity and circulate in the peripheral blood vessels, an adaptation to the biting habit of the potential mosquito vectors. Lok (1988) listed 26 species of mosquitoes, incriminated on the basis of field or laboratory studies, as vectors of *D. immitis*. Most vectors are found in the genera *Aedes/Ochlerotatus*, *Anopheles* and *Culex* (Eldrige and Edman, 2000). When ingested by a mosquito, the microfilariae penetrate the gut epithelium and enter the Malpighian tubes where a certain percentage grow and moult to the infective third larval stage within about two weeks, under temperate climatic conditions. The infective larvae migrate to the mosquito mouthparts and enter the labium from which they penetrate into the host skin when the

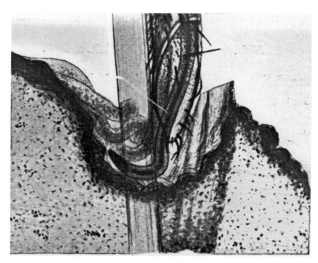

Figure 3.4. Worm larvae penetrating into the host skin (after Martini).

mosquito is biting (Fig. 3.4). The development to the adult worms in the heart and pulmonary artery of the vertebrate host takes several months, and after more than six months microfilariae are produced again. In addition, other dog infecting filarioids are *Dipetalonema* spp. or *Mansonella* spp. which have a similar biology as *D. immitis* (Aranda *et al.*, 1998).

3.4. CONCLUSIONS

To conclude it can be stated that in Europe, the risk for the re-introduction and transmission of mosquito-borne diseases has increased rapidly over recent decades, due to the large number of tourists arriving from the tropics and carrying parasites and pathogens, as well as aircrafts carrying vectors causing the so-called "airport malaria".

The trade with used tyres is responsible for the introduction of the so-called "Asian tiger mosquito", *Ae. albopictus*, into Europe, which has probably been present in Albania since the end of the 1970s (Adhami and Murati, 1987). Infestations were reported from Genoa in 1990 and from Orne (Basse-Normandie) and Vienne (Poitou-Charentes) in France in autumn 1999

(Schaffner *et al.*, 2001) and from Podgorica Montenegro in summer 2001 (Pelric *et al.*, 2001). Temperature and precipitation are the main limiting factors affecting the range extension of *Ae. albopictus*, which is nonetheless likely to spread throughout the total Mediterranean area and even to the north of the Alps (Mitchell, 1995).

A similar way of introduction of *Ae. albopictus* and *Ae. japonicus japonicus*, into the United States is well known. *Ae. albopictus* was found for the first time in the continental United States in used tyres and retrograde cargo returning from Asian ports (Pratt *et al.*, 1946; Eads, 1972). The first record of establishment of this species in the continental U.S. was reported by Sprenger and Wuithiranyagool (1986), when a large population was discovered breeding in used tyres (shipped from Japan) in Houston, Texas. Within a short period (1985–1999), widespread infestations of *Ae. albopictus* were reported from 26 states east of the Mississippi River (Moore, 1999). More recently, this species was introduced for the first time on the west coast of the United States, into California in June 2001. Developmental stages and adult specimens were discovered in shipments of *Dracaena* sp. ("lucky bamboo") plants packaged in standing water, arriving in refrigerated maritime containers at the Los Angeles and Long Beach Harbors. *Ae. albopictus* as container breeder has adapted very well to breeding in such man-made containers (Madon *et al.*, 2002).

Nowadays this mosquito has not only become a major pest species in some areas in Italy, but also raised considerable official concern because of the competence of this species as vector of arboviruses (CDC, 1987). The Asian tiger mosquito is a vector of dengue viruses in Asia and has been experimentally shown to transmit these viruses from infected human hosts to uninfected volunteers. *Ae. albopictus* is also capable of transmitting a variety of pathogens to humans and other mammals. Under experimental conditions *Ae. albopictus* has proven to be a competent vector of several other viruses of public health significance in the United States (Moore and Mitchell, 1997). To date, there have been no records of this species transmitting any diseases to humans in Europe and the United States.

Ae. j. japonicus is another exotic species which was recently introduced into the United States. It is a common mosquito subspecies in Korea and Japan. The first two recorded cases of introduction of this subspecies into the United States were reported by Peyton *et al.* (1999), when adults were trapped in CDC light traps in Southland, New York in August 1998, and in Colliers Mills, New Jersey in September 1998. The most likely mode of introduction of *Ae. j. japonicus* may have been via used tyres exported to the United States. This subspecies prefers to breed in a variety of natural and artificial containers. It is a daytime biter and is known to read-ily bite humans. Although little is known about the public health significance of *Ae. j. japonicus*, preliminary studies indicate that it has the vector potential to transmit Japanese encephalitis (Takashima and Rosen, 1989).

These are examples of vectors adapting to different environmental and climatical conditions. The change of the climate with increasing temperatures and unusual heavy rainfalls may lead to an extension of the range of mosquito vectors and mosquito-borne diseases. The occurrence of the Asian tiger mosquito *Ae. albopictus* in North America and South Europe and of the Mediterranean mosquito *Uranotaenia unguiculata* in Germany (Becker and Kaiser, 1995), may be seen as evidence of the increased risk of the introduction of new and maybe more virulent mosquito-borne diseases in Europe.

4

Mosquito Research

Basic knowledge about the distribution, abundance, phenology and ecology of different mosquito species is an essential requirement for a successful control campaign against these insects. For example, population dynamics and migration behaviour of the target organisms have a vital influence on the design of a control strategy. In parasitological and epidemiological studies, the interaction between the parasite/pathogen, or the vector and host, must be evaluated to suppress mosquito-borne diseases successfully. Therefore, at the beginning of all mosquito control campaigns, extensive entomological studies have to be carried out. In this chapter, the most important methods of mosquito research will be presented. A complete review of mosquito collecting methods and analysis of collected data is given by Service (1993).

4.1. SAMPLING OF MOSQUITO EGGS

Mosquitoes lay their eggs singly or in egg rafts in many different habitats, such as swamps, marshes or pools, as well as in a great variety of small, natural and artificial water collections, such as tree holes, rock pools or man-made containers. Some mosquito females lay their eggs on the water surface, whereas others lay them onto the moist soil of the breeding site, on the edge of the water or on the wall of natural or artificial containers. The determination of egg densities in natural habitats provides not only a better understanding of the egg-laying behaviour of the various mosquito species, but can also aid in predicting future larval populations and possible control areas.

Because of the various egg-laying behaviours and physical characteristics of the eggs, different collecting techniques have to be employed.

4.1.1. *Anopheles* Eggs

Anopheles females lay their eggs on the water surface where they float due to air-filled chambers formed from the outer egg layer, the exochorion. Often the eggs group themselves into net-like structures until the larvae hatch. Less than 2 mm in size, the eggs are hardly recognisable with the naked eye. Eggs can be sampled from the water surface using a light-coloured dipper. However, a dipper whose bottom surface has been cut out and replaced with a fine wire mesh, is preferred. This modification allows the dipper to be pulled through the water, skimming the surface and thus collecting the eggs from the surface. The mesh is then washed with water into a light-coloured plastic dish, where the eggs can be collected with a pipette. In the absence of a dipper, a metal ring (10 cm diameter) with a nylon mesh stretched across the surface can be used. A wooden handle can also be attached for easier use (WHO, 1975).

4.1.2. Egg Rafts

Species of the genera *Culex*, *Uranotaenia*, *Coquillettidia* and subgenus *Culiseta* of the genus *Culiseta* lay their eggs in rafts on the water surface. With a size of several millimetres, the egg rafts can be easily seen and sampled by feather forceps, pipettes or small nets.

4.1.3. *Aedes/Ochlerotatus* Eggs

Aedes and *Ochlerotatus* females lay their tiny eggs into the moist substrate of breeding sites. It is extremely difficult to recognise the eggs *in situ* even with the help of a magnifying lens, so soil samples from the breeding sites have to be taken. For the determination of the egg density the so-called "flooding method" or the "saltwater method" can be employed. For the estimation of the number of eggs per surface unit, it is important to standardise the soil samples (Becker, 1989; Service, 1993). For this purpose, a metal frame of angle iron (20 × 20 × 2.5 cm) is suitable. The frame can be driven into the ground with a hammer until the angled side is flush with the ground surface. Using a trowel, the soil along the bottom of the frame can be cut horizontally to a depth of 2.5 cm. The soil sample can then be carefully transferred to a plastic bag, and labelled with the location and date of sampling. In this manner, the *Aedes/ Ochlerotatus* eggs should remain in their natural state.

Females prefer particular sites for egg deposition. It is known that the moisture and quality of the soil, or the plant associations which indicate a special flooding sequence or degree of soil moisture, are important factors in determining where the female mosquito lays her eggs. This usually results in a heterogenous distribution of eggs. In order to secure the necessary data from different zones, it is therefore recommended to take the samples in transects, to determine the variation in egg densities in a potential breeding site. Along the bank of a breeding site, samples should be taken in equal distances beginning at the deepest point, and continuing to the upper margin of the breeding site. This ensures that the areas with the highest densities of *Aedes/Ochlerotatus* eggs will be recorded. Often the highest numbers

of eggs are laid in bands around pools subject to fluctuations in water level. The bands indicate the zones above the water edge with a high content of moisture during the time of mass oviposition by the gravid females. Until their further treatment, the soil samples should be stored out of the sun to avoid drying of the soil and damage to the eggs. If the samples are kept for several days or weeks, the soil should be regularly moistened.

When the "flooding method" is applied to calculate the relative egg density of a certain breeding site, the standardised soil samples should be flooded with water at a temperature which aids the hatching of the larvae (approximately 20°C). It is recommended to use water with an initial high content of oxygen. The decrease of oxygen caused by the metabolism of the microbes in the soil will stimulate the hatching of the larvae. Ascorbic acid, or yeast and sugar, can be added for further reduction of oxygen to increase the hatching stimulus. The hatched larvae can be collected a day or two days after flooding, counted and identified after being reared to the fourth larval instar or adult stage. Because of "hatching in installments" (batches of larvae hatching during consecutive floodings), samples with high egg densities must be repeatedly flooded. The standing water must be decanted so that the soil can dry out until the next flooding procedure takes place. The alternate flooding and drying raises the hatching stimulus of the eggs that remain in the soil. Even after several floodings, larvae will still continue to hatch.

As the *Aedes/Ochlerotatus* larvae may be in diapause and therefore unable to hatch even when suitable hatching conditions are provided, it is important to break the diapause and so condition the larvae for hatching, before the samples are treated. For instance, when eggs of *Ae. vexans* are sampled during winter time, the samples must be stored for at least two weeks at a temperature above 20°C, so that larvae are able to hatch. The higher the difference between the low and high temperature, the higher is the response to the hatching stimulus.

In contrast to the flooding method, the "saltwater method", described by Horsfall (1956b), is less time consuming and ensures close to a 90% discovery rate of the eggs. The principle behind this method is that the density of an aqueous solution is increased by adding salt, thereby causing the eggs with a specific weight lower than the saline solution, to float to the surface.

If the soil samples are taken from the field during summer, it is recommended to store the samples at 5°C for at least two weeks to make the larvae (*e.g. Ae. vexans*) unable to hatch. To wash the eggs out of the soil, the sample should be placed in a tub and flooded with cold water (below 10°C). The cold water further reduces the hatching stimulus. The flooded sample should be carefully and uniformly mixed, and lighter particles (leaves, wood) can be removed from the surface of the water (Butterworth, 1979). The water and the finer, suspended particles in the water have to be carefully decanted. This procedure can be repeated several times. Then salt (sodium chloride) is added to the water, until a 100% saturated salt solution has been achieved and the salt crystals no longer dissolve. This solution has a higher specific weight than the eggs. By thoroughly mixing the sample, the eggs rise to the surface, where they can be collected by means of a pipette or with a filter paper and counted. The eggs can be determined either by rearing the larvae, or in some species the specific pattern of the egg shell (chorion) can also be used for species determination.

Another method for collecting the eggs is the use of a sequence of meshes with a decreasing mesh-size to separate the eggs from the soil.

Due to the similar egg-laying behaviour and physical properties of the eggs of the sub-genus *Culicella* within the genus *Culiseta*, the same methods can be employed.

4.1.4. Eggs in Artificial Oviposition Sites

Mosquito species have preferred breeding habitats, which can differ very much in abiotic and biotic factors, such as water quality, light intensity, available food or vegetation. The knowledge of the critical factors in the choice of a breeding site by a certain mosquito species allows the construction of artificial oviposition sites or traps. The oviposition traps or so-called "ovitraps" can be useful tools in surveillance programmes of certain mosquito species, such as mosquitoes breeding in artificial containers, tree-holes or rock-pools.

In survey programmes *Ae. albopictus* and *Ae. aegypti* are mainly monitored by means of ovitraps (Fay and Eliason, 1966; Pratt and Jakob, 1967; Jakob and Brevier, 1969a,b; Evans and Brevier, 1969; Thaggard and Eliason, 1969; Chadee and Corbet, 1987, 1990; Freier and Francy, 1991; Service, 1993; Bellini *et al.*, 1996; Reiter and Nathan, 2001).

Population levels of both species can be determined by using a sufficient number of ovitraps in a certain area (Mogi *et al.*, 1990; Bellini *et al.*, 1996). Commonly used ovitraps consist of a dark plastic or glass jar which is painted black on the outside, about 8 cm in diameter at the top and 5 cm at the bottom, with a height of 12.5 cm. When the ovitraps are placed in the field, a strip of hardboard or masonite (2×12 cm) with a smooth and a rough surface is attached vertically with a paper clip to the inside of the jar so that the rough side is facing towards the centre of the jar (Service, 1993). Then, approximately 200 ml of dechlorinated tap water is poured in. A small hole in the plastic jar above the water line prevents water overflow during heavy rainfalls. Usually the females of *Ae. albopictus* and *Ae. aegypti* lay their eggs just above the water line on the rough side of the strip. At regular intervals (once a week) the strips are changed and the water replaced. The number of eggs can be counted on the strips using a dissecting microscope.

Plastic buckets, wash tubes or other receptacles containing several litres of infusion of hay, brewer's yeast, dog biscuit, alfalfa or sewage water can be used as oviposition traps for *Cx. p. pipiens* and *Cx. p. quinquefasciatus* (Yasuno *et al.*, 1973; Sharma *et al.*, 1976; Leiser and Beier, 1982; Reiter, 1983, 1986; O'Meara *et al.*, 1989; Becker, 1989). The *Culex* mosquitoes are apparently attracted by gaseous compounds such as ammonia, methane, or carbon dioxide, which are released when organic material decomposes in the breeding waters. Fatty acids and n-capric acid are also oviposition attractants for females of *Cx. p. pipiens*. In order to study the egg-laying behaviour, or the population density of *Culex* mosquitoes, egg rafts can be sampled at regular intervals. It can be assumed that in human settlements with limited numbers of natural breeding sites, the *Culex* population may be effectively reduced by destroying the egg rafts sampled in oviposition traps at regular intervals. In addition to attractants, insect growth regulators can be added to the oviposition traps in order to avoid the development of adult mosquitoes, or to sterilise them.

4.2. SAMPLING OF MOSQUITO LARVAE

Many sampling techniques are used to assess the population of the aquatic stages of mosquitoes. They are especially employed to study the population dynamics of mosquitoes, or to estimate the population densities before and after larvicide application. Although mosquito larvae occur in a wide range of habitats, some simple techniques can be employed for the assessment of the larval population. The most commonly used tool is the dipper, which can vary in size and shape. Often soup ladles, white plastic or enamel bowls or photographic trays with a capacity of a few hundred millilitres to 1 litre are inexpensive and easily used tools for collecting larvae. For comparative purposes it is recommended to use standardised dippers. In most programmes the so-called "standard pint dipper" is used, consisting of a white plastic container measuring 11 cm in diameter and with a capacity of 350 ml (Dixon and Brust, 1972; Lemanager *et al.*, 1986). The white background aids accurate counting of larvae. A wooden handle is attached to the dipper to reach water bodies from a distance, and to avoid disturbing the water, which causes larvae to dive during sampling (Fig. 4.1).

According to the size of the surface of the larval habitat, a sufficient number of dips have to be taken to estimate the number of developmental stages. In larger ponds (>20 m^2) 10–20 samples along the edges and in the centre should be taken. The number of various larval instars and pupae per dip should be separately recorded. By calculating the volume of water inhabited by larvae, a rough estimation of the aquatic mosquito population can be made (Papierok *et al.*, 1975; Croset *et al.*, 1976; Mogi, 1978). Service (1993) gives more precise

Figure 4.1. Assessment of larval densities by using a standard pint dipper.

methods for population estimates based on data evaluation by regression statistics.

In large habitats a plankton net attached to a handle can be used to catch a relatively large number of larvae within a short time. The net should be drawn through the water in a figure-of-eight pattern to sample the larvae.

In small collections of water such as tree-holes, sampling can be difficult. Larvae can be pipetted directly from the surface or can be collected by siphoning out the water.

A sufficient number of larvae/pupae have to be transported to the laboratory alive for species determination, or alternatively fourth-instar larvae can be preserved in 70% ethyl alcohol on the spot. In the larval stage only larvae in the fourth-instar are suitable for species determination. Earlier instars have to be reared to the fourth-instar or adult stage. For transportation to the laboratory, three-quarters of the volume of a glass or plastic jar with a close-fitting cap should be filled with pond water. The larvae can be pipetted into the jar, or the contents of a dipper or net catches can be released into the vessel. The jars must be marked carefully recording the date and location of sampling.

In the laboratory, larvae can be killed by hot water (60°C) and transferred into preserving media for further handling. Alternatively the larvae can be identified at the fourth-instar, or reared to the adult stage in a mosquito breeder for approval of the larval identification (see rearing of mosquitoes).

4.3. SAMPLING OF ADULT MOSQUITOES

Various sampling techniques to catch adult mosquitoes can be employed, such as direct human bait catches (HBC), netting, catching with aspirators, suction or attractant traps *e.g.* containing carbon dioxide (CO_2 traps). Several factors such as the weather conditions, activity pattern, host-seeking and resting behaviour, as well as the physiological stages of the mosquito, will determine the composition of the catch as regards species and sexes. Here, only the most common techniques will be discussed. Service (1993) provides more detailed information.

4.3.1. Bait Catches

The bait or host for a blood-meal (*e.g.* human) emits olfactory stimuli, so-called kairomones, which are attractants for the host-seeking mosquitoes. Female mosquitoes respond to compounds such as carbon dioxide from the exhaled breath, host odour (*e.g.* lactic acid) deriving from sweat, water vapour, body heat and vision aided by the movement of the host (Bar-Zeev *et al.*, 1977; McIver, 1982; Takken and Kline, 1989; Takken, 1991; Lehane, 1991). A traditional method of assessing the adult population is to count females landing on a human, and to express the number of females per unit of time. The catch can be performed in hourly or bi-hourly intervals over a 24-hour period to assess the activity pattern of the female mosquitoes in relation to abiotic factors such as temperature, relative humidity, light intensity or wind speed. For instance, the biting activity of many mosquitoes is often highest during sunset, when the temperature is still high and the humidity is increasing.

Depending on the number of mosquitoes, the collector might expose only a part of the body (leg or arm) or the whole body, to collect the mosquitoes by means of an aspirator. Widely used aspirators consist of a clear polystyrene tube, which can be fitted with a two-hole stopper to hold the intake and exhaust tubes, made from flexible clear plastic (Fig. 4.2b). The top of the intake tube should hold a small funnel to aid the sampling of landing mosquitoes, and the exhaust tube should be protected with a fine nylon net to limit inhalation of undesirable materials. The polystyrene tubes can be removed and capped by snap-on caps. Mosquitoes can be killed by placing a small

(a)

(b)

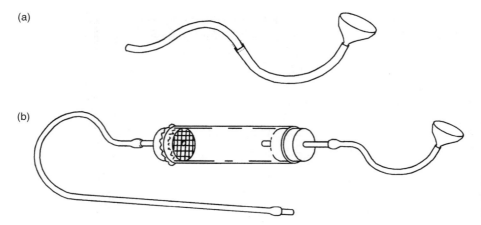

Figure 4.2. Simple (a) and chamber aspirator (b) for collecting adult mosquitoes.

piece of cotton wool which has absorbed a very small quantity of a killing agent (*e.g.* ethyl acetate), into the tube. A glass tube is recommended because it is impermeable to the killing agent. If plugs of cotton wool or paper strips are added to the tube, movement of the dead fragile mosquitoes and loss of scales are avoided.

A simple aspirator can be easily constructed by joining two flexible plastic tubes with slight differences in diameters, so that they fit closely together. A nylon net fixed between the joined tubes prevents the inhalation of mosquitoes by the collector (Fig. 4.2a). The catch can then be blown into a separate killing container.

The use of a drop-net is advisable when the mosquitoes are excessively numerous (Fig. 4.3). The trap consists of a bell-shaped net with 3 metal hoops which can be fixed on a branch of a tree above the collector. The two lower hoops have a diameter of 1 metre. With cords fixed on the hoops the collector can pull up the lower hoop for exposure to mosquitoes. After an exposure time, ranging from 2–10 min or longer, the net is dropped. The collector then samples all mosquitoes inside the drop-net, with an aspirator.

It is known that mosquito species respond differently to various kairomones and to individual attractiveness of the collectors, resulting in variations in catch sizes and species composition. To allow a more standardised monitoring

of adult mosquito populations in large areas without the need of numerous collectors, many types of mosquito traps have been developed to attract different target species. In particular, a wide range of mosquito light traps have been developed and tested. Some rely solely on a conventional incandescent filament light bulb as the main source of attraction, others use an ultra-violet light source, while others supplement the light source with carbon dioxide or another chemical attractant. For monitoring of adult floodwater mosquito populations, light traps with a carbon dioxide source are recommended. Traps may also incorporate a photosensitive switch which turns the light and motor off during daylight hours, and closes a valve to prevent the mosquitoes escaping. A variety of models are now commercially available, and for a full review of light traps, see Service (1993). The EVS model has been successfully used for some years and is described in detail below.

4.3.2. Carbon dioxide/Light Trap

The EVS (Encephalitis Vector Survey) trap was introduced in 1979 and mainly developed by the Orange County Vector Control District in California. The top part consists of a 3.5 litre plastic dry ice container. The walls and snap-on polyethylene lid are insulated with polyethylene foam to prevent rapid sublimation of the

Figure 4.3. Collecting mosquitoes with a drop-net.

dry ice that is placed in the container during use. In the lower part of the container, two to four holes of 0.5 cm in diameter allow the sublimated CO_2, which is the primary attractant, to escape and flow downwards over the lower part of the trap. The middle part of the trap consists of a plastic tube with holders for three 1.5 V dry cell batteries, which provide power for a fan and a subminiature 1.5 V, 70 mA lamp, which can be operated by an off/on switch. Female mosquitoes, attracted mainly by the carbon dioxide, enter through an opening and are sucked downward by airflow created by the plastic fan attached to a small motor into a 30-cm long nylon netting catch bag (Fig. 4.4).

In routine monitoring programmes the trap is baited with approximately 1 kg dry ice, which is enough to catch mosquitoes during one night.

A carrying handle is provided, along with a metal chain to facilitate hanging from a tree branch or other object. The traps are set up in the late afternoon and removed the next morning. The catch, including the net, can be transferred into a container with dry ice or other killing agents to kill the mosquitoes. In the laboratory the species composition is determined. Catches at regular intervals (*e.g.* each fortnight) yield valuable information about the phenology and population size of the adult mosquitoes. Furthermore, the comparison of the catches in controlled and uncontrolled areas allows an estimation of the reduction of the mosquito population by control operations. Under favourable conditions in areas with mass breeding sites more than 15 000 female *Ae. vexans* can be caught in one EVS-trap per night.

Figure 4.4. Carbon dioxide baited trap.

Figure 4.5. The Mosquito Magnet™.

Male mosquitoes are not attracted by CO_2 traps, but can respond to light traps. They are best collected in the field when they are swarming. The males of most species usually swarm above prominent landmarks where they can be caught by the use of a long-handled net.

4.3.3. The Mosquito Magnet™

In addition to the range of conventional mosquito light traps used by mosquito control organisations, a novel device, named the Mosquito Magnet™, has recently become available (Fig. 4.5). This is a free-standing machine, that is intended for more use as a local mosquito control device, than for monitoring and survey work. It uses a container of propane gas as an internal energy source. The propane is converted into carbon dioxide by a platinum catalyst, and the heat produced by the reaction is used to generate enough electricity to drive a small fan. The device incorporates a replaceable lure that releases 1-octen-3-ol into the carbon dioxide plume. The machine will run continuously for up to 21 days,

after which time the propane gas cylinder and the octenol lure need replacement.

Mosquitoes and other biting insects such as phlebotomine sandflies, are attracted by the plume of carbon dioxide and 1-octen-3-ol emitted by the machine, and are then sucked into a net cage by the fan.

Preliminary field tests of the machine indicate that it will catch a range of *Aedes/ Ochlerotatus*, *Culex* and *Anopheles* mosquitoes. The robust, weatherproof construction, and it's ability to run unattended for days and weeks, indicate that it may have potential as a monitoring tool.

4.4. PRESERVING MOSQUITOES

4.4.1. Larvae

The best method for rapidly killing larvae is to drop them into hot water (60°C), which leaves the specimens in a nicely distended condition. After a few minutes they are transferred to the preserving fluid, which can be MacGregor's solution (10 ml 5% borax, 80 ml distilled water, 10 ml formaldehyde and 0.25 ml glycerine) or

70% alcohol, with 1% glycerine added if specimens are to be prevented from drying out when in long-term storage. The specimens can be examined by means of a dissecting microscope in the preserving fluid, or stored in small glass containers with rubber-lined screw caps to avoid fast evaporation. For quick reference, the glass containers must be clearly labelled. Labels in pencil on a paper stripe can be put into the bottle; those in waterproof ink can be stuck to the outside of the bottle. The label should contain at least the location of collection, type of breeding site, date of sampling and name of the collector.

For permanent preparations, larvae are mounted in Caedax, Canada balsam, Eukit, Euparal or Histomount. For this purpose, the larvae should first be transferred with some fluid into a solid, small glass vessel. When very late fourth-instar larvae are mounted it may be necessary to macerate the larvae before dehydration. It is recommended to leave the larvae in a 5% solution of potassium or sodium hydroxide overnight, or as an alternative, to heat them below the boiling point in the same fluid for about 10 minutes. The caustic solution should be removed and 5% acetic acid added for several minutes to neutralise the liquid. Then the larvae should be dehydrated in graded ethyl alcohols (50%, 70%, 80%, 96%), iso propanol and xylene, with each step lasting at least 30 minutes. The larvae should be kept for 10 minutes in a 1 : 1 mixture of xylene and mounting media when Caedax or Canada balsam is used. Xylene is not needed when mounted in Euparal, which is the most widely used material. The fluids must be carefully pipetted off and then replaced with the next fluid by means of a Pasteur pipette to avoid damaging the delicate setae, which are important taxonomic characteristics. Finally, a large drop of viscous mountant is to be put on a slide before the larva is transferred by featherweight forceps.

When culicine larvae are mounted, it is necessary to cut off the last abdominal seg-

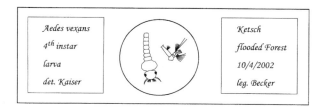

Figure 4.6.   Arrangement of culicine larvae on a slide.

ments and to arrange this part alongside the rest of the body. Using two micro-needles the specimen should be worked to the slide surface, dorsal side upwards, with the head and the end of abdomen (lateral side up) toward the long-side and across the slide (Fig. 4.6). Then, a coverslip is gently placed onto the specimen, starting from the abdominal end of the larva rather than the head, which will help keep the larva in position. If needed, more fluid mountant should be added from the side to apply the coverslip.

A label with data for the specimen, including information on the location of sampling and breeding site as well as date of sampling and name of the collector, should be placed to the right of the specimen. The label with the name of the specimen and instar should be placed on the left, with the specimen between the two labels. The same procedure can be used to preserve larval skins.

4.4.2. Pupae

Slide mounts of pupae and pupal skins can be made in the same way as described above. When adults are bred out from larvae and pupae it is useful to keep the adult, the fourth-instar larval and pupal skin together in the collection.

To mount the pupal skin, the cephalothorax should be cut from the abdomen and cut open ventrally to lay flat on the slide alongside with the abdomen (dorsal side up).

4.4.3. Adults

Adult mosquitoes are preferably killed with ether or ethyl acetate vapour. This is done by placing a small piece of cotton, which has absorbed several drops of the killing agent, into a glass vessel containing the mosquitoes.

Some field workers also use cigarette smoke in the absence of other killing agents, or a killing bottle with a layer of Plaster of Paris containing cyanide, on the bottom. The dead mosquitoes are pinned and kept dry. It is recommended to pin the mosquitoes soon after killing because the soft bodies can be better handled compared to the fragile, dry insect bodies, which tend to disintegrate. For this reason, dry specimens should be relaxed in a humidity chamber prior to pinning. The dried mosquitoes should be spread separate from one other on a filter paper in a petri-dish and kept for 12–24 hours in a desiccator containing water to soften the specimens. Care should be taken not to use humidity chambers in which drops of condensed water can fall on the specimens.

The mosquito should then be removed by fine-pointed forceps and pinned by the double-mounted method. In this method, a large stainless steel pin about 38 mm long, no. 2 in thickness, holds the stage where the mosquito is mounted. The mosquito is attached with a small (about 12 mm) stainless steel point (minute pin), and usually two labels with all necessary data similar to those for permanent preparations of larvae (Fig. 4.7).

Stage materials can be strips of cardboard, cork, polystyrene board, polyporous or expanded polyethylene foam. The narrow strips should be equal in size, usually 13 mm long and a few millimetres in width. First, the minute pin for the mosquito is put through the stage about 2 mm from one end. Then, the mosquito is placed upside down on a plate of polystyrene; the prepared stage holding the minute pin is picked up with entomological pinning forceps; and the end of the minute pin is inserted from the ventral side into the thorax of the specimen between the legs, pushing through until the point of the pin reaches the scutum. The head of the mosquito should face away from the end of

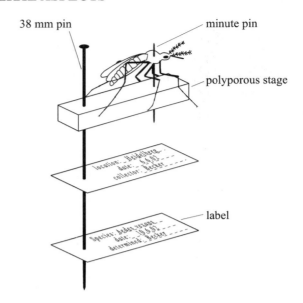

Figure 4.7.   Mounted adult mosquito.

the stage, and the abdomen and legs along the stage. It is recommended to pull the minute pin through the stage so that the abdomen of the mosquito remains horizontal to the stage, but not to arrange the legs, as they could be damaged. A moderate puff of air from behind the mounted specimen will spread wings into a position so that they do not cover the abdomen. The stage should be carefully held between the fingertips of one hand by the end in which the large pin will be inserted. About 1 mm from the end of the stage, the large pin should be pushed through until the upper third of the pin remains above the stage. This allows room for holding the mounted specimen and for putting the two labels below the stage. Labels should be of good quality card and written in Indian ink, using a fine pen of uniform size and always attached to the pin in the same position. On the label pushed next to the stage, the locality of sampling, date and collector's initials are recorded. The label with the identity of the specimen should be placed below this. A pinning block is useful for a uniform and neatly presented collection of specimens (Snow, 1990).

A more simple and widely used method guarantees also a good protection of the mosquito and easy handling of the mounted mosquito (Bohard and Washino, 1978). The freshly killed, frozen or newly relaxed mosquito is attached in a horizontal position by the left pleuron to a drop of clear fingernail polish on the tip of a triangular cardboard or cork strip, of at least 7 mm in length, which is located at the top of a 38 mm pin. The scutum should face away, and the legs towards the pin. Then, the cardboard or cork strip is pulled down to about 2/3 of the pin's length, to leave enough room for the two labels.

After drying for about 1–2 days (if possible in a refrigerator) the insects should be transferred into a wooden cork-lined storage box with an anti-pest fumigant (*e.g.* crystals of naphthalene) to avoid pest damage.

The external genitalia of males are usually sufficient, sometimes essential for identification. After the mosquito is killed the tip of the abdomen is cut between segment VII and VIII with optical scissors or fine forceps. The end of the abdomen is placed in a solution of warm sodium or potassium hydroxide for about 15 minutes for maceration. Then, the genitalia are transferred to glacial acetic acid in order to neutralise the maceration fluid and to dissolve any remaining fat droplets. They are dehydrated and mounted as described for larvae (see subsection 4.1.1). To obtain the best view of the genitalia, they should be placed with the ventral surface facing upwards before the cover slip is placed in position. The balsam should be viscous enough and in sufficient quantity to ensure that the specimen is not flattened. The remainder of the male is pinned, and both the slide preparation and the pinned male are labelled to indicate their relationship.

4.5. REARING MOSQUITOES

It is often necessary to breed mosquitoes to the adult stage in order to identify them. Frequently, the fourth larval instar, its exuviae, or the exuviae of the pupae or the adult are needed before a confident identification can be made. The breeding of mosquitoes is also essential for the assessment of the efficacy of insecticides, and the study of the bionomy of the species. Therefore, many different techniques have been developed for rearing various mosquito species in the laboratory (Gerberg, 1970).

It is recommended that larvae are kept in water obtained from the breeding sites, to assist successful development to the adult stage. If the developmental stages of the mosquitoes have to be transferred into clean water, either distilled or rainwater should be used. Usually it is not necessary to add food, as water containing debris from the original breeding site is used. However, food in the form of powdered fish food, dried yeast or liver powder should be added when distilled water is used, or when large numbers of larvae are reared. Only small quantities of food should be added to avoid fouling of the rearing medium. A regular change of water or supply of oxygen is necessary if excess food is accumulating, to avoid anaerobic conditions and scum formation. Before emergence of the adults, fourth-instar larvae or pupae are transferred into the so-called mosquito breeder. The mosquito breeder consists of two clear polystyrene containers. The water sample with larvae is placed into the bottom portion. A screw-on lid between the two sections contains a vinyl funnel through which the emerging adults can fly into the upper part. Refrigerating the apparatus stuns the adults to facilitate removal. They could also be kept in a small container such as a jam jar. As pupae develop, the container has to be covered with fine netting held by a rubber band, or alternatively a larger glass container may be put upside-down over the jar, so that the emerging mosquitoes cannot escape.

When larger numbers of larvae are reared they should be placed inside a mesh cage. These can be easily constructed by making a frame with a side of about 30 cm, covered with curtain netting. On one side a circular opening of about 10 cm should be left to allow access into the cage, *e.g.* to remove vessels or mosquitoes using an aspirator. A sleeve should be sewn in the opening to prevent the escape of mosquitoes. It can be knotted when access is not required. A density of not more than 1000 larvae per litre of water should be kept in shallow containers to avoid overcrowding. On a daily basis, pupae can be removed with pipettes and transferred into separate containers with clean water, also kept in the mesh cage.

The rearing temperature should be adapted to the species requirements. For snow-melt mosquitoes it should be about 20°C, and for summer species, about 25°C. Adult mosquitoes can be fed with sugar solutions soaked into cotton wool or with wet raisins.

Only a few species can be successfully reared in the laboratory for several generations, due to the fact that only a few species mate readily in artificial small cages. However, artificial techniques of mating are sometimes used. Males are glued on the edges of glass petri-dishes and decapitated with small scissors. Then the tip of the abdomen of the anaesthetised female is moved towards the tip of the abdomen of the male at an angle of about 45° for grasping and copulation. Afterwards the female is transferred to a cage and fed with blood. Some researchers feed the mosquitoes on themselves, or on small mammals (*e.g.* mice) or chickens. A few species are able to mate in small cages and to lay eggs without a blood-meal, such as *Cx. p. pipiens* biotype *molestus*. This mosquito is therefore very often kept in cultures where containers with water for laying the egg rafts and a sugar solution or raisins as food resource are provided.

Females can also be collected in the field from a host when they take their blood-meal. A glass tube which is covered with cotton wool at the bottom is carefully placed onto the female when it starts to suck blood. When the blood-meal is completed (about 2 minutes) the glass

container with the blood-fed mosquito is covered with a nylon mesh and kept for several days at about 20°C. Raisins can be placed on the nylon mesh as additional food. The cotton wool is kept wet to keep the humidity high, and when *Aedes/Ochlerotatus* species are captured the wet cotton could also serve for egg laying. Females of *Culex* and *Anopheles* need a small water body for egg laying. The time from the blood-meal until the deposition of the eggs can be measured. The number of eggs can be counted, the morphology of the eggs can be described, or hatching experiments can be conducted with defined egg batches.

The structure and colouration of the chorion of *Anopheles* eggs is of crucial importance in distinguishing certain species, *e.g.* within the Anopheles Maculipennis Complex. To enable the females to lay eggs, they need to be handled gently both during sampling and in the laboratory. The engorged females of the complex can be easily obtained within animal shelters in any part of a year, depending on the overwintering behaviour of the species. Test tubes (diameter 15 mm, length 75 or 150 mm) lined with wet 10 mm wide strips of filter paper can be used for direct sampling of single females within shelters. After the specimen has been captured, the tube is closed with a hydrophilic cotton ball and transferred to the laboratory. For successful egg laying, the filter paper needs to be kept damp until the end of the gonotrophic cycle.

4.6. METHODS FOR MEASURING THE PHYSIOLOGICAL STAGE

The vector capacity of a mosquito species depends to a great extent on the ability of the females to have several blood-meals, which in turn enables the vector mosquito to transmit pathogens or parasites from a previous infection. In this respect the knowledge of the physiological stage of a mosquito is of epidemiological importance. It can be determined by the assessment of the number of gonotrophic cycles through which a mosquito female has passed. The gonotrophic cycle for species that require a single blood-meal for each batch of eggs, is defined as the time period needed from the blood-meal until the oviposition of the eggs. After a blood-meal the basal oocytes within the ovarian tubes (ovarioles) in the ovaries develop into mature eggs. As a result, the epithelium surrounding a single ovariole is stretched by the expanding oocyte. When the egg is deposited, the epithelial sheet that surrounded the egg during its development shrinks, leaving a residual lump or a knotlike dilatation. Following the next blood-meal the next sequence of oocytes located more distally in the direction of the germarium of the ovariole, continue their development. After the deposition of the next batch of mature eggs a second series of dilatations remains in the ovarioles. Therefore, the number of dilatations corresponds to the number of previous ovipositions. This allows an estimation of the longevity of the individual mosquito, and its importance as a vector. The larger the number of blood-meals, the greater is the possibility for the mosquito to acquire and transmit pathogens.

Furthermore, the age or the physiological stage of a mosquito can also be determined by examining the changes in ovarian structure. During the first gonotrophic cycle the coiled ends of the tracheoles are unwound by the swelling of the ovaries. These uncoiled tracheoles indicate parity (eggs developed) whereas nulliparity (no eggs developed) is indicated by still coiled tracheoles.

For the examination of the ovaries the mosquito must be dissected. After the female mosquito is killed it is placed in a drop of water on a microscope slide. The mosquito is held with a fine mounted needle on the thorax, and with a second needle the intersegmental membrane between abdominal segments VII and VIII is cut. Then, the gut and the ovaries can be pulled out by means of forceps. The gut is cut and the ovaries are inspected. For the examination of the dilatations of the ovarioles, the ovarian sheet has to be removed and the ovarioles have to be separated with a needle. For examination using a microscope, the ovaries are covered with a coverslip.

The appearance of the abdomen can also be used to determine whether a mosquito has engorged blood, or to assess the stage of egg development. When the abdomen of the mosquito is thin it is unfed, or has passed through a gonotrophic cycle. In freshly fed females the abdomen (midgut) is filled with red blood, which gradually becomes dark red. As the eggs develop the ovaries increase in size, and the red colour disappears. In the gravid females, in most cases a trace of dark blood is left and ovaries occupy almost the whole abdomen.

4.7. MORPHOLOGICAL STUDIES

Although the morphological characteristics of most of the European mosquito species are described, there is still the need for further studies on character variation within individuals and populations of a species. Not only the number and size of scales and setae can vary but also the colouration, size or other features. In particular the comparison of species from different areas can reveal new results on variation. However, in some cases the morphological features are not sufficient for species identification, and therefore other techniques have to be employed.

4.8. CYTODIAGNOSTIC METHODS FOR THE IDENTIFICATION OF SIBLING SPECIES

The morphological similarities of the members of a mosquito species complex (so-called sibling species) in the larval and/or adult stages, made it necessary to develop tools for species identification which are not based on external morphological characters. Genetic mapping is of great importance to distinguish between species, subspecies or ecotypes, and for investigations on population genetics. Various types of maps can be constructed (Hillis, 1996):

(a) Preparations of polytene chromosomes made from larval salivary glands or ovarian nurse cells, provide the basis for physical maps where the specific banding patterns of polytene chromosomes are illustrated in photomicrographs or drawings. For several decades polytene chromosomes have been particularly instrumental in chromosomal studies, due to their specific properties.

(b) Genetic or linkage maps where the relative position of genes and the distances between them are plotted.

(c) Restriction maps and

(d) whole genome or molecular maps.

The typical karyotype of culicines consists of three pairs of homomorphic chromosomes. Typical for the karyotype of anophelines is one pair of heteromorphic sex chromosomes, as well as two pairs of autosomes. In certain mosquito tissues the chromosomes replicate repeatedly without mitosis-like events, so that hundreds of sister chromatids remain synapsed. This multiple DNA duplication is accompanied by alignment and condensation of the DNA strands, thus forming specific pattern of bands and interbands. These banded polytene chromosomes occur in larval and adult tissues, such as the salivary glands and midgut epithelium, the malpighian tube cells, as well as in the ovarian nurse cells of adult females. Banding sequences can change by rearrangements within or between chromosome arms *e.g.* by inversions or translocations. Chromosome aberrations are typical of certain mosquito populations. Banding patterns can therefore be used as markers for distinguishing species and ecotypes. Differences in banding *e.g.* due to inversions, have proved to be a useful tool in distinguishing members of sibling species complexes, and for resolving interrelationships between species.

The polytene or so-called giant chromosomes are easily seen using a microscope at high magnification (view at ×1000 with oil immersion lens and phase contrast). In the culicine mosquito *Cx. p. pipiens* and anopheline mosquitoes, high quality squashes of polytene chromosomes can be obtained from the salivary glands of fourth-instar larvae. The specimen is placed in 5% Carnoy's fixative on a glass slide (Carnoy's fixative: mixture of one part of glacial acetic acid and three parts of 95% ethanol). The abdomen is cut off and the thorax is slit open dorsally along the mid-line. By pulling on the head, the glands attached to the head are pulled out of the thorax. Then the salivary glands are cut free and transferred to a slide in a droplet of Carnoy's fixative, stained for about 5 minutes by adding two droplets of a mixture of acetic and lactic acids with orcein (2% by weight of orcein powder is dissolved in 1 part 85% lactic acid and 1 part glacial acetic acid, and this concentrated stain is then again diluted 1:3 with 45% acetic acid). The tissue is carefully squashed by placing a coverslip on top of the preparation and pressing it with a finger. Thus the cells are broken and the chromosomes are spread. Species, sex and karyotype composition of larvae from various habitats can be determined by the examination of the chromosome preparations with a microscope. This technique can be successfully employed for population studies on the Anopheles Maculipennis Complex. Methods for making polytene chromosome preparations have been described by French *et al.* (1962), Green (1972), Hunt (1973), Green and Hunt (1980) and Graziosi *et al.* (1990).

4.9. BIOCHEMICAL AND MOLECULAR METHODS IN STUDIES ON SYSTEMATICS

Taxonomy utilises homologous morphological characters for diagnosis of species and evaluation of relationships among taxa. The phenotype depends both on genotype and environmental conditions, so that even characters depending on a monomorphic gene locus, may vary between individuals or populations of a distinct species. Consequently, separation of closely related species may be difficult using variable, quantitative characters.

Molecular methods for analysing variation in DNA, and biochemical investigation of gene products, are widely used in support of taxonomic studies. Generally, investigations that incorporate both morphological and molecular approaches will provide the most substantiated descriptions and interpretations of diversity within and between taxa.

In taxonomy and evolutionary research, protein-electrophoresis is a widely used and highly efficient technique. For this, homogenates of animals or organ samples are applied to a gel (polyacrylamide, agarose or starch) and exposed to an electrical field under ionic buffer conditions. The migration of dissolved proteins is determined by their charge, size and shape. Depending on these factors, the proteins segregate to bands in the gel and then are visualised by unspecific or, in case of enzymes, by substrate transformation coupled staining. Differential migration of the same protein is the result of a change of its amino acid sequence. Proteins are primary gene products, so changes in their amino acid sequence are caused by changes in the underlying DNA sequence, *i.e.* there are different alleles on their gene locus. When various species possess different alleles exclusively, these fixed alleles may be used as genetic markers and enrich the diagnoses of species by morphological traits. By investigation of several proteins, the resulting data are suitable for the quantification of genetic variation within populations (*e.g.* degree of polymorphism and heterozygosity), detection of species boundaries, phylogenetic relationships and evolutionary processes (Harris and Hopkinson, 1976).

Investigation of DNA has been driven particularly by the development of the polymerase chain reaction (PCR) that allows the amplification of small amounts of DNA millionfold, and makes DNA accessible for many purposes. There are several methods of DNA analysis: DNA-DNA-hybridisation estimates the amount of sequence divergence between genomes, but tells us nothing about structural background of differences. Fragment and restriction analysis such as variation in fragment size (*e.g.* VNTR), restriction site variation (RFLPs, RAPDs, DNA fingerprinting) and others, provide some information about genome organisation, and are suitable for detection of individual and species relationships. By DNA sequencing, exact information about the genotype is obtained that allows solid conclusions to be drawn about molecular evolution and phylogeny (Hillis, 1986).

All these techniques provide important and interesting information about biodiversity and evolution, but the technique should always be selected according to the aim of the investigation, considering the complexity and cost efficiency of each particular method.

4.10. ECOLOGICAL AND BEHAVIOURAL STUDIES

The above mentioned techniques represent the tools for many ecological studies which are required for a better understanding of the biology of mosquitoes, and the development of appropriate strategies for mosquito control. Studies of the mosquito fauna at regular intervals in a wide range of habitats and in various geographic areas enhance our knowledge of species occurrence based on abiotic and biotic factors. The seasonal occurrence and the speed of development of larvae depend not only on biotic features of a breeding site, but also on water quality, temperature and light regimes. The availability of food and occurrence of predators influence the survival rate and speed of development of a population. Regular investigations on adults reveal information on the biting habit related to temperature, humidity or light intensity, and the horizontal and vertical dispersal of a species according to environmental conditions and population size. Service (1993) describes a number of mark-recapture techniques to study the population size, the dispersal and feeding behaviour, or the duration of the gonotrophic cycle and survival rates. Due to their wide distribution and excellent adaptation to their habitats, mosquitoes are not only a group of organisms which are suitable for basic ecological studies, but due to their nuisance and vector capacity, they are usually of interest to the general public.

Morphology of Mosquitoes

5.1. ADULTS

Mosquitoes differ from all other members of the Nematocera by having a long scaled proboscis (labium and stylets), always longer than the thorax, which projects forward together with the maxillary palps (Fig. 5.1). The latter are as long as or longer than the proboscis in males of most species, and females of the genus *Anopheles*. The head, thorax and abdomen are covered with scales, the extent of coverage of which is genus specific. The legs, wing margins and wing veins are typically clothed with scales. The closest resemblance is found within the families of slender, long-legged crane flies (Tipulidae) and non-biting midges (Chironomidae), the latter often being mistaken for mosquitoes especially around artificial light at night. However neither of these families have mouthparts for piercing and sucking. The mandibulate mouthparts of the tipulids are of the biting and chewing type, and articulated on the tip of a prolonged, beak-like gnathocephalon. However, the Chironomidae usually have a reduced gnathocephalon and biting mouthparts. In addition, the Chironomidae possess a conspicuously humped thorax, and often particularly long, forward-facing fore legs.

Males of most mosquito species clearly differ from the females by having plumose antennae and long, hairy maxillary palps (Fig. 5.2a, b). The first 12 flagellomeres of a male antenna bear dense, long setae, which are at least as long as the head capsule. The setae on the last flagellomere are shorter than the head capsule, and of similar length like those on the female antenna.

Unusually for Diptera, the integument of mosquitoes is quite extensively covered with scales. Essentially, scales are flattened setae containing pigment and often have a striate surface which produces optical effects, giving some mosquitoes physicochemical or physical colouration. The shiny, metallic blue, green and purple appearance is mostly found within tropical species. The colour of the scales may vary from white to almost black but is usually referred to as pale or dark in the description of the species. Scale colouration can change after long term storage and in different light sources. Pale and dark scales can be intermixed wherever they occur. They also may be grouped forming the specific contrasting patterns which will be referred to as rings on all legs, stripes on the scutum, and bands on the abdominal terga. The overall body colour can be influenced by the colour of the integument which may show through between the scales. The abdominal terga and sterna are densely covered by scales in the subfamily Culicinae, while the sterna, and usually also the terga, are wholly or largely devoid of scales in the subfamily Anophelinae. Like all other Diptera, the mosquito head, thorax and abdomen are covered with setae of different length, shape and colouration which can have significant genus and species specific taxonomic importance.

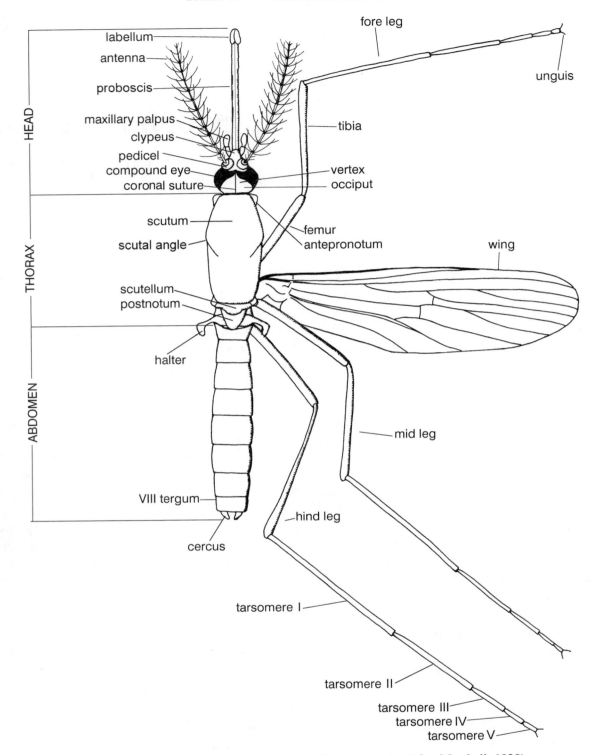

Figure 5.1.   General outline of a female culicine mosquito (after Marshall, 1938).

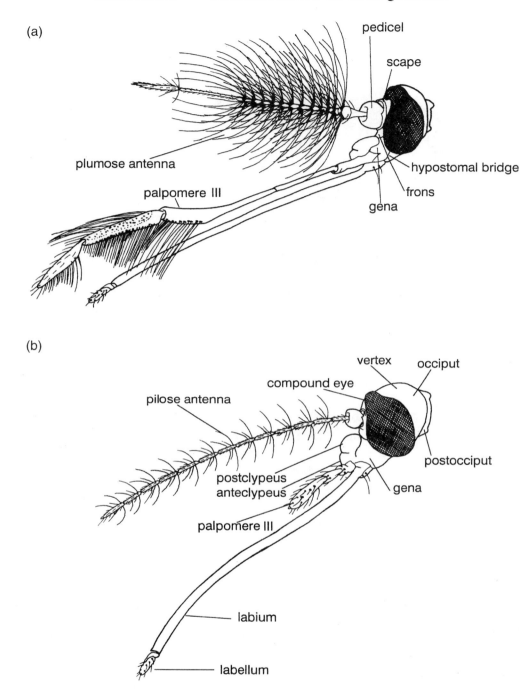

(a)

pedicel

scape

plumose antenna

hypostomal bridge

frons

gena

palpomere III

(b)

vertex

occiput

compound eye

pilose antenna

postocciput

postclypeus
anteclypeus

gena

palpomere III

labium

labellum

Figure 5.2. Head of (a) male; (b) female culicine mosquito (after Wood *et al.*, 1979).

5.1.1. Head (Figs 5.2 and 5.3)

Mosquitoes, like all Pterygota (winged insects) and Thysanura, have annulated antennae. Its basal segment, or scape, is collar shaped and hidden behind an enlarged, spherical second segment, the pedicel (Fig. 5.2a). In the Chironomidae and Culicidae the pedicel is specially enlarged to accommodate a highly developed mechano- and sound-receptor, the Johnston's organ. The remaining 13 divisions

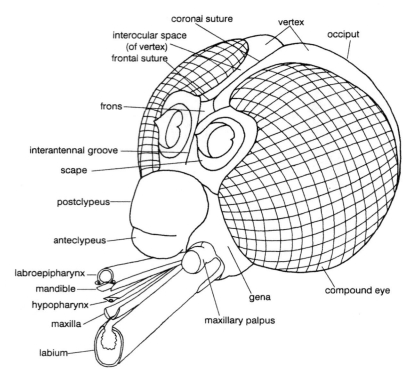

Figure 5.3. Head regions and mouthparts (after Clements, 1992).

of the antenna together constitute the flagellum and are entirely without intrinsic muscles. Consequently, those divisions are named annuli or flagellomeres instead of segments, the latter term being used only for the partitions of the insect body having intrinsic musculature. Setae are basally articulated at all flagellomeres in female mosquitoes, but only at the two most distal flagellomeres in males. The remaining 11 flagellomeres of the latter bear the setae subapically. The exterior of the head is composed of several sclerites amalgamated to form the head capsule (Figs 5.2 and 5.3). Compound eyes occupy a substantial portion of the head, and if they meet in the middorsal line, this condition is then called holoptic. Ocelli are absent. The epicranial suture is quite indistinct. Its median part is called the coronal suture and the two branches above the scape are called the frontal sutures. The region of the head below the frontal sutures is the frons, above them lies the vertex, though the precise limits of the former areas are not easily defined. A narrow strip

of the vertex between the compound eyes is the interocular space. At the back of the head, between the vertex and the cervix, or neck, the occiput and the narrow ring-like postocciput are located. The vertex and occiput may be covered with truncate or forked erect scales and narrow, curved decumbent scales. The nearest structure, found immediately anterior to the frons, is the bulge of the gnathocephalon, called the rostrum in some other orders of insects. The gnathocephalon is composed of the postclypeus and anteclypeus dorsally, of the genae laterally and of the hypostomal bridge ventrally.

The proboscis, which is articulated with the gnathocephalon, is made of six slender stylets (Fig. 5.3); the labroepipharynx or labro-palatum (a more or less fused structure made of an outer wall, the labrum, and an inner wall, the epipharynx), the paired mandibles, the hypopharynx and the paired maxillae. These are lodged in the groove of the elongated labium (prementum) which ends with a distally articulated pair of labella, thought to represent the modified labial palps

(Figs 5.1 and 5.2). Between the labella a short pointed lobe is situated, the ligula, and in its shallow groove lies the tip of the stylet fascicle. The mandibles and the maxillary laciniae are modified into long piercing stylets. The latter, being stronger and serrated apically, are the main piercing organs used to perforate the skin or integument of the host. The labroepipharynx, which bears three pairs of sensillae on its tip, and the other stylets form a fascicle which probes until it enters a blood capillary or haemocoel of a host. The saliva which acts as an anticoagulant, is injected through the hypopharynx. The blood is taken via a channel made mostly of the labroepipharynx. The mandibular and maxillary stylets are reduced or absent in females of *Toxorhynchites* sp. and *Malaya* sp. which feed on nectar. They have a functional labroepipharynx and hypopharynx. The labium, being only a protective sheath for the stylets, is looped backwards during feeding and does not penetrate the host tissue. The labium of *Toxorhynchites* sp. females is rigid.

In males, both mandibles and maxillary laciniae are reduced or lacking and cannot be used for piercing. Nectar and other sugary juices are simply imbibed through the tubular labroepipharynx.

The well developed maxillary palps are articulated below the clypeus, dorsolaterally to the proboscis (Figs 5.1 and 5.2). Palps are divided into five divisions which are clearly visible in both sexes of *Anopheles* and in the males of almost all the other genera. The females of other genera frequently have an atrophied basal segment and apical annuli, therefore their palps appear to have less segments. The three basal divisions are true segments, having intrinsic muscles (Clements, 1992). Although the two apical divisions are not true segments they will be referred to as segments or all five as palpomeres, for the convenience in species description. The palps of most female mosquitoes are less than half the length of the proboscis. Only in anopheline females (except

former *Bironella* sp.) are the palps of similar length to the proboscis. In almost all male mosquitoes the maxillary palps are as long as, or longer than the proboscis. Regarding the mosquitoes of Europe, only in males of the subgenus *Aedes* and in *Ur. unguiculata* are the palps of similar length as in females.

5.1.2. Thorax (Figs 5.4–5.7)

The three thoracic segments are known as the prothorax, mesothorax and metathorax. In mosquitoes and all other Diptera, where only the fore wings are used for flight, the mesothorax is the best developed segment. The hind wings are modified into a gyroscopic organ, called the halteres. The prothorax and metathorax are reduced to little more than leg bearing collars fore and aft. Each of the thoracic segments is divisible into four main regions, the dorsal tergum or notum, the ventral sternum and the two lateral pleurae. These regions are differentiated into separate sclerites which are termed as tergites, sternites and pleurites respectively. The sclerites of the thorax, according to the segments to which they belong, are denoted by the prefixes pro, mes(o) and met(a).

The pronotum is seemingly absent medially but well developed laterally. It is separated into an anterior lobe like division, the antepronotum, and a flattened posterior division, the postpronotum which appears to be part of the mesopleuron (Fig. 5.5).

The mesonotum of Diptera is clearly subdivided into prescutum, scutum and scutellum, the fourth tergite, the postnotum, being generally hidden beneath the scutellum. The prefix meso- is usually omitted for dipterous insects because those subdivisions do not occur in the weakly developed metanotum, and the pronotum is usually simple or bipartite. In mosquitoes (Figs 5.4 and 5.5), the transverse suture separating the prescutum and scutum is not fully developed, hence the term scutum is used for both mesotergites. The scutum is the principal

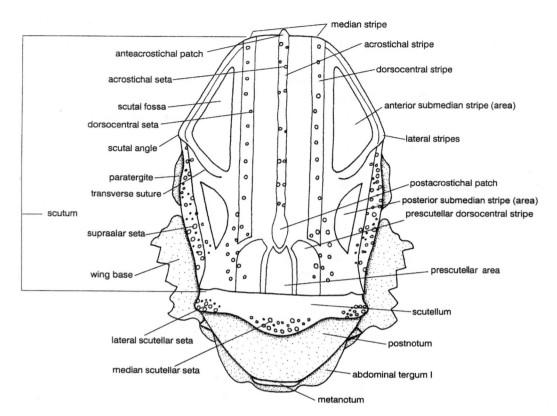

Figure 5.4.   Dorsal view of thorax.

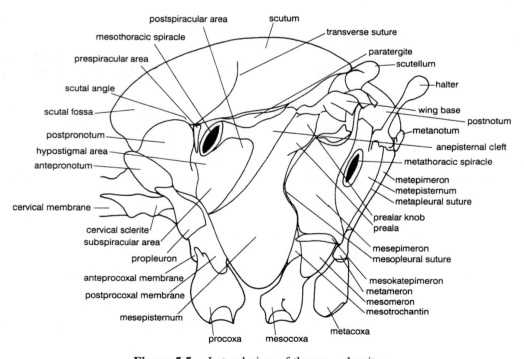

Figure 5.5.   Lateral view of thorax—pleurites.

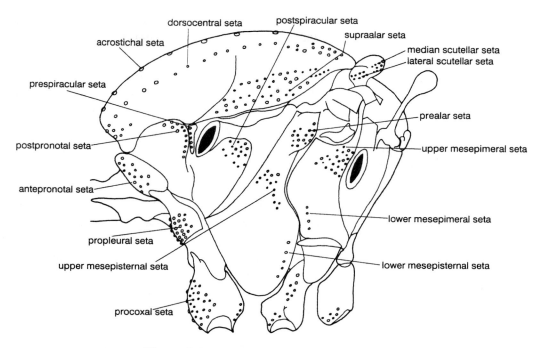

Figure 5.6. Lateral view of thorax—setation.

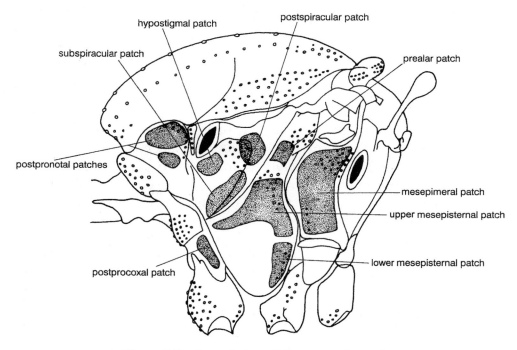

Figure 5.7. Lateral view of thorax—scale patches.

dorsal area of the thorax and is sometimes incorrectly referred to as the mesonotum in the mosquito literature. The lateral scutal margin above the mesothoracic spiracle bears a tubercle named the scutal angle from which the transverse suture extends. The scutal fossa is a slightly depressed area anterior to the scutal angle, which lies between the scutal margin and the dorsocentral row of setae. A small elongate sclerite of mesonotal origin, the paratergite, is situated anterior to the wing root and in some species it is not apparent from above. The scutum is followed by the scutellum (Figs 5.4 and 5.5). Its posterior margin may be evenly convex (genera *Anopheles* and *Toxorhynchites*) or three lobed (most genera of Culicinae and genus *Chagasia*). An additional, well developed sclerite of intersegmental origin, the postnotum, bears the phragma to which the dorsal longitudinal flight muscles are attached. The metanotum is reduced to a narrow, weakly sclerotised transverse band (Figs 5.4 and 5.5). The metapostnotum is usually not visible.

The postpronotum usually bears a vertical row of setae close to its posterior margin, the postpronotal setae (Fig. 5.6). Sometimes they may cover the mesothoracic spiracle and can easily be mistaken for the prespiracular setae (an important diagnostic characteristic for the genus *Culiseta*). It is therefore advisable to carefully focus on the point where the setae are inserted. Setae may be concentrated in several areas of the scutum or scattered over most of its surface. As far as the identification of the mosquitoes is concerned, the main setae are inserted in three pairs of, more or less dispersed, longitudinal rows (Fig. 5.4). The setae of the acrostichal rows are articulated close to each other and to the median line. The second pair of dorsocentral rows of setae is more dispersed towards the end of the scutum. Finally two rows of scattered supraalar setae occupy the posterior parts of the lateral margins of the scutum. The scutellum in species of the genus *Anopheles* bears a uniform row of backwardly

directed setae along its evenly rounded posterior margin. In the other genera that have a trilobed scutellum, the setae are arranged in three sets, a median and two groups of lateral scutellar setae. The postnotum is usually bare.

Narrow or broad scales may cover the whole scutum or only some of its areas, the amount of scaling varies between different genera. The scales may be all the same colour on the whole scutal surface, or differently coloured producing usually linear, longitudinal scale patterns which are usually species specific and consequently important for identification. Therefore it is important to catch, handle and mount specimens with special care to prevent scales of the central scutal area from being rubbed off, or to collect larvae/pupae and let adults emerge in the laboratory. It seems that, although a variety of colour patterns occur, most of them are based on a common template. The nomenclature devised here for the principal elements of these patterns, is derived from Berlin (1969) who regarded the general distribution of the setae mentioned above (Fig. 5.4). The acrostichal stripe is always a very narrow row of scales confined to the median line of the scutum. It may be absent, or often it is reduced to an anterior or anteacrostichal patch and a usually larger posterior or postacrostichal patch of scales. If present, the acrostichal stripe divides the much broader median stripe which is located between the dorsocentral rows of setae. The paired dorsocentral stripes lie in the dorsocentral areas. They can be subdivided into inner and outer dorsocentral stripes. For convenience, if only one of them is present it will be simply named the dorsocentral stripe. The lateral stripes border the scutum to varying degrees from the anterior border to the wing base, and may be broken into several prescutal or supraalar fragments. Between the dorsocentral and lateral stripes are the moderately broad, lanceolate or crescent shaped submedian stripes. The stripes are usually subdivided into anterior and posterior portions. Lateral to the prescutellar area lie the prescutellar dorsocentral stripes, which can be fused with the dorsocentral stripes and/or the postacrostichal patch of scales above (Fig. 5.4). The prescutellar area is often bare.

The pleuron in advanced insects has become an enlarged, flattened and sclerotised part of the thoracic wall which it rigidly supports (Fig. 5.5). It is in general transversely divided into two main pleurites, the anterior episternum and the posterior epimeron. These are separated by the pleural suture, and may be further divided into upper and lower pleurites which are denoted by the prefixes an- and kat-, thus anepisternum refers to the upper pleurite of episternum and katepimeron to the lower pleurite of epimeron. A small pleurite, the trochantin, is often present near the lower margin of the episternum. Another small pleurite, the meron (meso- and meta-) was originally the posterior part of the basicoxite which has become completely separated from the coxa and forms a part of the thoracic wall. In a wing bearing segment the pleuron develops a dorsal (alar) articular process for the wing.

In mosquitoes the elongated, lobe shaped propleuron is located between the antepronotum and the procoxa (Fig. 5.5). The membrane which articulates the procoxa with the prothorax is arbitrarily separated into an anterior part lying between the prosternum and procoxa, the anteprocoxal membrane, and the posterior part between the propleuron, mesepisternum and procoxa, named the postprocoxal membrane.

The mesopleural suture divides the mesopleuron transversely into a large anterior mesepisternum and a smaller posterior mesepimeron. The mesepisternum is further subdivided into two pleurites, an upper mesanepisternum and a lower mesokatepisternum. The former pleurite bears the mesothoracic spiracle. It is small, displaced towards the postpronotum and often divided into four areas (Fig. 5.5). The prespiracular area is very narrow, confined between the postpronotum and the anterior margin of the spiracle. A small area immediately below the spiracle is called the hypostigmal area. The smooth subspiracular area lies below it and is usually connected to the sclerotised postspiracular area. Because the names of the four mentioned areas are so widely accepted and used in mosquito morphology, there is no practical reason to use the designation mesanepisternum for the description of the whole pleurite. Consequently, the mesokatepisternum is referred to as the mesepisternum. The membranous anepisternal cleft lies above the mesepisternum, which is particularly well developed towards the ventral side. There it frequently unites with the opposite mesepisternum separating the fore and mid coxae from one another, thus concealing the mesosternum. The very narrow upper portion of the mesepisternum or preala, bears the lobe-like prealar knob. The mesepimeron, which lies behind the mesopleural suture, is also divided into a large upper mesanepimeron and a very narrow, bare, inconspicuous mesokatepimeron. The latter is not mentioned in the description of the species and consequently, the mesanepimeron will be referred to as the mesepimeron (Fig. 5.5). Below the mesopleural suture lies the small triangular mesotrochantin. Slightly above and immediately behind the mesocoxa and just below the mesokatepimeron, is the small triangular mesomeron located.

The metapleuron, like all other sclerites of the metathorax, is greatly reduced. It is divided by the metapleural suture in an anterior triangular or trapezoid metepisternum, which bears the metathoracic spiracle and the haltere, and the posterior stripe-like metepimeron (Fig. 5.5). The small metameron lies above the metacoxa.

Propleural setae are grouped above the lower border of the propleuron (Fig. 5.6). The prespiracular area bears the taxonomically important prespiracular setae, present in the genus *Culiseta* among European mosquitoes and some other genera worldwide, except *Aedes, Ochlerotatus, Coquillettidia*, and *Culex*.

The subspiracular area rarely supports setae, whereas the postspiracular area may bear the taxonomically significant postspiracular setae. Three groups of setae are typical for the mesepisternum. Displaced towards the wing base, inserted in the prealar knob, is the group of prealar setae. The upper mesepisternal setae are grouped at the level where the mesepisternum narrows into the preala, and the lower mesepisternal setae are located along the posterior margin, anterior to the mesotrochantin. The mesepimeron bears two groups of setae, the lower mesepimeral setae located at the anterior lower corner, behind the upper mesepisternal setae, and the upper mesepimeral setae at the posterior upper corner, anterior to the metathoracic spiracle (Fig. 5.6).

The postpronotum is usually covered with scales of various size. Sometimes the patch is divided into an upper and lower portion, called the upper postpronotal patch and lower postpronotal patch (Fig. 5.7). Both anteprocoxal and postprocoxal membranes may be covered with scales in some species, the patches are named anteprocoxal and postprocoxal patch respectively. A group of scales on the hypostigmal area is, if present, consequently named the hypostigmal patch. The subspiracular area may bear scales, the subspiracular patch. The postspiracular patch of scales can extend posteriorly onto the anepisternal cleft. Scales of the mesepisternum are usually grouped into two patches, the prealar one below the prealar knob, and the mesepisternal one blended to varying degrees with the upper and lower mesepisternal setae. The mesepisternal patch may be subdivided into an upper and lower portion (Fig. 5.7). The upper half of the mesepimeron is usually covered with scales, the mesepimeral patch, can extend downwards towards the lower margin of the mesepimeron to a varying degree.

The sternum in advanced insects is often greatly reduced in width and partly enfolded between the legs, while the two sternal apophyses are often united on a common base to form a Y-shaped furca. The prosternum in Culicidae is the region lying anteriorly between the procoxae, connected dorsolaterally with the propleuron. It bears a distinct median suture and sometimes setae or scales. The scaling of the prosternum is of taxonomic significance in some *Aedes* and *Ochlerotatus* species. The mesosternum is obscured by the mesepisterna, which are fused ventrally, and is restricted to the region between the mesocoxae, having a strong median suture. The metasternum is reduced to a narrow crescent shaped sclerite.

Mosquitoes, as all insects with the exception of apodous larval forms and a few specialised adults, have three pairs of legs, one pair on each of the thoracic segments (Fig. 5.1). They are referred to as fore legs, mid legs and hind legs. Each leg consists typically of six segments, the coxa, trochanter, femur, tibia, tarsus and pretarsus, the first five being almost entirely covered with scales. The tarsus is subdivided into five tarsomeres. These are, similarly to the flagellomeres, differentiated from true segments by the absence of intrinsic muscles. The basal tarsomere I articulates with the distal end of the tibia. Between the tarsomeres there is no articulation; they are connected by a flexible membrane so that they are freely movable (Chapman, 1982). The distal tarsomere V bears the pretarsus which consists of a pair of claws, or ungues, a ventral unguitractor plate where the tendon of the flexor muscle of the claws is inserted, the paired setose pulvilli, if present, one under each claw, and a median spine-like empodium. The tarsal claw may be simple or have a subbasal tooth. In females of most of the genera, except some of *Aedes/Ochlerotatus* and *Psorophora*, all tarsal claws are simple and of quite similar shape on all three pairs of legs. The shape of the tarsal claws of the fore legs can be used as a diagnostic characteristic, especially in the genus *Ochlerotatus*. In males the outer claws of the fore and mid legs are much bigger than the inner claws, modified for grasping the females. The two claws of the hind legs are similar and may be simple or denticulated.

The scaling of the coxa and trochanter is not usually used as a characteristic for species identification. However, pale scales can form specific patterns on a usually dark background of the femur, tibia and tarsus, generally in the shape of longitudinal stripes and/or rings. The surfaces of those segments (particularly those of the tibia) are described as if the leg was held stretched straight out, forming a right angle with the longitudinal body axis. Thus, it can be referred to as an anterior and posterior, dorsal and ventral, as well as an anterodorsal surface, and so on (Wood *et al.*, 1979). The femur, when characteristically coloured, has basally more pale scales than apically. The tibia and each of the five tarsomeres may be entirely dark scaled or they may bear pale longitudinal stripes or rings of varying width.

In all Diptera the mesothoracic wings are used for flight; the metathoracic are modified to form small vibrating organs known as halteres, which control the equilibrium during flight. The wings consist of the upper and lower epidermal layers, not fused along certain strengthening tubes, the wing veins. The complete system of wing veins is called the venation. Beginning from the anterior margin of the mosquito wing (Fig. 5.8), the first unbranched vein is the costa (C), which passes round the apex of the wing and forms its anterior margin. The subcosta (Sc) is located closely behind the costa and is also undivided. The radius (R) forks into an anterior branch R_1 and a posterior branch, or radial sector R_s, which branches again into R_{2+3} and R_{4+5}. The R_{2+3} divides once more into R_2 and R_3, while R_{4+5} remains unbranched. The fourth vein, or media (M), bifurcates apically into M_{1+2} and M_{3+4}. Likewise, the fifth vein, or cubitus (Cu), divides into Cu_1 and Cu_2. Finally, there is one anal vein (A) present. The longitudinal veins may be connected by six different cross veins. Two of them are situated close to the wing base, the humeral vein (h) stretches from C to Sc and the arculus (Ar) from R to M and Cu. The other four cross veins are displaced towards the wing apex. They are the subcostal-radial vein (sc-r) extending from Sc to R, the r_1–r_s from R_1 to R_s (apparent in Anophelinae—Fig. 5.8b), the radio-medial vein (r-m) from R_{4+5} to M and the medio-cubital vein (m-cu) from M to Cu_1. The veins divide the wing area into cells, the name for each cell is derived from that of the vein forming its anterior margin. Where two veins have fused or make one part of the anterior border, the cell is named after the posterior component. Thus, when veins R_4 and R_5 coalesce, as in mosquitoes, the area behind it is called cell R_5 (Davies, 1992).

Almost all wing veins are both dorsally and ventrally covered with scales. The cross veins are without scales except in the genus *Culiseta*. The wing margin, from the apex of the costa until the level of the base of the anal vein, is fringed with a row of scales. The veins and margin are usually covered with dark scales, which can aggregate to form distinct spots (some species of *Anopheles* and *Culiseta*). Pale scales may be scattered or grouped into alternating patches along the veins in a pattern of considerable taxonomic value. In males, the scales are

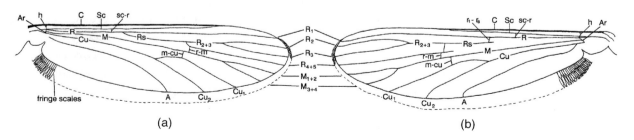

Figure 5.8. Wing of (a) Culicinae; (b) Anophelinae.

less dense and more easily rubbed off, even though of similar colouration and distribution as in the females. The same is true for the scaling pattern of other body parts. As a consequence the scaling pattern of the males will not be discussed in detail in the species description nor used in the keys.

5.1.3. Abdomen (Fig. 5.1)

Basically the insect abdomen consists of eleven segments and the nonsegmental telson (absent in most insects). An abdominal segment consists of a sclerotised tergum and sternum joined by membranous pleural regions. If a tergum or sternum is further subdivided each sclerite is named tergite or sternite respectively. Numerous reductions and amalgamations of segments can occur in the abdomen, especially in the last abdominal segments with the functions of mating, egg laying and disposal of faeces. In both sexes segments X and XI are generally represented by a dorsal lobe, or epiproct, and two lateroventral lobes, or paraprocts. In mosquitoes the first segment of the abdomen is somewhat reduced. The median part of tergum I is often not visible from above. In such cases only lateroterga are apparent (Fig. 5.4). The rest of the pregenital segments, II–VIII in males and II–VII in females, are well developed. The posterior part of each segment overlaps the anterior part of the next segment, and the segments are joined by an intersegmental membrane. The pleural membranes of segments I–VII bear a pair of spiracles anterolaterally (Knight and Laffoon, 1971). In males segment IX supports the reproductive opening and is highly modified to produce a well developed copulatory apparatus, the hypopygium. The postgenital segments X and XI are reduced, fused and partly telescoped into segment IX. Segment X, the cercal sclerites and the paraprocts are fused, forming the proctiger. In males both the proctiger and in some species of *Culiseta* the sexually differentiated abdominal segment VIII can be of taxonomic value. In females, genital segment VIII is well developed but shorter than segment VII. The opening of the internal reproductive organs lies at the VIII segment border. Segment IX is reduced to less than a third of the size of the preceeding segment and bears the rest of the postgenital segments, the proctiger (epiproct and paraprocts) and the cerci (Fig. 5.1), which support oviposition. Most of these structures are partly telescoped within segment VIII. The terga and sterna of the eight easily distinguishable segments (except sternum I) are entirely clothed with decumbent scales or covered with setae only.

Insect genitalia in both sexes are said to be derived from modified abdominal appendages, present in other arthropods and used for mating and egg laying. Basically, each genital appendage, the gonopod, has been regarded as consisting of a limb base, the gonocoxite, which shoulders an apical appendage, the gonostylus, and a median process, the gonapophysis, arising from its basal region. Some or all of these parts may be reduced, obliterated or fused, therefore the homologies of the various structures are not known with certainty for all orders of advanced insects (Davies, 1992).

The male genitalia are said to be derived from appendages of the genital segment. In some insects such as the Ephemeroptera these consist of a pair of lateral claspers, the gonocoxite with the gonostylus, which grip the female in copulation, a pair of parameres and of the median intromittent organ, the aedeagus. Parameres of many orders are most probably not homologous throughout the class of insects. But they and paired rudiments, the mesomeres, from which the aedeagal structures develop, are suggested to represent the divided gonapophyses of abdominal segment IX (Davies, 1992). For mosquitoes Horsfall and Ronquillo (1970) showed the development of the aedeagal complex from the genital primordia. Some authors claim that claspers of mosquitoes are parameres and not homologous with the claspers of mayflies (Seguy, 1967; Snodgrass, 1935). Thus parameres and claspers should not be confused, as the latter most probably are derivates of segment IX as in other nematocerans (Dahl, 1980). In the present book the widely accepted nomenclature of gonocoxite and gonostylus is applied for practical reasons and without implications of the true origin.

The uncertainty of origin of the various structures and consequently the variation in terminology are further affected by the fact that a few hours after emergence of the males, the abdominal segments VIII, IX, X and XI rotate through 180°. This inverts the terga, sterna and genitalia into an upside-down position. Thus, the aedeagus lies above the anus instead of being under it and the hindgut is twisted over the reproductive duct. However, it is unanimously accepted for all the structures to be referred to as though they were in their correct, prerotational, morphological position. All drawings of the genitalia are consequently of dorsal, prerotational view (Figs 5.10–5.12). The synonyms most frequently used for the genital structures are given in brackets.

The tergum, sternum and pleurae of segment IX are usually fused into a complete ring (Fig. 5.9). The tergum IX is often bilobed, each lobe bearing a various number of strong setae. These are not always shown in the drawings of the hypopygia, except in cases of taxonomic importance. The largest unit of the male genitalia is the gonocoxite (basimere, basistyle, coxite, sidepiece or paramere according to some authors) which articulates with sternum IX (Fig. 5.9). The gonocoxite may be a sclerotised, truncate cone or may have the dorsal and ventral surfaces mesally separated by a median membrane. It may lack lobes, as in most *Anopheles* and some *Aedes/Ochlerotatus* species, or be variously lobed. These lobes are named with the appropriate combination of position adjectives (*e.g.* basal lobe, apical lobe, subapical lobe) without intent to imply homologies (Knight and Laffoon, 1971). The gonocoxite and its lobes are covered with sparse or dense setae of different length, width and shape, sometimes with scattered scales. The movable, elongated appendage, the gonostylus (clasper, distimere or dististyle) is articulated with the gonocoxite at or close to its apex. The gonostylus usually bears one or more setae, the apical spines of gonostylus (Wood *et al.*, 1979), which are often short, spine or peg-like and situated apically or near the apex of the gonostylus. Dorsomesally to the gonocoxite and laterally to the aedeagus a paired sclerite, the paramere, is articulated with the gonocoxal and parameral apodemes, thus supporting the aedeagus (Fig. 5.9). Some authors apply the term phallosome to the

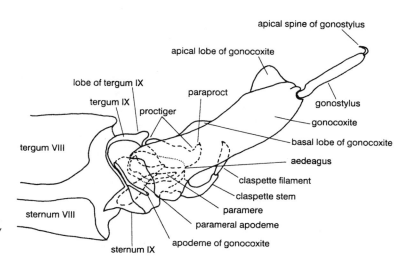

Figure 5.9. Abdomen of *Aedes/Ochlerotatus* sp.—male, lateral view.

complex of structures made up of the aedeagus, parameres and parameral apodemes while in this publication these structures will be referred to separately as in Harbach and Knight (1980).

The aedeagus (mesosome, phallosome), is the median structure of the genitalia. It may be quite variable in shape but is often bulb or flask shaped (Figs 5.10 and 5.11). It lies medially, above the ventral surfaces of the gonocoxites and below the proctiger (the position is switched after rotation). The lateral borders of the aedeagus are usually sclerotised and form two darkly coloured plates, the lateral aedeagal plates.

Dorsal to the aedeagus lies the proctiger (anal segment, anal cone, tenth segment), a complex formed by segment X, the cercal sclerite and the paraprocts (Figs 5.9–5.11). The proctiger bears the anus and is largely membranous. Usually the sclerotised parts of the proctiger are tergum X, a portion of the cercal sclerite and the paraprocts (tenth sternite— Figs 5.11 and 5.12). Tergum X is often developed as a pair of dorsally directed arm like tergites, arising laterobasally on the proctiger and sometimes confluent with the basal part of the paraproct. The paraproct is paired, usually a well sclerotised sclerite, situated laterally on the proctiger. Its strongly sclerotised apex can be denticulated or covered with numerous short spine-like setae (in *Culex*).

Largely membranous projections of the gonocoxites support variably shaped ventromesal lobes, the claspettes (harpes, harpagones), which are well developed in *Anopheles, Aedes/Ochlerotatus* and *Psorophora* but absent in *Culex* and other genera (Figs 5.9–5.12). The male genitalia of different genera and related species often differ considerably in detail and are therefore of great taxonomic value.

In most Anophelinae one or more strong, spine-like setae, the parabasal setae, are sunk

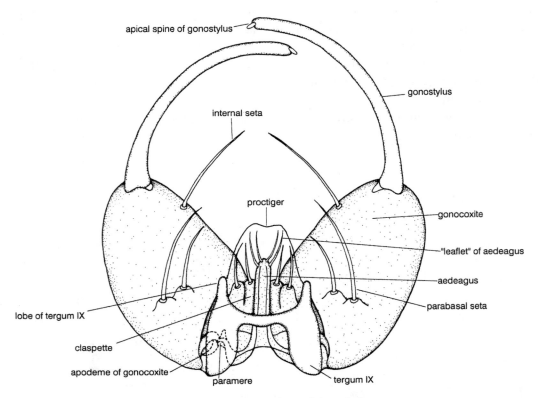

Figure 5.10. Hypopygium of *Anopheles* sp.

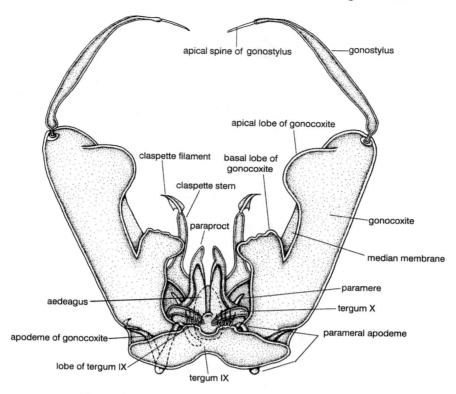

Figure 5.11. Hypopygium of *Aedes/Ochlerotatus* sp.

into a cuticular socket or elevated on a tubercle, both located dorsobasally on the gonocoxite (Fig. 5.10). Additionally, a strong spine-like seta, the internal seta (mesal seta), is mesally attached on the gonocoxite of many species. One or more pairs of spine-like or leaf-like setae, the leaflets of aedeagus (aedeagal setae), are often attached to the apex of the aedeagus. The proctiger is usually greatly reduced or membranous (cone shaped). It is hardly visible, especially after decolouration with KOH during slide preparation, so it is not shown in any drawing of anopheline hypopygia. Only some anophelines (subgenus *Nyssorhynchus*) have well sclerotised paraprocts. The claspettes of *Anopheles* may be divided longitudinally into two or three lobes, the inner, median and outer lobe. All of these can bear spine-like or flattened, spatula-like setae.

In the genera *Aedes* and *Ochlerotatus* the typical claspette is divided into a proximal, more or less cylindrical portion, the claspette stem,

and a single leaf shaped division, the claspette filament (Fig. 5.11). In the subgenus *Aedes* the apex of the claspette is bifurcate (claspetoid) with the two branches bearing a few setae, while in *Aedimorphus* and *Stegomyia* it is of variable shape, covered with numerous setae or bare.

In *Culex* (Fig. 5.12), the apex of the paraproct is called the paraproct crown. It may bear a group of dense, short setae, be denticulate or with a combination of both. The lateral aedeagal plates are particularly well developed and subdivided into a lateral portion, the outer division, and a mesal portion, the inner division. The outer division is often further differentiated into lateral arms or teeth, the inner division of some species is further divided into a dorsal arm and a ventral arm. Laterobasally from the paraproct originates a ventrally pointed process, the ventral arm of the paraproct (basal arm of tenth sternite, basal arm of anal segment).

Mosquito females do not possess a true appendicular ovipositor. The posterior segments

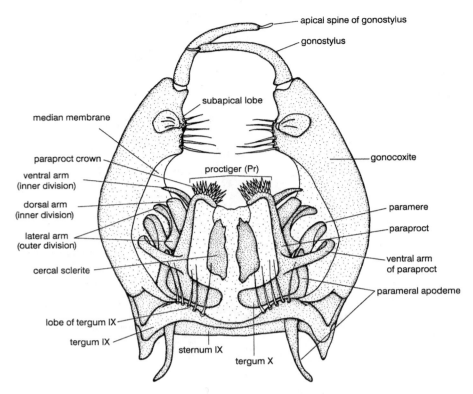

Figure 5.12. Hypopygium of *Culex* sp.

of the abdomen form a short apparatus modified for egg laying which acts as a functional ovipositor. The female genitalia are of little taxonomic value. The structures of the postgenital segments are greatly reduced. The best developed are the postgenital plate and cerci. The cerci (Fig. 5.1) are long and readily visible in most species of the genera *Aedes*, *Ochlerotatus* and *Psorophora* while in other genera the cerci are short and inconspicuous.

5.2. LARVAE

The body of the mosquito larva is divided into three principal parts, the completely sclerotised head capsule, the flattened thorax composed of three fused segments which is distinctly broader than the two other parts in fully grown instars, and the abdomen which consists of ten segments. Mosquito larvae are distinguished from all other dipterous larvae by the combination of the following characters: the presence of distinct labral brushes (exceptions may be found in carnivorous larvae, *e.g. Toxorhynchites* sp.), the expanded thorax and the tubular or cylindrical breathing tube, the siphon. The siphon is located on the dorsal surface of abdominal segment VIII in all genera except *Anopheles*. In the larvae of Dixidae, which possess labral brushes but lack a siphon, the thoracic segments are not fused or broadened. The larvae of the Chaoboridae, especially of *Mochlonyx* sp. which have a siphon, the mouthparts are converted for a predacious life and not used for filter feeding. They have long, ventrally curved prehensile antennae, adapted to capture aquatic prey. In addition, the chaoborid larvae possess conspicuous hydrostatic organs, one pair in the thorax, the other in abdominal segment VII, which allow the larvae to float in a horizontal position in the water.

Mosquito larvae pass through four larval instars. During their development various diagnostic characters change, *e.g.* the size of the head capsule, the number of pecten teeth or the number of branching of certain setae may increase. For this reasons most larval identification is based on the fourth-instar. This may require rearing of earlier instars until the fully grown fourth-instar is reached. The following description of the larval morphology is mainly confined to those characters which are of taxonomic importance among European species and used in the keys. Although the larvae of the anopheline and culicine subfamilies differ from one another in many respects, they are structurally similar. Therefore, they are described together, directing attention to those characters which are exhibited by one or the other subfamily.

The larval body is ornamented with 222 pairs of setae (Forattini, 1996). Their arrangement, called the chaetotaxy, and structure are important taxonomic features. The setae may be simple or variously branched (Fig. 5.13). A simple seta is undivided and usually cylindrical and apically attenuated. An aciculate seta

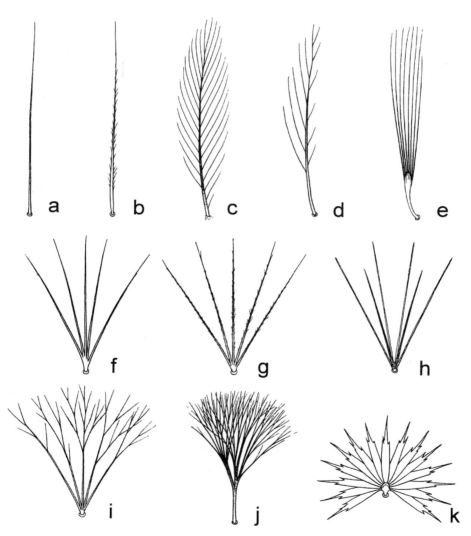

Figure 5.13. Types of larval setae: (a) simple; (b) aciculate; (c) pinnate or plumose; (d) lateral branched; (e) fan-like; (f, g) branched; (h) stellate; (i, j) dendriform; (k) palmate (after Marshall, 1938).

consists of a main stem with lateral secondary branches which are short and delicate. The branching may be sparse or dense. A branched seta consists of a long or short stem split into two or a few more divisions usually approximately equal in length, none of them can be recognised as the continuation of the main stem. A branched seta, which has no apparent stem and is divided practically at the base into several more or less equal divisions, is called a tuft. A stellate seta has a very short stem, is basally branched, the stiff branches radiating from the stem at various angles. A pinnate or plumose seta consists of a main stem with regularly arranged, long lateral branches, which usually shorten towards the apex. A lateral branched seta has a main stem with lateral branches, irregularly arranged and often irregular in length. A fan-like seta, which can be found in the ventral brush of the larva, consists of a very stout stem, usually apically widened, and many branches. A dendriform seta has various stems with branches that are divided and subdivided so that they resemble the branches of a tree. Specialised setae, the so called palmate setae, characteristic of the genus *Anopheles*, can be found on several abdominal segments. A palmate seta consists of a short stem from which leaf shaped, flattened branches, called the leaflets, extend. They may have smooth or serrated margins. The number of the leaflets, their shape and marginal serration can vary in different species. Sometimes the leaflets are narrowed or shouldered more or less abruptly at or beyond the middle, and thus divided into a wider proximal part or blade, and a narrower distal part, called the filament. Palmate setae support the larval body at the water surface film while feeding. When the anopheline larvae lie just beneath the water surface, the setae are held in an open position and the leaflets are spread to at least 180°. When the larvae dive, the leaflets become folded and hold an air bubble inside which is used to break the surface film at the next rise to the surface.

5.2.1. Head (Fig. 5.14)

Although the heads of culicine and anopheline larvae resemble each other in structural and

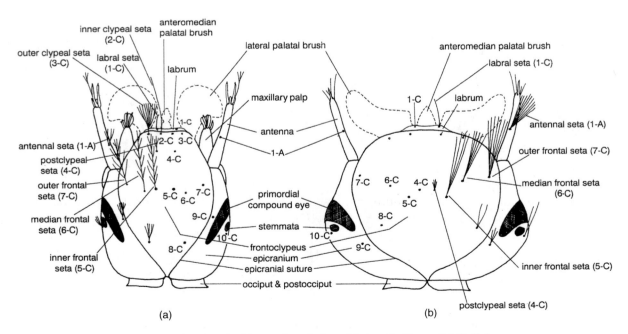

Figure 5.14.   Head of larva-dorsal view (a) Anophelinae; (b) Culicinae.

other features, they are noticeably different in shape. In most species of culicine larvae, the head is significantly broader than it is long, whereas the head of anopheline larvae is normally longer than broad, but the ratio of the length to the width may vary in different species.

The head capsule is formed by four sclerotised plates, the frontoclypeus (dorsal apodeme), a large sclerite forming the dorsal aspect of the head, and two epicranial plates (lateralia), covering the lateral surfaces as well as the ventral surface and the fused collar-shaped occiput and postocciput. The first three plates meet dorsally in the epicranial suture. In some species the frontoclypeus has areas of darker colouration which gives rise to a specific pattern, sometimes of diagnostic value (*e.g.* in *An. algeriensis*). To evaluate this character correctly, the larval head must be examined under low magnification. The crescent shaped labrum is located anterior to the frontoclypeus. Its well developed ventral surface or palatum carries the conspicuous labral brushes, called lateral palatal brushes. Each labral brush is composed of numerous long, sometimes sigmoid setae closely arranged on the palatum. The brushes can be moved rapidly and synchronously to produce currents in the surrounding water and thereby attract floating food particles. The presence or absence of serrations on the setae of the labral brushes may distinguish those larvae that brush food particles from those that sweep the water as filter feeders. In predatory larvae the brushes have only a few, stout, prehensile setae. The median part of the palatum bears a brush much less developed than the lateral palatal brushes, the anteromedian palatal brush. The larval mouthparts are rarely used for taxonomic purposes and they will not be discussed in detail here. Besides the labrum, they consist of two pairs of flattened sclerites, the mandibles and maxillae and the mentum (labium). The maxillary palps arise close to the base of each maxilla. They are much more developed in anophelines than in

culicines. For particular descriptions of the mouthparts see Dahl (2000), Clements (1992), Dahl *et al.* (1988), Harbach and Knight (1980) and Pucat (1965).

The antennae, which arise at the anterolateral corners of the head, are slender and slightly tapering sensory appendages. They may be shorter than the head and either straight or slightly curved (genera *Aedes* and *Ochlerotatus*, except *Oc. diantaeus*, and subgenus *Culiseta*) or as long as or longer than the head and evenly curved (genus *Culex* and subgenus *Culicella*). Each antenna bears six setae, which are numbered 1-A to 6-A. The antennal seta 1-A projects from the shaft of the antenna; it can be simple and inconspicuous (*e.g. Oc. geniculatus*) or mostly has multiple branches of large size (*e.g. Cs. morsitans*). Its position on the antennal shaft is often of diagnostic importance. Setae 2-A to 6-A are minute and located at the antennal apex. The antennal shaft is usually covered with small anteriorly projecting spines or spicules, only in the subgenus *Finlaya* and in *Oc. pulcritarsis* is the surface of the shaft smooth.

Two pairs of eyes are situated medio-laterally at the epicranial plates. The dark crescent shaped anterior patches are the primordial compound eyes of the future adult showing through the larval skin. The smaller simple larval eyes or stemmata are located just behind them. Posterior to the frontoclypeus and the epicranial plates lies the occipital foramen, the opening of the cranium to which the cervix (neck) is attached.

On the head are up to 18 symmetrically paired setae which are numbered 0-C to 17-C. The letter "C" is used to indicate that the setae are located on the head or caput. Two more setae affiliated with the letter "C", C-18 and C-19, are articulated to the cervical sclerite. The setae which are of diagnostic importance and used for identification in the keys, arise from the frontoclypeus and are numbered 2-C to 7-C (Fig. 5.14). They provide their diagnostic indication not only by their relative position but also by their length, degree of branching and other characteristics. One additional seta, 8-C also arises from the frontoclypeus. Setae 1-C are prominent, forwardly directed setae arising on the labrum. The inner and outer clypeal setae 2-C and 3-C are located close to the anterior margin of the frontoclypeus. In culicines they are either greatly reduced in size and not easily visible, or absent. In anophelines they are among the most conspicuous setae, and their position in relation to one another is used to separate the subgenus *Anopheles* from the subgenus *Cellia*. The postclypeal setae 4-C are located at some distance behind the clypeal setae in anophelines and they are more weakly developed. In most culicine species the postclypeal setae 4-C are very short, usually multiple branched and situated closer to the midline than the frontal setae. The frontal setae consist of three pairs, the inner (5-C), median (6-C) and outer (7-C) frontals. In *Anopheles* larvae the frontal setae arise more or less side by side, forming a transverse row. Usually they are plumose, except in *An. plumbeus*, which has very short and simple setae. In the other genera, except a few *Aedes* species, setae 6-C are displaced forward, sometimes even in front of 4-C, forming a triangle with its two companions 5-C and 7-C. The frontal setae are mostly well developed and branched. The outer frontal setae 7-C arise near the edge of the frontoclypeus close to the antennal bases. They are often more densely branched than the inner or median frontal setae. In larvae of *Ur. unguiculata* setae 5-C and 6-C are unique, they are long and stout. Setae 9-C and 10-C are located at the epicranium, below the primordial compound eyes, the latter being displaced more laterally. All other setae arise on the ventral side of the head.

5.2.2. Thorax (Fig. 5.15)

The thorax is the most conspicuous part of the larvae. Its cuticle is mainly or entirely membranous, and during the growth of the instars it becomes increasingly larger, relative to the head. Just before pupation of the fourth-instar larvae, it is much broader than the head. As in the adult, the thorax consists of three segments, the pro-, meso-, and metathorax. The segments are completely fused; their borders can only be determined by the arrangement of the setae in three distinct sets. The symmetrically paired setae are numbered 0-P to 14-P on the prothorax, 1-M to 14-M on the mesothorax and 1-T to 13-T on the metathorax. The numbering starts with the pair of setae closest to the middorsal line and ends with the one nearest the midventral line, the only exception being seta 0-P which is articulated lateral to 1-P and displaced towards the mesothorax. Many of the 42 pairs of thoracic setae may be useful for identification, but only setae 1-P to 3-P are used in the larval keys, because there are other convenient characteristics for identification, especially on the head and the last abdominal segments. Setae 1-P to 3-P usually arise very close to each other in a line and may, at a first glance, be mistaken for branches of a single seta. They are often situated on a sclerotised plate.

5.2.3. Abdomen (Figs 5.15–5.17)

The larval abdomen consists of ten segments, the first seven segments closely resemble each other. Abdominal segment I bears 13 pairs of setae and each of segments II to VII has 15 pairs. When referring to a seta, its number is followed by the segment number, for example 3-VI refers to seta 3 on abdominal segment VI. The numbering of the setae follows the same principle as described for the thorax. Of all pairs of abdominal setae available on segments I to VII, only a few are used for identification in the keys. In anophelines seta 1 is of palmate type in some or all abdominal segments (Fig. 5.15b). The number of segments with fully developed palmate setae, and the shape of the branches, varies between species. To distinguish

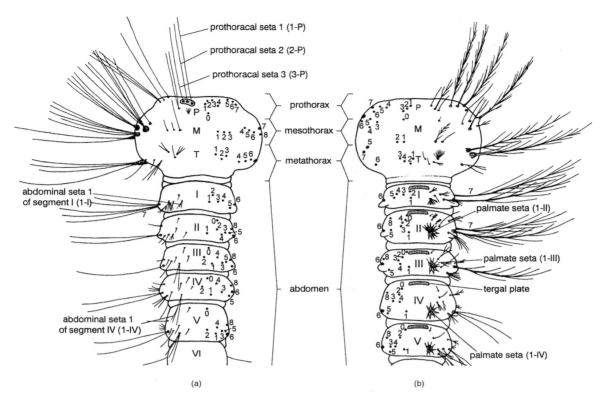

Figure 5.15. Thorax and abdomen—dorsal setation (a) Culicinae; (b) Anophelinae.

between the two species so far known in the Anopheles Claviger Complex, the branching of the antepalmate setae 2-IV and 2-V is a useful characteristic. These setae are located anterior to the palmate setae and closer to the dorsomedian line, than the setae 3 to 5 of the segments. The most laterally placed setae 6 are the longest and most conspicuous on segments I to VI. In European culicines no palmate setae are present (Fig. 5.15a). In the genus *Culex* setae 1 on segments III–V show a diagnostic number of branches useful in separating species of *Cx. p. pipiens* and *Cx. torrentium*. In anophelines, each of the abdominal segments I–VII bears a sclerotised tergal plate anteriorly, and may also have one or more smaller plates posterior to it near the centre of each segment. Usually these plates do not occur in culicine larvae, except *Or. pulcripalpis* which has sclerotised plates of

different size dorsally on segments VI–VIII in the fourth-instar larva.

Segment VIII is entirely different from the preceeding segments. It bears the only functional external openings of the metapneustic respiratory system, the spiracles, which are located posteriorly on the dorsal surface of the segment. In culicines the spiracles are situated at the tip of a long, tubular and cylindrical organ called the siphon (Fig. 5.16). In anophelines, the siphon is almost completely reduced to the spiracular plate and often is said to be absent (Fig. 5.17).

On each side of segment VIII a number of decumbent scales can be found. The whole structure is called the comb (Fig. 5.16). Each comb scale is directed posteriorly and fringed with small spines. All spines may be of the same length or the median, or terminal spine

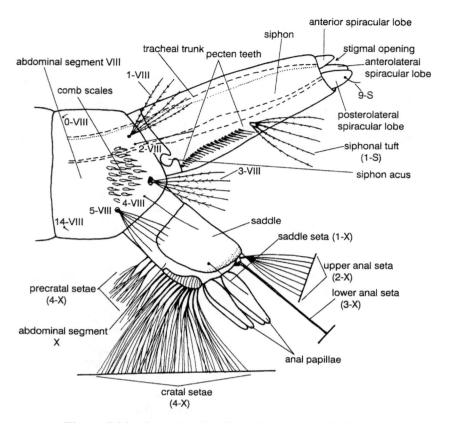

Figure 5.16.   Lateral view of larval abdomen—Culicinae.

can be exceptionally distinct and longer than the others. The shape of the spines can only be studied adequately under sufficient magnification, therefore this character is used in the keys only to separate a few species within the genera *Aedes/Ochlerotatus* and *Culex*. The number of comb scales varies greatly with the species, from 5–7 to more than 100 and they may be arranged in a single row, a double row or in an irregular pattern. The total number of comb scales, with its ranges of variability, is of diagnostic value. In *Ur. unguiculata* the comb scales arise from the posterior margin of a sclerotised lateral plate. In *Or. pulcripalpis* the comb scales are much more elongated than in the other genera and form two regular rows. In fourth-instar larvae of anophelines there is no structure which corresponds to the culicine comb (Fig. 5.17).

In culicines, of the seven pairs of setae on the lateral sides of segment VIII, five pairs are located posterior to the comb scales. They cannot be homologised with the setae on the first seven segments, except the ventral-most one which is consequently named 14-VIII instead of 6-VIII. All setae are numbered, successively from dorsal to ventral, 0-VIII to 14-VIII. Only 1-VIII to 5-VIII are distinct (Fig. 5.16) and in the fourth-instar larvae the setae 1-VIII, 3-VIII and 5-VIII are usually strongly branched, whereas the intermediate ones, 2-VIII and 4-VIII, are shorter and nearly always simple, except in *Coquillettidia*. Both 0-VIII and 14-VIII are minute and of no taxonomic importance.

At the base of the siphon, two small lateral projections arise close to its ventral margin, each called acus. They are the points of muscle attachment which enable the siphon to be bent

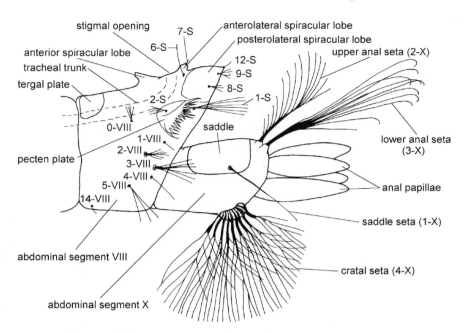

Figure 5.17. Lateral view of larval abdomen—Anophelinae.

posteriorly. In all species of the subgenus *Stegomyia* the acus is absent and in *Ae. vittatus* it is very indistinct. The siphon of the culicines is one of the most useful structures for identification. Its shape and proportions vary considerably. Very often the siphonal index is used for identification, its value is expressed as the ratio of the length of the siphon to its basal width. Harbach and Knight (1980) and Belkin (1962) suggested measuring the width of the siphon at its middle. In many cases such measurement is difficult and quite subjective. For this reason it is suggested to measure the width of the siphon at its base (Forattini, 1996; Gutsevich *et al.*, 1974). The index may vary from less than 2.0 in some species, such as *Oc. mariae*, to over 7.0 in others, like *Cx. hortensis*. In some *Culiseta* species the ratio of the siphonal length to the width of its apex is diagnostic. The siphon also carries a pair of ventrolateral rows of stout, sclerotised spines, the pecten teeth. Each row is known as the pecten. A single pecten tooth usually has 1–4 lateral denticles on its ventral margin. In culicines the pecten teeth commonly increase uniformly in size toward the apex. All teeth may be evenly spaced, but in some species

the distal pecten teeth may be more widely and irregularly spaced than the others. In the keys they are then referred to as being detached apically. The proportion of the siphon to which the pecten extends from the base, is of taxonomic importance. It may be short, with a few teeth which do not exceed the middle of the siphon, or it may extend almost to the apex of the siphon, as in *Oc. cataphylla*. In *Cs. longiareolata* the pecten consists of a row of teeth which are widely spaced along the entire siphon. In *Cs. fumipennis* the stout teeth continue on in an irregularly row of large spines. In the subgenus *Culiseta* typical pecten teeth are found only at the base of the siphon. They are then followed by a row of thin, hair-like setae. Among the European culicine species the pecten is absent only in the species of the genus *Coquillettidia* and in *Or. pulcripalpis*. In anophelines which lack an elongated siphon, the pecten is situated immediately below the spiracles and consists of a row of alternate long and short teeth arising from the posterior side of a triangular sclerite, the pecten plate (Fig. 5.17). The two plates on each side of segment VIII are connected by a U-shaped band. The setae 0-VIII to 14-VIII are located ventrally to the pecten and are less conspicuous than in culicines.

The siphon and spiracular plate bear 13 pairs of setae designated 1-S to 13-S. In culicine larvae the siphon is adorned with one or more pairs of tufts or simple setae on its ventral and/or lateral surface, the siphonal setae 1-S (Fig. 5.16). If there is more than one pair of setae 1-S present, the basalmost is named 1a-S, the next one towards the apex of the siphon 1b-S and so on, proceeding distally (Darsie and Ward, 1981). Larvae of the genus *Culex* have several pairs of tufts, they may be arranged symmetrically and sometimes the penultimate tuft is found to be situated dorsally out of line with the others, or they may arise in a more or less zig-zag row. The total number of siphonal tufts, their length compared to the width of the siphon at the point of articulation of each 1-S, and their position on the siphon, is often characteristic. A single paired siphonal tuft is usually present, either at the base, as in *Culiseta*, which is the principal diagnostic feature of this genus, or between the mid length and the apex of the siphon, as in other culicines. The position of 1-S with respect to the pecten is used to separate some species of *Ochlerotatus*. Usually 1-S is attached distal to the pecten, but sometimes it is attached basally to the distal pecten teeth, and is then described as being attached within the pecten. The number and length of the branches of 1-S may vary and are used for species identification. Occasionally, additional setae can be found on the dorsal surface of the siphon (subgenus *Rusticoidus*). In anopheline larvae seta 1-S is articulated behind the pecten plate (Fig. 5.17). Seta 2-S is located dorsoapically, closely to the spiracular apparatus in culicines and on the pecten plate in anophelines. The spiracles are surrounded by the spiracular plate. It is a 5-lobed valve which closes the openings during submersion of the larva. These lobes are the anterior spiracular lobe and two pairs of anterolateral and posterolateral spiracular lobes. They bear several setae, 3-S to 13-S, of which only 9-S, the second one below the apices of the posterolateral, largest valves, are used in the keys. It is elongated, thickened and hook-shaped in some *Ochlerotatus* species. In the genus *Coquillettidia* the spiracular plate is highly modified, bearing inner and outer spiracular teeth at the apex and a row of teeth on the anterior surface known as the saw, for piercing and penetrating submerged parts of plants in order to obtain oxygen. Similar modifications are not found in any other European genus, but may be characteristic of larvae of some *Culex* (subgenus *Lutzia*), *Mimomyia* and *Hodgesia* species occurring elsewhere.

Segment IX is reduced and partly fused with segments VIII and X. Its remnant is visible as a small ring at the base of segment X in many species, but in some others, like in *Ur. unguiculata*, this rudiment disappears entirely. It does not exist as a separate morphological unit and is of no taxonomic importance.

The most posterior, or anal segment X is narrower than the others, forms an angle to the ventral side of segment VIII and bears four pairs of setae, 1-X to 4-X. It possesses a curved, sclerotised plate called the saddle. Although in many species the plate is saddle-shaped and extends down the lateral sides of the segment to various degrees, *e.g.* almost reaching to the midventral line in *Oc. punctodes*, it completely encircles the anal segment in others. The shape of the saddle is of diagnostic value in many species. It bears a lateral saddle seta, 1-X. In the fourth-instar the saddle seta arises well within the saddle, closer to its posterior margin than to its ventral margin. Its length compared to the length of the saddle is frequently used for identification. The posterior margin of the saddle may bear denticles or spines of varying form. Dorsally at the distal end, the anal segment bears two long paired setae, the upper anal setae 2-X and the lower anal setae 3-X. In anophelines both pairs are composed of multiple branched setae (Fig. 5.17). In most culicines, seta 2-X is multiple branched and 3-X is a long stout seta which is usually either simple or only slightly branched (Fig. 5.16). Along the mid-ventral line of the anal segment, close to its apex, arise a number of long setae 4-X, which make up the ventral brush or fin. All setae of the ventral brush are fan-like and they act as a rudder during swimming. Some or all of these setae are attached to a heavily sclerotised network of bars which resembles a ladder or grid for greater support basally. In anophelines all setae arise from the grid, while in culicines the setae may be situated partly upon, and partly proximal, to the grid. In this latter case, those setae attached to the grid are called the cratal setae and those attached to the segment anterior to the cratal setae are called the precratal setae (Fig. 5.16). The number of precratal setae is an important feature to separate various *Ochlerotatus* species. Sometimes it is not easy to differentiate between the cratals and pre-cratals, especially when they are close together. In this case attention should be focused to the proximal end of the grid and the first tuft which is attached to it. Then the number of tufts anterior to it should be counted. The anal segment terminates with two pairs of flexible, papilli-form structures, the anal papillae which surround the anus and are involved in osmoregulation. The length of the anal papillae varies remarkably in different species. In salt marsh species and others which are associated with brackish or alkaline water, the anal papillae are extremely short. In some species the length of the papillae depends upon the physico-chemical conditions of the water in which the larvae develop, as is well known for *Oc. caspius*. Usually the two pairs are of the same length, but ocasionally one pair may be longer than the other. The length and shape of the anal papillae are regularly used for identification but often they are broken off or hardly visible, especially in mounted specimens.

5.3. PUPAE

The mosquito pupae provide less distinct characteristics for identification than the larvae or adults. Although there are differences in the external morphology and chaetotaxy of the pupae in the different genera and even at the species level, no attempt is made here to include the pupae in the keys. It is easier to rear pupae, which are collected in the field to the adult stage and then identify them. However, occasionally there may be difficulties in separating certain closely related species in the adult or larval stage, *e.g. Cs. annulata* and *Cs. subochrea*, or *An. claviger s.s.* and *An. petragnani*. In these cases some pupal characteristics are needed for proper identification. Thus, a brief overwiew of the external morphology and chaetotaxy of mosquito pupae is given.

The body of the pupa consists of a large globular anterior portion, the cephalothorax, and a narrower articulated abdomen, which is kept flexed under the cephalothorax and used to propel the individual while swimming (Fig. 5.18). Mosquito pupae, unlike the pupae of most other insects, are quite mobile and can rapidly dive from the water surface when disturbed. Usually they remain at the water surface most of the time, with the paired respiratory trumpets in contact with the air.

The general appearance of the mosquito pupa with its division into only two obvious parts is mainly due to the morphology of the pupal case. The head with the mouthpart sheaths, and the thorax with the sheaths of the wings and legs, together appear to form one structure, the cephalothorax. The flattened head with the mouthparts is located at the front of the cephalothorax. The mouthparts are bent beneath it along the ventral surface to its posterior part, like a keel. The pupal cuticle is transparent and the compound eyes of the adult are visible at the sides of the head and behind these, the stemata of the former larva. The antennae arise in front of the compound eyes in the upper part, and are directed backwards in a curved line over the sides of the thorax. The broad convex scutum of the mesothorax extends on the anterior dorsal surface of the cephalothorax. Along its median dorsal line a crest-like ridge or median keel is visible. This ridge forms the line of weakness, the ecdysial line, along which the cuticle splits before the emergence of the adult. The respiratory trumpets project from the sides of the scutum. In pupae of culicine mosquitoes the respiratory trumpets are long and cylindrical. An exception, as with the siphon of the larvae, can be found in the genera *Coquillettidia* and *Mansonia*. In this case, the trumpets taper apically and have a strong sclerotised hook, adapted for piercing submerged parts of aquatic plants. In pupae of anopheline mosquitoes the respiratory trumpets are shorter and broad, with a flap-like appearance. The base of the trumpet is very flexible allowing it to be easily moved into position when brought to the water surface. A tracheal trunk, which leads from the base of the trumpet to the developing mesothoracic spiracle of the imago, is visible through the cuticle. The wing bases from which the pupal wing sheaths pass downwards on the sides of the cephalothorax, are located posterior to the respiratory trumpets. Between the sheaths of the wings and mouthparts, the sheaths of the three legs are visible, the tarsi being curled up under the wing apex.

The abdomen of the pupa is flattened dorsoventrally and consists of nine segments, the last of which is very small and carries the terminal paddles (Figs 5.18 and 5.19). The segments are sclerotised, connected with intersegmental membranes and freely movable on each other in flexion and tension, but with little or no lateral movement. The first abdominal segment bears a pair of conspicuous palmate setae (1-I), which resemble those of the *Anopheles* larvae and have the same function in supporting

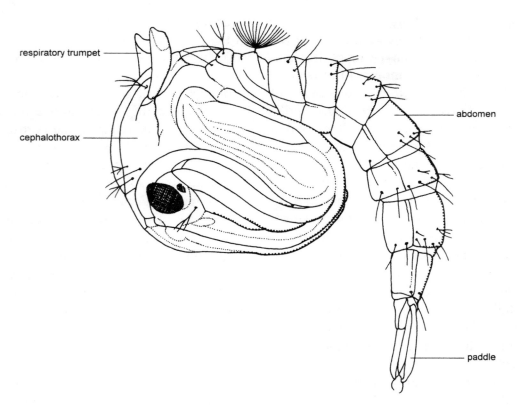

Figure 5.18.   Pupa—lateral aspect of the body.

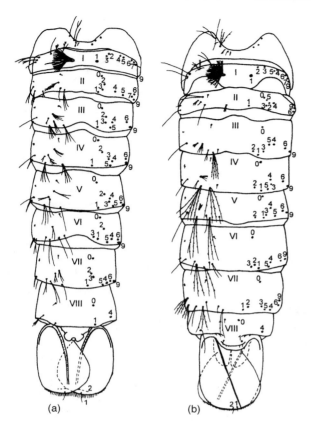

Figure 5.19. Abdominal setation of pupa-dorsal view, (a) Anophelinae; (b) Culicinae.

the pupal body at the water surface in the upright position. Again the genera *Coquillettidia* and *Mansonia* are exceptions. According to the permanently submerged life of the pupae, the seta 1-I is not of the palmate type, but short and simple. Both ventrally and laterally, the first abdominal segment is largely membranous. Abdominal segments II to VII are more or less the same size. They consist of completely sclerotised rings without pleural membranes between the terga and sterna. Segment VIII is smaller than the preceeding segments and bears on its dorsal aspect the only sclerite area of the IX segment that is usually visible, the median caudal lobe (IX tergum). The lateral lobes of the IX tergum are developed into the paddles, and the IX sternum is an indistinct transverse band. The paddles are irregularly oval in shape with a narrow base, and overlap by about half of their width.

They are the main organs of movement of the pupae. They are provided with a median longitudinal strengthening or midrib. Close to the end of the midrib arises an apical seta (1-P). In pupae of *Anopheles* a small accessory seta (2-P) is present above the apical seta on the ventral side (Fig. 5.19a), while in *Culex* the accessory seta arises dorsally side by side with 1-P (Fig. 5.19b). In the genera *Coquillettidia* and *Mansonia*, both the apical and accessory setae are absent. In many species the margin of the paddle is fringed with spicules or small spines which are of taxonomic significance.

Caudal and ventral to the median caudal lobe lies medially a small rudiment of the X segment, the anal lobe or proctiger. In female mosquitoes it is widened and carries a more or less distinct pair of cercal lobes (XI segment). Between the bases of the paddles, caudal and ventral to the anal lobe, arises a sheath, enclosing the developing genitalia of the adult, the genital lobe. In males the lobe is relatively large, conical and clearly divided, almost completely separated by a deep median fissure which extends closely to its base (Fig. 5.20a). In females the lobe is smaller and of similar size to the cercal lobes, with a more or less blunt end, and the median fissure is short and does not divide the lobe completely (Fig. 5.20b). The sex of the pupae can be readily determined by the shape and the size of the genital lobe in relation to the size of the paddles, and the extent of the median fissure.

Another separation technique may also be used in the laboratory. If larvae are reared in aerated water and overfed, the female pupae are distinctly larger than those of the males and the abdomen is almost of the same width as the cephalothorax, while in male pupae the abdomen is distinctly narrower than the cephalothorax. Bearing in mind that the development of male larvae into pupae is usually faster than that of females, sexing according to the size and general appearance of the pupae can be used with quite high accuracy. Both

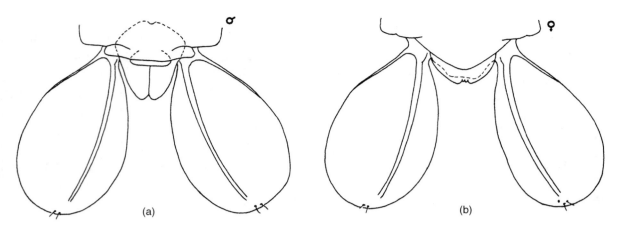

Figure 5.20. Abdomen of pupa, (a) male; (b) female.

separation techniques can be applied to living individuals and are useful in a number of laboratory experiments, *e.g.* for single pair rearing without the necessity of handling the adults.

The abdominal segments bear paired dorsal and ventral setae. Except for segments I and VIII, they resemble those of the larvae. Together with the form of the respiratory trumpets and the characteristics of the paddles, the chaetotaxy of the pupal abdomen is mainly used for description and identification of species. The notation used in this publication follows that of Harbach and Knight (1980) and is given in Fig. 5.19. Characteristic of the pupae of the subfamily Anophelinae are setae 9 of the abdominal segments III to VII. They are located at the lateral posterior corner of the segments and have the form of stout spines (Fig. 5.19a). In pupae of the Culicinae, the setae do not arise quite at the corner but a little anterior to it and

they are usually small and simple, rarely branched (Fig. 5.19b). Within the Culicinae, the genera *Coquillettidia* and *Mansonia* again show an exception in the chaetotaxy by having scarcely developed abdominal setae.

The shape and structure of the terminal appendages or paddles may be used to separate mosquito pupae from those of the closely related families of Chaoboridae, Chironomidae and Dixidae. As mentioned above, the paddles of culicid pupae are provided with a median longitudinal strengthening or midrib. Chaoborid pupae have paddles quite similar to those of mosquitoes, but they are stiffened by three veins or ribs. Despite great variations within the family, the paddles of chironomid pupae lack a midrib and they are never completely separated. In Dixidae the pupal abdomen ends with sharply tapered caudal lobes (Fig. 5.21).

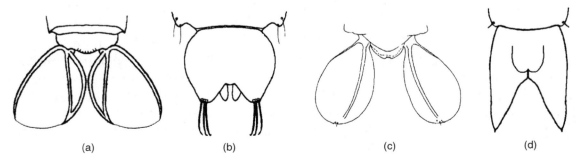

Figure 5.21. Abdomen of pupa, (a) Chaoboridae; (b) Chironomidae; (c) Culicidae; (d) Dixidae (after Cranston *et al.*, 1987).

II

The Identification Keys

Key to Female Mosquitoes

GENERA

1 Palps as long as proboscis. Scutellum evenly rounded and uniformly setose
(Fig. 6.1a) . . . . *Anopheles* (p. 88)
Palps distinctly shorter than proboscis. Scutellum trilobed, setae arranged in
three sets (Fig. 6.1b). Abdominal terga and sterna densely covered with scales 2

2 (1) Anal vein (A) sharply bent apically, ending slightly before or at the same level as
furcation of cubitus (Cu) (Fig. 6.2a) . . . . *Uranotaenia unguiculata* (p. 339)
Anal vein evenly curved, ending distinctly beyond furcation of cubitus
(Fig. 6.2b) 3

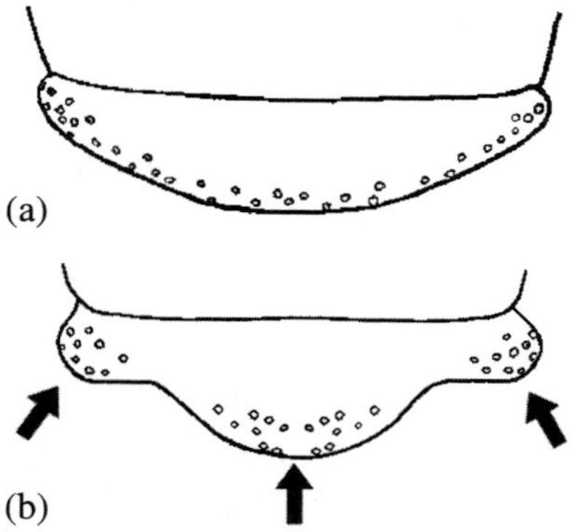

Figure 6.1. Scutellum of:
(a) *Anopheles* sp.; (b) *Aedes* sp.

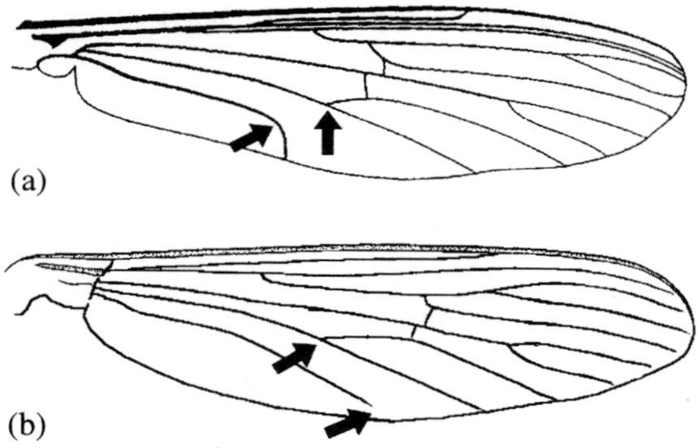

Figure 6.2. Wing of:
(a) *Ur. unguiculata*; (b) *Ae. vexans*

3 (2) Prespiracular setae present (Fig. 6.3a) . . . . *Culiseta* (p. 105)
Prespiracular setae absent (Fig. 6.3b) 4

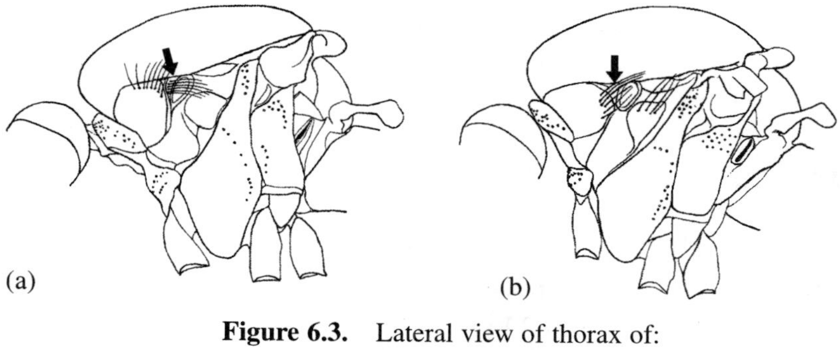

Figure 6.3. Lateral view of thorax of:
(a) *Cs. annulata*; (b) *Oc. geniculatus*

4 (3)　　Tarsomere I of fore legs longer than tarsomeres II to V together. Tarsomere IV of fore legs reduced, not longer than broad (Fig. 6.4a) . . ***Orthopodomyia pulcripalpis*** (p. 336) Tarsomere I of fore legs usually shorter than tarsomeres II to V together. Tarsomere IV of fore legs not reduced, distinctly longer than broad (Fig. 6.4b) 5

5 (4)　　Postspiracular setae present. Claws usually with subbasal tooth. Abdomen tapering apically, cerci long, easy visible (Fig. 6.5a) . . . . ***Aedes*** and ***Ochlerotatus*** (p. 92) Postspiracular setae absent. Claws simple, without subbasal tooth. Abdomen rounded apically, cerci short, hardly visible (Fig. 6.5b) 6

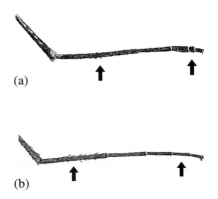

Figure 6.4.　Fore tarsus of: (a) *Or. pulcripalpis*; (b) *Aedes* sp.

Figure 6.5.　Dorsal view of abdomen of: (a) *Aedes* sp.; (b) *Culex* sp.

6 (5)　　Pulvilli present. Wing scales narrow (Fig. 6.6a) . . . . ***Culex*** (p. 103) Pulvilli absent. Wing scales usually broad and conspicuous (Fig. 6.6b) . . . . ***Coquillettidia*** (p. 108)

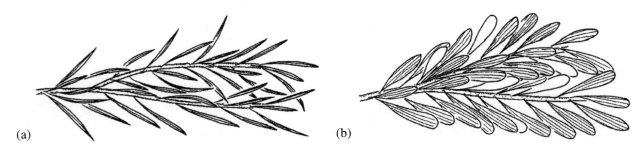

Figure 6.6.　Wing scales of: (a) *Culex* sp.; (b) *Coquillettidia* sp.

6.1. GENUS *ANOPHELES*

1　　　　Wing veins entirely dark scaled (Fig. 6.7a) . . . . 2 Wing veins covered with dark and pale scales, forming contrasting spots at least on costa (C), radius (R) and R_1 (Fig. 6.7b) 7

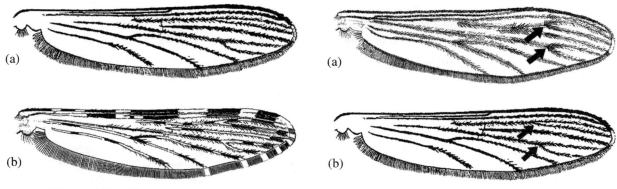

Figure 6.7. Wing of:
(a) *An. marteri*; (b) *An. hispaniola*

Figure 6.8. Wing of:
(a) *An. maculipennis s.l.*; (b) *An. marteri*

2 (1) Dark scales on wing veins more dense in some areas forming spots, mainly on cross veins and furcations. Furcation of R_{2+3} and M situated at the same distance from base of wing (Fig. 6.8a) 3
Dark scales evenly distributed, not aggregated into spots. Furcation of R_{2+3} usually situated slightly closer to base of wing than furcation of M (Fig. 6.8b) 4

3 (2) Scutum more or less pale brown, without a pale longitudinal stripe, dark spots on wing usually inconspicuous, scales of wing fringe unicolourous dark, without a pale patch at the apex (Fig. 6.9a) . . . . ***An. sacharovi*** (p. 178)
Scutum dark brown with a broad pale longitudinal stripe, dark spots on wing conspicuous, scales of wing fringe with a pale patch at the apex
(Fig. 6.9b) . . . . . all other members of **Anopheles Maculipennis Complex** (p. 172)

4 (2) All erect scales on vertex dark brown. Scutum unicoloured brown, without a median stripe of pale scales (Fig. 6.10a) . . . . ***An. algeriensis*** (p. 165)
Erect scales on median part of vertex white or cream coloured, lateral parts dark. Scutum with median stripe of pale scales (Fig. 6.10b) 5

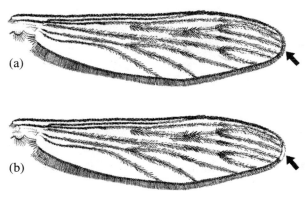

Figure 6.9. Wing of:
(a) *An. sacharovi*; (b) *An. maculipennis s.l.*

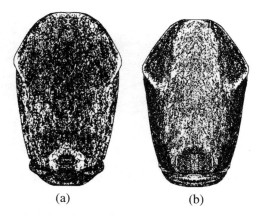

Figure 6.10. Scutum of:
(a) *An. algeriensis*; (b) *An. plumbeus*

5 (4)   Apex of proboscis pale scaled (Fig. 6.11a) . . . . ***An. marteri*** (p. 180)
        Apex of proboscis dark scaled (Fig. 6.11b) 6

6 (5)   Body blackish-grey with a leaden tinge. Anteacrostichal patch well developed,
        distinct, snow white (Fig. 6.12a) . . . . ***An. plumbeus*** (p. 182)
        Body yellowish brown or brown. Anteacrostichal patch weakly developed,
        yellowish (Fig. 6.12b) . . . . . . ***An. claviger s.s.*** and ***An. petragnani*** (pp. 167, 169)

(a)   (b)   (a)   (b)

Figure 6.11.   Head of: **Figure 6.12.**   Scutum of:
(a) *An. marteri*; (b) *An. plumbeus*   (a) *An. plumbeus*; (b) *An. claviger s.l.*

7 (1)   Costal margin of wing with 2 pale spots in the apical half. Base of fore femur
        distinctly swollen (Fig. 6.13a, b) 8
        Costal margin of wing with more than 3 pale spots along its entire length.
        Base of fore femur not swollen (Fig. 6.13c, d) 9

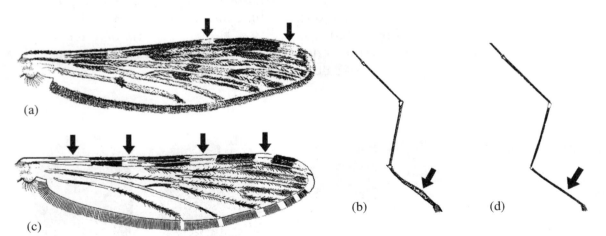

Figure 6.13.   Wing and fore leg of:
(a, b) *An. hyrcanus*; (c, d) *An. superpictus*

8 (7)   Tarsomere IV of hind leg mostly dark, pale only at apex
        (Fig. 6.14a). . . . . ***An. hyrcanus*** (p. 170)
        Tarsomere IV of hind leg entirely pale scaled
        (Fig. 6.14b) . . . . ***An. hyrcanus*** var. ***pseudopictus*** (p. 172)

Figure 6.14. Hind tarsus of:
(a) *An. hyrcanus*; (b) *An. hyrcanus* var. *pseudopictus*

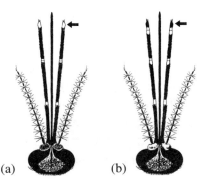

Figure 6.15. Head of:
(a) *An. sergentii*; (b) *An. multicolor*

9 (7) Palps pale at apex (Fig. 6.15a) 10
Palps dark at apex (Fig. 6.15b), rarely with a small white spot 11

10 (9) Dark scales predominate on wing veins R_s to Cu. Vein R_{4+5} almost entirely dark
scaled. Pale spot at the apex of wing fringe small (Fig. 6.16a) . ***An. sergentii*** (p. 188)
Pale scales predominate on wing veins R_s to Cu. Vein R_{4+5} almost
entirely pale scaled. Pale spot at the apex of wing fringe large
(Fig. 6.16b) . . . . ***An. superpictus*** (p. 190)

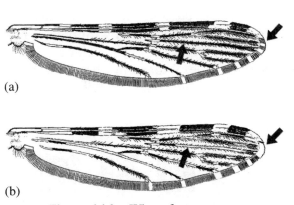

Figure 6.16. Wing of:
(a) *An. sergentii*; (b) *An. superpictus*

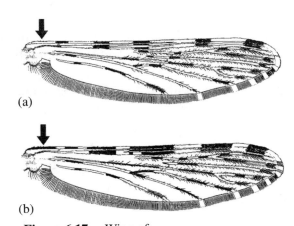

Figure 6.17. Wing of:
(a) *An. multicolor*; (b) *An. cinereus hispaniola*

11 (9) Pleurites with a few pale scales. Submedian and lateral areas of scutum with
scattered well developed, pale scales. Base of costa pale scaled
(Fig. 6.17a) . . . . ***An. multicolor*** (p. 186)
Pleurites without scales. Submedian and lateral areas of scutum without
scales, sometimes a few hair-like pale scales present, but confined to the
extreme anterior margin of scutum. Base of costa dark scaled
(Fig. 6.17b) . . . . ***An. cinereus hispaniola*** (p. 185)

6.2. GENERA *AEDES* AND *OCHLEROTATUS*

1   Tarsomeres with pale rings, usually more distinct on hind legs, rings sometimes very narrow (pale rings better visible against a dark background and with a blue light filter) (Fig. 6.18a) 2

Tarsomeres without pale rings (Fig. 6.18b) 20

2 (1)   Each pale ring embraces two tarsomeres, the apex of one and the base of the following tarsomere (Fig. 6.19a) 3

Pale rings present only at base of tarsomeres (Fig. 6.19b) 6

(a)

(b)

Figure 6.18.   Hind tarsus of:
(a) *Oc. cantans*; (b) *Ae. rossicus*

(a)

(b)

Figure 6.19.   Hind tarsus of:
(a) *Oc. caspius*; (b) *Ae. vexans*

3 (2)   Wing veins uniformly dark scaled (in *Oc. berlandi* isolated pale scales sometimes present on wing veins and legs). Tarsomeres V of all legs entirely pale scaled (Fig. 6.20a) . . . . *Oc. berlandi* and *Oc. pulcritarsis* (pp. 229, 270)

Wing veins with pale and dark scales. Only tarsomere V of hind legs entirely pale scaled (Fig. 6.20b) 4

4 (3)   Terga with a median longitudinal light band (in *Oc. caspius* rarely restricted to tergum II). The last terga sometimes almost completely covered with pale scales (Fig. 6.21a) 5

Terga without a median longitudinal light band. Pale scales form narrow basal transverse bands of varying pattern (Fig. 6.21b, c, d) . . *Oc. mariae* and *Oc. zammitii* (pp. 255, 257)

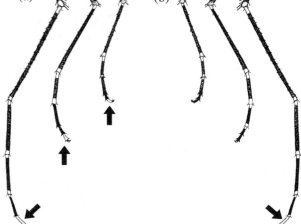

Figure 6.20.   Fore, mid and hind tarsi of:
(a) *Oc. berlandi*; (b) *Oc. dorsalis*

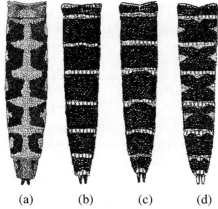

(a)   (b)   (c)   (d)

Figure 6.21.   Dorsal view of abdomen of:
(a) *Oc. caspius*; (b, c, d) *Oc. mariae s.l.*

5 (4) Scutum fawn coloured, with 2 narrow dorsocentral white stripes, reaching to the posterior margin of scutum. Wing veins with dark and pale scales more or less evenly mixed (Fig. 6.22a, b) . . . . ***Oc. caspius*** (p. 227)
Scutum with a dark brown median stripe, reaching to the prescutellar dorsocentral area. Median stripe posteriorly ornamented with a pair of narrow white lines. Posterior submedian area usually dark brown scaled. Lateral parts of scutum greyish white. Base of C, Sc, and R predominantly pale scaled (Fig. 6.22c, d) . . . . ***Oc. dorsalis*** (p. 239)

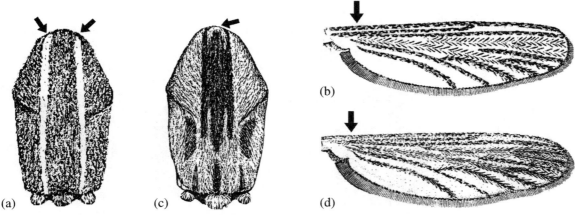

Figure 6.22. Scutum and wing of:
(a, b) *Oc. caspius*; (c, d) *Oc. dorsalis*

6 (2) Proboscis as long as fore femur or slightly shorter. Scutellum with broad, white scales (Fig. 6.23a) 7
Proboscis distinctly longer than fore femur. Scutellum with narrow, yellowish or pale and curved scales (Fig. 6.23b) 10

7 (6) Scutum with 2 to 3 pairs of small white spots, distributed along the dorsocental area (Fig. 6.24a). Tibia of hind legs with white rings in the middle . . . ***Ae. vittatus*** (p. 204)
Scutum with one or more longitudinal white stripes (Fig. 6.24b). Tibia of hind legs entirely dark scaled 8

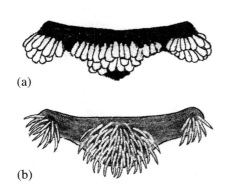

Figure 6.23. Scutellum of:
(a) *Ae. aegypti*; (b) *Ae. vexans*

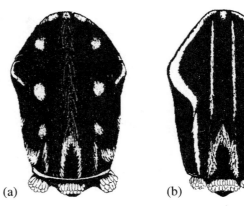

Figure 6.24. Scutum of:
(a) *Ae. vittatus*; (b) *Ae. aegypti*

8 (7)   Scutum without acrostichal stripe on anterior part, but with two narrow white dorsocentral stripes separated from anterior margin. Lateral white stripes broad, continuing over transverse suture to the end of scutum, lyre shaped (Fig. 6.25a) . . . . *Ae. aegypti* (p. 207)
Scutum with a white acrostichal stripe extending from the anterior margin to the beginning of the prescutellar area, where it forks to end at the anterior margin of scutellum. If lateral stripes are present, they are narrow and do not continue over transverse suture, never lyre shaped (Fig. 6.25b) 9

9 (8)   Acrostichal stripe broad. Posterior dorsocentral white stripes narrow, short, not reaching to middle of scutum (Fig. 6.26a) . . . . *Ae. albopictus* (p. 210)
Acrostichal stripe narrow. Posterior dorsocentral white stripes narrow, long, reaching to middle of scutum, slightly posterior of the level of scutal angle (Fig. 6.26b) . . . . *Ae. cretinus* (p. 212)

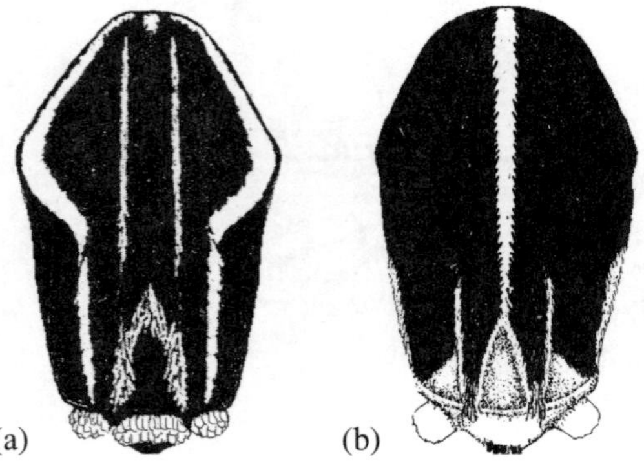

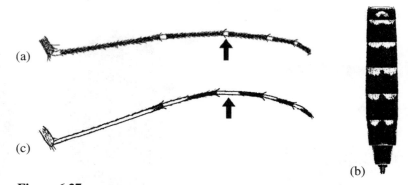

(a)   (b)   (a)   (b)

Figure 6.25.   Scutum of: **Figure 6.26.**   Scutum of:
(a) *Ae. aegypti*; (b) *Ae. albopictus*   (a) *Ae. albopictus*; (b) *Ae. cretinus*

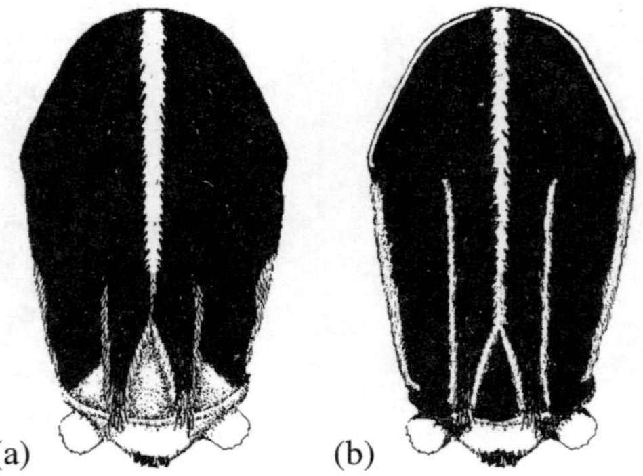

10 (6)   Pale rings on tarsi very narrow, usually not exceeding more than 1/4 the length of the tarsomeres. Terga with white transverse basal bands constricted in the middle giving them a bilobed pattern (Fig. 6.27a, b). . . . . . . *Ae. vexans* (p. 201)
Pale rings on tarsi broad. Tarsomere III of hind legs with the pale ring embracing at least 1/3 of its length (Fig. 6.27c) 11

Figure 6.27.
Tarsus (a) and abdomen (b) of *Ae. vexans* and tarsus (c) of *Oc. flavescens*

11 (10) Dorsal surface of terga with light scales, sometimes mixed with isolated dark scales (Fig. 6.28a) 12

Dorsal surface of terga with scattered dark scales (not isolated). Sometimes dark scales predominate (Fig. 6.28b) 13

12 (11)   Scutum covered with copper or golden brown scales. Scales on terga straw coloured, distinctly paler than scutal scales. Pleural scales also paler compared to scutal scales. Lower mesepimeral setae absent (Fig. 6.29a) . . . . ***Oc. flavescens*** (p. 245)
Scutum covered with golden yellowish scales. Scales on terga usually ochre yellowish, not distinctly differing in colour from the scales of the scutum. Pleurites with cream coloured scales, colouration similar to scutum. Lower mesepimeral setae present (Fig. 6.29b) . . . . ***Oc. cyprius*** (p. 233)

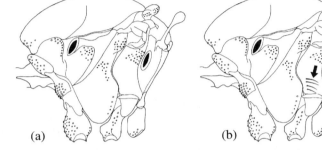

(a)   (b)   (a)   (b)

Figure 6.28. Dorsal view of abdomen of: **Figure 6.29.** Lateral view of thorax of:
(a) *Oc. flavescens*; (b) *Oc. cantans*   (a) *Oc. flavescens*; (b) *Oc. cyprius*

13 (11)   Palps, proboscis and wing veins uniformly dark scaled, occasionally isolated pale scales may be present at tip of palps or at base of costa (C) (Fig. 6.30a) . . . . ***Oc. mercurator*** (p. 257)
Palps, proboscis and wing veins, or at least one of them, with scattered or grouped pale scales (Fig. 6.30b) 14

14 (13)   Terga almost completely covered with dark scales dorsally, without pale transverse bands. Pale scales usually forming diffuse patches in the midline of the terga (Fig. 6.31a). Scutum usually uniformly coloured with small, bronze or rust coloured scales . . . . ***Oc. behningi*** (p. 221)

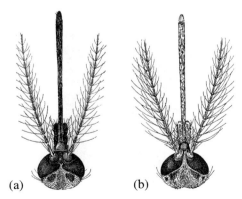
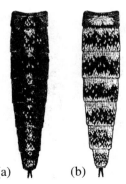

(a)   (b)   (a)   (b)

Figure 6.30. Head of: **Figure 6.31.** Dorsal view of abdomen of:
(a) *Oc. mercurator*; (b) *Oc. excrucians*   (a) *Oc. behningi*; (b) *Oc. excrucians*

Terga with scattered pale scales not forming patches or with more or less distinct pale transverse bands dorsally (Fig. 6.31b). Scutum with a dark median stripe or with diffuse pale patches 15

15 (14) Claw strongly and abruptly bent beyond the base of subbasal tooth. Subbasal tooth nearly parallel enclosing an angle of less than 25°. Claw more or less sinuous beyond bend (Fig. 6.32a) . . . . *Oc. excrucians* (p. 243)
Claw evenly curved. Subbasal tooth clearly diverging, enclosing an angle of at least 30°. Apex of claw not sinuous (Fig. 6.32b) 16

(a)   (b)

Figure 6.32. Claws of:
(a) *Oc. excrucians*; (b) *Oc. riparius*

16 (15) Scutum with a more or less well defined median stripe formed by brown scales (Fig. 6.33a) 17
Scutum without a well defined median dark brown stripe (Fig. 6.33b) 18

17 (16) Median stripe of scutum dark brown or bronze. Integument reddish brown. Terga with distinct white or pale basal transverse bands, sometimes interrupted in the middle on terga II–V forming indistinct pale triangular patches laterally. Apical bands of pale scales present at least on segments VI–VIII (Fig. 6.34a). Lower postpronotal patch with sickle shaped pale scales . . . . *Oc. riparius* (p. 270)
Median stripe of scutum chocolate brown, golden brown or fawn coloured (sometimes not so distinct). Integument brownish, mesepimeral sclerite honey coloured. Terga with distinct yellowish basal transverse bands, scattered light scales might be present at the apex (Fig. 6.34b). Lower postpronotal patch with broad white scales . . . . *Oc. annulipes* (p. 219)

Note: To separate *annulipes* from *cantans*, the hind supraalar setae may be considered. They are straw coloured in the former and dark coloured in the latter. The general body colouration should also be considered, see species description.

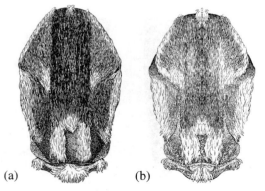

(a)   (b)

Figure 6.33. Scutum of:
(a) *Oc. riparius*; (b) *Oc. euedes*

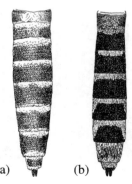

(a)   (b)

Figure 6.34. Dorsal view of abdomen of:
(a) *Oc. riparius*; (b) *Oc. annulipes*

18 (16)   Scutum covered with dark brown or bronze brown scales. Usually a pair of white submedian patches present just beyond scutal angle (Fig. 6.35a). White basal transverse bands on terga of variable width, sometimes indistinct. More or less numerous white scales scattered among the dark distal part of the terga . . . . **Oc. cantans** (p. 225)
Scutum covered with reddish brown, golden or bronze scales. Submedian pale patches absent (Fig. 6.35b). At least anterior terga with narrow basal transverse bands. Numerous scattered white scales apically, sometimes forming narrow transverse apical bands 19

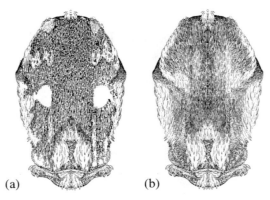

Figure 6.35.   Scutum of:
(a) *Oc. cantans*; (b) *Oc. euedes*

19 (18)   Dorsocentral stripes of pale scales start at some distance from anterior margin of scutum, extending to level of transverse suture, white lateral stripes continuing over transverse suture and reaching the end of dorsocentral stripes . . . **Oc. euedes** (p. 241)
Dorsocentral stripes of pale scales usually start at about the level of transverse suture reaching close to the posterior margin of scutum. White lateral stripes sometimes indistinct, transverse suture with well visible stripes of white scales . . . . **Oc. surcoufi** (p. 244)

20 (1)   Proboscis not longer than fore femur (Fig. 6.36a) 21
Proboscis distinctly longer than fore femur (Fig. 6.36b) 22

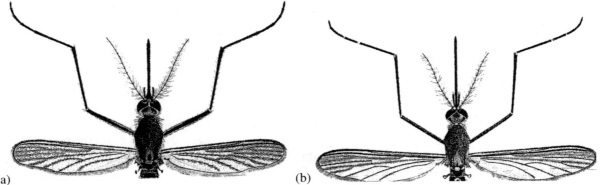

Figure 6.36.   Head and thorax of:
(a) *Ae. cinereus*; (b) *Ae. vexans*

21 (20)   Scales of scutum fawn brown, scales of terga dark brown. Colouration distinctly different. Scales of pleurites yellowish white, scales of sterna yellowish . . . . ***Ae. cinereus*** and ***Ae. geminus*** (pp. 195, 198) Scales of scutum and terga unicolourous, dark brown. Scales of pleurites and sterna greyish white . . . . ***Ae. rossicus*** (p. 199)

22 (20)   Cerci short, blunt (Fig. 6.37a). Pale patches on terga silvery, with metallic sheen . . . . 23 Cerci long, tapering (Fig. 6.37b). Pale scales on terga, if present, without silvery sheen 24

23 (22)   Scutellum with narrow yellowish scales . . . . ***Oc. geniculatus*** (p. 216) Scutellum with broad whitish scales . . . . ***Oc. echinus*** (p. 214)

24 (22)   Pale scales present or predominating in apical half of terga (Fig. 6.38a) . . . . . . 25 Apical half of terga with dark scales; pale scales forming basal bands or lateral spots (Fig. 6.38b) 30

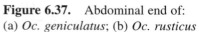

(a)   (b)   (a)   (b)

Figure 6.37. Abdominal end of: (a) *Oc. geniculatus*; (b) *Oc. rusticus*

Figure 6.38. Dorsal view of abdomen of: (a) *Oc. detritus*; (b) *Oc. cataphylla*

25 (24)   Upper scales of postpronotum broad, straight and black (Fig. 6.39a). Scutum with a dark median stripe, which can be divided into 2 stripes by a narrow acrostichal stripe of paler scales . . . . 26 Upper scales of postpronotum often narrow, curved (Fig. 6.39b). If broad and straight, the colour is yellowish or light brown, not black. Scutum usually without dark median stripe, if present, the colour of the stripe is bronze or yellowish . . 28

26 (25)   Basal transverse bands on terga distinct, usually widened middorsally. Often at least the apical 2–3 terga with a continuous median longitudinal band (Fig. 6.40a) . . . . 27 Basal transverse bands on terga present, sometimes not well defined and not widened in the middle. Pale scales scattered on apical part of terga (sometimes predominating) and occasionally forming narrow apical bands (Fig. 6.40b) . . . . ***Oc. refiki*** (p. 277)

27 (26)   Subcosta with scattered pale scales, most numerous at its apex   ***Oc. rusticus*** (p. 279) Subcosta without pale scales . . . . ***Oc. quasirusticus*** (p. 276)

28 (25)   Terga with transverse bands of pale scales. Scattered pale scales usually present in apical part of terga (Fig. 6.38a) . . . . ***Oc. detritus*** (p. 235)

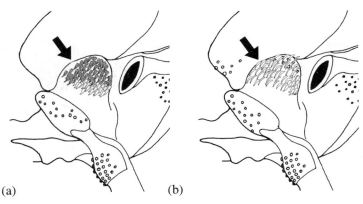

Figure 6.39. Lateral view of thorax of:
(a) *Oc. refiki*; (b) *Oc. detritus*

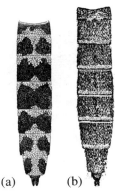

Figure 6.40. Dorsal view of abdomen of:
(a) *Oc. rusticus*; (b) *Oc. refiki*

Pale transverse bands absent. Terga completely covered with pale scales or with scattered dark scales sometimes forming diffuse spots 29

29 (28)   Postnotum with a group of scales . . . . **Oc. lepidonotus** (p. 274)
Postnotum without a group of scales . . . . **Oc. subdiversus** (p. 281)

30 (24)   Scutum with dense, long and black setae. Postpronotal setae scattered in several rows (Fig. 6.41a) 31
Setae of scutum not that long and fewer, often brown or golden. Postpronotal setae only present along posterior margin, sometimes with a few setae confined to dorsal margin of pospronotum (Fig. 6.41b) 32

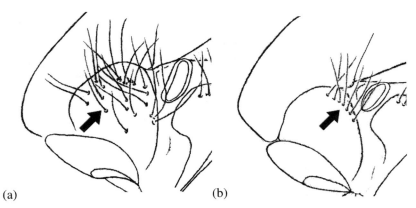

Figure 6.41. Lateral view of thorax of:
(a) *Oc. impiger*; (b) *Oc. cataphylla*

31 (30)   Tarsal claws sharply bent in the middle. Apical part of claw nearly parallel to long subbasal tooth (most distinct on fore legs). Postspiracular area with 10 or less setae (Fig. 6.42a, b) . . . . **Oc. impiger** (p. 250)
Tarsal claw evenly curved. Apical part of claw not parallel to short subbasal tooth (enclosing an angle of more than 45°). Postspiracular area with 14 or more setae (Fig. 6.42c, d) . . . . **Oc. nigripes** (p. 260)

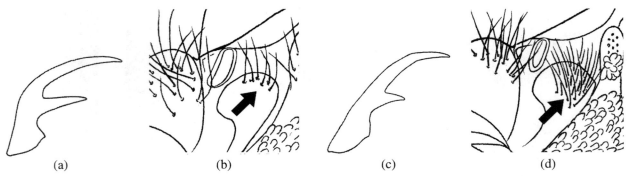

Figure 6.42.   Claw and lateral view of thorax of:
(a, b) *Oc. impiger*; (c, d) *Oc. nigripes*

32 (30)   Wing veins with pale and dark scales intermixed, especially on C and R$_1$ 33
          Wing veins without pale scales, if present, pale scales restricted to basal part of
          veins 34

33 (32)   Proboscis uniformly dark scaled . . . . **Oc. cataphylla** (p. 230)
          Proboscis speckled with more or less numerous pale scales, especially in the
          middle . . . . **Oc. leucomelas** (p. 253)

34 (32)   Hypostigmal patch of scales present, postprocoxal patch of scales absent
          (Fig. 6.43a) 35
          Hypostigmal patch of scales absent, postprocoxal patch of scales present or
          absent (Fig. 6.43b) 36

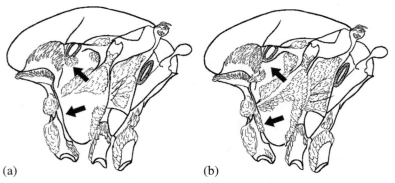

Figure 6.43.   Lateral view of thorax of:
(a) *Oc. pullatus*; (b) *Oc. punctor*

35 (34)   Mesepimeral patch of white scales reaching lower margin of mesepimeron.
          Lower mesepimeral setae present (Fig. 6.44a) . . . . **Oc. pullatus** (p. 265)
          Mesepimeral patch of white scales not reaching lower margin of mesepimeron.
          Lower mesepimeral setae usually absent (Fig. 6.44b) . . . . **Oc. intrudens** (p. 252)

36 (34)   Upper mesepisternal patch of scales not reaching anterior angle of mesepisternum,
          or the lower edge ending above the level of anterior angle (Fig. 6.45a) 37
          Upper mesepisternal patch of scales reaching anterior angle of mesepisternum or
          at least a few scales situated close to it or below the level of anterior angle
          (Fig. 6.45b) 38

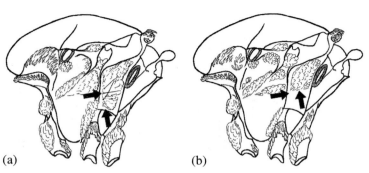

Figure 6.44. Lateral view of thorax of:
(a) *Oc. pullatus*; (b) *Oc. intrudens*

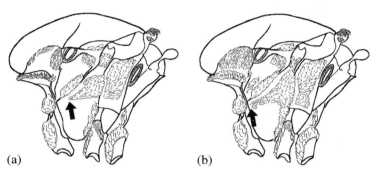

Figure 6.45. Lateral view of thorax of:
(a) *Oc. diantaeus*; (b) *Oc. communis*

37 (36) Upper mesepisternal patch of scales large, not divided into two portions (Fig. 6.45a). Abdominal terga with triangular patches of pale scales, patches usually connected by a basal transverse band on the last segments . . . . ***Oc. diantaeus*** (p. 237)
Upper mesepisternal patch of scales small, divided into two or more portions (Fig. 6.44b). All abdominal terga with pale transverse basal bands . . . . ***Oc. intrudens*** (p. 252)

38 (36) Mesepimeral patch of scales reaches lower margin of mesepimeron (Fig. 6.46a) 39
Mesepimeral patch of scales ends distinctly above lower margin of mesepimeron (Fig. 6.46b) 42

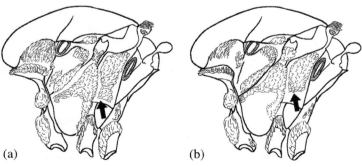

Figure 6.46. Lateral view of thorax of:
(a) *Oc. communis*; (b) *Oc. sticticus*

39 (38) Postprocoxal scales present (Fig. 6.47a) 40
Postprocoxal scales absent (Fig. 6.47b) . . . . ***Oc. communis*** (p. 232)

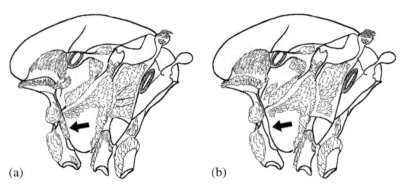

(a)   (b)

Figure 6.47.   Lateral view of thorax of:
(a) *Oc. punctor;* (b) *Oc. communis.*

40 (39) Pale bands of terga II–V distinctly confined in the middle
(Fig. 6.48a) . . . . ***Oc. punctor*** and ***Oc. punctodes*** (pp. 268, 267)
Pale bands of terga II–V of uniform width or only slightly confined in the middle
(Fig. 6.48b) 41

41 (40) Scutum with dark brown median stripe, occasionally divided into two dorsocentral stripes by an acrostichal row of yellowish scales (Fig. 6.49a). Base of costa covered with dark scales, rarely a few pale scales present. . . . . ***Oc. pionips*** (p. 262)
Scutum more or less uniformly covered with yellowish brown scales
(Fig. 6.49b). Base of costa with numerous pale scales often forming a large patch . . . . ***Oc. hexodontus*** (p. 247)

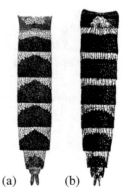

(a)   (b)

Figure 6.48.   Dorsal view of abdomen of:
(a) *Oc. punctor*; (b) *Oc. hexodontus*

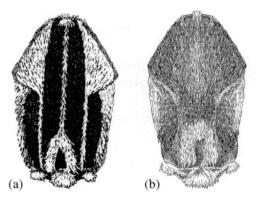

(a)   (b)

Figure 6.49.   Scutum of:
(a) *Oc. pionips*; (b) *Oc. hexodontus*

42 (38) Hind tibia generally covered with light scales on anterior surface 43
Hind tibia with dark scales on anterior surface . . . . ***Oc. hungaricus*** (p. 249)

43 (42) Wing veins usually entirely dark scaled. Terga II–IV with pale basal bands distinctly constricted in the middle, on following terga basal bands interrupted in the middle forming triangular pale patches at lateral sides (Fig. 6.50a).

Figure 6.50.   Dorsal view of abdomen of:
(a) *Oc. sticticus*; (b) *Oc. nigrinus*

First flagellomere yellowish at the base, second and third not distinctly
shortened . . . . ***Oc. sticticus*** (p. 272)
Wing veins with scattered pale scales at base of C, entire Sc and M proximal to
the cross veins. Terga with broad pale basal bands, slightly, if at all, constricted in
the middle (Fig. 6.50b). First flagellomere entirely black, second and third
distinctly shortened . . . . ***Oc. nigrinus*** (p. 259)

6.3. GENUS *CULEX*

1   Tarsomere I of hind legs distinctly shorter than hind tibia (Fig. 6.51a) 2
           Tarsomere I of hind legs as long as or slightly longer than hind tibia
           (Fig. 6.51b). 3
2 (1)   Terga with continuous lateral longitudinal bands of pale scales. Pale scales
           sometimes form more or less developed, triangular patches, which are usually
           connected . . . . ***Cx. modestus*** (p. 284)
           Terga without continuous lateral longitudinal bands. More or less developed patches
           of pale scales at the lateral basal margins of terga present . . . . ***Cx. pusillus*** (p. 286)
3 (1)   Tarsi with pale rings, which are sometimes narrow. Wings with large, prominent
           spots of pale scales, especially on C (Fig. 6.52a) . . . . ***Cx. mimeticus*** (p. 292)

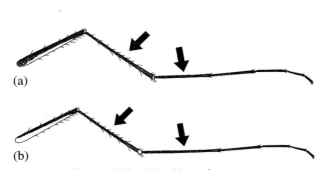

Figure 6.51.   Hind leg of:
(a) *Cx. modestus*; (b) *Cx. p. pipiens*

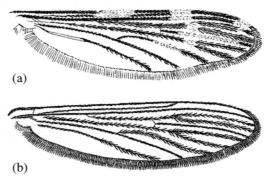

Figure 6.52.   Wing of:
(a) *Cx. mimeticus*; (b) *Cx. p. pipiens*

Tarsi without pale rings. Wing veins mainly covered with dark scales (in *theileri* a few pale scales may be present on C) (Fig. 6.52b) 4

4 (3)   Terga uniformly covered with reddish brown scales, without pale transverse bands . . . . ***Cx. martinii*** (p. 307)
Terga with more or less developed transverse bands formed by white or yellowish scales 5

5 (4)   Terga with pale basal bands, sometimes reduced to lateral triangular patches . . 6
Terga with pale apical bands, sometimes reduced to lateral spots or completely absent 10

6 (5)   Femora and tibiae of fore and mid legs with a distinct anterior pale longitudinal stripe, rarely only the fore femur with a pale stripe (Fig. 6.53a). Pale basal bands on abdominal terga usually triangularly extended posteriorly ***Cx. theileri*** (p. 300)
Femora and tibiae of fore and mid legs without anterior pale longitudinal stripe (Fig. 6.53b) 7

7 (6)   Proboscis with an indistinct pale median ring . . . . ***Cx. brumpti*** (p. 289)
Proboscis dark scaled 8

8 (9)   Basal bands on abdominal terga broad and white, covering 1/2 to 2/3 of each tergum (Fig. 6.54a). Tibia of hind leg with a prominent apical pale spot . . . . ***Cx. laticinctus*** (p. 290)
Basal bands on abdominal terga narrow, usually covering less than 1/2 of each tergum (Fig. 6.54b). Tibia of hind leg without apical pale spot 9

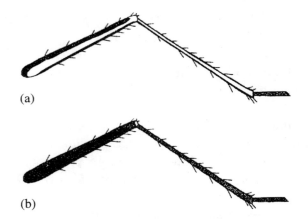

(a)

(b)

Figure 6.53. Fore femur and tibia of: (a) *Cx. theileri*; (b) *Cx. p. pipiens*

(a)   (b)

Figure 6.54. Dorsal view of abdomen of: (a) *Cx. laticinctus*; (b) *Cx. p. pipiens*

9 (8)   Basal bands on terga formed by white scales. Hind tibia with more or less distinct longitudinal anterior pale stripe (Fig. 6.55a) . . . . ***Cx. perexiguus*** (p. 294)
Basal bands on terga formed by yellowish scales. The bands are sometimes reduced or pale scales form patches at the sides of the terga. Hind tibia without longitudinal pale stripe (Fig. 6.55b) . . . . ***Cx. p. pipiens*** and ***Cx. torrentium*** (pp. 296, 299)

10 (5)   Apical bands narrowed or interrupted in the middle on some terga . . . . ***Cx. impudicus*** (p. 305)
Apical bands well developed, not interrupted in the middle 11

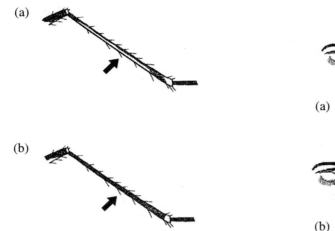

Figure 6.55. Hind tibia of:
(a) *Cx. perexiguus*; (b) *Cx. p. pipiens* and *Cx. torrentium*

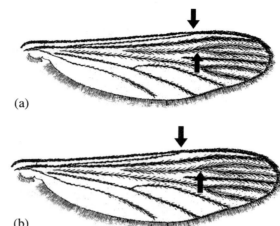

Figure 6.56. Wing of:
(a) *Cx. hortensis*; (b) *Cx. territans*

11 (10) Palps with dark and pale scales. End of Sc nearly alined with furcations of R_{2+3} and M (Fig. 6.56a). Pale bands on terga relatively broad, distinctly widened in the middle of some segments. Apex of hind tibia with a pale spot ...... ***Cx. hortensis*** (p. 303)
Palps with dark scales. End of Sc distinctly displaced towards the wing base compared to furcations of R_{2+3} and M (Fig. 6.56b). Pale bands on terga relatively narrow, without widening in the middle. Apex of hind tibia without a pale spot. Abdomen usually greenish ventrally ...... ***Cx. territans*** (p. 308)

6.4. GENUS *CULISETA*

1 Scutum with distinct longitudinal pale stripes (resemble a lyre in shape). Femora and tibiae with well defined pale spots and stripes. Costa mainly covered with pale scales .... ***Cs. longiareolata*** (p. 311)
Scutum without distinct longitudinal stripes. Femora and tibiae without pale stripes, uniformly dark scaled or with scattered light scales. Costa completely or mainly dark scaled 2

2 (1) Cross veins r-m and m-cu well separated. The distance between them usually at least the length of m-cu (Fig. 6.57a). Wings usually without dark spots, otherwise with an indistinct dark spot at the base of R_{4+5} 3

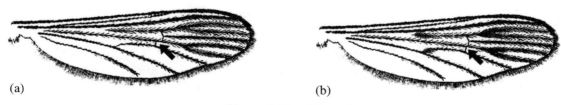

Figure 6.57. Wing of:
(a) *Culicella*; (b) *Culiseta*

Cross veins r-m and m-cu alined or slightly separated. If separated, the distance between them is usually not longer than m-cu (Fig. 6.57b). Wings weakly or distinctly spotted (spots sometimes absent in *Cs. glaphyroptera*) 6

3 (2)   Proboscis with scattered pale scales, mainly in its middle part. Sterna usually with dark scales arranged in a pattern of an inverted "V" (Fig. 6.58a, b) 4
Proboscis uniformly dark scaled, rarely with a few pale scales in the middle (in *Cs. morsitans*), or pale scales predominating (in *Cs. ochroptera*). Sterna without a pattern of an inverted "V", pale and dark scales diffused (Fig. 6.58c, d) 5

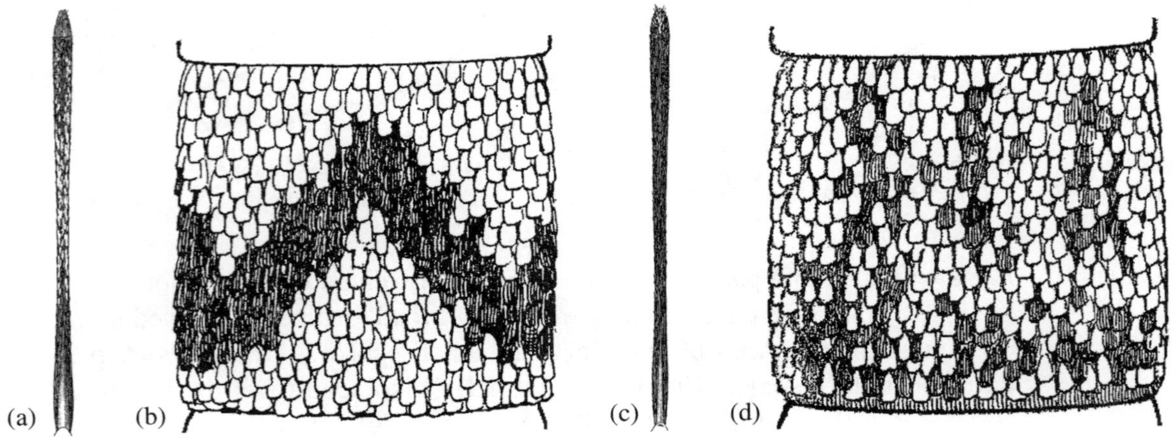

Figure 6.58.   Proboscis and sternum of:
(a, b) *Cs. fumipennis*; (c, d) *Cs. morsitans*

4 (3)   On fore legs the pale rings include apical and basal parts between tarsomeres III–IV and IV–V (Fig. 6.59a) . . . . ***Cs. fumipennis*** (p. 313)
On fore legs apical parts of tarsomeres III and IV entirely dark scaled (Fig. 6.59b) . . . . ***Cs. litorea*** (p. 315)

5 (3)   Terga with narrow basal pale bands. Scales on wing veins evenly distributed, not forming spots. Tibia of fore leg mainly dark brown scaled . . ***Cs. morsitans*** (p. 317)
Terga with narrow indistinct basal and apical pale bands, sometimes absent.
Tergum VIII completely covered with pale scales. Indistinct dark spot at the base of R_{4+5} may be present. Tibia of fore leg mainly yellowish scaled ***Cs. ochroptera*** (p. 319)

6 (2)   Tarsi dark (Fig. 6.60a) 7
Tarsi with white rings (Fig. 6.60b) 8

Figure 6.59.   Fore tarsus of:
(a) *Cs. fumipennis*; (b) *Cs. litorea*

Figure 6.60.   Hind tarsus of:
(a) *Cs. glaphyroptera*; (b) *Cs. annulata*

7 (6) Palps dark with scattered pale scales, eyes distinctly bordered with light scales. Spots on wing veins always present, sometimes indistinct. Usually not more than 15 prespiracular setae and not more than 10 lower mesepisternal setae present (Fig. 6.61a) . . . . *Cs. bergrothi* (p. 326) Palps entirely dark scaled, eyes not bordered with light scales. Spots on wing veins absent or indistinct. Number of prespiracular setae 16–22 and of lower mesepisternal setae 12–18 (Fig. 6.61b) . . . . *Cs. glaphyroptera* (p. 328)

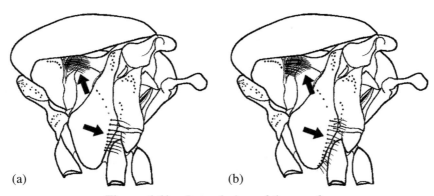

(a)   (b)

Figure 6.61. Lateral view of thorax of:
(a) *Cs. bergrothi*; (b) *Cs. glaphyroptera*

8 (6) Femora with subapical pale rings. Tarsomere I of hind legs with a median white ring (Fig. 6.62a) . . . . 9 Femora without subapical pale rings. Tarsomere I of hind legs without a median white ring (Fig. 6.62b) . . . . *Cs. alaskaensis* (p. 322)

9 (8) Costa (C) usually completely dark scaled or isolated pale scales could be present on C, Sc and R. Cu entirely dark scaled. Dark spots on wings distinct. Terga with distinct white basal bands, pale scales absent in apical half of terga. Cross veins r-m and m-cu alined (Fig. 6.63a) . . . . *Cs. annulata* (p. 324) Costa, Sc and R with scattered pale scales. Cu with more or less numerous pale scales. Dark spots on wings indistinct. Terga with indistinct pale basal bands formed by yellowish scales (not white), pale scales are also present among the dark scales in the apical half of the terga. Cross veins r-m and m-cu slightly separated (Fig. 6.63b) . . . . *Cs. subochrea* (p. 330)

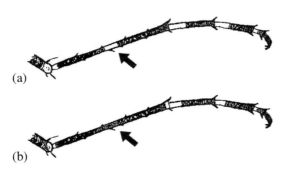

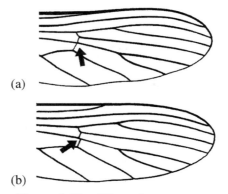

Figure 6.62. Hind tarsus of:
(a) *Cs. annulata*; (b) *Cs. alaskaensis*

Figure 6.63. Wing of:
(a) *Cs. annulata*; (b) *Cs. subochrea*

6.5. GENUS *COQUILLETTIDIA*

1   Tarsomeres with pale rings. Proboscis and palps with numerous pale scales. Wing veins with broad dark and pale scales intermixed   . . . . ***Cq. richiardii*** (p. 333)
      Tarsomeres without pale rings. Proboscis and palps uniformly blackish brown. Wing veins with uniformly dark scales . . . . ***Cq. buxtoni*** (p. 333)

7

Key to Male Mosquitoes

GENERA

1   Gonocoxite without lobes. Gonostylus about as long as gonocoxite. Sclerotised paraproct absent. Proctiger, if present, membranous, cone shaped and hardly visible (Fig. 7.1a) . . . . *Anopheles* (p. 111)
Gonocoxite with 1 or 2 lobes. Lobes sometimes rudimentary or absent (subgenus *Finlaya*). Gonostylus shorter than gonocoxite. Proctiger different, sclerotised paraproct present (Fig. 7.1b) 2

 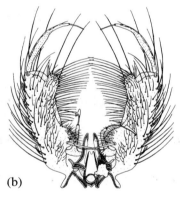

(a)   (b)

Figure 7.1.   Hypopygium of:
(a) *An. maculipennis s.l.*; (b) *Oc. mercurator*

2 (1)   Gonocoxite with one subapically located lobe covered with spines and setae. Apex of paraproct with abundant spines or rows of numerous denticles (paraproct crown) (Fig. 7.2a) . . . . *Culex* (p. 125)
Gonocoxite with 2 lobes (basal and apical), if only one lobe is present, it is situated at the base of the gonocoxite, lobes absent in subgenus *Finlaya*. Apex of paraproct without spines or rows of numerous denticles (Fig. 7.2b) 3
3 (2)   Apical spine of gonostylus longer than the maximum width of gonostylus (Fig. 7.3a). If the spine is shorter than the maximum width of the gonostylus, it is articulated subapically (Fig. 7.3b). If the spine is absent, the gonostylus is divided at its base (Fig. 7.3c) . . . . *Aedes* and *Ochlerotatus* (p. 114)
Apical spine of gonostylus shorter than or equal to the maximum width of the gonostylus (Fig. 7.3d) 4

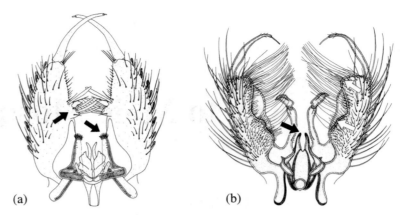

Figure 7.2.   Hypopygium of:
(a) *Cx. modestus*; (b) *Oc. annulipes*

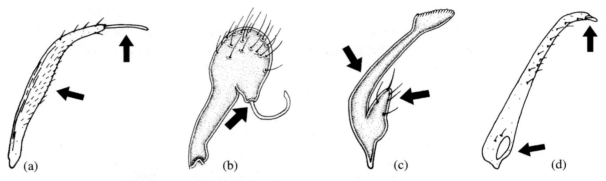

Figure 7.3.   Gonostylus of:
(a) *Oc. riparius*; (b) *Ae. vittatus*; (c) *Ae. rossicus*; (d) *Cs. morsitans*

4 (3)   Basal lobe of gonocoxite with one strongly sclerotised rod like spine
(Fig. 7.4a) . . . . ***Coquillettidia*** (p. 131)
Basal lobe of gonocoxite with at least 2 spines or strong setae (Fig. 7.4b) 5

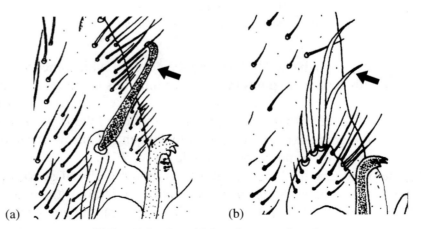

Figure 7.4.   Basal lobe of gonocoxite of:
(a) *Cq. richiardii*; (b) *Cs. fumipennis*

5 (4) Gonocoxite broad and short, plumpy in appearance with a small, flattened basal lobe. Gonostylus broad, flattened dorsoventrally. Proctiger lobe shaped, membranous (Fig. 7.5a) . . . . *Uranotaenia unguiculata* (p. 339)
Gonocoxite more elongated, with a distinct basal lobe. Gonostylus narrow, not flattened dorsoventrally (Fig. 7.5b) 6

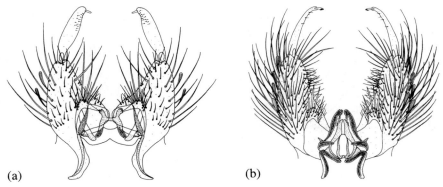

(a)   (b)

Figure 7.5.   Hypopygium of:
(a) *Ur. unguiculata*; (b) *Or. pulcripalpis*

6 (5) Apical spine of gonostylus fingered at the apex. Spine as long as the maximum width of the gonostylus (Fig. 7.6a) . . . . *Orthopodomyia pulcripalpis* (p. 336)
Apical spine of gonostylus not fingered (there may be 2 spines at the apex of the gonostylus in *Cs. longiareolata*, Fig. 7.6b). Spine distinctly shorter than the maximum width of the gonostylus (except in *Cs. glaphyroptera*, Fig. 7.6c) . . . . *Culiseta* (p. 129)

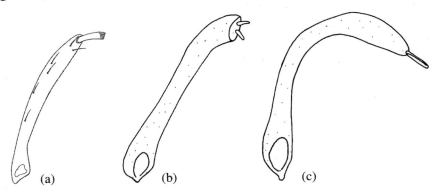

(a)   (b)   (c)

Figure 7.6.   Gonostylus of:
(a) *Or. pulcripalpis*; (b) *Cs. longiareolata*; (c) *Cs. glaphyroptera*

7.1. GENUS *ANOPHELES*

1   Base of gonocoxite with 1–3, usually 2, parabasal setae. At least one of them originated from a tubercle (tubercles weakly developed in *An. plumbeus*). Internal seta present (subgenus *Anopheles*, Fig. 7.7a) 2
Base of gonocoxite with 4–7, usually 6, parabasal setae. None of them elevated on a tubercle. Internal seta absent (subgenus *Cellia*, Fig. 7.7b) 7

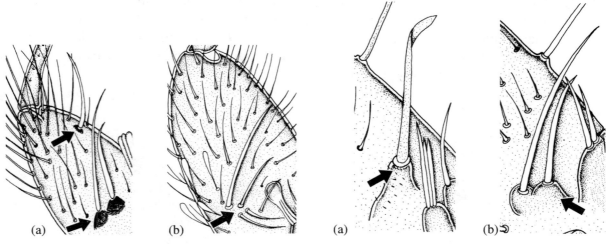

Figure 7.7.   Gonocoxite of:
(a) *An. marteri*; (b) *An. superpictus*

Figure 7.8.   Base of gonocoxite of:
(a) *An. algeriensis*; (b) *An. maculipennis s.l.*

2 (1)   Base of gonocoxite with one parabasal seta (Fig. 7.8a) . . . . ***An. algeriensis*** (p. 165)
Base of gonocoxite with 2–3 parabasal setae (Fig. 7.8b) 3

3 (2)   Base of gonocoxite with 3 parabasal setae, inner one simple, the 2 outer setae
apically branched (Fig. 7.9a) . . ***An. claviger s.s.*** and ***An. petragnani*** (pp. 167, 169)
Base of gonocoxite with 2 simple parabasal setae (Fig. 7.9b) 4

4 (3)   Aedeagus with leaflets. Parabasal setae of different length, outer seta longer than
inner seta (Fig. 7.10a) 5
Aedeagus without leaflets. Parabasal setae of approximately the same length,
tubercles weakly sclerotised (Fig. 7.10b) . . . . ***An. plumbeus*** (p. 182)

5 (4)   Parabasal setae arising from a strongly sclerotised base (Fig. 7.7a). Outer and
inner claspette lobes bear flattened spatula like or longer sabre-like setae
(Fig. 7.11a) . . . . ***An. marteri*** (p. 180)

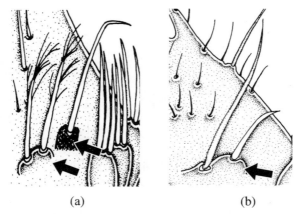

Figure 7.9.   Base of gonocoxite of:
(a) *An. claviger s.l.*; (b) *An. hyrcanus*

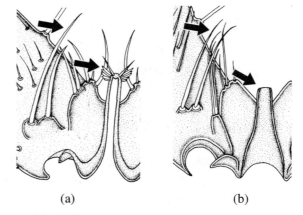

Figure 7.10.   Aedeagus and parabasal setae of:
(a) *An. maculipennis s.l.*; (b) *An. plumbeus*

Parabasal setae arising from a weakly or not sclerotised base (Fig. 7.8b). At least
the inner claspette lobe bears spine-like setae (Fig. 7.11b) 6

6 (5) Claspette lobes not well defined, bearing only spine-like setae of variable length and
shape (Fig. 7.12a) . . . . **Anopheles Maculipennis Complex** (p. 172)
Claspette lobes well defined. Setae of outer lobe fused into broad, spatula like
processes, inner lobe with 2 spine-like setae (Fig. 7.12b) . . . ***An. hyrcanus*** (p. 170)

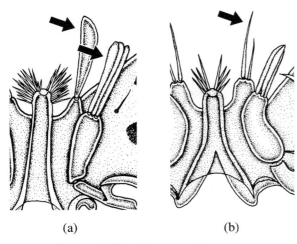

(a)   (b)

Figure 7.11.   Claspette of:
(a) *An. marteri*; (b) *An. hyrcanus*

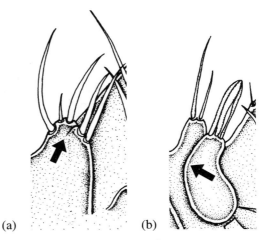

(a)   (b)

Figure 7.12.   Claspette of:
(a) *An. maculipennis s.l.*; (b) *An. hyrcanus*

7 (1) Leaflets of aedeagus present (Fig. 7.13a) 8
Leaflets of aedeagus absent (Fig. 7.13b) . . . . ***An. multicolor*** (p. 186)

8 (7) Leaflets of aedeagus short, distinctly shorter than half the length of the aedeagus
(Fig. 7.14a) . . . . ***An. superpictus*** (p. 190)
Leaflets of aedeagus well developed, longest leaflet almost half as long as the
aedeagus (Fig. 7.14b) 9

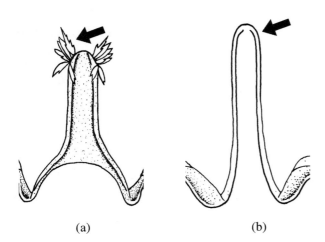

(a)   (b)

Figure 7.13.   Aedeagus of:
(a) *An. superpictus*; (b) *An. multicolor*

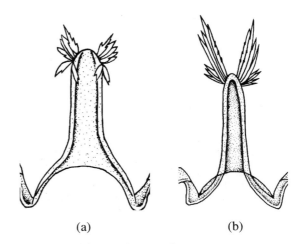

(a)   (b)

Figure 7.14.   Aedeagus of:
(a) *An. superpictus*; (b) *An. cinereus hispaniola*

9 (8)   Longest spine-like seta of claspette distinctly longer than spatula-like seta. The longest leaflet of aedeagus is less than 1/3 longer than the second longest. Leaflets weakly serrated (Fig. 7.15a) . . . . **An. cinereus hispaniola** (p. 185)
Longest spine-like seta of claspette of similar length as spatula-like seta. The longest leaflet of aedeagus is distinctly longer than the second longest. Leaflets strongly serrated (Fig. 7.15b) . . . . **An. sergentii** (p. 188)

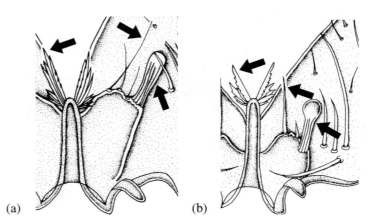

Figure 7.15.   Claspette and aedeagus of:
(a) *An. cinereus hispaniola*; (b) *An. sergentii*

7.2. GENUS *AEDES* AND *OCHLEROTATUS*

1   Apex of gonocoxite exceeding articulation point of gonostylus, which is divided into two branches (Fig. 7.16a). Palps several times shorter than the proboscis, as in females 2
Gonostylus arising at the apex of gonocoxite, simple, not divided into two branches (Fig. 7.16b). Palps about as long as the proboscis 4
2 (1)   Longer branch of gonostylus forked into 2 prongs at the apex (Fig. 7.17a) 3
Longer branch of gonostylus not forked at the apex, denticulated at outer apical margin (Fig. 7.17b) . . . . **Ae. rossicus** (p. 199)

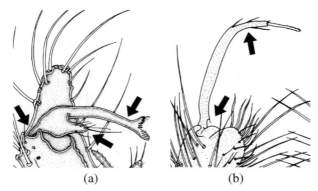

Figure 7.16.   Apex of gonocoxite and gonostylus of:
(a) *Ae. cinereus*; (b) *Oc. communis*

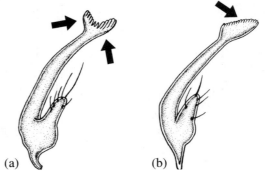

Figure 7.17.   Gonostylus of:
(a) *Ae. cinereus*; (b) *Ae. rossicus*

3 (2) Outer prong of the fork usually longer than or equal to the inner one
(Fig. 7.18a, b) . . . . ***Ae. geminus*** (p. 198)
Inner prong of the fork usually longer than the outer one
(Fig. 7.18c, d) . . . . ***Ae. cinereus*** (p. 195)

Note: Transitional forms present, specific status of *geminus* uncertain.

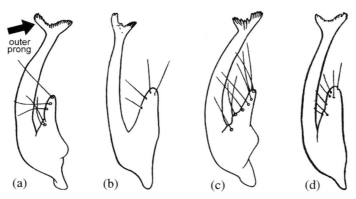

Figure 7.18. Gonostylus of:
(a, b) *Ae. geminus*; (c, d) *Ae. cinereus*

4 (1) Typical claspettes (divided into stem and filament) present (Fig. 7.19a) 9
Typical claspettes absent (Fig. 7.19b) 5
5 (4) Gonostylus distinctly expanded apically (Fig. 7.20a). Claspettes of different shape,
elongated, well separated from basal part of gonocoxite 6
Gonostylus slightly expanded apically or evenly tapering (Fig. 7.20b, c). Claspettes
lobe-like, seem to be inner basal part of gonocoxite and covered with dense setae,
some of them may be enlarged, spine-like 7

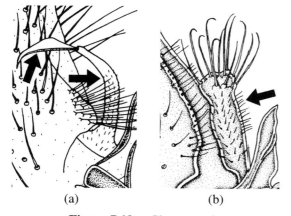

Figure 7.19. Claspette of:
(a) *Oc. nigripes*; (b) *Ae. vexans*

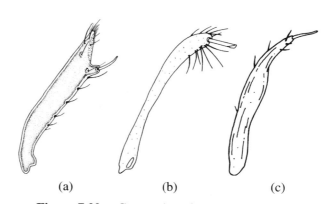

Figure 7.20. Gonostylus of:
(a) *Ae. vexans*; (b) *Ae. albopictus*; (c) *Ae. aegypti*

6 (5) Gonostylus gradually expanded to apex. Spine of gonostylus articulated subapically,
straight (Fig. 7.21a) . . . . ***Ae. vexans*** (p. 201)
Gonostylus abruptly expanded apically, flask shaped. Spine of gonostylus
articulated subapically, strongly curved (Fig. 7.21b) . . . . ***Ae. vittatus*** (p. 204)

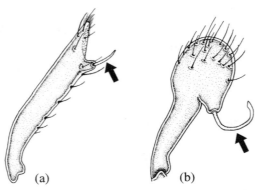

Figure 7.21.   Gonostylus of:
(a) *Ae. vexans*; (b) *Ae. vittatus*

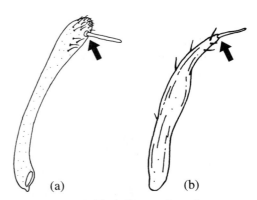

Figure 7.22.   Gonostylus of:
(a) *Ae. cretinus*; (b) *Ae. aegypti*

7 (5)   Apical part of gonostylus slightly broader than median part. Spine of gonostylus articulated subapically (Fig. 7.22a) 8
          Apical part of gonostylus distinctly narrower than median part, tapering. Spine of gonostylus articulated at the apex (Fig. 7.22b) . . . . *Ae. aegypti* (p. 207)

8 (7)   Setae below subapical spine of gonostylus usually scattered (Fig. 7.22a). Claspette lobes covered with short setae. Median part of tergum IX evenly rounded (Fig. 7.23a) . . . . *Ae. cretinus* (p. 212)
          Setae below subapical spine of gonostylus usually forming a row (Fig. 7.20b). Claspette lobes covered with long setae, several of them stronger, spine-like, curved at the apex. Median part of tergum IX pointed (Fig. 7.23b) . . . . *Ae. albopictus* (p. 210)

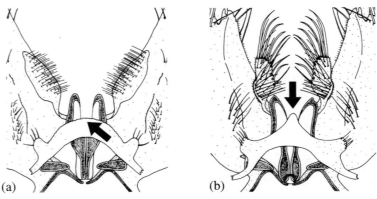

Figure 7.23.   Claspette lobes and tergum IX of:
(a) *Ae. cretinus*; (b) *Ae. albopictus*

9 (4)   Gonocoxite without lobes. Several tubercles bearing thin setae may be present at the inner base of gonocoxite (Fig. 7.24a) 10
          Gonocoxite with more or less distinct basal and apical lobes, or at least basal lobe present (Fig. 7.24b) 11

10 (9)   Setae of gonocoxite very long and dense (Fig. 7.25a) . . . . . . *Oc. echinus* (p. 214)
          Setae of gonocoxite short and less dense (Fig. 7.25b) . . . *Oc. geniculatus* (p. 216)

Note: Males of the two species are difficult to separate.

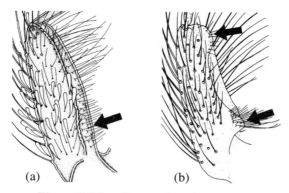

Figure 7.24. Gonocoxite of:
(a) *Oc. geniculatus*; (b) *Oc. nigripes*

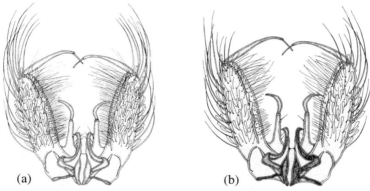

Figure 7.25. Hypopygium of:
(a) *Oc. echinus*; (b) *Oc. geniculatus*

11 (9)   Basal lobe of gonocoxite usually divided, one lobe with several long, lanceolate, flattened setae which may be slightly or strongly curved (Fig. 7.26a) 12
Basal lobe of gonocoxite usually undivided, without a row of several lanceolate, flattened setae, but 1 or 2 could be present (Fig. 7.26b) 16

12 (11) Apical spine of gonostylus distinctly S-shaped (Fig. 7.27a)   ***Oc. rusticus*** (p. 279)
Apical spine of gonostylus straight or slightly curved (Fig. 7.27b) 13

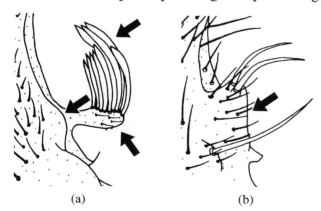

Figure 7.26. Basal lobe of gonocoxite of:
(a) *Oc. rusticus*; (b) *Oc. intrudens*

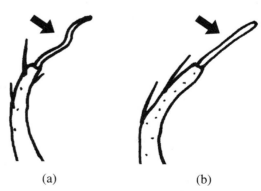

Figure 7.27. Apex of gonostylus of:
(a) *Oc. rusticus*; (b) *Oc. refiki*

13 (12)   Division of basal lobe which bears lanceolate setae elongated, at least in basal part. Claspette filament distinctly transversely striated (Fig. 7.28a) 14
Division of basal lobe which bears lanceolate setae arising gradually from gonocoxite, more or less conical. Claspette filament not distinctly striated (Fig. 7.29a) 15

14 (13)   Claspette stem broader at the apex than at the base. Number of slightly curved lanceolate setae on basal lobe 15–16, arranged in several rows
(Fig. 7.28a) . . . . ***Oc. refiki*** (p. 277)
Apex and base of claspette stem of more or less equal width. Number of strongly curved lanceolate setae on basal lobe 6–8, located at the distal margin
(Fig. 7.28b) . . . . ***Oc. quasirusticus*** (p. 276)

15 (13)   Base of gonocoxite with one lobe, bearing a group of lanceolate, flattened setae
(Fig. 7.29a) . . . . ***Oc. lepidonotus*** (p. 274)
Base of gonocoxite with two or more lobes. Only one lobe bears lanceolate, flattened setae, the others with hair-like setae (Fig. 7.29b) . . ***Oc. subdiversus*** (p. 281)

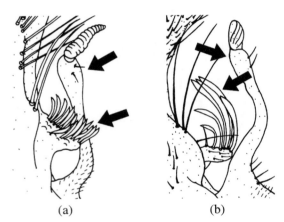
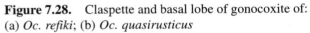
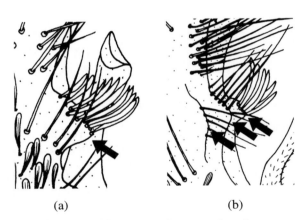

Figure 7.28.   Claspette and basal lobe of gonocoxite of:
(a) *Oc. refiki*; (b) *Oc. quasirusticus*

Figure 7.29.   Base of gonocoxite of:
(a) *Oc. lepidonotus*; (b) *Oc. subdiversus*

16 (11)   Basal lobe of gonocoxite with 3 setae distinctly larger than the others
(Fig. 7.30a) 17
Basal lobe of gonocoxite with 0–2 setae which are larger than the others
(Fig. 7.30b) 19

17 (16)   Basal lobe of gonocoxite with all 3 large setae spine-like. Claspette stem with a thorn shaped process beyond the middle (Fig. 7.31a) 18
Basal lobe of gonocoxite with 1 or 2 of the larger setae flattened and lanceolate. Claspette stem without a thorn shaped process, bent and swollen in the middle
(Fig. 7.31b) . . . . ***Oc. pullatus*** (p. 265)

18 (17)   Apical half of gonocoxite with very dense setae directed more inwardly
(Fig. 7.32a) . . . . ***Oc. diantaeus*** (p. 237)
Only small subapical zone with dense setae directed more distally
(Fig. 7.32b) . . . . ***Oc. intrudens*** (p. 252)

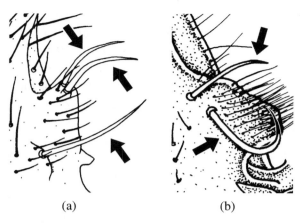

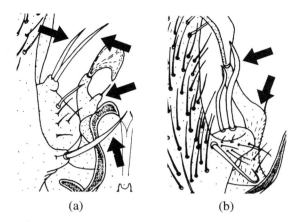

Figure 7.30. Basal lobe of gonocoxite of:
(a) *Oc. intrudens*; (b) *Oc. caspius*

Figure 7.31. Basal lobe of gonocoxite and claspette of:
(a) *Oc. diantaeus*; (b) *Oc. pullatus*

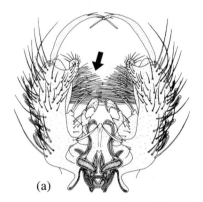

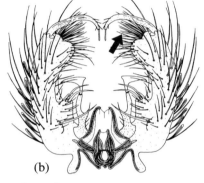

Figure 7.32. Hypopygium of:
(a) *Oc. diantaeus*; (b) *Oc. intrudens*

19 (16)   Basal lobe of gonocoxite with 2 spine-like setae (Fig. 7.33a) 20
          Basal lobe of gonocoxite with 0-1 spine-like seta (Fig. 7.33b) 22
20 (19)   Larger spine of basal lobe hook shaped (Fig. 7.34a) 21
          Larger spine of basal lobe not hook shaped, slightly curved in the middle
          (Fig. 7.34b) . . . . ***Oc. hungaricus*** (p. 249)

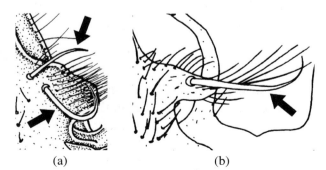

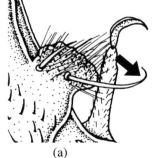

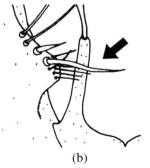

Figure 7.33. Basal lobe of gonocoxite of:
(a) *Oc. caspius*; (b) *Oc. impiger*

Figure 7.34. Basal lobe of gonocoxite of:
(a) *Oc. dorsalis*; (b) *Oc. hungaricus*

21 (20)   Basal lobe gradually arising from gonocoxite, spines situated close together, larger spine strongly curved apically (usually tip extending backwards to almost the middle of the spine) (Fig. 7.35a) .... *Oc. caspius* (p. 227)
Basal lobe of gonocoxite slightly constricted at base, spines widely separated; larger spine slightly curved apically (usually tip extending backwards to not more than one third of the spine) (Fig. 7.35b) .... *Oc. dorsalis* (p. 239)

Note: The structure of the hypopygium of the two species is very similar and difficult to distinguish, intermediate forms are common.

22 (19)   Apical lobe of gonocoxite small, weakly developed or absent (Fig. 7.36a) 23
Apical lobe of gonocoxite well developed (Fig. 7.36b) 26

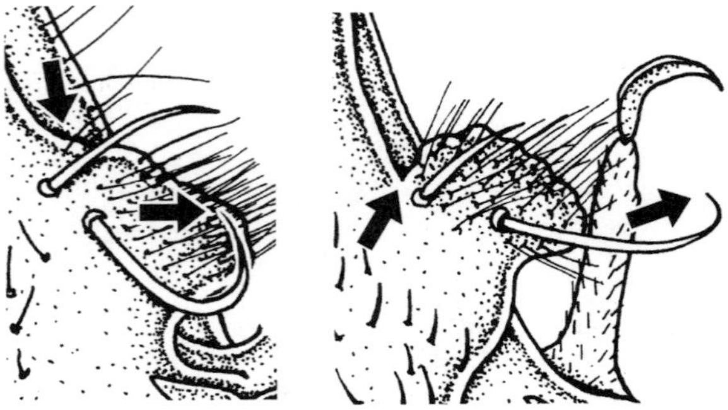

(a)   (b)   (a)   (b)

Figure 7.35.   Basal lobe of gonocoxite of: **Figure 7.36.**   Apical lobe of gonocoxite of:
(a) *Oc. caspius*; (b) *Oc. dorsalis*   (a) *Oc. pulcritarsis*; (b) *Oc. cataphylla*

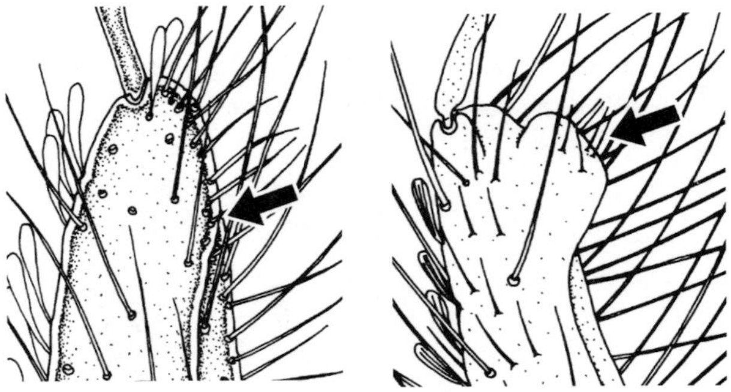

23 (22)   Basal lobe of gonocoxite indistinct, weakly developed (Fig. 7.37a) 24
Basal lobe well developed (Fig. 7.37b) 25

24 (23)   Longer setae on basal lobe thin, of equal length and thickness. Strong spine-like setae absent (Fig. 7.38a) .... *Oc. mariae* and *Oc. zammitii* (pp. 255, 257)
Longer setae on basal lobe distinctly differ in length and thickness, 1 strong spine-like seta present (Fig. 7.38b) ..... *Oc. pulcritarsis* and *Oc. berlandi* (pp. 264, 223)

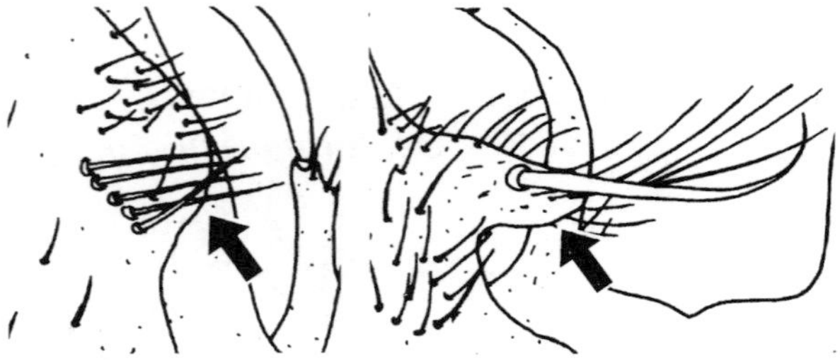

(a)   (b)   (a)   (b)

Figure 7.37.   Basal lobe of gonocoxite of: **Figure 7.38.**   Basal lobe of gonocoxite of:
(a) *Oc. mariae s.l.*; (b) *Oc. impiger*   (a) *Oc. mariae s.l.*; (b) *Oc. berlandi*

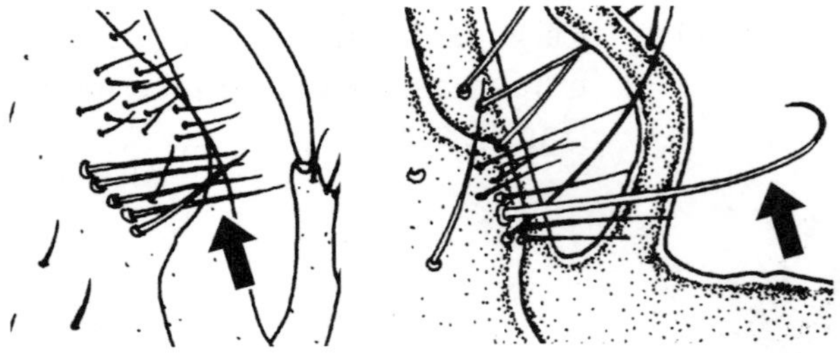

25 (23)   Gonocoxite with long setae predominating on inner side. Basal lobe with setae of more or less equal thickness (Fig. 7.39a) .... *Oc. nigripes* (p. 260)
Gonocoxite with short setae predominating on inner side. Basal lobe with one seta distinctly stronger than the others (Fig. 7.39b) .... *Oc. impiger* (p. 250)

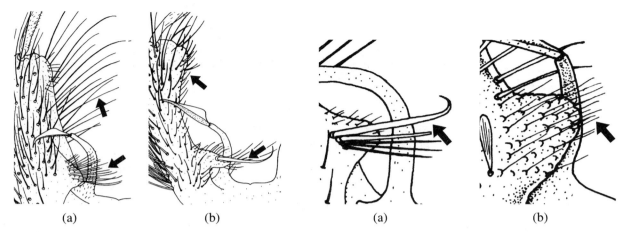

(a) (b)   (a) (b)

Figure 7.39. Gonocoxite of: **Figure 7.40.** Basal lobe of gonocoxite of:
(a) *Oc. nigripes*; (b) *Oc. impiger*   (a) *Oc. cataphylla*; (b) *Oc. behningi*

26 (22) Basal lobe of gonocoxite with one spine or with at least one enlarged seta among thinner setae (Fig. 7.40a) 27
Basal lobe without a spine or enlarged seta. All setae on basal lobe with more or less the same length and width (Fig. 7.40b) 40

27 (26) Claspette filament evenly sclerotised, without transparent wings (Fig. 41a) . . . 28
Claspette filament differentiated into a well sclerotised ridge and 1 or 2 weakly sclerotised, transparent wings (Fig. 7.41b) 29

28 (27) Claspette filament relatively short, shorter than the stem, strongly sclerotised (Fig. 7.42a) . . . . **Oc. hexodontus** and **Oc. punctor** (pp. 247, 268)
Claspette filament relatively long, of almost the same length as the stem, weakly sclerotised (Fig. 7.42b) . . . . **Oc. punctodes** (p. 267)

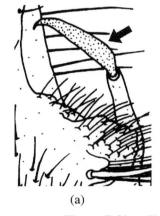

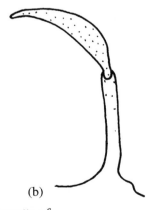

(a) (b)   (a) (b)

Figure 7.41. Claspette filament of: **Figure 7.42.** Claspette of:
(a) *Oc. punctor*; (b) *Oc. communis*   (a) *Oc. hexodontus*; (b) *Oc. punctodes*

29 (27) Wing narrow, of more or less similar width along the whole claspette filament (Fig. 7.43a) 30
Wing broad, distinctly widening the claspette filament at any section between its base and apex (Fig. 7.43b) 32

Figure 7.43.   Claspette filament of:
(a) *Oc. leucomelas*; (b) *Oc. cataphylla*

30 (29)   Upper part of basal lobe with a row of apically strongly curved, sometimes
hooked setae (Fig. 7.44a) . . . . **Oc. communis** (p. 232)
Upper part of basal lobe with straight or slightly curved setae, never hooked
(Fig. 7.44b) 31

31 (30)   Claspette stem straight or slightly curved (Fig. 7.45a) . . . . . **Oc. pionips** (p. 262)
Claspette stem strongly curved (Fig. 7.45b) . . . . **Oc. leucomelas** (p. 253)

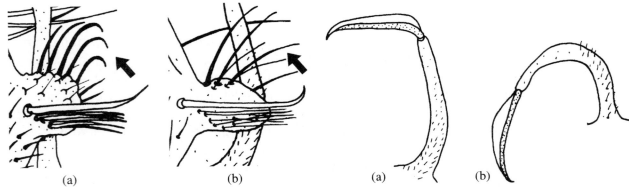

Figure 7.44.   Basal lobe of gonocoxite of:
(a) *Oc. communis*; (b) *Oc. pionips*

Figure 7.45.   Claspette of:
(a) *Oc. pionips*; (b) *Oc. leucomelas*

32 (29)   Setae above basal lobe of gonocoxite long, usually overlapping in the middle. Claspette
stem strongly curved in the middle (Fig. 7.46a, b) . . . . **Oc. cataphylla** (p. 230)
At least several setae immedeately above basal lobe of gonocoxite short, not
overlapping in the middle, or if all setae are long, claspette stem more or less
straight or slightly curved in the apical part (*Oc. detritus*) (Fig. 7.46c, d) 33

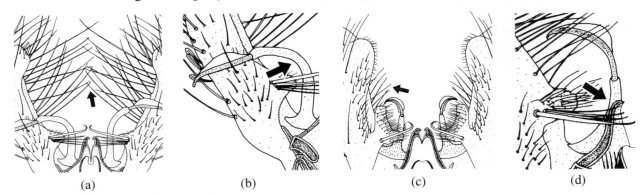

Figure 7.46.   Gonocoxite and claspette of:
(a, b) *Oc. cataphylla*; (c) *Oc. sticticus*; (d) *Oc. detritus*

33 (32) Basal lobe of gonocoxite constricted at its base. (Fig. 7.47a) 34
Basal lobe gradually arising from gonocoxite (Fig. 7.47b) 35

34 (33) Basal lobe of gonocoxite more or less crescent shaped, its upper part slender
(Fig. 7.48a) . . . . **Oc. sticticus** (p. 272)
Upper part of basal lobe of gonocoxite broad and rounded
(Fig. 7.48b) . . . . **Oc. nigrinus** (p. 259)

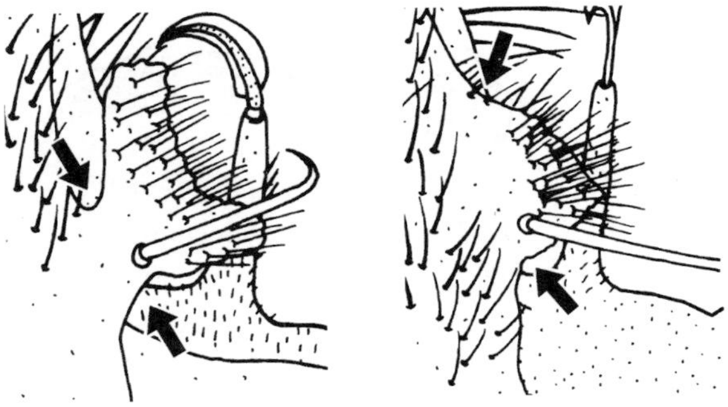

Figure 7.47. Basal lobe of gonocoxite of:
(a) *Oc. sticticus*; (b) *Oc. mercurator*

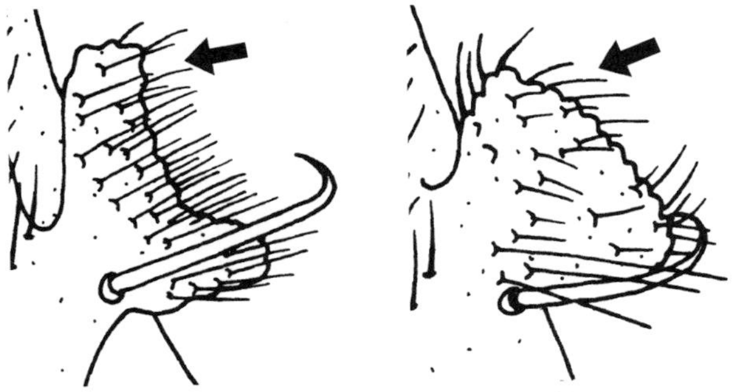

Figure 7.48. Basal lobe of gonocoxite of:
(a) *Oc. sticticus*; (b) *Oc. nigrinus*

35 (33) Basal lobe of gonocoxite longer than broad at its base (Fig. 7.49a) 36
Basal lobe distinct but not longer than broad at its base (Fig. 7.49b) 37

36 (35) Basal lobe of gonocoxite narrow and markedly elongated. Claspette filament
extremely broad, at least half as wide as long (Fig. 7.50a) . . . **Oc. cantans** (p. 225)
Basal lobe of gonocoxite slightly longer than broad at its base. Claspette filament
moderately broad, less than 1/4 of its length (Fig. 7.50b) . . . **Oc. riparius** (p. 270)

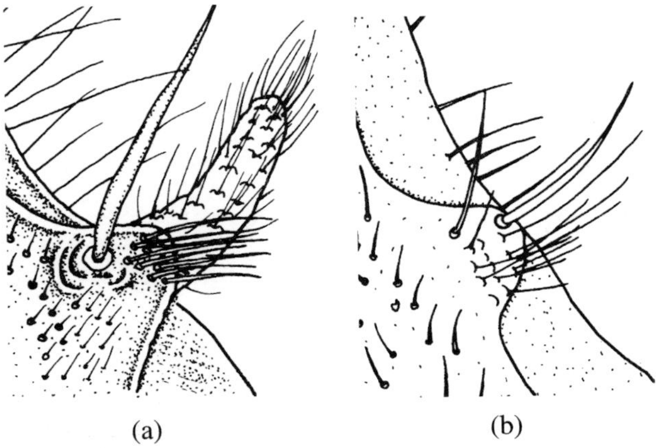

Figure 7.49. Basal lobe of gonocoxite of:
(a) *Oc. cantans*; (b) *Oc. cyprius*

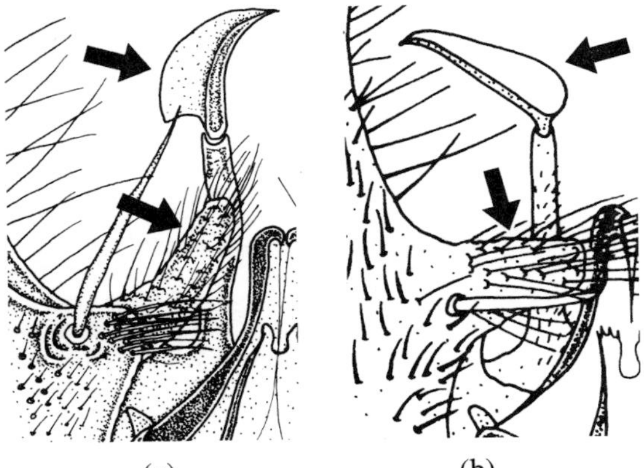

Figure 7.50. Basal lobe of gonocoxite and claspette of:
(a) *Oc. cantans*; (b) *Oc. riparius*

37 (35) Basal lobe of gonocoxite more or less conical, with 1 spine and long setae
(Fig. 7.51a) 38
Basal lobe of gonocoxite more flattened, with 1 spine and short, dense setae
(Fig. 7.51b) . . . . **Oc. flavescens** (p. 245)

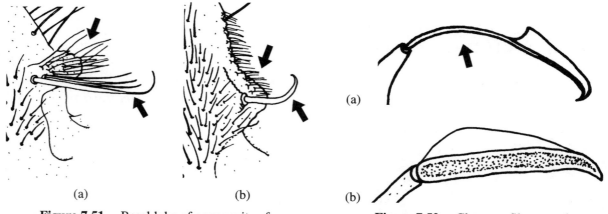

Figure 7.51. Basal lobe of gonocoxite of:
(a) *Oc. detritus*; (b) *Oc. flavescens*

Figure 7.52. Claspette filament of:
(a) *Oc. mercurator*; (b) *Oc. cyprius*

38 (37)   Claspette filament with long stalk, widened from about or beyond the middle
(Fig. 7.52a) 39
Claspette filament without stalk, widened from its base
(Fig. 7.52b) . . . . *Oc. cyprius* (p. 233)

39 (38)   Claspette filament evenly widened into a wing from about its middle. Lobe of
tergum IX with 3–8 spine-like setae (Fig. 7.53a, b) . . . . *Oc. detritus* (p. 235)
Claspette filament abruptly widened into a wing beyond the middle. Lobe of
tergum IX with 6–12 spine-like setae (Fig. 7.53c, d) . . . *Oc. mercurator* (p. 257)

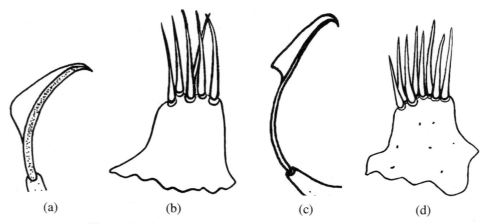

Figure 7.53. Claspette filament and lobe of tergum IX of:
(a, b) *Oc. detritus*; (c, d) *Oc. mercurator*

40 (26)   Basal lobe of gonocoxite conical, more or less as long as broad at the base
(Fig. 7.54a) . . . . *Oc. behningi* (p. 221)
Basal lobe of gonocoxite flattened, indistinct or absent (Fig. 7.54b) 41

41 (40)   Claspette stem stout, slightly swollen at the apex. Claspette filament slightly
swollen beyond the middle (Fig. 7.55a) . . . . *Oc. annulipes* (p. 219)
Claspette stem slender, tapering apically. Claspette filament tapering gradually
towards the apex (Fig. 7.55b) 42

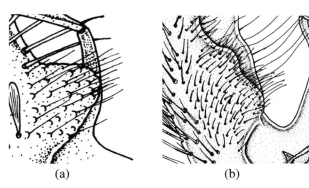

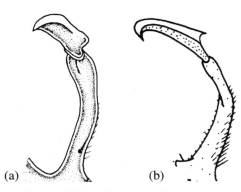

Figure 7.54. Basal lobe of gonocoxite of:
(a) *Oc. behningi*; (b) *Oc. annulipes*

Figure 7.55. Claspette of:
(a) *Oc. annulipes*; (b) *Oc. excrucians*

42 (41)   Apical lobe small, not reaching the level of gonostylus articulation
(Fig. 7.56a) . . . . *Oc. excrucians* and *Oc. surcoufi* (pp. 243, 244)
Apical lobe prominent, protruding above the level of gonostylus articulation
(Fig. 7.56b) . . . . *Oc. euedes* (p. 241)

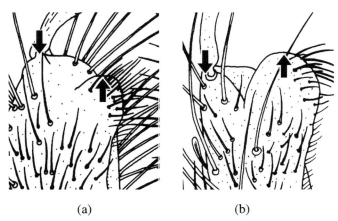

Figure 7.56. Apical lobe of gonocoxite of:
(a) *Oc. excrucians s.l.*; (b) *Oc. euedes*

7.3. GENUS *CULEX*

1   Gonocoxite with a broad, flattened, sclerotised process at the apex, extending
distinctly beyond the base of the gonostylus (Fig. 7.57a) . . . *Cx. hortensis* (p. 303)
Gonocoxite without a process at the apex (Fig. 7.57b) 2
2 (1)   Gonocoxite with small scales on the outer surface. Lobe of gonocoxite located
slightly beyond the middle (Fig. 7.58a) 3
Gonocoxite without scales on the outer surface. Lobe of gonocoxite located well
beyond the middle (Fig. 7.58b) 4
3 (2)   Gonostylus relatively long, more or less half as long as gonocoxite. Ventral
arm of aedeagus short, its apex not extending beyond apex of paraproct
(Fig. 7.59a) . . . . *Cx. modestus* (p. 284)

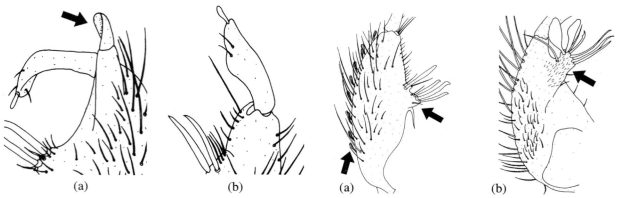

Figure 7.57.   Apex of gonocoxite of:
(a) *Cx. hortensis*; (b) *Cx. martinii*

Figure 7.58.   Gonocoxite of:
(a) *Cx. modestus*; (b) *Cx. brumpti*

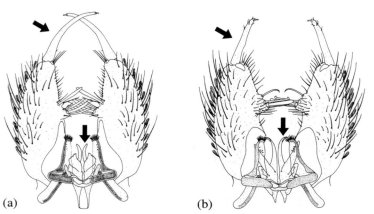

Figure 7.59.   Hypopygium of:
(a) *Cx. modestus*; (b) *Cx. pusillus*

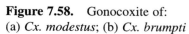

Gonostylus short, distinctly shorter than half the length of gonocoxite. Ventral arm of aedeagus long, its apex extending beyond apex of paraproct (Fig. 7.59b) . . . . ***Cx. pusillus*** (p. 286)

4 (2)   Lobe of gonocoxite with one or several transparent, broad, oval or lanceolate scale-like setae. Apex of paraproct with several rows of spines (Fig. 7.60a) 5
Lobe of gonocoxite with distinctly sclerotised broad and narrow setae. Transparent scale-like setae absent. Apex of paraproct with a row of large denticles, sometimes in addition to the denticles several rows of short spines could be present (Fig. 7.60b) 11

5 (4)   Gonostylus expanded beyond the middle (Fig. 7.61a) 6
Gonostylus not expanded beyond the middle, tapering apically (Fig. 7.61b). 7

6 (5)   Ventral arm of aedeagus slender, extended apically into a fan shaped process which bears 5 spines. The spines do not exceed the upper border of the fan (Fig. 7.62a) . . . . ***Cx. brumpti*** (p. 289)
Ventral arm of aedeagus stout, with a concave apex not bearing any spines (Fig. 7.62b) . . . . ***Cx. perexiguus*** (p. 294)

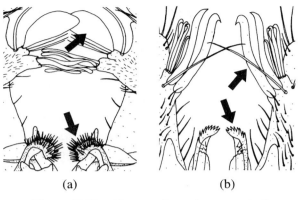

Figure 7.60. Gonocoxite and paraproct of:
(a) *Cx. perexiguus*; (b) *Cx. territans*

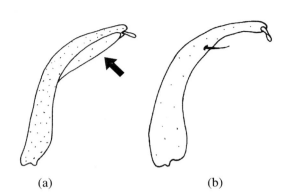

Figure 7.61. Gonostylus of:
(a) *Cx. perexiguus*; (b) *Cx. mimeticus*

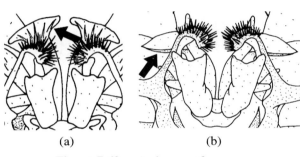

Figure 7.62. Aedeagus of:
(a) *Cx. brumpti*; (b) *Cx. perexiguus*

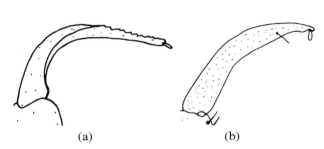

Figure 7.63. Gonostylus of:
(a) *Cx. laticinctus*; (b) *Cx. theileri*

7 (5) Gonostylus sharply bent (Fig. 7.63a) . . . . ***Cx. laticinctus*** (p. 290)
       Gonostylus gradually curved (Fig. 7.63b) 8

8 (7) Ventral arm or both, dorsal and ventral arms of aedeagus with denticles or teeth
       (Fig. 7.64a) 9
       Dorsal and ventral arms of aedeagus without denticles or teeth (Fig. 7.64b) . . . 10

9 (8) Ventral arm of aedeagus with 1–2 denticles, dorsal arm usually with 3 finger-like
       processes (Fig. 7.65a) . . . . ***Cx. mimeticus*** (p. 292)
       Ventral arm of aedeagus with 2–4 strong lateral teeth, dorsal arm simple,
       slender, pointed (Fig. 7.65b) . . . . ***Cx. theileri*** (p. 300)

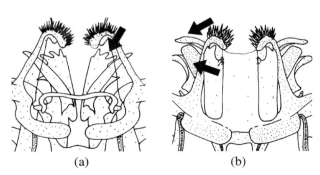
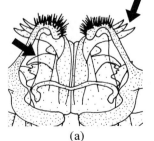

Figure 7.64. Aedeagus of:
(a) *Cx. theileri*; (b) *Cx. p. pipiens*

Figure 7.65. Aedeagus of:
(a) *Cx. mimeticus*; (b) *Cx. theileri*

10 (8)   Dorsal arm of aedeagus tube-like, truncate apically. Ventral arm of paraproct usually weakly developed, transparent (slightly sclerotised) (Fig. 7.66a) . . . . ***Cx. p. pipiens*** (p. 296)
Dorsal arm of aedeagus twisted and pointed apically. Ventral arm of paraproct well developed, strongly scerotized (Fig. 7.66b) . . . . . . ***Cx. torrentium*** (p. 299)

11 (4)   Gonocoxite with many long and conspicuous setae on its outer surface. Gonostylus constricted subapically, then expanding into a "T"-shaped apical part (Fig. 7.67a) . . . . ***Cx. impudicus*** (p. 305)
Gonocoxite with less conspicuous setae on its outer surface. Apex of gonostylus not "T"-shaped, gradually tapering from the middle (Fig. 7.67b) 12

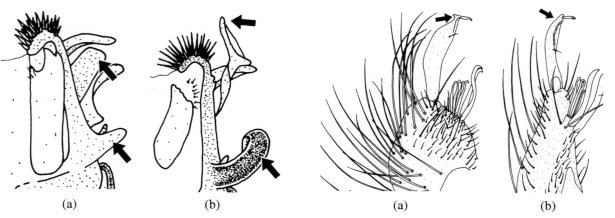

(a)   (b)   (a)   (b)

Figure 7.66.   Aedeagus and paraproct of: **Figure 7.67.**   Gonocoxite of:
(a) *Cx. p. pipiens*; (b) *Cx. torrentium*   (a) *Cx. impudicus*; (b) *Cx. territans*

12 (11)   Gonostylus short, markedly widened beyond the middle. Apex of paraproct widened with a convex row of denticles (Fig. 7.68a) . . . . . ***Cx. martinii*** (p. 307)
Gonostylus elongated, usually evenly tapering from base to apex. Apex of paraproct not widened, with an inwardly curved row of denticles (Fig. 7.68b) . . . . ***Cx. territans*** (p. 308)

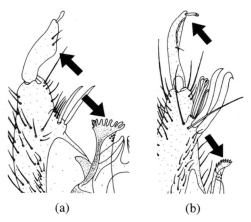

(a)   (b)

Figure 7.68.   Gonocoxite and paraproct of:
(a) *Cx. martinii*; (b) *Cx. territans*

7.4. GENUS *CULISETA*

1   Tergum IX laterally expanded into two long and slender, sclerotised lobes bearing tiny spine-like setae at the apex. Gonostylus broadened apically, blunt, with 2 short, pointed apical spines (Fig. 7.69a, b) .. ***Cs. longiareolata*** (p. 311)
Lobe of tergum IX small or indistinct, bearing long setae. Gonostylus tapering apically, apical spine simple (Fig. 7.69c, d) 2

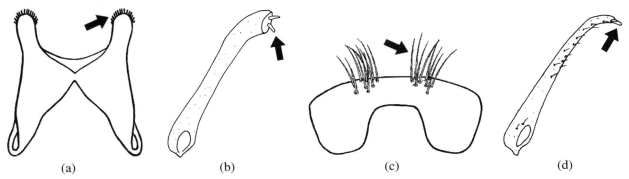

       (a)   (b)   (c)   (d)

Figure 7.69. Tergum IX and gonostylus of:
(a, b) *Cs. longiareolata*; (c, d) *Cs. morsitans*

2 (1)   Aedeagus usually oval, slightly sclerotised (Fig. 7.70a), (in *Cs. bergrothi* the oval shaped aedeagus is slightly sclerotised at least in the lateral parts) 3
Aedeagus usually elongated, conical. Lateral plates strongly sclerotised, pointed and well separated at the apex (Fig. 7.70b) 7

3 (2)   Basal lobe of gonocoxite well developed, elongated. Median lobe of tergum VIII with a row of less than 10 strong spine-like setae (Fig. 7.71a) 4
Basal lobe of gonocoxite weakly developed. Median lobe of tergum VIII with a row of usually more than 10 (4–18) strong spine-like setae
(Fig. 7.71b) .... ***Cs. bergrothi*** (p. 326)

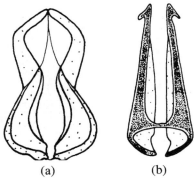

      (a)   (b)

Figure 7.70. Aedeagus of:
(a) *Cs. litorea*; (b) *Cs. alaskaensis*

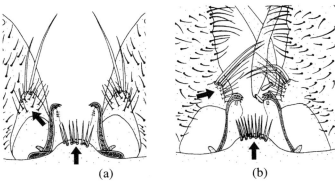

      (a)   (b)

Figure 7.71. Hypopygium and tergum VIII of:
(a) *Cs. litorea*; (b) *Cs. bergrothi*

4 (3)   Gonocoxite plumpy in appearance, more or less twice as long as its basal width. Basal lobe of gonocoxite with 2 stout setae, longer seta reaching apex of gonocoxite (Fig. 7.72a) . . . . ***Cs. litorea*** (p. 315)
      Gonocoxite elongated, at least 2.5 times as long as its basal width. Basal lobe of gonocoxite with 2 or more (usually 3–8) stout setae, no seta reaching apex of gonocoxite (Fig. 7.72b) 5

5 (4)   Basal lobe of gonocoxite with 5–8 strong setae (Fig. 7.73a) . . ***Cs. ochroptera*** (p. 319)
      Basal lobe of gonocoxite with 2–4 strong setae (Fig. 7.73b) 6

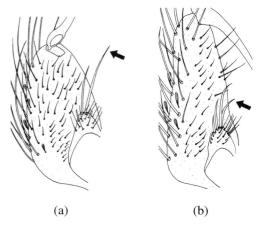

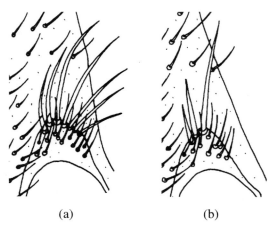

(a)   (b)   (a)   (b)

Figure 7.72.   Gonocoxite of: **Figure 7.73.**   Basal lobe of gonocoxite of:
(a) *Cs. litorea*; (b) *Cs. fumipennis*   (a) *Cs. ochroptera*; (b) *Cs. morsitans*

6 (5)   Gonostylus long and slender, abruptly constricted shortly beyond the base (Fig. 7.74a) . . . . ***Cs. fumipennis*** (p. 313)
      Gonostylus more stout, without abrupt constriction (Fig. 7.74b) . . . . ***Cs. morsitans*** (p. 317)

7 (2)   Apical lobe of gonocoxite with several long lanceolate setae (Fig. 7.75a) . . . . ***Cs. glaphyroptera*** (p. 328)
      Apical lobe of gonocoxite with thin short setae or absent (Fig. 7.75b) 8

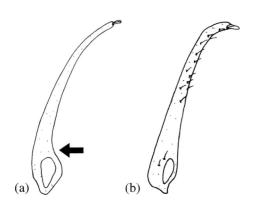

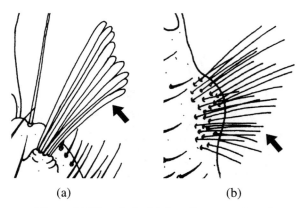

(a)   (b)   (a)   (b)

Figure 7.74.   Gonostylus of: **Figure 7.75.**   Apical lobe of gonocoxite of:
(a) *Cs. fumipennis*; (b) *Cs. morsitans*   (a) *Cs. glaphyroptera*; (b) *Cs. alaskaensis*

8 (7) Apical lobe of gonocoxite well developed and slightly convex, densely covered with short setae (Fig. 7.76a) .... *Cs. alaskaensis* (p. 322)
Apical lobe of gonocoxite indistinct or absent (Fig. 7.76b) 9

9 (8) Basal lobe of gonocoxite with 2 (rarely 3) setae distinctly stouter than the others (Fig. 7.77a). Median lobe of tergum VIII usually without stout setae, rarely a few present (1–4) .... *Cs. annulata* (p. 324)
Basal lobe of gonocoxite with 3–5 setae distinctly stouter than the others (Fig. 7.77b). Median lobe of tergum VIII with several stout setae (usually 4) *Cs. subochrea* (p. 330)

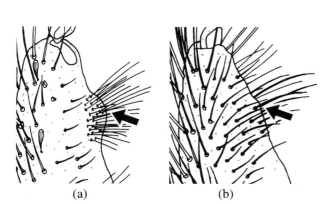

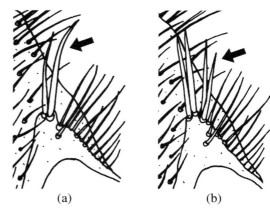

Figure 7.76. Apical lobe of gonocoxite of: (a) *Cs. alaskaensis*; (b) *Cs. annulata*

Figure 7.77. Basal lobe of gonocoxite of: (a) *Cs. annulata*; (b) *Cs. subochrea*

7.5. GENUS *COQUILLETTIDIA*

1   Base of gonostylus slightly constricted below expanded portion, then sharply narrowed in the middle part (Fig. 7.78a) .... *Cq. richiardii* (p. 333)
Base of gonostyle broad, continuation stem like, expanded apically into a bulbous structure (Fig. 7.78b) .... *Cq. buxtoni* (p. 333)

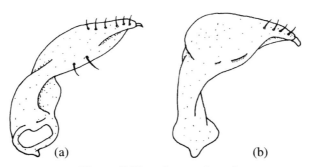

Figure 7.78. Gonostylus of: (a) *Cq. richiardii*; (b) *Cq. buxtoni*

8

Key to Mosquito Fourth-Instar Larvae

GENERA

1   Abdominal segment VIII without elongate siphon (Fig. 8.1a) . . ***Anopheles*** (p. 135)

        Abdominal segment VIII with elongate siphon (Fig. 8.1b) 2

2 (1)   Siphon short, apex strongly sclerotised and pointed, with saw-like apparatus for cutting and piercing plant tissues (Fig. 8.2a) . . . . ***Coquillettidia*** (p. 160)

        Siphon longer, apex not pointed, without such an adaptation (Fig. 8.2b) 3

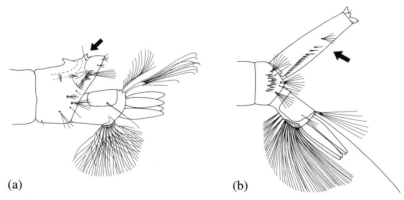

Figure 8.1. Abdominal segment VIII of:
(a) *Anopheles* sp.; (b) *Aedes* sp.

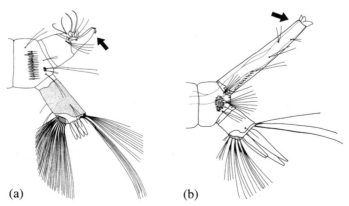

Figure 8.2. Siphon of:
(a) *Cq. richiardii*; (b) *Cx. p. pipiens*

133

3 (2)   Siphon with several pairs of siphonal tufts (1-S) (Fig. 8.3a) . . . . ***Culex*** (p. 152)
        Siphon with one pair of siphonal tufts (1-S) (Fig. 8.3b) 4

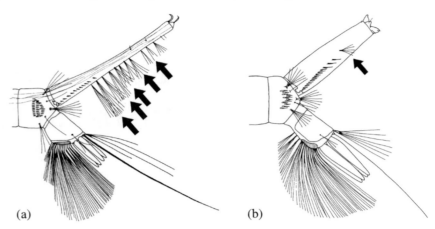

(a)   (b)

Figure 8.3.   Siphon of:
(a) *Cx. hortensis*; (b) *Ae. vexans*

4 (3)   Pecten absent. Abdominal segments VI–VIII with more or less developed
        sclerotised plates dorsally (Fig. 8.4a) . . . . ***Orthopodomyia pulcripalpis*** (p. 336)
        Pecten present. Abdominal segments VI–VIII without sclerotised plates dorsally
        (Fig. 8.4b) 5

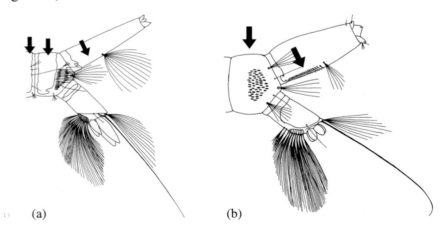

(a)   (b)

Figure 8.4.   Larva of:
(a) *Or. pulcripalpis*; (b) *Oc. detritus*

5 (4)   Siphonal tuft (1-S) inserted at the base of the siphon
        (Fig. 8.5a) . . . . ***Culiseta*** (p. 157)
        Siphonal tuft (1-S) inserted close to the middle or near the apex of the siphon
        (Fig. 8.5b) 6

6 (5)   Abdominal segment VIII with sclerotised plates laterally, comb scales arising from
        the posterior margin of the plates (Fig. 8.6a) . . . ***Uranotaenia unguiculata*** (p. 339)
        Abdominal segment VIII without sclerotised plates laterally. Only comb scales are
        present and they never arise from sclerotised plates
        (Fig. 8.6b) . . . . ***Aedes*** and ***Ochlerotatus*** (p. 139)

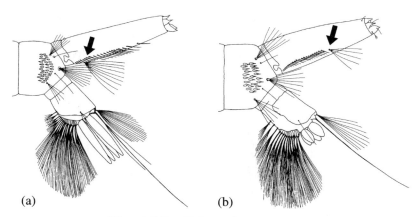

Figure 8.5. Siphon of:
(a) *Cs. annulata*; (b) *Oc. caspius*

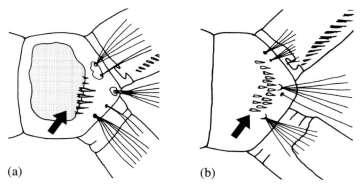

Figure 8.6. Abdominal segment VIII of:
(a) *Ur. unguiculata*; (b) *Ae. cinereus*

8.1. GENUS *ANOPHELES*

1 (2) Frontal setae (5-C to 7-C) long, pinnate. Antenna covered with spicules, at least on its inner surface (Fig. 8.7a) 2
Frontal setae short, simple. Antenna not covered with spicules
(Fig. 8.7b) . . . . *An. plumbeus* (p. 182)

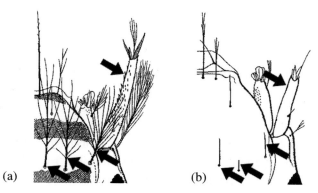

Figure 8.7. Head of:
(a) *An. algeriensis*; (b) *An. plumbeus*

2 (1)   Inner clypeal setae (2-C) situated close together, closer to each other than to outer clypeal setae (3-C) (Fig. 8.8a) 3
Inner clypeal setae widely separated, closer to outer clypeal setae than to each other (Fig. 8.8b) 8

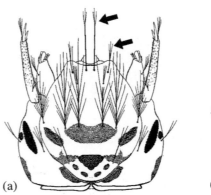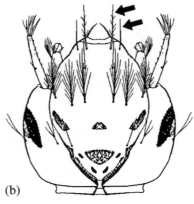

Figure 8.8.   Head of:
(a) *An. claviger s.s.*; (b) *An. superpictus*

3 (2)   Outer clypeal seta (3-C) simple or aciculate (Fig. 8.9a) (2–3 apical branches in *An. claviger s.l.*) 4
Outer clypeal seta dendriform (Fig. 8.9b) 7

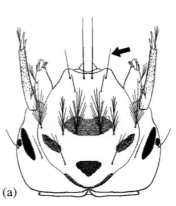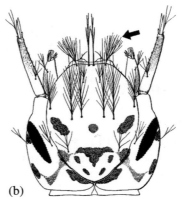

Figure 8.9.   Head of:
(a) *An. marteri*; (b) *An. maculipennis s.l.*

4 (3)   Frontoclypeus with 3 dark transverse bands. Clypeal setae (2-C and 3-C) aciculate (Fig. 8.10a) . . . . ***An. algeriensis*** (p. 165)
Frontoclypeus spotted but not banded. Clypeal setae simple or with 2–3 apical branches (Fig. 8.10b) 5

5 (4)   Postclypeal seta (4-C) simple, sometimes with 2 branches (Fig. 8.11a). Antepalmate setae on abdominal segments IV and V (2-IV and 2-V) with 1–3 branches 6
Postclypeal seta with 2–5 branches (Fig. 8.11b). Antepalmate setae on abdominal segments IV and V (2-IV and 2-V) with 3–5 branches. Palmate setae on abdominal segment II (1-II) with 10–15 leaflets . . . . . . ***An. claviger s.s.*** (p. 167)

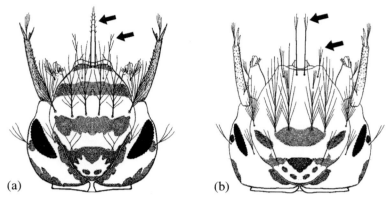

Figure 8.10. Head of:
(a) *An. algeriensis*; (b) *An. claviger s.s.*

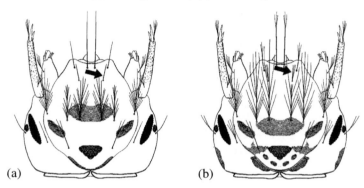

Figure 8.11. Head of:
(a) *An. marteri*; (b) *An. claviger s.s.*

6 (5)   Leaflets of palmate setae terminating in a long filament which is 1/3 as long
as the blade (Fig. 8.12a). Antepalmate setae on abdominal segments IV and V
(2-IV and 2-V) with 1, rarely 2 branches . . . . ***An. marteri*** (p. 180)
Leaflets of palmate setae with slightly elongated apex but without terminal
filaments (Fig. 8.12b). Antepalmate setae on abdominal segments IV and V
(2-IV and 2-V) with 2–3 branches. Palmate setae on abdominal segment II (1-II)
with more than 15 leaflets . . . . ***An. petragnani*** (p. 169)

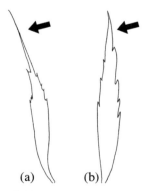

Figure 8.12. Leaflet of palmate seta of:
(a) *An. marteri*; (b) *An. petragnani*

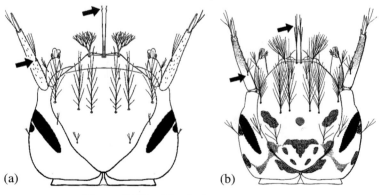

Figure 8.13. Head of:
(a) *An. hyrcanus*; (b) *An. maculipennis s.l.*

7 (3)   Inner clypeal seta (2-C) with short apical branches. Antennal seta (1-A) inserted in the middle or slightly below the middle of antenna (Fig. 8.13a) .. ***An. hyrcanus*** (p. 170)
Inner clypeal seta with long apical branches. Antennal seta inserted in basal 1/4 to 1/3 of antenna (Fig. 8.13b) . . . . **Anopeles Maculipennis Complex** (p. 172)

8 (2)   Inner frontal setae (5-C) slightly longer than median frontal setae (6-C) (Fig. 8.14a) 9
Inner frontal setae distinctly longer than median frontal setae (Fig. 8.14b) 10

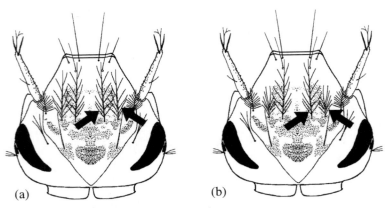

Figure 8.14.   Head of:
(a) *An. sergentii*; (b) *An. multicolor*

9 (8)   Sclerotised tergal plate on abdominal segment VII broad, broader than distance between palmate setae (1-VII). Filament of abdominal palmate setae long, more than half as long as the blade (Fig. 8.15a, b) . . . . ***An. sergentii*** (p. 188)
Sclerotised tergal plate on abdominal segment VII relatively small, smaller than distance between palmate setae (1-VII). Filament of abdominal palmate setae short, about half as long as the blade (Fig. 8.15c, d) . . . . ***An. cinereus hispaniola*** (p. 185)

10 (8)   Inner clypeal seta (2-C) simple (Fig. 8.16a). Typical palmate setae on metathorax (1-T) absent . . . . ***An. multicolor*** (p. 186)
Inner clypeal seta with short branches (aciculate) (Fig. 8.16b). Typical palmate setae on metathorax present . . . . ***An. superpictus*** (p. 190)

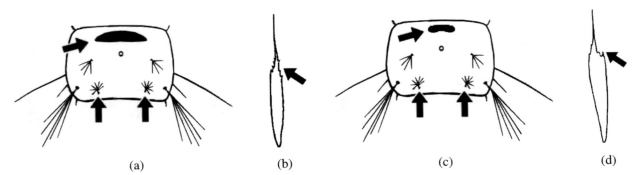

Figure 8.15.   Abdominal segment VII and leaflet of palmate seta of:
(a, b) *An. sergentii*; (c, d) *An. cinereus hispaniola*

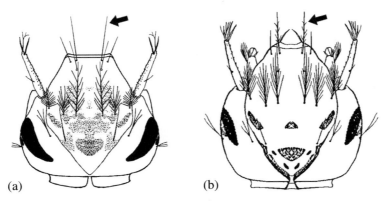

Figure 8.16. Head of:
(a) *An. multicolor*; (b) *An. superpictus*

8.2. GENERA *AEDES* AND *OCHLEROTATUS*

1 (2) Antenna distinctly longer than the head (Fig. 8.17a) . . . . . . **Oc. diantaeus** (p. 237)
        Antenna as long as or shorter than the head (Fig. 8.17b) 2
2 (1) Siphon extremely long and slender, siphonal index at least 5.5
        (Fig. 8.18a) . . . . **Oc. berlandi** (p. 223)
        Siphon otherwise, siphonal index never exceeds 5.5 (Fig. 8.18b) 3

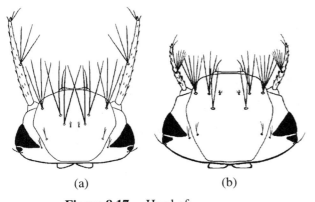

Figure 8.17. Head of:
(a) *Oc. diantaeus*; (b) *Oc. riparius*

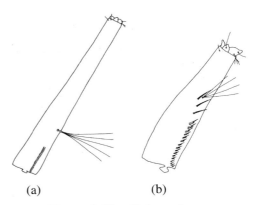

Figure 8.18. Siphon of:
(a) *Oc. berlandi*; (b) *Oc. euedes*

3 (2) Entire body surface covered with dense rows of spicules
        (Fig. 8.19a) . . . . **Oc. cyprius** (p. 233)
        Body surface without spicules. Some indistinct spicules may be present on the last
        abdominal segments (Fig. 8.19b) 4
4 (3) Siphonal tuft (1-S) small, about half as long as the width of the siphon at the point
        of its origin, or shorter. Tuft inserted well beyond the middle of siphon
        (Fig. 8.20a) 5
        Siphonal tuft large, at least 2/3 as long as the width of the siphon at the point of
        its origin, or longer. Tuft may be inserted before or beyond the middle of siphon
        (Fig. 8.20b) 7

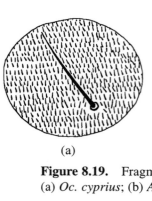
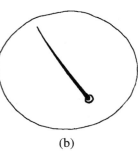
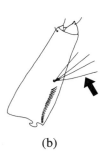

(a)   (b)   (a)   (b)

Figure 8.19.   Fragment of integument of: **Figure 8.20.**   Siphon of:
(a) *Oc. cyprius*; (b) *Ae. vexans*   (a) *Ae. vexans*; (b) *Oc. hexodontus*

5 (4)   Frontal setae (5-C to 7-C) arranged in an triangular pattern (Fig. 8.21a). Median
           setae of labral brush serrated apically . . . . ***Ae. vexans*** (p. 201)
           Frontal setae arranged in an arc-like row (Fig. 8.21b). All setae of labral brush
           simple 6

6 (5)   Antennal seta (1-A) inserted in the middle of antenna. Prothoracal seta 4-P with
           4 branches, 7-P with 5–6 branches (Fig. 8.22a) . . . . ***Ae. rossicus*** (p. 199)
           Antennal seta (1-A) inserted slightly before the middle, at 2/5 of the length of
           antenna. Prothoracal seta 4-P with 2 branches, 7-P with 3 branches
           (Fig. 8.22b) . . . . ***Ae. cinereus*** and ***Ae. geminus*** (pp. 195, 198)

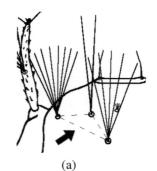
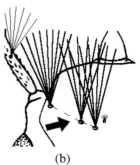
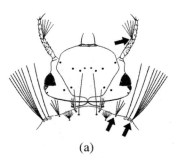
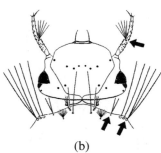

(a)   (b)   (a)   (b)

Figure 8.21.   Head of: **Figure 8.22.**   Head and prothorax of:
(a) *Ae. vexans*; (b) *Ae. cinereus*   (a) *Ae. rossicus*; (b) *Ae. cinereus*

7 (4)   Base of siphon with acus (Fig. 8.23a) 8
           Base of siphon without acus (indistinct in *Ae. vittatus*) (Fig. 8.23b) 45

8 (7)   Antennae covered with more or less numerous spicules (Fig. 8.24a) 9
           Antennae without spicules (Fig. 8.24b) 43

9 (8)   Dorsal surface of siphon with several pairs of additional setae (Fig. 8.25a) . . . 10
           Dorsal surface of siphon without additional setae (Fig. 8.25b) 14

10 (9)   Siphonal tuft (1-S) attached within the distal pecten teeth (Fig. 8.26a) 11
           Siphonal tuft attached beyond the distalmost pecten tooth (Fig. 8.26b) 13

11 (10)   Siphonal seta (1-S) simple. 3–4 distal pecten teeth atypical, spine-like,
           widely spaced, almost reaching apex (Fig. 8.27a) . . . . . ***Oc. subdiversus*** (p. 281)
           Siphonal seta with 5–8 branches. 1–3 distal pecten teeth widely spaced, but not
           reaching to apical third of siphon (Fig. 8.27b) 12

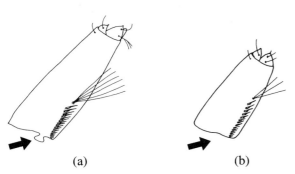

Figure 8.23. Siphon of:
(a) *Oc. punctor*; (b) *Ae. aegypti*

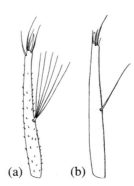

Figure 8.24. Antenna of:
(a) *Oc. rusticus*; (b) *Oc. geniculatus*

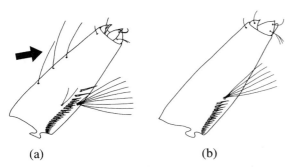

Figure 8.25. Siphon of:
(a) *Oc. rusticus*; (b) *Oc. punctor*

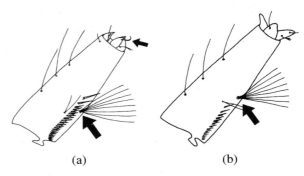

Figure 8.26. Siphon of:
(a) *Oc. rusticus*; (b) *Oc. refiki*

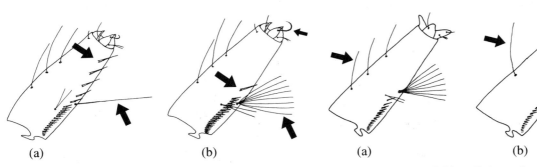

Figure 8.27. Siphon of:
(a) *Oc. subdiversus*; (b) *Oc. quasirusticus*

Figure 8.28. Siphon of:
(a) *Oc. refiki*; (b) *Oc. lepidonotus*

12 (11) Prothoracic seta 1-P simple. Seta 9-S small (Fig. 8.26a) . . . *Oc. rusticus* (p. 279)
 Prothoracic seta 1-P with 3 branches. Seta 9-S large, hook shaped
 (Fig. 8.27b) . . . . *Oc. quasirusticus* (p. 276)

13 (10) Dorsal surface of siphon with 3 pairs of additional setae
 (Fig. 8.28a) . . . . *Oc. refiki* (p. 277)
 Dorsal surface of siphon with 2 pairs of additional setae
 (Fig. 8.28b) . . . . *Oc. lepidonotus* (p. 274)

14 (9) Saddle completely surrounding anal segment or ventral margins of the saddle are
 separated by a very narrow gap (Fig. 8.29a) 15

Saddle extending to the lateral parts of the anal segments to a various degree,
but leaving the ventral part of segment uncovered (Fig. 8.29b) 18

15 (14)   Distal pecten teeth (1–3) detached (Fig. 8.30a) . . . . *Oc. nigripes* (p. 260)

Pecten teeth evenly spaced, close together (Fig. 8.30b) 16

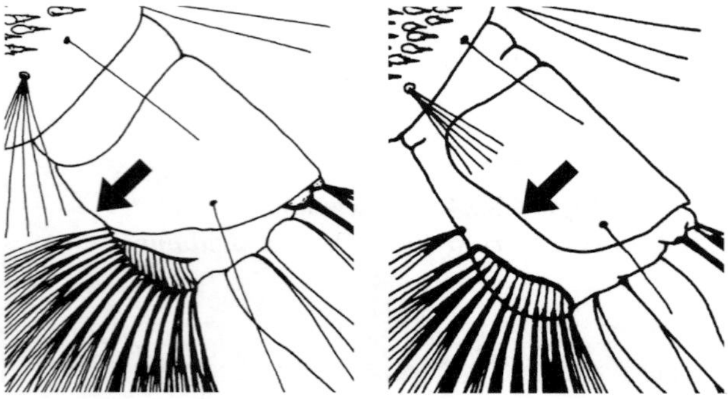

(a)   (b)

Figure 8.29.   Anal segment of:
(a) *Oc. punctor*; (b) *Oc. cataphylla*

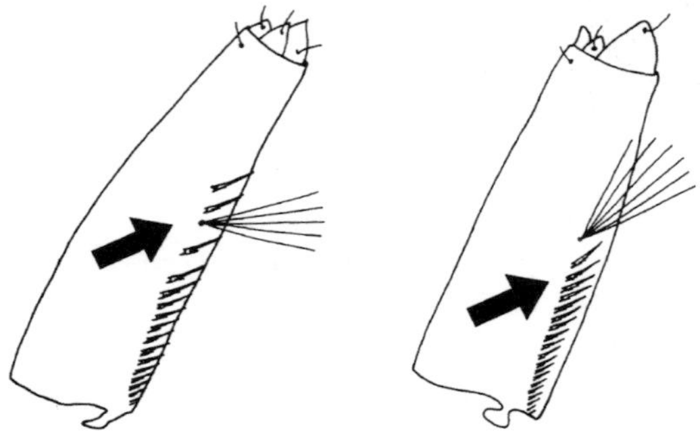

(a)   (b)

Figure 8.30.   Siphon of:
(a) *Oc. nigripes*; (b) *Oc. punctodes*

16 (15)   Number of comb scales 6–9 (Fig. 8.31a) . . . . *Oc. hexodontus* (p. 247)

Number of comb scales 10–30 (Fig. 8.31b) 17

17 (16)   Saddle completely surrounding anal segment. Inner (5-C) and median (6-C)
frontal setae with 1–3 branches (Fig. 8.32a) . . . . *Oc. punctor* (p. 268)

Lower margins of the saddle are separated by a very narrow gap. Inner and
median frontal setae simple (Fig. 8.32b) . . . . *Oc. punctodes* (p. 267)

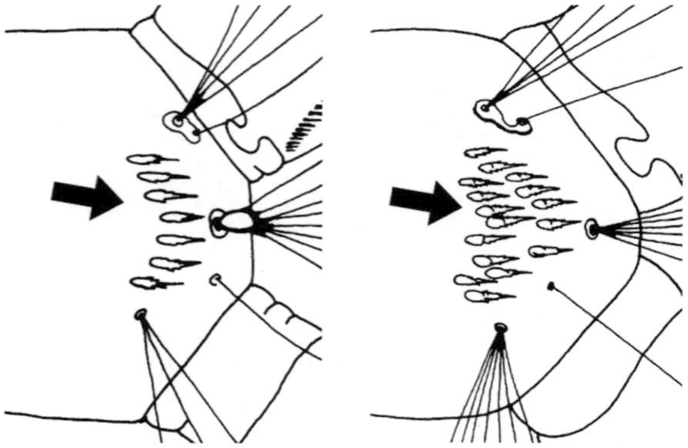

(a)   (b)

Figure 8.31.   Abdominal segment VIII of:
(a) *Oc. hexodontus*; (b) *Oc. punctor*

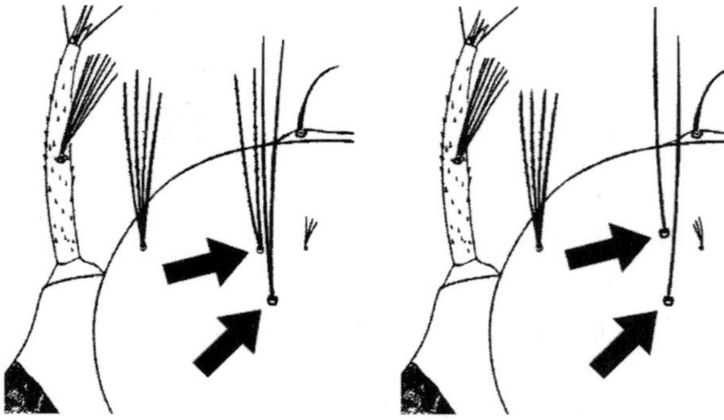

(a)   (b)

Figure 8.32.   Head of:
(a) *Oc. punctor*; (b) *Oc. punctodes*

18 (14)   Anal segment with 1–3 precratal setae (4-X) (Fig. 8.33a) 19

Anal segment with 4–6 precratal setae, sometimes up to 10 (Fig. 8.33b) 33

19 (18)   Distal pecten teeth (1–4) detached (Fig. 8.34a) 20

Pecten teeth evenly spaced, close together (Fig. 8.34b) 21

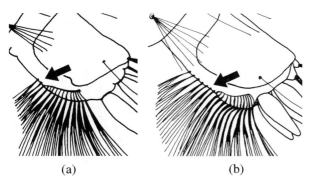

Figure 8.33. Anal segment of:
(a) *Oc. intrudens*; (b) *Oc. annulipes*

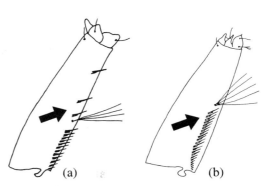

Figure 8.34. Siphon of:
(a) *Oc. cataphylla*; (b) *Oc. caspius*

20 (19) Inner (5-C) and median (6-C) frontal setae simple. All detached pecten teeth
(2–4) located beyond the insertion point of the siphonal tuft (1-S)
(Fig. 8.35a, b) . . . . ***Oc. cataphylla*** (p. 230)
Inner and median frontal setae with 3–5 branches. Detached pecten teeth usually
located below insertion point of the siphonal tuft, the distalmost tooth may be
located slightly beyond it (Fig. 8.35c, d) . . . . ***Oc. intrudens*** (p. 252)

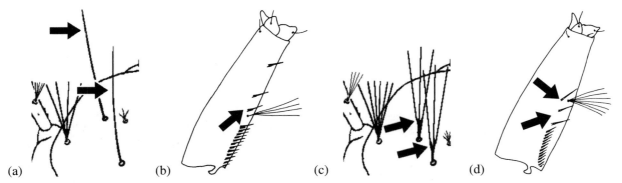

Figure 8.35. Head and siphon of:
(a, b) *Oc. cataphylla*; (c, d) *Oc. intrudens*

21 (19) Anal papillae usually much shorter than the saddle, rarely as long as or up to
1.3 times longer than the saddle (Fig. 8.36a) 22
Anal papillae distinctly longer than the saddle, at least 1.3 times longer than the
saddle (Fig. 8.36b) 25

22 (21) Comb with more than 40 scales. Comb scales blunt ended (median spine is not
distinctly longer than the others) (Fig. 8.37a). Inner frontal seta (5-C) with
2–3 branches . . . . ***Oc. detritus*** (p. 235)
Comb with less than 35 scales. Comb scales pointed (median spine always
distinctly longer than the others, at least in some scales) (Fig. 8.37b). Inner
frontal seta usually simple, sometimes with 2 branches 23

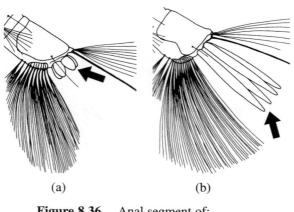

Figure 8.36.   Anal segment of:
(a) *Oc. detritus*; (b) *Oc. communis*

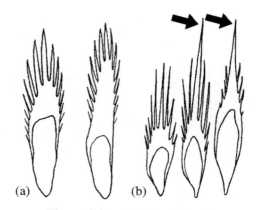

Figure 8.37.   Comb scales of:
(a) *Oc. detritus*; (b) *Oc. caspius*

23 (22)   Siphonal tuft (1-S) situated beyond the middle of siphon
(Fig. 8.38a) . . . .*Oc. caspius* (p. 227)
Siphonal tuft usually situated below, rarely slightly beyond the middle of
siphon (Fig. 8.38b) 24

24 (23)   Anal papillae tapering. Saddle seta (1-X) long, nearly as long as the saddle
(Fig. 8.39a) . . . . *Oc. leucomelas* (p. 253)
Anal papillae rounded. Saddle seta half as long as the saddle
(Fig. 8.39b) . . . . *Oc. dorsalis* (p. 239)

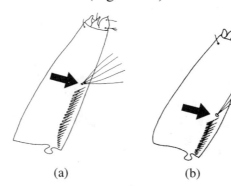

Figure 8.38.   Siphon of:
(a) *Oc. caspius*; (b) *Oc. leucomelas*

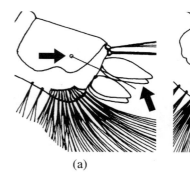

Figure 8.39.   Anal segment of:
(a) *Oc. leucomelas*; (b) *Oc. dorsalis*

25 (21)   Comb with more than 40 scales (Fig. 8.40a) 26
Comb with less than 40 scales (Fig. 8.40b) 28

26 (25)   Inner (5-C) and median (6-C) frontal setae simple, rarely one seta with
2 branches (Fig. 8.41a) . . . . *Oc. communis* (p. 232)
Inner and median frontal setae multiple branched, with 3 or more branches
(Fig. 8.41b) 27

27 (26)   Antennae long, about 2/3 as long as the head or slightly longer. General
appearance of comb scales blunt. Elongated median spine absent, all spines of
similar length (Fig. 8.42a, b) . . . . *Oc. pionips* (p. 262)
Antennae shorter, about half as long as the head. General appearance of comb
scales pointed. At least some lateral scales with the median spine distinctly
longer than the others (Fig. 8.42c, d) . . . . *Oc. pullatus* (p. 265)

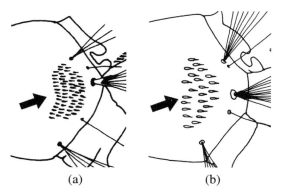

Figure 8.40. Abdominal segment VIII of:
(a) *Oc. communis*; (b) *Oc. sticticus*

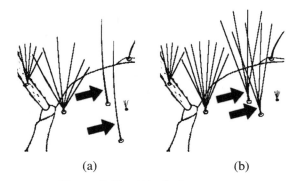

Figure 8.41. Head of:
(a) *Oc. communis*; (b) *Oc. pionips*

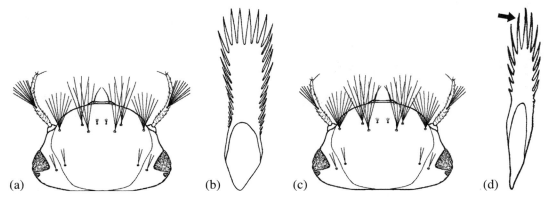

Figure 8.42. Head and comb scale of:
(a, b) *Oc. pionips*; (c, d) *Oc. pullatus*

28 (25) Number of comb scales 6–16 (Fig. 8.43a) 29
         Number of comb scales more than 16 (Fig. 8.43b) 30

29 (28) Saddle almost completely covering lateral parts of anal segment
         (Fig. 8.44a) . . . . ***Oc. nigrinus*** (p. 259)
         Saddle more plate shaped, extending slightly beyond lateral half of
         anal segment (Fig. 8.44b) . . . . ***Oc. impiger*** (p. 250)

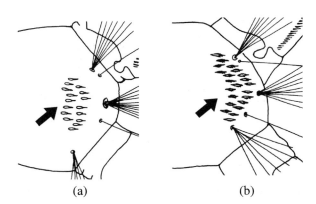

Figure 8.43. Abdominal segment VIII of:
(a) *Oc. nigrinus*; (b) *Oc. hungaricus*

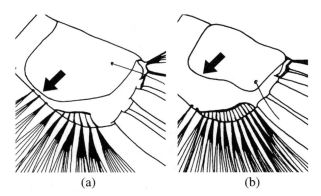

Figure 8.44. Anal segment of:
(a) *Oc. nigrinus*; (b) *Oc. impiger*

30 (28)   Inner frontal seta (5-C) simple. Anal segment with 3 precratal setae (4-X), saddle
           extending at most to half of lateral sides (Fig. 8.45a) . . ***Oc. hungaricus*** (p. 249)
           Inner frontal seta (5-C) with 2 or more branches. If one of the pairs of setae is
           simple, anal segment with 1–2 precratal setae and saddle extending to
           at least 2/3 of lateral sides (atypical *Oc. sticticus*) (Fig. 8.45b) 31
31 (30)   Siphonal tuft (1-S) short, never longer than the width of the siphon at the
           insertion point of 1-S (Fig. 8.46a) . . . . ***Oc. sticticus*** (p. 272)
           Siphonal tuft long, distinctly longer than the width of the siphon at the insertion
           point of 1-S (Fig. 8.46b) 32

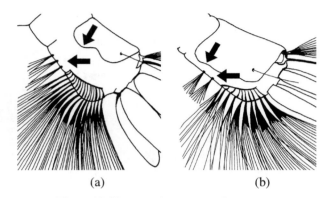

(a)   (b)

Figure 8.45.   Anal segment of:
(a) *Oc. hungaricus*; (b) *Oc. mercurator*

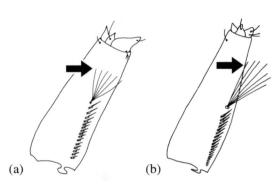

(a)   (b)

Figure 8.46.   Siphon of:
(a) *Oc. sticticus*; (b) *Oc. pullatus*

32 (31)   Prothoracic setae 2-P and 3-P distinctly shorter and thinner than 1-P
           (Fig. 8.47a) . . . . ***Oc. mercurator*** (p. 257)
           Prothoracic setae 2-P and 3-P nearly as long and strong as 1-P
           (Fig. 8.47b) . . . . ***Oc. pullatus*** (p. 265)
33 (18)   Number of comb scales 6–12, rarely 15–17 in *Oc. nigrinus* (Fig. 8.48a) 34
           Number of comb scales 15–45 (Fig. 8.48b) 35

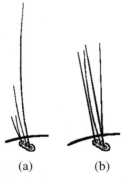

(a)   (b)

Figure 8.47.   Prothoracic setae 1-P to 3-P of:
(a) *Oc. mercurator*; (b) *Oc. pullatus*

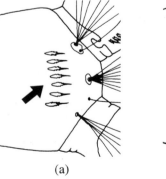
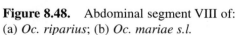

(a)   (b)

Figure 8.48.   Abdominal segment VIII of:
(a) *Oc. riparius*; (b) *Oc. mariae s.l.*

34 (33)   Inner (5-C) and median (6-C) frontal setae with 2–3 branches. Comb scales arranged
           in one row (Fig. 8.49a, b). Siphonal index 3.5–4.0 . . . . ***Oc. riparius*** (p. 270)

Inner and median frontal setae simple (rarely one seta with 2 branches). Comb scales arranged in 2 (rarely 3) rows (Fig. 8.49c, d). Siphonal index never exceeds 3.0, usually 2.0–2.5 .... ***Oc. nigrinus*** (p. 259)

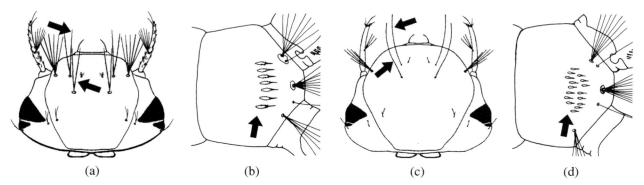

(a)   (b)   (c)   (d)

Figure 8.49. Head and abdominal segment VIII of:
(a, b) *Oc. riparius*; (c, d) *Oc. nigrinus*

35 (33)   Siphonal index less than 2.0. Anal segment with weakly developed saddle, lateral part triangular. Anal papillae very short and spherical (Fig. 8.50a) .... ***Oc. mariae*** and ***Oc. zammitii*** (pp. 255, 257)
Siphonal index more than 2.0. Anal segment with well developed saddle, lateral part more or less rectangular. Anal papillae shorter or longer than the saddle, tapering (Fig. 8.50b) 36

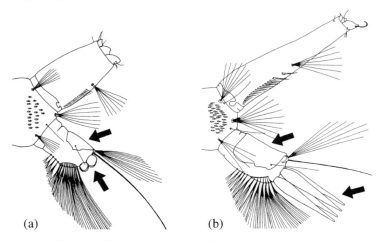

(a)   (b)

Figure 8.50. Larva of: (a) *Oc. mariae s.l.*; (b) *Oc. excrucians*

36 (35)   Number of comb scales less than 20. Siphonal tuft (1-S) short, inserted distinctly beyond the middle of siphon (Fig. 8.51a) .... ***Oc. euedes*** (p. 241)
Number of comb scales more than 20. Siphonal tuft (1-S) usually long, inserted at about the middle of siphon. (Fig. 8.51b) 37

37 (36)   Seta 9-S on posterolateral flap of stigmal plate strong and hooked or curved (Fig. 8.52a) 38
Seta 9-S on posterolateral flap of stigmal plate relatively weak, slightly curved (Fig. 8.52b) 40

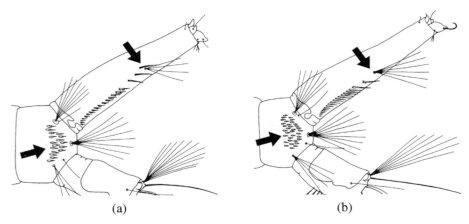

Figure 8.51.   Larva of:
(a) *Oc. euedes*; (b) *Oc. excrucians*

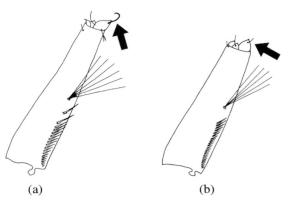

Figure 8.52.   Siphon of:
(a) *Oc. excrucians*; (b) *Oc. mercurator*

38 (37)   Postclypeal seta (4-C) with 6–8 branches. Comb with less than 30 scales
(Fig. 8.53a, b) . . . . *Oc. behningi* (p. 221)
Postclypeal seta (4-C) with 2–3 short, thin branches. Comb with more than 30 scales
(Fig. 8.53c, d) 39

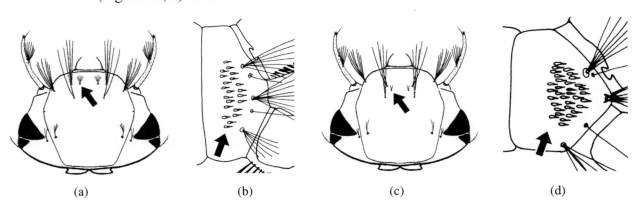

Figure 8.53.   Head and abdominal segment VIII of:
(a, b) *Oc. behningi*; (c, d) *Oc. surcoufi*

39 (38) Abdominal seta 6 with 2 branches on segments I and II, simple on
segments III to VI (Fig. 8.54a) . . . . ***Oc. excrucians*** (p. 243)
Abdominal seta 6 with 2 branches on segments I to VI
(Fig. 8.54b) . . . . ***Oc. surcoufi*** (p. 244)

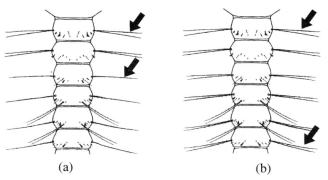

(a)   (b)

Figure 8.54.   Abdominal segments I–VI of:
(a) *Oc. excrucians*; (b) *Oc. surcoufi*

40 (37) Saddle seta (1-X) short, distinctly shorter than the saddle. Siphonal tuft (1-S)
distinctly longer than the width of the siphon at the point of its insertion
(Fig. 8.55a) . . . . ***Oc. mercurator*** (p. 257)
Saddle seta long, about as long as the saddle. Siphonal tuft (1-S) about as long
as the width of the siphon at the point of its insertion (Fig. 8.55b) 41

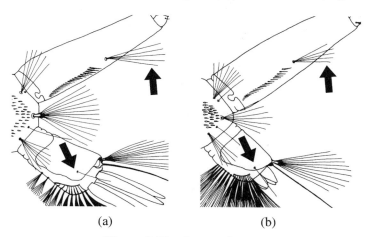

(a)   (b)

Figure 8.55.   Larva of:
(a) *Oc. mercurator*; (b) *Oc. flavescens*

41 (40) Anal papillae short, about half as long as the saddle. Siphonal index more
than 3.0 (Fig. 8.56a) . . . . ***Oc. flavescens*** (p. 245)
Anal papillae long, about as long as the saddle or longer. Siphonal index usually
less than 3.0 (Fig. 8.56b) 42

42 (41) Ventral brush with 4–6 precratal setae (4-X) and 15–20 cratal setae (4-X)
(Fig. 8.57a) . . . . ***Oc. cantans*** (p. 225)
Ventral brush with 6–10 precratal setae (4-X) and 10–15 cratal setae (4-X)
(Fig. 8.57b) . . . . ***Oc. annulipes*** (p. 219)

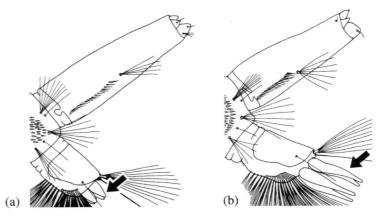

Figure 8.56.   Larva of:
(a) *Oc. flavescens*; (b) *Oc. cantans*

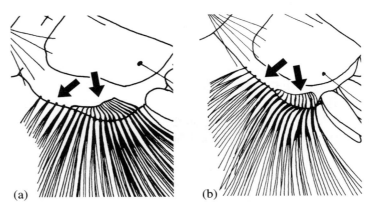

Figure 8.57.   Anal segment of:
(a) *Oc. cantans*; (b) *Oc. annulipes*

43 (8)   Numerous stellate setae on thorax and abdomen. Antennal seta (1-A) simple, pecten teeth long and spine-like (Fig. 8.58a, b) 44
Stellate setae absent. Antennal seta with 3–4 short branches, pecten teeth short, not spine-like, with a broad base (Fig. 8.58c, d) . . . . . . ***Oc. pulcritarsis*** (p. 264)

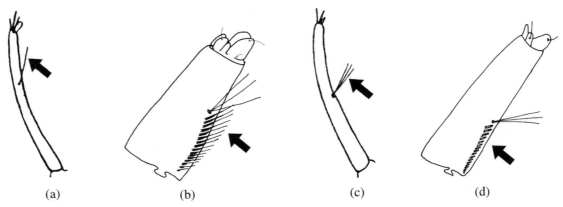

Figure 8.58.   Antenna and siphon of:
(a, b) *Oc. echinus*; (c, d) *Oc. pulcritarsis*

44 (43)   Branches of stellate setae on abdominal segment I longer than length of the segment. Pecten at least half as long as the siphon (Fig. 8.59a, b) . . . . . . ***Oc. echinus*** (p. 214)
Branches of stellate setae on abdominal segment I about the same length as the segment. Pecten 1/4–2/5 as long as the siphon (Fig. 8.59c, d) . . . . ***Oc. geniculatus*** (p. 216)

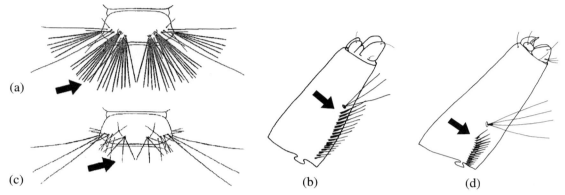

Figure 8.59.   Abdominal segment I and siphon of: (a, b) *Oc. echinus*; (c, d) *Oc. geniculatus*

45 (7)   Outer frontal seta (7-C) simple (Fig. 8.60a) 46
Outer frontal seta (7-C) usually with 2 or more branches, rarely simple (Fig. 8.60b) 47

46 (45)   In addition to the siphonal tuft (1-S), a simple seta is inserted laterally within the apical third of the siphon (Fig. 8.61a) . . . . ***Ae. cretinus*** (p. 212)
Additional lateral seta absent (Fig. 8.61b) . . . . ***Ae. aegypti*** (p. 207)

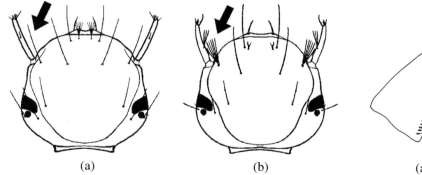

Figure 8.60.   Head of: (a) *Ae. aegypti*; (b) *Ae. vittatus*

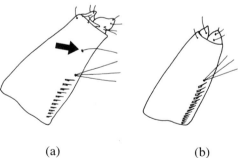

Figure 8.61.   Siphon of: (a) *Ae. cretinus*; (b) *Ae. aegypti*

47 (45)   Antennal seta (1-A) with 2–3 branches. Siphonal tuft (1-S) inserted at about 2/3 the length of the siphon and within the pecten. Distalmost pecten tooth spine-like and apically detached (Fig. 8.62a, b) . . . . ***Ae. vittatus*** (p. 204)

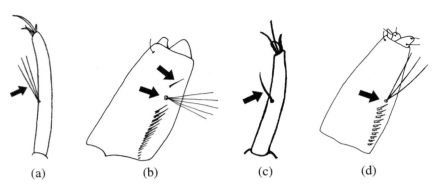

Figure 8.62.   Antenna and siphon of:
(a, b) *Ae. vittatus*; (c, d) *Ae. albopictus*

Antennal seta simple. Siphonal tuft inserted slightly beyond the middle of the siphon, distal to the pecten (Fig. 8.62c, d) . . . . ***Ae. albopictus*** (p. 210)

8.3. GENUS *CULEX*

1   Siphon long and slender, siphonal index 6.5 or more (Fig. 8.63a) 2
           Siphon shorter, siphonal index usually less than 6 (Fig. 8.63b) 7

           Note: The siphonal index in specimens of *Cx. perexiguus, Cx. brumpti* and *Cx. territans* may vary between the above defined groups, consequently they are treated under both branches of the key.

2 (1)   Basal siphonal tuft (1a-S) shorter than the width of siphon at point of its origin (Fig. 8.64a) 3
           Basal siphonal tuft (1a-S) equal or longer than the width of siphon at point of its origin (Fig. 8.64b) 4

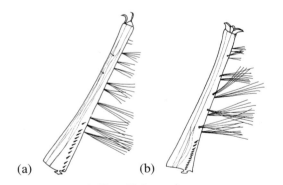

Figure 8.63.   Siphon of:
(a) *Cx. hortensis*; (b) *Cx. mimeticus*

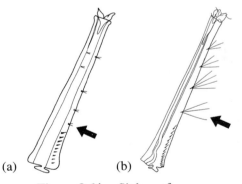

Figure 8.64.   Siphon of:
(a) *Cx. brumpti*; (b) *Cx. territans*

3 (2)   Upper anal seta (2-X) with 2 branches (Fig. 8.65a) . . . . ***Cx. perexiguus*** (p. 294)
           Upper anal seta (2-X) with 3–4 branches (Fig. 8.65b) . . . . . ***Cx. brumpti*** (p. 289)
4 (2)   Prothoracal seta 3-P nearly as long as 1-P, always longer than the half of 1-P. Basal siphonal tuft (1a-S) more or less 3 times longer than the width of siphon at the point of its origin. At least one siphonal tuft clearly inserted within the pecten (Fig. 8.66a, b) . . . . ***Cx. hortensis*** (p. 303)

Prothoracal seta 3-P never exceeding half the length of 1-P. Basal siphonal tuft (1a-S) of variable length, usually not more than twice as long as the width of siphon at the point of its origin. Usually no siphonal tuft inserted within the pecten, rarely one tuft may be inserted close to the last pecten tooth (Fig. 8.66c, d) 5

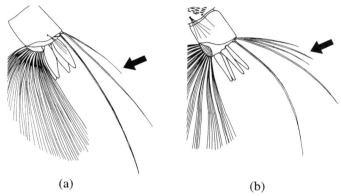

(a)   (b)

Figure 8.65.   Anal segment of:
(a) *Cx. perexiguus*; (b) *Cx. brumpti*

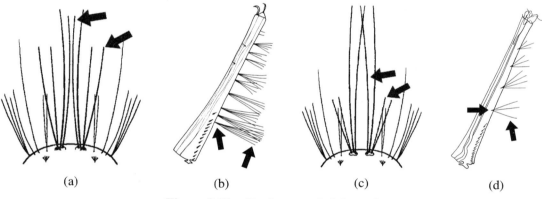

(a)   (b)   (c)   (d)

Figure 8.66.   Prothorax and siphon of:
(a, b) *Cx. hortensis*; (c, d) *Cx. territans*

5 (4) Siphon usually evenly tapering towards the apex. More than 1, usually 2, apical siphonal tufts inserted laterally. Pecten occupying 1/5 of the siphon length. Anal papillae half as long as saddle (Fig. 8.67a) . . . . ***Cx. martinii*** (p. 313) Siphon distinctly widened at the apex. Only 1 apical siphonal tuft inserted laterally. Pecten occupying more than 1/4 of siphon length. Length of anal papillae variable (Fig. 8.67b) 6

6 (5) Anal papillae nearly as long or longer than saddle, pointed (Fig. 8.68a) . . . . ***Cx. territans*** (p. 308) Anal papillae about half as long as saddle, blunt ended (Fig. 8.68b) . . . . ***Cx. impudicus*** (p. 305)

7 (1) Siphon short, siphonal index at most 3.0. Siphonal tufts (1-S) arranged in a zigzag row on ventral side (Fig. 8.69a) . . . . ***Cx. pusillus*** (p. 286) Siphon longer, siphonal index at least 4.0 (Fig. 8.69b) 8

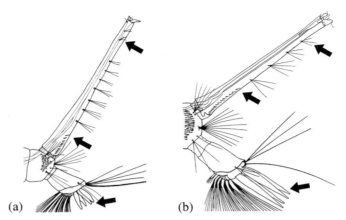

Figure 8.67.   Larva of:
(a) *Cx. martinii*; (b) *Cx. territans*

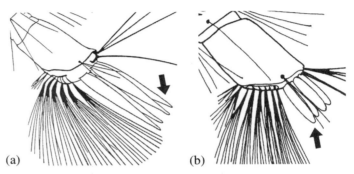

Figure 8.68.   Anal segment of:
(a) *Cx. territans*; (b) *Cx. impudicus*

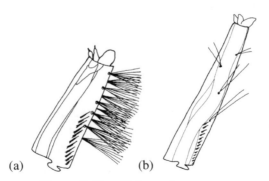

Figure 8.69.   Siphon of:
(a) *Cx. pusillus*; (b) *Cx. p. pipiens*

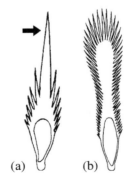

Figure 8.70.   Comb scale of:
(a) *Cx. mimeticus*; (b) *Cx. territans*

8 (7)   Comb scales with a distinct median spine longer than the others (Fig. 8.70a) 9
Comb scales without a distinct median spine. All spines at the apex of more or less the same length (Fig. 8.70b) 10

9 (8)   Subapical setae (2-A, 3-A) inserted at about 2/3 the distance between antennal seta (1-A) and tip of antenna. Main tracheal trunks narrow, less than half as wide as the siphon at its apical third (Fig. 8.71a, b) . . . . *Cx. mimeticus* (p. 292)

Subapical setae (2-A, 3-A) inserted close to the tip of antenna. Main tracheal trunks broad, at least half as wide as the siphon at its apical third (Fig. 8.71c, d) . . . . *Cx. theileri* (p. 300)

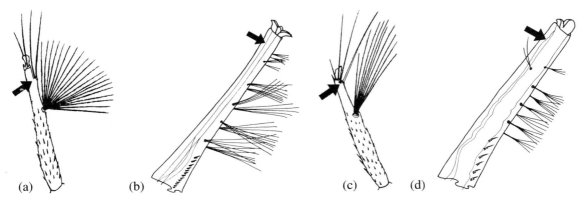

Figure 8.71. Antenna and siphon of:
(a, b) *Cx. mimeticus*; (c, d) *Cx. theileri*

10 (8)   Several basal siphonal tufts not paired, inserted at the ventral surface forming a zigzag row. If paired, inserted very close together, near the ventral midline (Fig. 8.72a) 11
All siphonal tufts paired. The pairs more widely separated, inserted ventrolaterally or laterally (Fig. 8.72b) 12

11 (10)   All siphonal tufts (1-S) arranged in a ventral zigzag row. The length of the tufts suddenly drops towards the apex of the siphon. Usually one tuft inserted within the pecten. Saddle seta (1-X) 2–3 branched (Fig. 8.73a) . . . . *Cx. modestus* (p. 284)
Not all siphonal tufts arranged in a ventral zigzag row, penultimate tuft arising from the lateral surface of siphon. The length of the ventral tufts gradually decreases towards the apex of the siphon. Usually 3 tufts inserted within the pecten. Saddle seta (1-X) simple, sometimes 2-branched (Fig. 8.73b) . . . . *Cx. laticinctus* (p. 290)

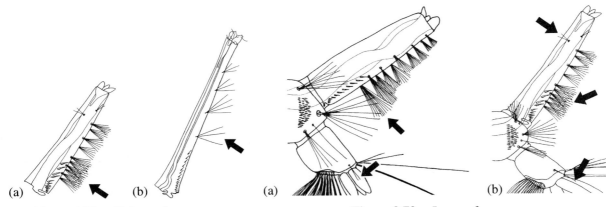

Figure 8.72. Siphon of:
(a) *Cx. laticinctus*; (b) *Cx. territans*

Figure 8.73. Larva of:
(a) *Cx. modestus*; (b) *Cx. laticinctus*

12 (10) All siphonal tufts (1-S) shorter than or equal to width of the siphon at point of 1-S insertion (Fig. 8.74a) 13

At least some siphonal tufts distinctly longer than the width of the siphon at point of 1-S insertion (Fig. 8.74b) 14

13 (12) Upper anal seta (2-X) with 2 branches (Fig. 8.75a) . . . . ***Cx. perexiguus*** (p. 294)

Upper anal seta (2-X) with 3–4 branches (Fig. 8.75b) . . . . . ***Cx. brumpti*** (p. 289)

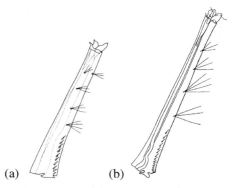

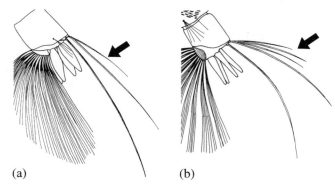

(a)   (b)   (a)   (b)

Figure 8.74. Siphon of: **Figure 8.75.** Anal segment of:
(a) *Cx. perexiguus*; (b) *Cx. territans*   (a) *Cx. perexiguus*; (b) *Cx. brumpti*

14 (12) All siphonal tufts longer than the width of the siphon at the point of their insertion. Basalmost tuft usually inserted apart from the last pecten tooth (Fig. 8.76a) . . . . ***Cx. territans*** (p. 308)

Apicalmost siphonal tuft as long as or shorter than the width of the siphon at its point of insertion. Basalmost tuft usually inserted close to the last pecten tooth (Fig. 8.76b) 15

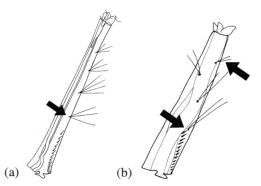

(a)   (b)

Figure 8.76. Siphon of:
(a) *Cx. territans*; (b) *Cx. p. pipiens*

15 (14) Seta 1-T longer than half of the length of 2-T. Seta 1 on abdominal segments III–V (1-III to 1-V) usually with 4 or 5 branches (sum of the branches on one side usually greater than 10). Saddle seta (1-X) usually with 2 branches (Fig. 8.77a, b) . . . . ***Cx. torrentium*** (p. 299)

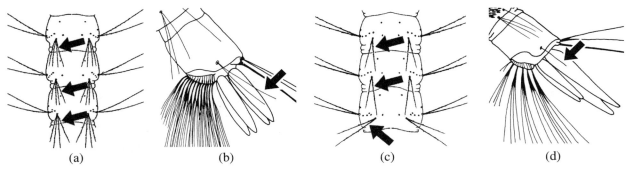

Figure 8.77. Abdominal segments III–V and anal segment of:
(a, b) *Cx. torrentium*; (c, d) *Cx. p. pipiens*

Seta 1-T shorter than half of the length of 2-T. Seta 1 on abdominal segments III–V usually with 1 or 2 branches (sum of the branches on one side usually 6 or less). Saddle seta (1-X) usually simple (Fig. 8.77c, d) . . . . **Cx. p. pipiens** (p. 296)

8.4. GENUS *CULISETA*

1   Antenna shorter than the head, seta 1-A weakly developed. Siphon short, siphonal index less than 4.0 (Fig. 8.78a, b) 2
     Antenna longer than the head, seta 1-A well developed. Siphon long and slender, siphonal index more than 5.0 (Fig. 8.78c, d) 7

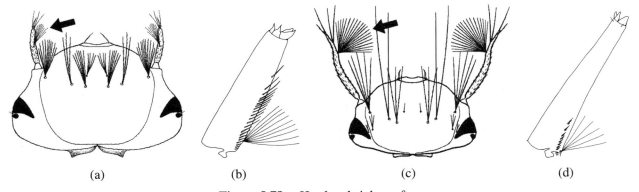

Figure 8.78. Head and siphon of:
(a, b) *Cs. annulata*; (c, d) *Cs. morsitans*

2 (1)   Inner (5-C) and median (6-C) frontal setae simple. Siphonal index 2.0 or less. Saddle plate shaped, not surrounding anal segment (Fig. 8.79a, b) . **Cs. longiareolata** (p. 311)
        Inner and median frontal setae multiple-branched. Siphonal index more than 2.0. Saddle completely surrounding anal segment (Fig. 8.79c, d) 3

3 (2)   Antenna less than half as long as the head. Median frontal seta (6-C) with 1–3 branches (Fig. 8.80a). Number of comb scales usually less than 50 4
        Antenna at least half as long as the head. Median frontal seta with 4–8 branches (Fig. 8.80b). Number of comb scales usually more than 60 6

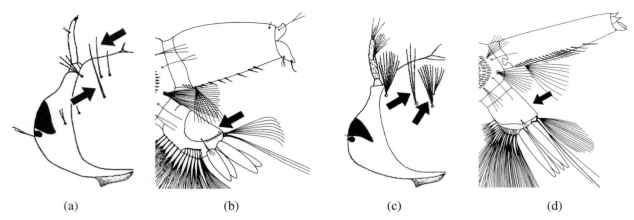

Figure 8.79. Head and abdomen of:
(a, b) *Cs. longiareolata*; (c, d) *Cs. annulata*

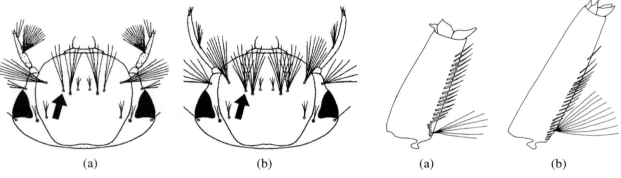

Figure 8.80. Head of:
(a) *Cs. alaskaensis*; (b) *Cs. glaphyroptera*

Figure 8.81. Siphon of:
(a) *Cs. alaskaensis*; (b) *Cs. annulata*

4 (3)   Siphonal index less than 3.0, siphon slightly tapering apically.
(Fig. 8.81a) . . . . ***Cs. alaskaensis*** (p. 322)
Siphonal index 3.0–4.0, siphon distinctly tapering apically (Fig. 8.81b) 5

5 (4)   Distance between postclypeal setae (4-C) equal to or longer than distance between
inner frontal setae (5-C) (Fig. 8.82a) . . . . ***Cs. annulata*** (p. 324)
Distance between postclypeal setae distinctly shorter than distance between inner
frontal setae (Fig. 8.82b) . . . . ***Cs. subochrea*** (p. 330)

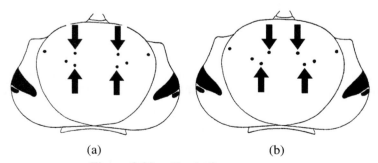

Figure 8.82. Head of:
(a) *Cs. annulata*; (b) *Cs. subochrea*

6 (3)   Antenna 2/3 as long as the head. Pecten spread over 3/4 the length of siphon. Ventral brush with 5 precratal setae (4-X) (Fig. 8.83a, b) . . . . ***Cs. glaphyroptera*** (p. 325)
Antenna half as long as the head. Pecten spread over 2/3 the length of siphon. Ventral brush with 3–4 precratal setae (4-X) (Fig. 8.83c, d) . . . . ***Cs. bergrothi*** (p. 326)

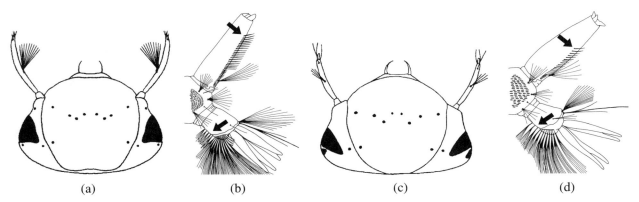

(a) (b) (c) (d)

Figure 8.83. Head and abdomen of:
(a, b) *Cs. glaphyroptera*; (c, d) *Cs. bergrothi*

7 (1)   In addition to the typical pecten teeth, the siphon bears spine-like setae irregularly scattered on its ventrolateral surface (Fig. 8.84a) . . . . ***Cs. fumipennis*** (p. 313)
Siphon with typical pecten teeth only (distal 1–2 teeth may be detached and atypical, spine-like in *Cs. ochroptera*) (Fig. 8.84b) 8

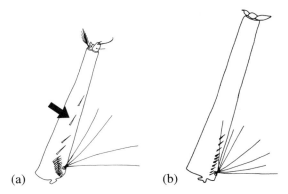

(a) (b)

Figure 8.84. Siphon of:
(a) *Cs. fumipennis*; (b) *Cs. ochroptera*

8 (7)   Inner frontal seta (5-C) with 5–9 branches. Anal papillae 1.5–2 times longer than the saddle (Fig. 8.85a, b) . . . . ***Cs. ochroptera*** (p. 319)
Inner frontal seta with 2–4 branches. Anal papillae shorter than the saddle (Fig. 8.85c, d) 9

9 (8)   Pecten confined to basal 1/4 of siphon. Length of siphonal tuft (1-S) usually 1/3 or less than length of siphon (Fig. 8.86a) . . . . ***Cs. morsitans*** (p. 317)

Pecten confined to basal 1/3 of siphon. Length of siphonal tuft usually more than 1/3 length of siphon (Fig. 8.86b) . . . . *Cs. litorea* (p. 315)

Note: Both species show variation and overlapping in this characters and are difficult to distinguish.

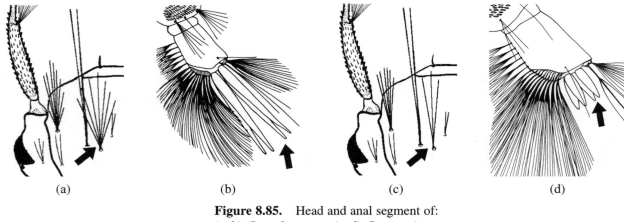

Figure 8.85.   Head and anal segment of:
(a, b) *Cs. ochroptera*; (c, d) *Cs. morsitans*

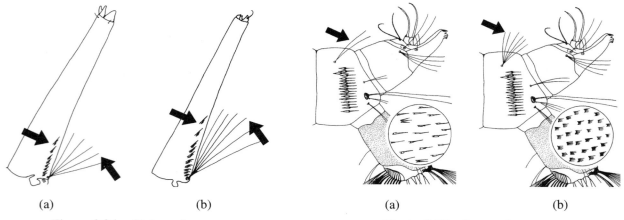

Figure 8.86.   Siphon of:
(a) *Cs. morsitans*; (b) *Cs. litorea*

Figure 8.87.   Larva of:
(a) *Cq. richiardii*; (b) *Cq. buxtoni*

8.5. GENUS *COQUILLETTIDIA*

1   Seta 1-VIII with 2–4 branches. Saddle completely covered with short and stout, usually single spicules, rarely 2–3 grouped on a common base (Fig. 8.87a) . . . . *Cq. richiardii* (p. 333)
Seta 1-VIII with 5–7, usually 6, branches. Spicules on saddle grouped in rows of 2–8 (usually 5–6) on a common base (Fig. 8.87b) . . . . *Cq. buxtoni* (p. 333)

III

Morphology, Ecology and Distribution of European Species

9

Subfamily Anophelinae

In adults of this subfamily, at least the first abdominal segment (tergum I) is devoid of any scales. In general the development of scaling has not reached the same level as in the subfamily Culicinae, often the abdomen is covered with fine setae only. The palps of both sexes are approximately the same length as the proboscis. The larvae have no discernible respiratory siphon and seta 1 of most abdominal segments is usually of the palmate type.

The subfamily comprises only two genera. *Chagasia* Cruz is a small and rare genus; its four species are exclusively distributed in the Neotropical region. Adults of this genus are mainly characterised by the somewhat trilobed scutellum with a set of setae on each lobe and the large claws on the fore and mid legs of the males. The larvae of *Chagasia* have the anterior flap of the spiracular apparatus produced into a long spine-like process and the palmate setae on the abdominal segments are characteristically shaped. Most of the species of the subfamily, including all European species, belong to the genus *Anopheles* Meigen that has a worldwide distribution with more than 400 described species, species complexes, subspecies and varieties. The former genus *Bironella* Theobald of the subfamily, including its three subgenera *Bironella, Brugella* Edwards and *Neobironella* Tenorio was recently synonymised with the genus *Anopheles* Meigen and redefined as an informal group within the subgenus *Anopheles* (Sallum *et al.*, 2000). The seven described species of *Bironella* are confined to the Australian region.

Adult anophelines are usually recognised at once by their attitude when at rest on walls or other objects. The proboscis is held straight out in line with the body axis, not at an angle as in culicine mosquitoes, and the body is strongly tilted downwards to the head end. This causes the abdomen to point away from the surface and the whole body to form an angle with the surface. The majority of species form angles with the surface on which they rest of 30° to 45°, but in some species, such as *An. hyrcanus*, this angle may even approach 90°. The tip of the labium and the palps are usually brought close to the surface rested upon, almost being in contact with it. In addition, adult anophelines generally have longer legs than culicines. In females the palps are elongated, about the same length as the proboscis. They are held closely adjacent to the proboscis (except during feeding), so that the palps and the proboscis appear to be a single organ. When the insect is dead and the tissue somewhat dried, the elongated palps separate and are easily recognised. In males, as in nearly all other mosquitoes, the palps are also elongated. The two apical segments of the palps are swollen and laterally flattened, giving them a club-like appearance. The larvae of anophelines are distinguishable from all other mosquitoes by their feeding behaviour. They usually rest under the water surface in a horizontal position and feed on particles from the surface film. The head can be rotated through 180° toward both sides, so that the ventral side is upward and the mouthparts are in contact with the surface. At the sides of the anterior margin of the thorax are two lobed, reversible organs (notched organs) which can be retracted and support, together with the palmate setae on the abdominal segments, the body when attached in a horizontal position at the surface film.

9.1. GENUS *ANOPHELES* MEIGEN

Members of the genus *Anopheles* are usually medium sized to small mosquitoes with long and slender legs and relatively narrow wings. Their general colouration may be variable from grey, brown or almost black to whitish or pale, but without a metallic shine. The proboscis is straight, long and slender. The palps are usually as long as the proboscis in both sexes, rarely slightly shorter in some tropical females. The palps of females consist of five slender palpomeres, except in *An. plumbeus* that has a slightly thickened palpomere V. The clypeus is usually longer than it is broad, and triangular in shape. The occiput is covered with erect, forked scales. The scutum is slightly convex and not strongly arched, and clothed with setae of various colours. The scutellum is evenly rounded with a more or less regular row of long setae. Prespiracular setae are usually present. Pulvilli are absent, and the tarsal claws of females are without a subbasal tooth. The wings are relatively narrow, often with dark or pale scales forming characteristic spots or evenly dark scaled and without spots, cross veins usually well separated. The integument of the terga is usually dark, covered with various coloured setae and usually without scales. The abdomen is parallel sided, with a blunt end. The cerci of the females are short, rounded, and inconspicuous. Only one spermatheca is present.

The male hypopygium with a conical gonocoxite, is about twice as long as it is broad, usually without an apical and basal lobe, but with prominent parabasal setae of variable number and shape in place of a basal lobe. The gonostylus is approximately as long as the gonocoxite, curved, with the apical spine of the gonostylus short and blunt ended. The claspettes are divided into two or more lobes, each lobe bearing long, various shaped setae at its apex. The aedeagus is tubular, bare or with apical aedeagal leaflets of various shape.

The head of the larvae is as long as or longer than it is broad, and the antennae are shorter than the head, and rod-shaped. Antennal seta (1-A) is small and simple or more prominent and multiple-branched. Two pairs of clypeal setae, the inner (2-C) and outer (3-C), are situated at the anterior margin of the head. The frontal setae (5–7 C) are of pinnate or plumose type, except in *An. plumbeus* that has short and simple frontal setae. Most setae of the thorax are usually plumose, varying in length, and arranged in three sets, corresponding to the thoracic segments. Seta 1 of abdominal segments I–VII is usually of the palmate type, it is often rudimentary on segments I and II, and well developed on segments III–VII. The lateral setae of the first three segments (6-I to 6-III) are long and plumose. Each segment has 1 or 2 dorsal sclerotised plates positioned medially. The spiracular apparatus is situated at the posterior margin of the dorsal side of abdominal segment VIII flush with its surface. On the sides of segment VIII pecten plates are located, each with a posterior row of conspicuous teeth. Both plates are connected by a narrow sclerotised band, bordering the spiracular apparatus posteriorly. The saddle is plate-shaped, and does not encircle the anal segment. The ventral brush consists of a group of fan-like setae, but without precratal setae. The anal papillae are usually blunt ended, and about the length of the anal segment.

The genus *Anopheles* is divided into four subgenera. Members of the two subgenera *Kerteszia* Theobald and *Nyssorhynchus* Blanchard have a mainly neotropical distribution. Only species of the subgenera *Anopheles*

Meigen and *Cellia* Theobald are represented in the Palaearctic region, which includes Europe. The former subgenera *Lophopodomyia* Antunes and *Stethomyia* Theobald were recently synonymised with subgenus *Anopheles* (Sallum *et al.*, 2000).

The eggs of *Anopheles* are commonly laid singly and directly on the water surface, where they lie horizontally; several of them often adhering to one another and forming characteristic configurations of triangles, stars or ribbons due to the surface tension and the shape of the eggs. The egg of anophelines is usually boat shaped with pointed ends and has a flattened or slightly concave upper surface or deck and a more convex lower surface. The frill or fringe surrounds the whole, or parts, of the upper surface. At the sides of the eggs are the floats or air cells, that may interrupt the fringe in the middle. The number of air cells is characteristic for different species. In species of the Anopheles Maculipennis Complex the ornamentation of the upper surface of the egg and the form of the intercostal membrane (surface of the air cells) are an important diagnostic tool for identification.

Larvae of *Anopheles* are usually found in natural water bodies and rarely in artificial collections of water. The preferred breeding places include permanent and semipermanent ponds, pools, puddles or ditches with a growth of emergent aquatic vegetation. They can also be found in rice fields, swamps or margins of rivers and lakes, where the water flow is reduced through shelters. Only a few species show exceptions, namely *An. plumbeus*, which is a typical tree hole breeder, and *An. multicolor*, which exhibits a preference for brackish water.

Although adult females seek their blood-meal from early evening on and then continue to be active throughout the night, they are considered predominantly biting during the crepuscular period after sunset and during dawn (Bates, 1949).

9.1.1. Subgenus *Anopheles* Meigen

The wing is often entirely without pale spots; if pale spots are present (*An. hyrcanus*), the costal margin is covered with one or two of them in its apical half. The cross veins and bifurcations of R_{2+3} and M are dark scaled. In the male genitalia, 1–3, usually 2, parabasal setae are situated at the base of gonocoxite with at least one of them arising from a more or less distinct tubercle. Internal setae are present. The larvae of the subgenus *Anopheles*, except those of *An. plumbeus*, are characterised by the branched antennal seta (1-A) which arises from the inner surface of the antennal shaft, and the inner clypeal setae (2-C) arising close together, their bases often nearly touching. In *An. plumbeus*, seta A-1 is simple and the pair of setae 2-C is more widely separated, the distance between them is nearly the same as the distance between 2-C and the outer clypeal setae (3-C). In all species of the subgenus, the leaflets of the palmate setae are lanceolate, not abruptly narrowed or shouldered, and have a more or less developed terminal filament. The subgenus *Anopheles* is widely distributed and occurs with approximately 150 species in all zoogeographical regions. In Europe, 14 species of the subgenus are recorded with a distribution range throughout the whole region.

Anopheles (Anopheles) algeriensis Theobald 1903

Female: *An. algeriensis* differs from all other European species of the subgenus *Anopheles* which have unspotted wings by the absence of white or cream coloured scales on the vertex, and by the scutum which is unicoloured and may vary from reddish-brown to brown or dark brown, showing no trace of darkening at the sides (Fig. 6.10a). The characteristic median stripe of pale scales on the scutum, that stretches at least along the anterior half in *An. marteri*, is also absent. The proboscis and palps are brown to dark brown, the latter nearly as long as the former. The erect head scales are broad, and dark brown throughout. The scutum and scutellum are brown with blackish setae, and the lobes of the antepronotum are dark brown. The colour of the pleurites is uniformly dark to greyish brown with 2 prespiracular setae, 3 prealar setae, 2 upper mesepisternal setae, and 10 upper mesepimeral setae. The lower mesepisternal setae and lower mesepimeral setae are absent. The legs are uniformly dark, very rarely the hind tarsi show indistinct pale ringing. The wing veins are covered with dark scales that are uniformly distributed. The abdominal terga are mainly dark brown without transverse bands.

Male: The genitalia are characterised by the gonocoxite with only one apically curved parabasal spine-like seta situated on a distinct lobe (Figs 9.1 and 7.8a). Above it another prominent pointed seta, the internal seta, arises from the inner side of the gonocoxite near the apex. The claspettes are 3-lobed, the outer lobe with 2–3 spine-like pointed setae located close together, the middle lobe with 2 flattened setae inwardly curved at apex and the inner lobe with 3 long, thin and pointed setae. The aedeagus has 2–3 pairs of long narrow leaflets.

Larva: The antenna is slightly curved, dark and spiculous throughout its inner surface. The antennal seta (1-A) is short with 5–6 branches arising close to the base of the antenna. The inner clypeal seta (2-C) is long, with a few minute branchlets along the apical half of the shaft. The outer clypeal seta (3-C) is about half as long as the inner clypeals (2-C), with 2–3 branches near the apex. The postclypeal seta (4-C) is simple, rarely with 2 branches at the apex. One of the most suitable features to separate larvae of *An. algeriensis* from those of the closely resembling *An. claviger* is the pattern of dark markings on the head. In *An. algeriensis*, at least three transverse dark bands run across the frontoclypeus (Fig. 8.10a). The most anterior band is located behind the postclypeal setae, the median band behind the frontal setae and the posterior band is situated close to the base of the frontoclypeus. In *An. claviger* the frontoclypeus is dark spotted, but never banded. The palmate setae on abdominal segments III to VII (1-III to 1-VII) each consist of 16–18 leaflets with a weakly developed terminal filament and a conspicuous sawed margin. The sclerotised tergal plate on segment VIII is as broad as or broader than the distance between the palmate setae on segment VII (1-VII). The saddle seta (1-X) arises well within the margin of the saddle.

Biology: The larvae are found in natural or artificial, well shaded, and more or less fresh, water bodies, such as ditches, canals or flooded pools with a rich overgrowth of vegetation (*Phragmites* sp.). They often occur in shaded permanent pools among thick sedge or in swamps and marshes. Due to their tolerance of a slight salinity, they are occasionally found in high numbers in brackish water. They are associated with larvae of *An. maculipennis* s.l. and *Ur. unguiculata*, but rarely with *An. claviger, Cx. hortensis* or *Cx. theileri*. In central Europe, adults of *An. algeriensis* first occur in early summer and the species usually hibernates in the larval stage, but in its southern range in North Africa, both adults and larvae can be found during the winter months. In this area the larvae can also be found in clear, cool, mountain streams or in wells and cisterns. (Senevet and Andarelli, 1956). *An. algeriensis* is regarded to

Figure 9.1. Hypopygium of *An. algeriensis*.

be an exophylic mosquito; the adults rest outside in dense grassy vegetation and readily attack their hosts, man and animals, in the open preferably at dusk and dawn. They rarely enter houses or stables.

Distribution: *An. algeriensis* is widely distributed throughout the Mediterranean region and North Africa. In central Europe, it is recorded from England, Germany, northern France, Hungary and Bulgaria. In the eastern part of Europe there is one record from the west coast of Estonia (Gutsevich *et al.*, 1974). The eastwards extension of *An. algeriensis* includes the costal plain area of Turkey to the northern slopes of the Caucasus and through Middle Asia.

Medical importance: Although the vectorial capacity of *An. algeriensis* is high (*e.g.* it can easily be infected with *P. falciparum* in the laboratory), it is considered as a secondary vector due to its exophily. *An. algeriensis* is rarely encountered in villages and the abundance of its populations, even in the open field, is rarely high.

Anopheles Claviger Complex

It has been shown by Coluzzi (1960, 1962) and Coluzzi *et al.* (1965) that *An. claviger* is a species complex with at least two distinct members, *An. claviger s.s.* (Meigen) and *An. petragnani* Del Vecchio. These two sibling species are distributed in the western Mediterranean subregion and differ distinctly in their larval and pupal morphology and larval and adult behaviour. The question remains unsolved if other described forms, *e.g. An. missiroli* Del Vecchio from Italy and *An. pollutus* Torres Canamares from Spain, so far treated as synonyms of *An. claviger*, are also distinct species or varieties of the nominative form.

Anopheles (Anopheles) claviger s.s. (Meigen) 1804

Female: *An. claviger s.s.* is distinguished from *An. marteri* by its entirely dark palps without pale scales at the tip and from the similar *An. plumbeus* by its decidedly larger size and its

general brownish colouration. The pleurites and lateral parts of the scutum are fawn brown or light brown, whereas *An. plumbeus* is a small mosquito with dark brown or bluish pleurites with the lateral parts of the scutum forming a distinct contrast to the pale scales of the median scutal area. In *An. claviger s.s.* the median pale stripe is always visible, but it is not so distinctly contrasted. In old and dry specimens with an indistinct colouration the different length of the palpomeres may be of more help in the separation of the two species. In *An. claviger s.s.* the last segment of palps is less than half as long as the penultimate segment, whereas in *An. plumbeus* it exceeds half the length of the penultimate segment. *An. claviger s.s.* has a uniformly dark brown proboscis and palps. Its antennae are brown, the vertex has a tuft of whitish and cream coloured scales and the setae are anteriorly directed. The occiput has narrow pale scales. The integument of the scutum is brown, with narrow to moderately broad pale scales forming the median stripe, which is broad and covers more than half of the scutum. The anteacrostichal patch of pale scales is weakly developed (Fig. 6.12b). The scutellum is brown, and darker in the middle with dark setae on its posterior margin. The pleurites of the thorax are brown, and the legs are brown or dark brown with a few pale scales at the articulations. The scales on the wing veins are dark, evenly distributed and do not form dark spots. The abdomen is brown, with indistinct lighter apical bands and long, pale brown setae.

Male (Fig. 9.2): The gonocoxite has 3 parabasal setae, the inner one simple, tapering and arising on a distinct lobe, the two outer setae are located close together and apically branched. The Internal seta is inserted on the inner side of the gonocoxite near the apex, and the gonostylus is relatively thick with a short apical spine. The claspettes are made up of three distinct lobes, the outer lobe carries 2–3 lanceolate blades, the middle one 2 and the inner one 2–3 pointed spines. The aedeagus has 2–3 pairs of leaflets at its apex.

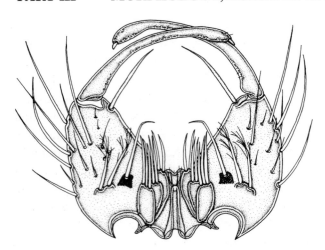

Figure 9.2.   Hypopygium of *An. claviger*.

Larva: The antenna is about half as long as the head with small spicules on the inner surface, decreasing in size towards the tip, and rarely absent from the distal third. The antennal seta (1-A) is short, with 4–7 branches, and located near the base of the antenna. The inner clypeal seta (2-C) is long, nearly as long as the antenna; the setae are situated close together and are simple or sometimes with 2–3 branches at the apex. The outer clypeal seta (3-C) is rarely simple, more often with 2–3 apical branches, and the postclypeal seta (4-C) is short and thin, with 2–4, rarely 5 branches (Fig. 8.11b). The frontal setae (5-C to 7-C) are strongly branched. Dark markings on the frontoclypeus are restricted to spots on the posterior part of it, and not banded as in *An. algeriensis*. The palmate setae on the abdominal segment II (1-II) have 10–15 leaflets. The palmate setae on the abdominal segments III–VII (1-III to 1-VII) have lanceolate leaflets with an elongated apex but without a long filament. The antepalmate setae on segments IV and V (2-IV and 2-V) have 4–5 branches, rarely 3-branched, and if so, the branches are of the same length. The sclerotised tergal plate on the abdominal segment VIII is not as broad as the distance of the palmate setae on segment VII (1-VII). The

saddle seta (1-X) arises just outside the margin of the saddle.

Biology: The larvae are found in a wide variety of habitats, but show a preference for clean water bodies with a more or less permanent character in rather shady situations with cool water, *e.g.* weedy pools sheltered by trees or among growth of reeds at the edges of ponds or lakes, where they are often associated with larvae of *An. atroparvus*, *Cs. annulata*, *Cx. impudicus* and *Cx. theileri*. The larvae are also common in ditches overgrown with weed; in the Mediterranean region they are frequently found in wells and water containers. In mountain areas the larvae breed among the aquatic vegetation of the margins of small well shaded mountain streams. In the northern range of its distribution, the larvae are often found together with *Cs. morsitans* and *Oc. punctor*. The larvae hibernate, and the adults appear in early spring. They can be found during the summer months until late autumn. In southern England maximum biting activity of adult females was observed in May and September showing that *An. claviger* is bivoltine in this area (Service, 1973b). The females deposit their eggs not directly on the water surface, like other *Anopheles* species, but above the water level into the wet soil. They die off before the end of the year. Hibernation takes place in the 3rd or 4th larval stage in water bodies that do not entirely freeze. In the southern range of its distribution the larval development is not interrupted in winter, but is very slow at that time. The larvae are very sensitive to disturbance and promptly descend to the bottom of the breeding sites after sensing the slightest movement of the water. *An. claviger s.s.* is a more exophilic species; the adult females do not readily enter houses or stables, but bite in the open outside villages (exophagy). *An. claviger s.s.* is a zoophilic species, and its preferred hosts are large domestic animals. Sampling carried out in human dwellings and stables revealed that more females of *An. claviger s.s.* fed on humans (3.3%) than was the case for *An. maculipennis s.l.* (2.3%) (Tovornik, 1974). Autogenous populations are reported from Italy (Rioux *et al.*, 1975).

Distribution: The species is widely distributed in the Palaearctic region. Its northern range runs through central Scandinavia, Estonia and the St. Petersburg area. It is found throughout nearly all of Europe from central Sweden and Norway to northern Africa and in the Caucasus and Crimea area. In the mountains of middle Asia, it is found at an altitude of up to 2000 m and its range stretches to Iraq, Iran and Pakistan. In the western Mediterranean area it is sympatric with *An. petragnani*, but less common (Ramos *et al.*, 1978).

Medical importance: *An. claviger s.s.* is a potential vector of malaria. Although its epidemiological importance is not large due to its usually small populations, it was well known as a principal malaria vector in the eastern Mediterranean region (Postiglione *et al.*, 1973).

Note on systematics: *Anopheles claviger* was first described by Meigen in 1804 as *Culex bifurcatus*. After the generic change, the name *Anopheles bifurcatus* was used for a long time, but this usage was inadmissible because *bifurcatus* was a name given earlier by Linnaeus for the males of *Culex pipiens*.

Anopheles (Anopheles) petragnani Del Vecchio 1939

Female: *An. petragnani* as a member of the Anopheles Claviger Complex can only be separated from the nominative form by larval and pupal characters. Coluzzi (1960) reported the adult females to have a darker colouration than *An. claviger* s.s., but this character is of less value for taxonomic separation and the females of the two species are particularly difficult to distinguish between.

Male: In the male hypopygium, as in the adult females, no distinct differences can

be found to distinguish *An. petragnani* from *An. claviger s.s.*

Larva: In *An. petragnani*, the postclypeal setae (4-C) are usually simple, rarely bifid. The palmate setae of abdominal segment II (1-II) usually have more than 15 leaflets with slightly elongated apices (Fig. 8.12b). The antepalmate setae on abdominal segments IV and V (2-IV and 2-V) have 2 or 3 branches, and if 3-branched the median one is shorter than the outer branches. In *An. claviger s.s.*, the postclypeal setae (4-C) are short and thin, with 2–4, rarely 5 branches. The palmate setae on abdominal segment II (1-II) have 10–15 leaflets. The antepalmate setae on segments IV and V (2-IV and 2-V) have 4–5 branches, rarely 3-branched, and if so they are all of the same length.

Pupa: A reliable morphological characteristic, which facilitates the separation of the two species, is found in the length of seta 9 on abdominal segments IV and V (9-IV and 9-V). In all anophelines, seta 9 on abdominal segments III-VII is developed into a short, stout, dark spine arising from the extreme posterior corner of the segment. According to Coluzzi (1960), in *An. claviger s.s.* the spine of abdominal segment IV (9-IV) is less than half as long as the spine of abdominal segment V (9-V), whereas 9-IV is more than half as long as 9-V in *An. petragnani*.

Biology: The larvae of *An. petragnani* are able to tolerate slightly higher water temperatures than *An. claviger s.s.* In Sardinia they are found from February through July and from October to December in freshwater rockholes, ditches, drainage canals, and the edges of streams and rivers preferably in shady situations (Marchi and Munstermann, 1987). Pires *et al.* (1982) reported larvae breeding in southern Portugal along shaded mountain streams with submerged green algae at an altitude of 600 m. The water surface was covered with *Lemna* sp. and the larvae were associated with those of *An. atroparvus* and *Cx. impudicus*. Ribeiro *et al.* (1989) found *An. petragnani* in Portugal from about sea level up to an altitude of 1750 m, and in central Portugal, 87% of its recorded sightings are above 500 m. They also reported a marked host preference of the adult females for livestock such as pigs and cows.

Distribution: *An. petragnani* seems to be restricted to the western Mediterranean subregion, where it is largely sympatric with *An. claviger s.s..* It is the prevalent species of the Anopheles Claviger Complex in southern Italy, Sardinia and Corsica, but has not been recorded from east of the Italian peninsula. Its whole range of distribution has not yet been fully clarified.

Medical importance: *An. petragnani* seems to be a strongly zoophilic species, which often appears in small numbers only. It apparently plays no role in malaria transmission.

Anopheles (Anopheles) hyrcanus (Pallas) 1771

Female: The proboscis is dark brown. The palps are nearly as long as the proboscis, and dark brown with a pale apex and 3 narrow pale bands at the joints of segments II–III, III–IV and IV–V. The clypeus is dark brown with a large tuft of dark scales positioned laterally on each side. The antennae are dark brown, and the basal 5–7 flagellomeres have a few white scales. The vertex has a tuft of anteriorly directed white narrow scales between the eyes. There are white erect scales on the dorsal surface of the occiput and blackish brown upright scales at its sides. The scutum is brown with a stripe of greyish narrow scales in the middle. Often the median stripe is divided by dark longitudinal stripes into two or four narrow greyish stripes. In addition, a few pale scales are present on the anterior margin of the scutum. The postnotum is light brown, and the pleurites are brown. The legs are brown, but lighter on the ventral surface of the femora and the inner surface of the tibiae. The bases of the front femora are distinctly enlarged (Fig. 6.13b). The tarsi are dark brown, and the fore and mid tarsi have white rings at the apices of tarsomeres I–III. Tarsomere IV of the hind tarsi is also pale at the apex (Fig. 6.14a), and entirely pale in var. *pseudopictus* (Fig. 6.14b). Sometimes the white ring extends to parts of the hind tarsomere V. The wing veins are covered with dark and pale scales, forming contrasting dark and pale spots. The dark costal margin of the wing is interrupted by 2 pale areas in the apical half; the more proximal spot includes C and R$_1$, and the spot near the apex extends from C to R$_2$ (Fig. 6.13a). White scales particularly dominate on Cu and A, and sometimes these veins are almost entirely white scaled. The fringe scales of the wing are white at the apex, but otherwise brownish. It has to be mentioned that *An. hyrcanus* exhibits a considerable amount of intraspecific variation in the wing markings. The dark to white area ratio may vary as well as the discreteness and vividness of the spotting. The abdomen is dark with long, dense, brown or golden setae, and sternum VII has a tuft of dark greyish scales at its apical half.

Male: Tergum IX has long lateral lobes which are enlarged at their apices. The gonocoxite has 2 parabasal setae, unequal in length, the outer seta being longer than the inner, and neither is inserted in a sclerotised base (Fig. 9.3). The internal seta is inserted near the middle of the gonocoxite, and its length slightly exceeds that of the tip of the gonocoxite. The outer lobe of the claspettes has setae fused into

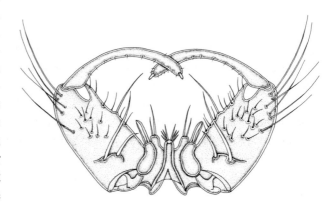

Figure 9.3. Hypopygium of *An. hyrcanus.*

a flattened spatula like process, and the inner lobe has 2 simple pointed setae. The aedeagus has several pairs of leaflets at the apex. These leaflets are delicate, and about 1/3 the length of the aedeagus.

Larva: The larvae of *An. hyrcanus* resemble those of the Anopheles Maculipennis Complex, but can easily be distinguished by the characters given in the keys. The head is longer than it is wide. The antennae are straight with small spiculae located on the inner aspect and a conspicuous antennal seta (1-A), which is multiple-branched (7–8 branches) and at least half as long as the antenna. It arises about midway on the antennal shaft or slightly below it. The inner clypeal setae (2-C) are situated close together, with short apical branches (Fig. 8.13a). The outer clypeal seta (3-C) is multiple-branched, and dendriform. The frontal setae (5-C to 7-C) are long. The palmate setae on abdominal segments I and II are rudimentary, but well developed on segments III–VII, with 17–24 leaflets ending in indistinct terminal filaments. The sclerotised tergal plates on the abdominal segments are not very long, but fairly broad.

Biology: The larvae of *An. hyrcanus* can be usually found in more or less clean, stagnant sun exposed water bodies, rich in aquatic vegetation. They are found especially in rice fields and associated irrigation systems, in swamps and similar locations in the open, *e.g.* pools, margins of lakes or ditches with grassy edges. Elsewhere the larvae breed at the edges of slowly moving waters such as grassy streams and canals or road side ditches. They exhibit a tolerance to a slight degree of salinity and can also be found in coastal and inland marshes. In rice fields in northern Greece the larvae are associated with those of *An. sacharovi* and *Cx. modestus*. From the coastal plains of Turkey, Postiglione *et al.* (1973) reported the larvae in association with those of *An. maculipennis s.l., An. superpictus* and *An. algeriensis*. At a temperature of about 20°C the

larval development of *An. hyrcanus* lasts 14–16 days (Senevet and Andarelli, 1956). The adults occur in small numbers in late April and May but the size of the population increases towards autumn. *An. hyrcanus* produces two to four generations per year. The resting behaviour of the adult females is predominantly exophilic (outdoors), but the degree of exophily is known to differ between places. Usually they rest outdoors in bushes and other dense vegetation during the day time. They rarely enter houses, but are common in cattle sheds or shelters, from where they return into the open after blood-feeding. Livestock, or when these are not available as hosts, humans are readily attacked in the open field at dusk or in the night. Occasionally feeding is observed during the day time in shaded situations. Autogenous populations have been reported from Kirgisia (Rioux *et al.*, 1975).

Distribution: In Europe *An. hyrcanus* is widely distributed throughout the northern Mediterranean countries, from Spain, southern France, Italy, and Greece to Turkey. Its northern range reaches the Pannonean Plain, Ukraine and southern Russia. It is distributed in southern Kazakhstan, the Caucasus and middle Asia. Together with closely related species it is common in South East Asia and its eastern distribution range includes China, Japan and Korea.

Note on systematics: *An. hyrcanus* is one of the most widespread and common species of the genus *Anopheles*, and certainly one of the most variable. It has an enormous range throughout the Palaearctic from the Atlanic in the west to the Pacific in the east and the Oriental region in the south. Due to its variability, a number of different forms have been described from different localities as variations or subspecies of the nominative form. Some of these forms from south-east Asia are now considered as distinct species within the *An. hyrcanus* sibling species group, *e.g. An. sinensis* Wiedemann, *An. nigerrimus* Giles, *An. paeditaeniatus* (Leicester) and others (Reid, 1953;

Harrison, 1972). The Palaearctic variations include the form *mesopotamiae* from western Asia with a lighter, diffuse colour pattern of the wing caused by more pale scales intermixed on the wing veins and a western Palaearctic form *pseudopictus* with tarsomere IV of the hindlegs entirely white, which is mainly found in southern and southeastern Europe. Glick (1992) treated *An. pseudopictus* as a distinct species well separated from *An. hyrcanus* by its sympatric distribution throughout Turkey, Iran and Afghanistan with apparently no evidence of hybridisation. However, Gutsevich (1976) reported a wide variation of *An. hyrcanus* in the extent of the pale ringing on the hind legs as well and found intermediate forms. Because Glick's judgement was solely based on one character of female morphology and no further investigations on the morphology of the male genitalia, the developing stages or cross mating experiments have been carried out so far, the opinion not to treat the European form *pseudopictus* as a distinct species nor a subspecies of *An. hyrcanus* is favoured here.

Medical importance: Because of its exophilic behaviour, *An. hyrcanus* has never been regarded as a dangerous vector of malaria in the Mediterranean region. With regard to the probability of changes in human behaviour (*e.g.* increased mobility of humans or increase in the number of seasonal workers in the rice and cotton fields), its role as a potential malaria vector should not be ignored.

Anopheles Maculipennis Complex

The mosquitoes of the Anopheles Maculipennis Complex are the classical example of a so called "species complex", comprising various sibling species. Before 1925, it was reported that malaria was transmitted by the malaria mosquito "*An. maculipennis*". Further research on the distribution and ecology of this mosquito uncovered considerable irregularities. It was found that the distribution of malaria and the distribution of *An. maculipennis* were not closely correlated. In some areas where individuals of *An. maculipennis* were abundant, the incidence of malaria was low or absent. This situation was characterised as anophelism without malaria (Bates, 1940). Furthermore, differences in the biology and behaviour of various populations were discovered. In some regions the larvae were restricted to fresh water, in others to brackish water; the adult females preferred to feed on humans in some areas and elsewhere they mainly fed on livestock. Exceptional differences in the swarming and mating behaviour (stenogamy, eurygamy) between certain populations were also observed. The first evidence for the existence of a complex of sibling species was brought up by Falleroni (1926), who described distinct morphological differences of the chorion pattern of the eggs in populations with a different biology. Unfortunately, Falleroni's observations laid dormant for five years and were not rediscovered until 1931. Research of Martini *et al.* (1931), Van Thiel (1933) and Hackett and Missiroli (1935) gave more evidence of the presence of a complex and by the end of the 1930s, most of the present species of the *An. maculipennis* complex were recognised (Bates, 1940). Studies using cross breeding experiments, cytotaxonomic methods or enzyme electrophoresis confirmed the existence of the different species within the complex (Stegnii and Kabanova, 1976; Bullini and Coluzzi, 1978; Suzzoni-Blatger *et al.*, 1990). Up to now the complex has comprised of at least a dozen reproductively isolated but morphologically similar species in the northern hemisphere (White, 1978) and it can be expected that more members will be found with the use of the above mentioned advanced techniques of identification.

Once the existence of a complex of sibling species was established, some morphological differences between adults and larvae were also found. The females may be separated by the form and shape of scales on certain wing veins, the males differ in the form and length of the spines on the outer claspette lobe and larvae may be distinguished by the overall number of branches of the four antepalmate setae on abdominal segments IV and V (2-IV and 2-V, four setae together). Unfortunately there exists an intraspecific variation of the characters and overlapping between the species is not uncommon. Therefore it is necessary to investigate a series of specimens with statistical analysis of the results, which is not applicable for individual identification. Thus the morphological identification is still based largely on characters of the egg (markings of the dorsal exochorion, presence of floats and their size, position and texture). Engorged females can be directly sampled into vials, *e.g.* from animal shelters, and allowed to lay their eggs for individual identification.

Identification key for European species of the Anopheles Maculipennis Complex based on characters of the eggs (after White, 1978):

1   Egg without floats (but rudimentary floats may develop at low temperatures), egg surface uniformly pale (frosty-white) from pole to pole (Fig. 9.4a) . . . . *sacharovi*
        Egg with fully developed floats, egg surface dark, barred or mottled 2

2 (1)   Intercostal membranes of floats (surface of the air cells) smooth 3
        Intercostal membranes of floats rough (finely corrugated) 5

3 (2)   Upper surface of egg entirely dark (Fig. 9.4b) . . . . *melanoon*
        Upper surface of egg otherwise, barred or mottled 4

4 (3)   Upper surface of egg softly patterned with wedge-shaped black marks on a pale background, ends of deck pale almost to the tip (Fig. 9.4c) . . . . *atroparvus*
        Upper surface of egg with pattern of 2 transverse dark bars near the ends of floats, poles dark and remainder of the upper egg surface irregularly mottled (Fig. 9.4d) . . . . *subalpinus*

5 (2)   Upper surface of egg richly patterned with wedge-shaped dark marks on frosted pale backround, tip of poles dark and narrow (Fig. 9.4e) . . . . *labranchiae*
        Upper surface of egg marked with two transverse dark bars near the end of the floats, with or without other pattern (mottled background) 6

6 (5)   Transverse dark bars on upper surface forming part of diffuse mottled pattern (Fig. 9.4f) . . . . *messeae*
        Transverse dark bars on upper surface sharply contrasted with unmottled pale background colour 7

7 (6)   Tips of eggs less acutely pointed, chorion of upper egg surface relatively rough, width of egg between floats about 17% of egg length (Fig. 9.4g) . . . . *maculipennis s.s.*
        Tips of eggs more acutely pointed, chorion of upper egg surface relatively smooth, width of egg between floats about 12% of egg length (Fig. 9.4h) . . . *beklemishevi*

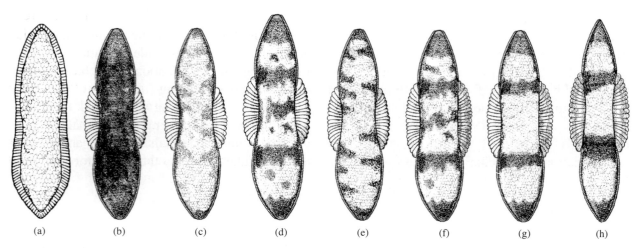

Figure 9.4. Eggs of Anopheles Maculipennis Complex:
(a) *sacharovi*; (b) *melanoon*; (c) *atroparvus*; (d) *subalpinus*; (e) *labranchiae*; (f) *messeae*; (g) *maculipennis s.s.*;
(h) *beklemishevi*

Morphological characteristics of adults and larvae shared between the members of the complex:

Female: Dark or medium brown in colour, although there is a wide variation in colouration and size. Individuals of southern origin are usually lighter and smaller. The characteristic features, which differ from all other European *Anopheles* species are the wings with an aggregation of dark scales forming several distinct spots (Fig. 6.8a). The proboscis is dark brown, and the palps are nearly as long as the proboscis, and of the same colour. The antennae are brown coloured. The vertex has a tuft of long, whitish, anteriorly directed narrow scales and setae. The occiput has erect dark brown scales. The scutum has a broad greyish median stripe tapering anteriorly and usually 2–3 indistinct brownish stripes in its anterior half. The lateral parts of the scutum are brown anteriorly and blackish brown posteriorly (for different colouration patterns of *An. sacharovi* see under species description). The anteachrostichal patch is made of pale, long and thin scales. The scutellum is brown with golden narrow scales, the postnotum is pale brown, and the pleurites are dark brown. The femora are dark brown on the dorsal side and pale brown on the ventral side, the tibiae are brown but slightly paler at their apices, and the tarsi are dark brown. The wings have dark scales unevenly distributed forming several dark spots close to the cross veins, base of R_s and furcations of R_{2+3} and M. The furcations of R_{2+3} and M are situated at about the same distance from the wing base. The fringe of the wing has a patch of pale scales at the apex of the wing (Fig. 6.9b); *An. sacharovi* has dark spots on the wing veins which are more indistinct and the fringe of the wing is uni-coloured dark (Fig. 6.9a). The abdomen is brown or blackish brown with long golden brown narrow scales.

Male (Fig. 9.5): The gonocoxite has 2 simple parabasal setae positioned on small, but distinct tubercles, and the setae are unequal in length, with the outer one being longer. The internal seta is inserted near the middle of the gonocoxite. The gonostylus is well developed,

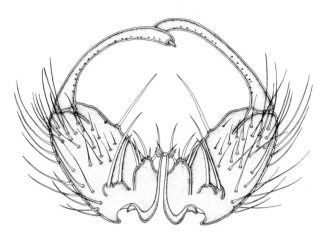

Figure 9.5. Hypopygium of *An. maculipennis s.l.*

and is longer than the gonocoxite, with a short apical spine. The claspette lobes are not well defined, and have spine-like setae of variable length and shape situated close together but not fused. The aedeagus is long and narrow with leaflets at the apex.

Larva: Very variable in pigmentation and size acording to their different habitats. Larvae from the northern parts of Europe are usually larger with a darker head capsule. The head is longer than it is wide, and the antenna is nearly straight, sparsely spiculous, and about 2/3 as long as the head. The antennal seta (1-A) is small, with 4–6 branches arising at the basal 1/3 to 1/4 of the antennal shaft (Fig. 8.13b). The inner clypeal setae (2-C) are situated close together, with long apical branches. The outer clypeal seta (3-C) is dendriform (Fig. 8.9b). The frontal setae (5-C to 7-C) are long and plumose. The palmate setae on abdominal segments I and II are rudimentary, but well developed on segments III–VII, with 16–24 leaflets which are slightly wider in the middle. The terminal filament is nearly 1/3 as long as the leaflet.

Anopheles (Anopheles) atroparvus
Van Thiel 1927

Biology: The larvae can be found in a variety of stagnant, semi-permanent or permanent, clean breeding sites, both in saline and fresh water, but they show a slight preference for brackish water. They may occur in canals, ditches, marshes in coastal areas, river margins, pools in river beds, rice fields and are sometimes even found in septic tanks. The water bodies are usually sun exposed and carry a considerable amount of filamentous green algae and other floating and submerged vegetation. In their southern distribution range the larvae are often found together with those of *Cx. theileri*, *Cx. impudicus* and *Cx. pipiens* (Ramos *et al.*, 1978). *An. atroparvus* hibernates as the adult female and usually shows incomplete diapause. After seeking shelter in stables or dwellings in autumn, the females remain active during winter time and may irregularly take blood-meals without subsequent oviposition. This habit mainly contributed to the indoor transmission of winter malaria in Great Britain, Netherlands and other parts of Europe at the beginning of the 20th century. The problem disappeared in the late 1940s, due to improved social and economic conditions. The duration of the diapause depends on day length as well as on temperature and thus varies with the latitude of the distribution of *An. atroparvus*. It may last from September until April in northern Europe or from November until February in southern Europe. The females are mainly zoophilic and prefer different domestic animals according to their availability in different areas, but also readily feed on humans. They usually rest indoors, predominantly in stables and man-made shelters. Although swarming of the males before mating has been observed on several occasions, it is considered to be a vestigial characteristic, playing only a minor part in the actual mating behaviour of the sexes (Cambournac and Hill, 1940). Usually the adults of *An. atroparvus* do not need to swarm before mating (stenogamy) and mate almost entirely indoors. Flight ranges of *An. atroparvus* females of at least 3 km have been reported (Cambournac and Hill, 1938).

Distribution: It is largely a littoral species occuring from southeastern Sweden to Portugal along the coasts of the Atlantic, Baltic and Mediterranean Sea. In southern and southeastern Europe it has a patchy distribution, *e.g.* from northern Italy and inland areas of central Italy through southwestern parts of Russia and the coastal area of the Black Sea. In Serbia and Macedonia it is widespread in the lowlands, but conspicuously dominant only in areas with saline/alkaline soils in the Pannonean Plain (Adamovic, 1980). In Portugal it is the most common, most abundant and most widely distributed *Anopheles* species throughout the country (Ribeiro *et al.*, 1988). *An. atroparvus* was more common in the past, but has a diminished distribution today. In the Netherlands, industrial pollution is considered to be one reason for the decrease of the species abundance (Jetten and Takken, 1994).

Anopheles (Anopheles) beklemishevi Stegnii and Kabanova 1976

Biology: The larvae may occur in breeding sites quite typical for *An. messeae*. Stegnii and Kabanova (1976) reported a finding of larvae in a pond that was heavily polluted with organic material and aquatic vegetation was absent. The larvae could be found only along the edges of the pond. They seem to be more cold tolerant than larvae of *An. messeae*, and thus show an adaptation to the continental climate. So far little is known about the biology and the biting behaviour of the females. Many characteristics of this species seem to be similar to those of *An. maculipennis s.s.*, *e.g.* its zoophily (preference for feeding on animals). The adults usually rest indoors. In northern Sweden, Jaenson *et al.* (1986a) collected females from livestock shelters. *An. beklemishevi* was found resting in cattle sheds, horse stables and pigsties without showing a preference for a particular species of domestic animal. It is an eurygamous species, *e.g.* swarming is needed before mating and undergoes complete

winter diapause of a duration that is inversely correlated with photoperiod (White, 1978).

Distribution: *An. beklemishevi* is endemic to the cooler highlands and northern latitudes of Russia and the Baltics (White, 1978). It can be found in Siberia, northern Sweden and Finland (Korvenkontio *et al.*, 1979; Jaenson *et al.*, 1986a) and is also recorded in several localities in southern Finland (Utrio, 1979).

Note on systematics: *An. beklemishevi* was the first mosquito species ever to be named and described primarily from cytogenetic evidence, rather than from a morphological description (White, 1978).

Anopheles (Anopheles) labranchiae Falleroni 1926

Biology: Larvae can be found in sunlit, stagnant fresh and brackish water containing horizontal vegetation in coastal areas in Europe and in some inland regions (Jetten and Takken, 1994). They may occur in coastal marshes, sheltered edges of flowing streams, rice fields, grassy pools and a number of similar breeding sites. In Europe, larvae of *An. labranchiae* are usually associated with brackish water, where the salinity of the breeding sites can reach 10 g/l (Weyer, 1939). In Sardinia, where other members of the complex are virtually absent, the larvae are almost entirely limited to fresh water breeding sites. On the island they can be found in ponds, ground-flooded pools, streams, ditches, drainages or canals (Aitken, 1954; Marchi and Munstermann, 1987). *An. labranchiae* larvae are better adapted to warmer waters than those of *An. atroparvus*. Hibernation takes place in the adult female stage in dark sheltered situations, *e.g.* in stables and/or natural cavities. Diapause may be complete or incomplete, *e.g.* occasional blood-feeding is possible during the winter months. The females principally feed on humans and they persistently try to enter bedrooms at night (Hackett and Missiroli, 1935), but domestic animals may also serve as hosts. *An. labranchiae* is predominantly an endophilic species; indoor resting sites include human dwellings, stables, animal shelters, *etc.* Occasionally they may be found resting outdoors in various natural shelters, such as tree cavities. Swarming before mating was reported by Horsfall (1955), but in the laboratory adults were found to mate in small cages of $50 \times 50 \times 100$ cm (Hackett and Bates, 1938). The flight range of the species is considered to be 2–5 km (Senevet and Andarelli, 1956).

Distribution: *An. labranchiae* has a restricted distribution in southern and southeastern Europe. It has been reported in the south east of Spain, Corsica, coastal areas of Italy, Sardinia, Sicily and the Dalmatian coast. In north Africa the species occurs in Morocco, Algeria and Tunisia.

Anopheles (Anopheles) maculipennis s.s. Meigen 1818

Biology: The larvae mainly occur in cold clear waters of upland areas, but they can also be found in plains and coastal zones (Hackett and Missiroli, 1935). Typical breeding sites are sheltered areas in running streams, river edges, rice fields or artificial pools. In mountainous regions in central Europe, the species can be found at altitudes of 1000 m and more, and occur there as the only member of the complex. Altitudes of 2190 m and up to 2300 m are reported from Bulgaria and Turkey (Postiglione *et al.*, 1973; Bozkov, 1966). *An. maculipennis s.s.* shows a better adaptation to breeding in small water bodies without vegetation and artifical collections of water with a wide fluctuation of the daily temperature than *An. messeae* (Mohrig, 1969). Therefore, this species seems to be more suited to survival in cultivated land that was widely created in Europe in the past decades through water management campaigns and the use of extensive agriculture techniques. The larvae can frequently be found in waters with a high content of organic matter together

with those of *Cx. p. pipiens*. Hibernation of adult females is complete, but it can be of relative short duration in warmer climates (Senevet and Andarelli, 1956). The winter months are usually passed in abandoned buildings without a potential host. The adults normally have little contact with humans; the females are considered to be strongly zoophilic feeding mainly on cattle (Weyer, 1939), but also on pigs and chicken (Tovornik, 1980). Nonetheless, in case of a shortage of livestock, *An. maculipennis s.s.* may feed on humans as well, both outdoors and indoors (Barber and Rice, 1935). It is usually an endophilic species; the daytime resting sites are stables and dwellings. The adults are eurygamous; they need to swarm before mating.

Distribution: *An. maculipennis s.s.* is widely distributed throughout Europe. With the exception of the southern parts of the Iberian peninsula it is recorded in nearly every European country. The range extends eastwards to southwest Asia and the Persian Gulf. It is considered to be a more continental species with demands for humidity considerably lower than those of *An. messae* and *An. atroparvus*.

Anopheles (Anopheles) melanoon Hackett 1934

Biology: Larvae of *An. melanoon* are typically found in sunlit stagnant water bodies with a large surface area and some vegetation (Jetten and Takken, 1994). They may occur in marshes, edges of rivers and lakes, ponds, and ground pools. In Sardinia, larvae can be found in fresh-water ground pools, streams and water holes with a sunny exposure in the springtime and more shady situations during the summer months (Aitken, 1954). Elsewhere the species is reported to breed in rice fields (Hackett and Missiroli, 1935). Hibernation takes place in the adult female stage with complete diapause. The winter quarters are not well known, but may be the same as in other members of the complex, *e.g. An. maculipennis s.s.*, *An. messeae* and *An. subalpinus*. *An. melanoon* females reveal a zoophilic preference; they mainly feed on cattle and have only occasionally been recorded as feeding on humans, both indoors and outdoors. Cattle stables are reported as daytime resting shelters (Ribeiro *et al.*, 1980), but others described the species as semi-exophilic (Artemiev, 1980). The adults are eurygamous. *An. melanoon* is a rare species with a limited distribution range. So far, it has never been found in the same numbers as other members of the complex.

Distribution: *An. melanoon* seems to be confined to southwestern and southern Europe. It has been found in Portugal and Spain (one record each), Corsica, Sardinia, and Italy. More eastern records probably refer to *An. subalpinus*.

Anopheles (Anopheles) messeae
Falleroni 1926

Biology: The larvae are typically found in cool, fresh, stagnant water bodies with an abundant growth of submerged vegetation. They occur at the edges of rivers and lakes, in swamps, flood plains, ponds and ditches. In central Europe the larvae are restricted to inland areas and fresh water habitats and avoid breeding sites with a high content of organic matter. *An. messeae* prefers larger water bodies which are available mainly in lowlands with poorly regulated ground water level (Mohrig, 1969). It is the predominant species of the Anopheles Maculipennis Complex in inundation areas and flood plains of larger rivers, *e.g.* Danube, Save, Rhone or Rhine rivers. It is rare or almost completely absent along the coastlines and in mountainous regions. Hibernation takes place in the adult female stage; the diapause is complete. The winter months are usually passed in abandoned buildings. The females are essentially zoophilic species feeding almost exclusively on domestic animals, thus contact with humans is largely suppressed in an agricultural area with much livestock (Jetten and Takken, 1994). Blood-meals are taken from humans only when the density of

An. messeae is very high and there is a shortage of livestock, but they may also attack humans in houses (Barber and Rice, 1935; Dahl, 1977). Artemiev (1980) described the species as endophilic as it was found resting during the daytime in stables, barns and cellars as well as in human buildings. The adults are eurygamous.

Distribution: *An. messeae* is the most widespread member of the complex. Its range stretches in the northern Palaearctic region from the Atlantic coast in the west, through Scandinavia, across northern and central Europe and Asia as far as into China. The species is virtually absent in southern Europe, it cannot be found on the Iberian peninsula, southern Italy and in the eastern Mediterranean region. *An. messeae* is regarded as being more susceptible to high temperatures and low humidities than *An. atroparvus* and this behaviour might limit the southern distribution of the species.

Anopheles (Anopheles) sacharovi
Favre 1903

Adults of this species are most readily distinguished from the other members of the complex by the lighter colouration of the mesonotum and some wing characters. The pale median stripe on the scutum, characteristic of the other members of the complex is absent. The lateral parts of the scutum are yellowish brown, more or less the same as in the middle part. The scales of the fringe of the wing are unicoloured dark, without a pale patch at the wing apex (Fig. 6.9a). The dark spots on the wings, particularly in males, are less conspicuous than in the other members of the complex and barely distinguishable in old worn out specimens. The eggs of *An. sacharovi* are unique in the lack of the air floats (Fig. 9.4a), but in the south of its distribution range, where development continues throughout the year, eggs deposited in winter may have rudimentary floats. Usually the larvae of *An. sacharovi* are smaller than the larvae of the other members of the complex, and the outer clypeal setae are relatively longer.

Biology: Larvae can be generally found in open, sun exposed shallow water bodies with abundant surface vegetation, both in fresh and saline waters. They are more tolerant of salinity than other members of the complex (*e.g. An. maculipennis s.s.*), and usually occur in coastal swamps and marshes, lagoons and nearby streams, irrigation, drainage and roadside ditches, rice fields, grassy ponds, pools or seepages. In rice fields the larvae are often found together with those of *Cx. modestus*. The larvae are not very mobile and rarely leave the water surface, and if so, they return after a short time (Gutsevich and Dubitzky, 1987). *An. sacharovi* is the most thermophilic species of the Anopheles Maculipennis Complex, and water temperatures of the breeding sites up to 39°C during daytime are not uncommon. Hibernation takes place in the adult female stage, and the individuals preferably seek stables and other animal shelters for overwintering. When the temperatures are suitable, they may take blood-meals occasionally during winter. Hibernation starts earlier and lasts longer than in other members of the complex (Martini, 1931). In spring, adults are not very abundant; the population maximum is usually reached between July and August. The females predominantly feed on humans but also on cattle when available. Even during the daytime they may be persistent biters in sheltered situations (Postiglione *et al.*, 1973). In some places they become very numerous and attack humans in large numbers (Gutsevich and Dubitzky, 1987). *An. sacharovi* is usually an endophilic species; adults are frequently found in stables and human dwellings during the day, leaving their daytime shelters during the hours of darkness. Occasionally they can also be found resting outdoors; recorded outdoor shelters include hollows and cavities under bridges and earth banks, hollow trees or rock cavities (Postiglione *et al.*, 1973). Saliternik (1957) caught marked mosquitoes at a distance of 3.5 km from the release point. The flight range of *An. sacharovi* may exceed this distance, if the blood source is far from the breeding places.

Distribution: In southern Europe, *An. sacharovi* is mainly a coastal species and distributed in the eastern Mediterranean subregion. It can be found in Corsica, Sardinia, Sicily, in coastal areas of Italy, Greece, former Yugoslavia and Albania, south of the Balkan peninsula and in western and southern coastal plains of Turkey. The range extends eastwards from the Near East through middle Asia, Iran, and Afghanistan to China. In its continental distribution range, *An. sacharovi* is a characteristic species for areas with a dry and hot climate.

Anopheles (Anopheles) subalpinus Hackett and Lewis 1935

Biology: Larvae can be found in various types of sunlit, stagnant water bodies where aquatic vegetation is present. They occur in swamps, edges of lakes, flooded rice fields or ponds and avoid more shaded places and small water collections. Regular breeding sites can be found up to an altitude of about 1200 m (Postiglione *et al.*, 1973). Hibernation of adult females is complete, and caves and abandoned buildings are chosen as winter quarters (Senevet and Andarelli, 1956). *An. subalpinus* is regarded as being a strongly zoophilic species, which rarely feeds on humans. Females are endophilic and can be found during the day in large numbers in stables and barns, and more rarely in human dwellings or buildings. The adults are eurygamous.

Distribution: *An. subalpinus* is a southern European mosquito with a range from the Iberian Peninsula, through the northern Mediterranean countries to the lowlands around the Caspian Sea.

Note on systematics: According to genetic studies and widespread surveys, *An. subalpinus* was formerly regarded as a subspecies of *An. melanoon*. It should merely represent an alternative egg phenotype of *An. melanoon*, the two forms of egg apparently being intergrading conspecific varieties that occur in pure populations in limited geographical areas (Bates, 1940; White, 1978). Based on material from Portugal, *An. melanoon* and *An. subalpinus* were treated as two distinct species on account of the differences of their eggs and the apparent area of sympatry of both forms in southwestern Europe (Ribeiro *et al.*, 1980; Ramos *et al.*, 1982). Furthermore, evidence was provided of reproductive isolation between sympatric populations of the two species (Cianchi *et al.*, 1987). Subsequently, regarding *An. subalpinus* as a distinct species of the Anopheles Maculipennis Complex was accepted by others (Ward, 1992).

Medical importance of the species of the Anopheles Maculipennis Complex: The anthropophilic species of the complex, such as *An. atroparvus* in coastal Europe, *An. labranchiae* around the western Mediterranean region and *An. sacharovi* around the eastern Mediterranean regions are the three main malaria vectors in the complex. All three species tend to be most prevalent in relation to saltwater habitats. *An. labranchiae* and *An. sacharovi* are important vectors throughout their distribution range, the medical importance of *An. atroparvus* depends on local conditions, *e.g.* presence of the species in large numbers, or transmission of winter malaria (availability of human hosts in or close to overwintering shelters). The more zoophilic species, *e.g. An. maculipennis s.s.*, *An. beklemishevi*, *An. messeae*, *An. melanoon* and *An. subalpinus* which are usually associated with freshwater habitats, play a minor role in malaria transmission. Nevertheless, at least *An. maculipennis s.s.* and *An. messeae* are responsible for malaria transmission in limited regions of Europe where high incidences of the species go along with a shortage of domestic animals.

Note on systematics: After the detection of a new species within the Anopheles Maculipennis Complex with cytotaxonomic techniques by Stegnii and Kabanova (1976) and evidence of the presence of other hitherto unknown members, White (1978) resurrected two members of the complex, previously regarded as synonyms, to specific rank, namely *An. martinius* Shingarev 1926, with an eastern distribution in central Asia, from synonymy with *An. sacharovi*, and *An. sicaulti* Roubaud 1935, found in northern Africa, from synonymy with *An. labranchiae*. Several North American mosquitoes clearly belong to the Anopheles Maculipennis Complex, *e.g. An. aztecus* Hoffmann 1935, *An. earlei* Vargas 1943, *An. freeborni* Aitken 1939, *An. occidentalis* Dyar and Knab 1906.

Anopheles (Anopheles) marteri Senevet and Prunnelle 1927

Female: Resembles *An. claviger* in general appearance; most easily distinguished from it by the whitish apex of the proboscis and a usually present pale apical spot on the fringe of the wing. The proboscis is without pale rings, and its apex is distinctly paler than the brown basal part (Fig. 6.11a). The palps are subequal to the proboscis, with scattered pale scales at the base. The terminal segment of the palps is at least half as long as the penultimate one. The occiput bears a tuft of whitish scales. The scutum has pale scales which form a median stripe; the sublateral and lateral areas are dark and bare. An anteacrostichal patch of pale scales is present. The lateral scutellar setae are longer than the median ones. The prespiracular setae are brown and often absent. The femora, tibiae and tarsi are blackish brown with a blue metallic shine. The femora and tibia have a pale scaled apex and the tarsi are without rings. Tarsomere I of the fore leg is distinctly longer than tarsomeres II to V combined. The scales on the wing veins are dense and all dark, and the wings are without spots (Fig. 6.7a). The wing fringe is made of lanceolate scales of unequal length, with a white apical spot.

Male: The palps are as long as or slightly shorter than the proboscis. The gonocoxite has two strong parabasal setae, curved at the end, arising from a strongly sclerotised base (Fig. 9.6). The outer seta is longer and thinner than the inner seta. A crescent shaped internal seta is inserted at the apical third of the gonocoxite, and numerous long thin setae cover the same portion of its dorsal surface. The claspettes have two lobes. The outer lobe bears three flattened, spatula-like setae and the inner lobe has a simple, longer, sabre-like seta. The aedeagus has 8–10 or more long leaflets on each side.

Larva: The antennae have a dark, spiculous apex. The antennal seta (1-A) has 3–4 moderately long branches inserted in the basal half of the antenna, close to the middle. The frontoclypeus is spotted but not banded. The clypeal setae are all simple. The inner clypeal setae (2-C) are closer to each other than to the outer (3-C), and more than twice as long as the outer setae. The postclypeal setae (4-C) are of variable length, and as long as the outer clypeal setae (3-C) which extend beyond the anterior margin of the frontoclypeus (frequent character of larvae from Portugal, Greece, Turkey, Iran

and Tadzhikistan populations), or are short and do not surpass the margin (larvae from Spain, Algeria, Tunisia, Israel and Jordan). The frontal setae (5-C to 7-C) are plumose, with 6–7 pairs of lateral branches. The end of the inner frontal seta (5-C) hardly reach the base of the postclypeal seta (4-C). The prothoracal seta 1-P is well developed, with a strong shaft, pinnately branched, similar to 2-P and 4-P. Seta 3-P is simple, rarely 2-branched. Seta 1-T on the metathorax of palmate type, has 10–11 leaflets. The abdominal seta 1-I is short and spine-like. The palmate setae 1–II to 1-VII are well developed, with 15–19 leaflets, sometimes more. The leaflets are narrowed in 2–3 irregular steps beyond the middle, forming a filament that is at least 1/3 as long as the blade (Fig. 8.12a). The antepalmate setae on abdominal segments IV and V (2-IV and 2-V) are simple, rarely with 2 branches. The abdominal setae 6-I to 6-III are pinnately branched.

Biology: *An. marteri* is a polycyclic species often found in mountainous regions. In Sardinia and Corsica, the species was collected from sea level up to 1000 m (Aitken, 1954), and in Tadzhikistan from 900 m to 1600 m (Keshishian, 1938). Females hibernate in natural shelters. Spring is the preferred season for larval development but a second, autumnal peak has been registered as well (Logan, 1953; Senevet and Andarelli, 1956). Larvae have been found from March until September in shaded, clear and cool water in rock pools and mountain streams and springs, with rocky bottom and scarce vegetation. They usually breed in acid water (pH 5.5–6.0) within a temperature range of 15° to 22°C (Senevet and Andarelli, 1956; Ribeiro *et al.*, 1987). The larvae can be found in association with those of *An. atroparvus, An. claviger s.s., An. petragnani, Cx. p. pipiens* and *Cx. territans* (Gutsevich *et al.*, 1974; Ribeiro *et al.*, 1987). *An. marteri* females are sylvatic, exophagic and markedly zoophilic.

Distribution: It is a southern Palaearctic species, which is distributed from Portugal and

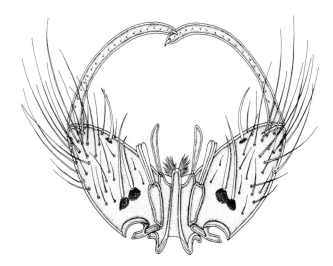

Figure 9.6. Hypopygium of *An. marteri.*

north Morocco to Tadzhikistan. In Europe it is registered in Portugal, Spain Corsica, Sicily, Sardinia, Albania, Greece and Bulgaria.

Note on systematics: Since 1927 when Senevet and Prunnelle *described An. marteri* as a new species, the closely related *An. sogdianus* was described in Tadzhikistan (Keshishian, 1938) and *An. marteri* var. *conquensis* in Spain (Torres Canamares, 1946). Later, *An. sogdianus* was given subspecific status under *An. marteri* after Beklemischev (in Boyd, 1949). Ribeiro *et al.* (1987) stated that *An. marteri* is a polymorphic, monotypic species, and that the names *sogdianus* and *conquensis* were only the morphs that have to be treated as junior synonyms of *marteri*. The authors hypothesize that the clinal distributions of the morphs is temperature dependent, with a January isoterm of +10°C separating *marteri/conquensis* in the south from *sogdianus* in the north. The proposed status was accepted by Ward (1992).

Medical importance: According to its strong preference for animal hosts, *An. marteri* is most probably of minor importance as a vector of human diseases (Ribeiro *et al.*, 1988).

Anopheles (Anopheles) plumbeus Stephens 1828

Female: *An. plumbeus* can be distinguished from the similar *An. claviger* by its smaller size and its general darker, leaden colouration. The pleurites of the thorax and the lateral parts of the scutum are blackish brown, forming a distinct contrast to the pale or ashy-grey median part of the scutum. Furthermore the wings are more densely scaled and darker in appearance than those of *An. claviger*. The proboscis and palps are black, with the palps being approximately the same length as the proboscis with the apical segment more than half as long as the penultimate one. The pedicel is brown, and the flagellum is blackish brown with black setae. The vertex has a tuft of narrow, pure white scales, which is directed anteriorly, and yellowish long setae. The occiput is covered in its median part with whitish lanceolate and erect forked scales, laterally with black erect forked scales. Lateral parts of the scutum are blackish brown, with a median grey stripe which covers at least 1/3 of the width of the scutum. The anterior margin of the scutum has a well developed anteachrostichal tuft of pure white, narrow scales (Fig. 6.12a). The scutellum is brown with dark setae, its posterior margin is slightly concave at the sides, and the postnotum is dark brown. The pleurites are blackish brown with 5–6 prespiracular setae. The legs are black or blackish brown, but the coxae and ventral surface of the tibiae are slightly paler. The wings are not spotted, and densely covered with dark brown, lanceolate scales, and the cross veins are well separated. The abdomen is black, and covered with pale brown or dark setae with a golden tinge.

Male (Fig. 9.7): The gonocoxite has 2 strong parabasal setae of approximately the same length. They arise directly from the surface of the gonocoxite, not elevated on a tubercle. The internal setae are inserted near the middle of the gonocoxite. The apical spine of the gonostylus is short. The claspettes are divided into 2 lobes; the outer lobe bears 3

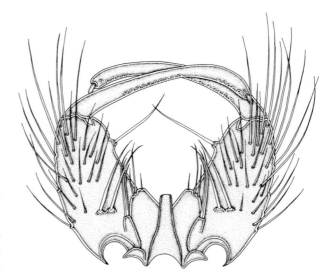

Figure 9.7. Hypopygium of *An. plumbeus*.

setae, which might be slightly flattened and are situated close together but are not fused. The inner lobe bears 1 short hair like seta and 2–3 spine-like setae of variable length, at least one of which is slightly bent at the apex. The aedeagus is short and broad, without spines or leaflets.

Larva: Larvae of *An. plumbeus* are at once distinguished from all other European species of the genus *Anopheles* by the frontal setae (5-C to 7-C) which are reduced in size and simple (Fig. 8.7b). The head is more or less oval, uniformly dark brown, and the primordial compound eyes are weakly developed. The antennae are approximately 1/3 of the length of the head, straight and smooth, and without spicules. The antennal seta (1-A) is very short and simple, and situated close to the base. The clypeal setae (2-C, 3-C) are thin and sparsely branched. The distance between the pair of inner clypeal setae (2-C) is smaller or almost the same as the distance between the inner (2-C) and outer clypeal setae (3-C). The post-clypeal setae (4-C) are short and simple, and situated wide apart. The distance between them is larger than the distance between the outer clypeal setae. Seta 1 of abdominal segment I (1-I) is short and simple. The palmate setae on abdominal segments II to VI (1-II to 1-VI) are conspicuous, but inconspicuous or rudimentary on segment VII. Each palmate seta consists of 14–15 lanceolate leaflets with a pointed apex, but without a terminal filament; their margin may be slightly serrated in the apical half. The lateral setae on segments I to VI (6-I to 6-VI) are large and pinnately branched. Each abdominal segment carries ventrally two pairs of stellate setae. The pecten plate is usually composed of teeth of more or less the same length. The saddle is covered with numerous spicules. The ventral brush has 17–19 fan-like setae arising from a common base. The anal papillae are shorter than the saddle.

Biology: Larvae of *An. plumbeus* develop almost exclusively in tree holes. The breeding water is usually dark brown due to the dissolved tannins and pigments derived from the wood and has a high concentration of salts in combination with a deficiency of oxygen (Mohrig, 1969). The larvae are ordinarily found in tree holes in beech (*Fagus sylvatica*), ash (*Fraxinus excelsior*), elm (*Ulmus* sp.), sycamore (*Acer pseudoplatanus*), lime (*Tilia* sp.), oak (*Quercus* sp.), birch (*Betula* sp.), horse-chestnut (*Aesculus hippocastanum*) and others, often together with larvae of *Oc. geniculatus*. Further associates may be larvae of *Or. pulcripalpis, Oc. berlandi, Oc. echinus* and *Oc. pulcritarsis*. The eggs of *An. plumbeus* are not laid on the water surface but on the side of the tree hole and hatching occurs when the hole is flooded. Thus, the number of generations per year depends on the hydrological situation. Hibernation takes place in the egg or larval stage. The larvae which hatch in autumn do usually grow into the second and third-instar by the end of the year, but do not pupate until the next spring. They spend most of the time at the bottom of the hole and can survive long periods when the water surface is frozen, but high mortality can be observed when the breeding water and the mud at the bottom are entirely frozen during a long period (Mohrig, 1969). In spring a major proportion of the larvae hatch from hibernated eggs. Occasionally, especially in periods of drought, larvae of *An. plumbeus* may also occur in artificial containers, rock holes or in ground depressions in shaded situations containing a rich infusion of fallen leafs (Aitken, 1954). In central Europe the adults usually occur from late spring on and are present until the end of September. They can be found from sea level up to an altitude of 1200 m (Senevet and Andarelli, 1956), and in the southern part of its range the species occurs in forests and in mountainous areas up to an altitude of 1600–2000 m (Gutsevich *et al.*, 1974). Females are persistent biters and they are most active during the crepuscular period, feeding principally on mammalian blood, including that of man (Service,

1971a). Occasionally they have been observed to attack humans during the day in shaded situations along forest edges. Some populations have shown a strong anthropophilic preference (Petric, 1989). Due to its preferred larval habitats, An. plumbeus is mainly found in forests and rural areas, but considerable populations may also be found in urban situations, where the larvae develop in tree holes in gardens or parks. However, it has adapted its breeding habit to widely available artificial breeding sites below ground, such as water catch basins and septic tanks with water contaminated with organic waste. Therefore, in central Europe An. plumbeus has increased in numbers during the last decades, and can be a major nuisance species in human dwellings especially when unused septic tanks support mass breeding.

Distribution: An. plumbeus is widely distributed throughout Europe wherever there are deciduous trees in which rot holes can be found. It is also distributed in the northern Caucasus, in the Middle East south to Iran and Iraq and in north Africa. A similar species which is found in India and Pakistan is considered by most experts to be the species An. barianensis James, but Gutsevich et al. (1974) pointed out that there are no distinct differences between An. plumbeus and An. barianensis. In North America, An. plumbeus is replaced by an allied species, An. barberi Coquillett, which closely resembles it (Marshall, 1938).

Medical importance: Although laboratory studies have shown that An. plumbeus can successfully be infected with P. vivax and P. falciparum (Weyer, 1939; Marchant et al., 1998) and that the species is an efficient carrier of malaria, it is considered to be of minor epidemiological importance at the present time due to its ecology. In the past, An. plumbeus played a major role as a vector in forests in Caucasian resorts (Gutsevich et al., 1974) and probably has been responsible for two recorded cases of locally transmitted malaria in the area of London, United Kingdom (Shute, 1954).

9.1.2. Subgenus *Cellia* Theobald

Members of this subgenus are characterised as follows: In the adults the cross vein areas and furcations of R_{2+3} and M are pale scaled, and C has four or more pale spots. In males the base of the gonocoxite bears 4–7, usually 6, parabasal setae, situated close together and not arising from distinct tubercles. The internal seta is absent. The larvae of the subgenus *Cellia* have a simple and small antennal seta (1-A) which is situated on the outer side of the antennal shaft. The inner clypeal setae (2-C) are wide apart, situated closer to the outer clypeal setae (3-C) than to each other. The leaflets of the palmate setae are not lanceolate, but always abruptly narrowed or shouldered, thus divided into a blade and a terminal filament. The subgenus *Cellia* is mainly distributed in the Oriental and Ethiopian regions and is not found in the Nearctic. In Europe, the distribution range of the subgenus is confined to the Mediterranean region, where three species and one subspecies of An. cinereus can be found. Therefore, An. cinereus is described here, despite the complicated situation of its real distribution.

Anopheles (Cellia) cinereus Theobald 1901

Female: The proboscis and palps are extraordinarily long and slender. The palps commonly have four pale bands. The basal three bands are broad, subequal, and extend to both segments at the articulations of palpomeres II–III, III–IV and IV–V. The distalmost band is often very narrow, sometimes indistinct or absent. When present and well developed, it occupies the apical third of palpomere V. Palpomere IV and V are long and when combined, distinctly longer than palpomere III. Glick (1992) described the palps of An. cinereus as having three bands and a dark apex which is usually true for ssp. hispaniola. The first flagellomere is speckled with white scales. The occiput has a well developed tuft of white scales. The scutum has a median stripe of long, usually narrow pale scales, and the submedian and lateral areas are avoid of scales. The femora and tibiae are dark with distinct pale spots on the femorotibial and tibiotarsal articulations. The fore and mid tarsi are entirely dark or have very narrow pale apical rings, but usually only on tarsomere I. The hind tarsi have very narrow white apical rings on tarsomeres I to IV. Tarsomere V is entirely dark scaled. The costa has five large dark spots, and its very base is dark scaled. R_{4+5} are mainly pale scaled, dark scaling of Cu is quite variable, and the anal vein usually has three dark spots with no pale fringe spot at its end.

Male: Palpomeres IV and V are very long; their lengths combined exceed that of palpomere III. The palps have three or four pale bands; the basalmost band is sometimes indistinct, and the apicalmost band is distinct. The gonocoxite has 3–6 subequal parabasal setae. The claspette lobe is undivided, and the inner seta is nearly twice as long as the lobe. Two additional outer setae may be present. The aedeagus has 7–10 pairs of long leaflets, some of them distinctly broadened, and serrated. The longest leaflet is almost half as long as the aedeagus.

Larva: The frontoclypeus is marked with dark spots of variable size and shape. The antenna is sparsely spiculous. The antennal seta (1-A) is simple, and inserted at about 1/3 of the length of the shaft; nearly as long as the width of the antenna at the point of insertion. The clypeal setae are all simple. The inner clypeal setae (2-C) are closer to the outer (3-C) than to each other, and 1.5–2.0 times as long as the outer setae. The length of the postclypeal setae (4-C) is variable, often subequal to the outer clypeal setae (3-C). The frontal setae (5-C to 7-C) are strong and pinnately branched (3–5 pairs of branches are evenly distributed over the entire main stem). The prothoracal seta 1-P is well developed, pinnately branched, and similar to 2-P and 4-P but distinctly shorter. Seta 3-P is short and simple. The abdominal seta 1-I

is short, variable in shape, and bifurcated or plumose. The palmate setae on abdominal segment II (1-II) are small, with 6–9 leaflets. The palmate setae on abdominal segments III to VII (1-III to 1-VII) have 13–15 leaflets abruptly narrowed in one or several irregular steps forming a filament which is about half as long as the blade (Fig. 8.15d). The abdominal setae 6-IV to 6-VI are pinnately branched.

Biology: *An. cinereus* is a polycyclic species. The most preferred breeding sites are edges of swamps, streams, irrigation ditches, rock pools and marshy pools. The larvae prefer moderately shaded water bodies with a slightly alkaline pH ranging from 7.7 to 8.3 (Evans, 1938). The females are zoophilic and frequently enter shelters of domestic animals. Occasionally they can be encountered inside human dwellings.

Distribution: Eastern Afrotropical region from Sudan and Ethiopia to South Africa, Arabian Peninsula.

Anopheles (Cellia) cinereus hispaniola (Theobald) 1903

Female: Similar to the nominative form in all stages, but easily distinguished from the other European members of the subgenus *Cellia*. It differs from *An. multicolor* by having a dark base of costa (Fig. 6.17b) and a bare submedian scutal area, from *An. sergentii* by having a mainly pale scaled R_{4+5} and from *An. superpictus* by having the apex of the palps dark and three relatively short dark spots on the anal vein. The palps usually have three pale bands, on both sides of articulations between palpomeres II–III, III–IV and IV–V, with a dark apex. Rarely the very tip may be white but the middle of palpomere V remains dark scaled. The hind tarsomeres have indistinct pale apical rings, sometimes reduced to a few scattered white scales. Other characteristics are similar to *An. cinereus*.

Male (Fig. 9.8): The gonocoxite has 4–7 (usually 6) short parabasal setae of similar size and length. The inner seta on the claspette lobe

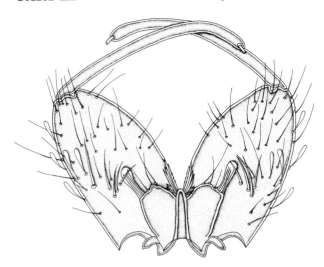

Figure 9.8. Hypopygium of *An. cinereus hispaniola.*

is about as long as the lobe, and one or two shorter median setae may be present. Several outer setae are fused to form a conspicuous broad spatula-like process, which is distinctly shorter than the inner seta. The aedeagus has 7–8 pairs of long, broad weakly serrated leaflets (Fig. 7.14b).

Larva: The frontal setae (5-C to 7-C) are poorly developed, with a few branches arising near the base of the main stem. The inner frontal seta (5-C) is slightly longer than the median frontal seta (6-C). The lateral abdominal setae 6-I to 6-V are pinnately branched, weakly pinnated on segments IV and V.

Biology: A polycyclic, orophilic (preference for higher altitudes) species, which can be found from sea level to 2334 m (Logan, 1953). The larvae breed on hilly grounds most numerous from midsummer on in the mountains of Sardinia (Aitken, 1954). The preferred breeding sites are sandy and gravelly edges of streams in shallow, clear water, but larvae can also be found in irrigation ditches and swamps (Senevet and Andarelli, 1956). According to the same authors, the larvae were found in waters with salinity ranging from 0 to 2.9% and a pH value ranging from 5.0 to 7.0. When disturbed, the larvae may remain below the surface of the water for one hour or more (Aitken, 1954). Often they breed in association with larvae of *An. labranchiae.* The females express zoophily and exophily, but readily bite humans in the open (Ribeiro *et al.*, 1988).

Distribution: Mediterranean region, North Africa, French Equatorial Africa, Sinai Peninsula and Transjordan (Ribeiro *et al.*, 1980; Knight and Stone, 1977). In Europe the species is registered in Portugal, Spain, Italy and Greece.

Note on systematics: The status of the name *hispaniola* is still rather undefined. Knight and Stone (1977) and Knight (1978) treated both *An. cinereus* and *An. hispaniola* as valid species of the subgenus *Cellia* Theobald. The status of *An. hispaniola* was changed to synonymy of *An. cinereus* after Dahl and White (1978) with no further explanation, which was acknowledged by others (Ward, 1984). Finally, the status of *An. hispaniola* was changed from species (former change to synonymy ignored) to subspecies of *An. cinereus* (Ward, 1992) after Ribeiro *et al.* (1980). The authors stated that due to a very low degree of morphological differences and absence of information concerning hybridisation between the forms, it is most advisable to treat *An. hispaniola* as a subspecies. Nevertheless, Ramsdale (1998) stated that the name *An. hispaniola* should be regarded as a junior primary synonym of *An. cinereus*, until further evidence is given.

Medical importance: The species is a potential vector of malaria in southern Europe (MacDonald, 1957). However, due to its exophilic and exophagic behaviour it is regarded as a vector of minor importance (Jetten and Takken, 1994).

Anopheles (Cellia) multicolor
Cambouliu 1902

Female: The proboscis is dark brown, and the palps are nearly as long as the proboscis with three pale bands at the joints of palpomeres II–III, III–IV and IV–V. The proximal band is very narrow with pale scales mainly at the apex of palpomere II and sometimes with a few scales at the base of palpomere III. The median and distal bands are broader than the proximal band. Half of palpomere V is dark. The clypeus is light brown, the flagellum brown, and the pedicel and the first 3–4 flagellomeres have a few pale scales. The vertex has long pale scales and setae forming a frontal tuft; the occiput in the upper half has pale scales, and dark scaled at the sides. The integument of the scutum is brown, the lateral areas somewhat darker than the median. The submedian and lateral areas are covered with narrow creamy scales, and the setae on the scutum are black. The scutellum is light brown, the pleurites brown with some pale scales, and about 3–5 prespiracular setae. The tibiae are prominently pale apically, the tarsi minutely pale apically, with pale scales not forming definite pale rings, and the hind tarsi may be completely dark. The wing veins are covered with narrow to moderately broad scales, and the cross veins are well separated. The wing is extensively pale scaled, and not well ornamented due to the paleness of the dark scales. The costa is largely pale, with dark areas reduced in some specimens, and pale at the extreme base (Fig. 6.17a). The anal vein is mainly pale with 3 small dark spots. There are pale fringe spots at the end of all the veins except the anal vein. The abdomen is brown, entirely devoid of scales, but with dark setae, and the first abdominal segment is more densely covered with setae.

Male (Fig. 9.9): The gonocoxite is slightly longer than it is broad. The base of the gonocoxite has 5 rather slender parabasal setae. The outermost seta is long and straight, arising from slightly above the other setae. The claspette is not clearly divided into distinct lobes, has a long and slender spine-like seta, several smaller setae and a broad, spatula-like seta situated at the outer border. The aedeagus is distinctly shorter than half the length of the gonocoxite, and leaflets and spines of the aedeagus are absent.

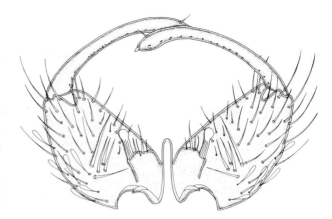

Figure 9.9.   Hypopygium of *An. multicolor.*

Larva: The larva of *An. multicolor* closely resembles that of *An. superpictus.* The antenna has numerous spicules along the inner border, and the antennal seta (1-A) is usually situated close to the base of the shaft and is simple. The apex of seta 1-A is sometimes divided into 2 or 3 extremely fine branches. The inner clypeal setae (2-C) have widely separated bases. Both inner and outer clypeal setae (2-C and 3-C) are simple (Fig. 8.16a); the outer seta is about 2/3 the length of the inner seta. The postclypeal setae (4-C) are simple, as long as or longer than the outer clypeals, with their distal ends reaching beyond the bases of the clypeal setae. The frontal setae are pinnately branched, and the inner frontals (5-C) are usually distinctly longer than the median frontals (6-C) (Fig. 8.14b). The palmate setae are undeveloped or rudimentary on abdominal segments I and II, and variably developed on segments III to VII, or sometimes undifferentiated. If fully developed, the leaflets are more or less uniformly pigmented and narrow. The shoulder is well marked, with a long filament about 2/3 the length of the blade, rather broad at the base and sharply pointed distally. The pecten plate has 7–8 long teeth usually alternating with 1 or 2 short teeth, finely serrated on the basal half. The arrangement is sometimes regular, but often irregular. The sclerotised tergal plates are small, and on abdominal segment V the width of the plate is about 1/3 of the distance between the palmate setae. A small, single accessory plate is present near the centre of the segment V. The saddle seta is long and simple, and the anal papillae are very short.

Biology: The larvae are generally to be found in inland or coastal breeding places in semi-arid regions in brackish and freshwater, but they show a preference for saline desert waters. They occur in salt pans, oases, small pools with or without aquatic vegetation, and sometimes in unused shallow wells. They were never found in rice fields (Christophers, 1933). The larvae are able to tolerate a high content of salinity, which may occasionally reach the point of saturation. In north Africa they were found in association with larvae of *Oc. caspius*, *Cs. longiareolata* and *An. cinereus hispaniola* (Senevet and Andarelli, 1956). The authors reported larval findings throughout the year with two peaks, the first between April and July and a second peak from November to December. Although *An. multicolor* has a wide distribution range, very little is known about the adult behaviour. The females are considered to feed on humans. In Egypt they readily enter houses and bite at night (Kirkpatrick, 1925). The same author reported considerable flight ranges. Adults have been found up to 13 km from the nearest possible breeding places.

Distribution: *An. multicolor* is mainly a desert species distributed in north Africa and much of Southwest Asia. Its range extends from Cyprus, Egypt, Palestine and Anatolia eastwards to west Pakistan. It has been reported in Spain in the province of Murcia and the Canary Islands (Romeo Viamonte, 1950).

Medical importance: The role of *An. multicolor* as a vector of malaria is still uncertain. Several authors have considered it to be an important vector on epidemiological grounds, but no naturally infected females have

been found. Other authors regard its role as doubtful even in areas of mass breeding (Gillies and de Meillon, 1968).

Anopheles (Cellia) sergentii (Theobald) 1907

Female: Resembling *An. superpictus*, but differs from the latter by its smaller size and the wing ornamentation. The proboscis is dark brown, long and slender. The palps are nearly as long as the proboscis, of uniform thickness from the apex to the base, and the last palpomere is very short and entirely pale scaled (Fig. 6.15a). In addition to the pale tip of the palps, two more narrow, but well marked pale bands are present at the joints of palpomeres II–III and III–IV. The clypeus is dark brown. The pedicel is dark brown, and without scales. The flagellum is brown with black setae and some narrow pale scales on the first and often on the succeeding flagellomeres. The vertex has pale lanceolate scales, which are directing forwards, and the frontal tuft is not well developed but is distinct. The occiput in the upper part has pale forked scales and at the sides has dark forked scales and setae. The eyes are bordered with broad scales. The scutum is brown, the lateral areas and transverse suture are somewhat darker than the median area, but not distinct and varies between the specimens. The scutum is covered with lanceolate hair-like scales and dark setae. The scutellum is brown with dark setae. The pleurites are entirely devoid of scales, with 2 prespiracular setae, and 8 upper mesepimeral setae. The coxae are without scales, and the femora are uniformly dark except for pale markings at the tips of the femora and tibiae of all the legs. The tarsi are entirely dark, without pale ringing. The wing has predominating dark scales on veins R_s to Cu, and vein R_{4+5} is usually almost entirely dark scaled, but may show an indistinct pale area about the middle. The extreme base of the radius (R) is entirely pale. Pale spots are also present at the furcations of R_{2+3} and M, at the cross veins, at the bases of Cu and A and at the middle of Cu$_2$. The wing fringe is well marked with pale spots at the ends of all veins except the anal vein (A) (Fig. 6.16a). The abdomen has no scales, is blackish brown, and covered with light setae.

Male (Fig. 9.10): The gonocoxite is slightly longer than it is broad. The base of the gonocoxite has 4–5 parabasal setae of varying length irregularly scattered over the surface and sometimes slightly curved at their apices. The claspette is not subdivided into lobes, and presents a massive unit, with a spatula-like seta on the outer border and a stout spine-like seta of similar length on the inner border. Between them is a second spine-like seta of half the length of the others. The aedeagus is long and slender, with 3–5 pairs of more or less broad leaflets. The longest leaflet is distinctly longer than the second longest, and all the broad leaflets are serrated through most of their length.

Larva: The antenna is covered with spicules along its inner border, and the antennal seta (1-A) is inserted at the basal third of the antennal shaft, and is short and simple. The clypeal setae are usually simple; rarely do the outer clypeals (3-C) have 2 branches. 3-C is slightly longer than half the length of the inner clypeals (2-C). The bases of the inner clypeal setae are not that much separated as in the other members of the subgenus *Cellia*, but the distance between them is nearly twice the distance between the inner and outer clypeals. The post-clypeal setae (4-C) are usually simple, about the same length as the outer clypeals, with their distal ends reaching well beyond the bases of 2-C and 3-C. Occasionally 4-C may be divided into 2–4 branches. The frontal setae (5-C to 7-C) are pinnately branched, and the inner frontals (5-C) are slightly longer than the median frontals (6-C) (Fig. 8.14a). The typical palmate setae are well developed on abdominal segments III to VII, with 17–18 large leaflets, which are more or less uniformly pigmented. The filament is long, slender and pointed, and more than half the length of the blade (Fig. 8.15b). The sclerotised tergal plates are broad; nearly as broad as the distance between the palmate setae on the first five abdominal segments and broader than the distance between the palmate setae on segments VI to VIII (Fig. 8.15a). The last tergal plate occupies more than one third of segment VIII. The pecten plate has 6–8 long and 4–8 short spine-like teeth. The saddle seta is long and simple.

Biology: The breeding places of *An. sergentii* are quite variable. Larvae can be found in rice fields, irrigation channels, slow moving streams, small pools in river beds or ponds with a rich aquatic vegetation and grassy swamps from overrunning irrigation ditches or streams. In the Canary Islands, it was found in pools in ravines, especially in small rock pockets (Christophers, 1929). *An. sergentii* often occurs together with larvae of *An. sacharovi* and *An. superpictus*, and other associates are *An. multicolor, Cx. p. pipiens* and *Cs. longiareolata* (Senevet and Andarelli, 1956). Adults are rarely found from spring to early summer and are most prevalent in late summer and autumn (Martini, 1931). Hibernation may take place in both the larval and adult stage. Adult females commonly

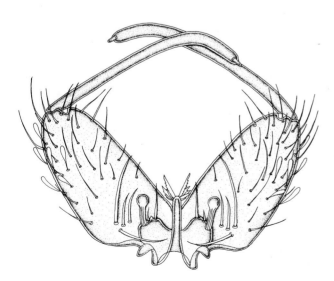

Figure 9.10.   Hypopygium of *An. sergentii.*

enter any kind of habitation, including houses, and readily bite man soon after dark, with the main activity period being during the night. They appear to be moderate flyers, and have been found in large numbers 2.5 km from the nearest possible breeding place (Christophers, 1933).

Distribution: *An. sergentii* has a wide distribution in the southern Mediterranean region. Its range stretches from north Africa, through the Middle East to Pakistan. It is recorded in the Canary Islands (Romeo Viamonte, 1950) and in Europe in Sicily. Senevet and Andarelli (1956) reported a doubtful record in Bulgaria.

Medical importance: Although natural infections were rarely found, the species is considered as a potential vector of malaria on epidemiological grounds. In Palestine it was regarded as the principal malaria vector (Saliternik, 1955).

Anopheles (Cellia) superpictus Grassi 1899

Female: A variable species, especially in the scaling of the palps, with white scales on the wings and legs. The four white scale patches on the costa and subcosta may be of variable length. The proboscis is dark; palpomeres III and IV have basal and apical white bands, and palpomere V is nearly completely covered with white scales. The vertex is dark, and the occiput has light, erect scales and an anterior patch of flat white scales. The scutum has a median stripe of pale greyish scales, but is otherwise dark, and the scutellum has some white scales. The pleurites are dark with some greyish tinge, pale setae and no distinct patches of scales. The costa (C) and subcosta (Sc) have four white areas, somewhat variable in length. Predominating pale scales are present on veins R_s to Cu, and vein R_{4+5} is usually almost entirely pale scaled. The pale spot at the apex of the wing fringe is large (Fig. 6.16b). Due to the numerous pale scales the whole wing has a light impression. The legs are dark, the coxae are without scales, the femora have a lighter ventral side, the knee spot is indistinct or absent and the tarsomeres have some white basal scales, sometimes forming indistinct rings. The abdomen has long yellowish setae.

Male (Fig. 9.11): The gonocoxite is rounded and short with very few scales and five strong parabasal setae; the two innermost setae are the shortest and the outermost one the longest. The gonostylus is long, bent at the apex, with a small apical spine. The claspette is not subdivided into lobes, but is broad, bearing several setae. The outer seta is spatula-like, the inner seta is long and spine-like. The aedeagus is short, less than half as long as the gonocoxite, and bearing several very short leaflets (Fig. 7.13a).

Larva: The inner clypeal seta (2-C) has short branches (Fig. 8.16b), and the outer clypeal seta (3-C) is simple and about half as long as 2-C. The inner frontal seta (5-C) is distinctly longer than the median frontal seta (6-C). Typical palmate setae on the metathorax are present. The palmate setae on abdominal segments II to VII (1-II to 1-VII) have 13–19 leaflets, with well developed, serrated and pointed filaments. The pecten plate has four long strong, and five smaller teeth.

Figure 9.11. Hypopygium of *An. superpictus.*

Biology: The larvae are found in slow running, warm and not polluted waters, *e.g.* in pools at the edges of rivers or in irrigated rice fields. They also occur in shallow pools in dry river beds (Gutsevich *et al.*, 1974). They are able to tolerate some amount of salinity but avoid eutrophic and muddy waters. They are found in smaller numbers early in the summer (first generation) and can occur in great numbers in autumn (further generations). The larval development may be rapid in high temperatures but usually takes nearly a month. The females may survive a cold climate, and can remain active throughout the cold season. They bite man and animals, and biting activity reaches its peak at sunset. They are both endo- and exophilic.

Distribution: In Europe the species is distributed in the Mediterranean region and occurs also in adjacent areas in Asia Minor and northern Africa. It has a southern Palaearctic distribution throughout Transcaucasia and middle Asia and is recorded as far as Pakistan and northwestern India.

Medical importance: The species is reported as being an important vector of malaria in middle Asia (Gutsevich *et al.*, 1974).

Note on systematics: Knight and Stone (1977) listed seven subspecies, described from Greece, Taschkent and Baluchistan. There are two more subspecies mentioned in the European literature (Martini, 1931; Peus, 1967). Thus the variability of the species needs to be further investigated.

10

Subfamily Culicinae

Most of the mosquito species of the world belong to this subfamily, which is subdivided into 11 tribes. The adults exhibit a high morphological variability ranging from species with little scaling to those with explicit patterns of scales of different colours, from white to black and even a metallic appearance, as well as from small to large species. Many species have prominent patterns of scales and setae on the scutum. The scutellum is trilobed with setae grouped on the lobes, except in Toxorhynchitini, which have an evenly rounded scutellum. In females the palps are short, and together with the trilobed scutellum make an elementary difference compared to members of the subfamily Anophelinae. The legs are often scaled in a characteristic pattern and the claws in some tribes have a subbasal tooth which can be species specific. The wings are often broader than those of Anophelinae and have the cross veins r-m and m-cu well expressed. They have usually three spermathecae (*receptaculum seminis*). The structure of the male hypopygium, especially the aedeagal apparatus is often more complex and in some tribes the gonocoxite may be enriched with well developed lobes. The head capsule of the larvae is more or less squared or rounded, and the antennae are of variable length. The larval thorax and abdomen are ornamented with long but less spiculate setae and lack, in European species, palmate setae. The elongated siphon with the plate-like, or in Mansoniini piercing, spiracular apparatus distinguishes the larvae of the Culicinae from those of the Anophelinae. The pupae have long trumpets, the opening of them being less wide than they are in Anophelinae. The chorionic pattern of the eggs varies with the mode of egg laying and can be used for genus, and in some cases, for species identification.

The culicine females belong to the fiercest biters especially in the cold temperate regions. They stand more parallel to their resting surface in contrast to most of the females of Anophelinae. Some species have developed autogeny, the ability to lay eggs without a previous blood-meal, usually in extreme cold regions and also when they inhabit confined spaces. The eggs can be laid either on the water surface for direct development in rafts (*Culex, subgenus Culiseta, Coquillettidia*) or into the substrate as single eggs. These may be drought resistant and able to overwinter for up to several years (most *Aedes* and *Ochlerotatus* species).

In tropical and subtropical regions the main portion of the mosquito fauna belongs to the tribes of Culicini, Sabethini, Mansoniini and others not occuring in Europe. In the more temperate zones most of the Aedini species belonging to the genera *Ochlerotatus* and *Aedes* are distributed, and they dominate in the extreme northern cold areas of the Holarctic region.

10.1. GENUS *AEDES* MEIGEN

The scaling and setation pattern of the adults is extensive and variable. The colour of the integument varies from brownish to blackish. The proboscis is long, straight, and always entirely scaled. The palps of the female are usually very short but in some species can extend to half the length of the proboscis. The vertex is scaled, and the occiput has erect scales. The scutum has a more or less species specific pattern of broad and/or narrow scales. The lobes of the scutellum are scarcely scaled, but have groups of setae. The prespiracular area is without setae, but post-spiracular setae are present. The postpronotum and sometimes also the postprocoxal membrane has pale scale patches. The mesepisternum, mesepimeron and mesomeron have groups of setae and patches of more or less broad scales. The coxae are scaled, and the femora and tibiae have distinct light and dark scaling patterns. Tarsomere IV is always distinctly longer than tarsomere V. The pulvilli are usually barely developed, and setose. The claws usually have a subbasal tooth; the European species of the subgenus *Aedes* have simple claws on all legs. The wing veins are covered with numerous dark scales, and sometimes scattered pale scales are present, most frequently on the costa (C), subcosta (Sc) and radius (R). The abdomen is covered with flat and more or less broad scales. The end of the abdomen differs from other Culicinae by its tapered last segments, and usually distinct cerci, which are elongated and rarely rounded. The scaling patterns of males is similar to those of females but often with less scaling on the pleurites. The palps of males are as long as or longer than the proboscis, and only rarely shorter (subgenus *Aedes*). The antenna has numerous whorls of setae. Tergum IX usually has two lateral lobes bearing a variable number of setae. The gonocoxite has more or less developed basal and apical lobes, but the lobes are sometimes absent. The gonostylus is of variable length, and can be simple or divided. The paraproct is sclerotised, and usually not fused, with a more or less acute tip. The claspettes are present but of variable shape and structure. The aedeagus is pear shaped, rounded or elongated. The head of the larva is oval, usually wider than it is long, and with

a rounded frontal margin. The antenna is about half as long as the head, occasionally longer than the head, and usually covered with more or less numerous spicules. The antennal seta 1-A is usually multiple-branched, inserted at about the middle of the antenna, and the inner frontal seta (5-C) is usually inserted in front of the median frontal seta (6-C). Branching of the prothoracic setae 1-P to 7-P is often species specific. Abdominal segment VIII has between a few to numerous comb scales arranged in one or more irregular rows. The siphon is usually moderately long, and only one siphonal tuft (1-S) is present. It is never located at the base of the siphon, but often inserted distal to the last pecten tooth, at about or beyond the middle of the siphon. The saddle does not usually encircle the anal segment, and seta 1-X is always located on the saddle, and is often simple. The ventral brush has a variable number of precratal and cratal tufts (4-X). The anal papillae are of variable length and shape.

The genus *Aedes* comprises more than 40 subgenera. Some of them are strictly tropical, confined to the Neotropical, Afrotropical, Oriental and/or Australian regions. In the European region, species of the subgenera *Aedes*, *Aedimorphus*, *Fredwardsius* and *Stegomyia* can be found. Some of the most feared vectors belong to the genus *Aedes*. They can transmit diseases such as Yellow Fever and Dengue as well as other arboviruses from several different families causing *e.g.* encephalitis in man and equines and transfer dog heart worm, filariae and bacteria.

10.1.1. Subgenus *Aedes* Meigen

Members of the subgenus *Aedes* are characterised by the palps which are short in both sexes and the proboscis which is usually about as long as the fore femur. The vertex is covered with narrow curved scales, the occiput with flat broad scales. The patches of scales on the pleurites are weakly developed and the lower mesepimeral setae are absent. The wings of the adults are dark scaled. The most characteristic features are found in the male genitalia. The gonostylus is subapically inserted, unequally bifurcated and does not support an apical spine. The shorter branch of the gonostylus bears several setae on its distal half. Typical claspettes, divided in a more or less slender stem and a flattened appendage or filament, are absent. Instead of this a process which arises from the dorsal part of the basal lobe is developed. It can be simple or unequally bifurcated. If present, both branches bear distinct setae on their apices. The larvae closely resemble those of the subgenus *Aedimorphus* and the genus *Ochlerotatus*.

Knight and Stone (1977) included more than 20 species into the subgenus, but earlier Reinert (1974) transferred nearly half of it, mostly from the Oriental region, into the subgenus *Verallina*. This transfer was acknowledged by Knight (1978). At present 12 species are listed under the subgenus *Aedes*; 3 from the Oriental, 1 from the Australian and 8 species from the Holarctic region, but none are found in the Ethopian or Neotropical regions.

General remarks on the systematics: The taxonomic status of the species of the subgenus *Aedes* found in the Palaearctic region is still uncertain. Gutsevich *et al.* (1974) treated *Ae. cinereus*, *Ae. rossicus* and *Ae. esoensis* as subspecies of the nominative form *Ae. cinereus*. This opinion is not shared, because there is a large overlap between *Ae. cinereus* and *Ae. rossicus* in Europe; very often the larvae occur in the same breeding sites and transitional forms are not known. There is no doubt that the Palaearctic species can be divided into two groups. First, the *cinereus*-group with *Ae. cinereus*, *Ae. geminus* and *Ae. sasai* described from Japan. Second, the *esoensis*-group with the two eastern species *Ae. esoensis* and *Ae. yamadai* and *Ae. rossicus* from Europe. The members of each group closely resemble each other. Whether the two other species of the

subgenus, *Ae. dahuricus* (Danilov, 1987) and *Ae. mubiensis* (Luh and Shih, 1958), belong to one of these groups remains unresolved so far.

Peus (1972) treated *Ae. rossicus* as a subspecies of the eastern *Ae. esoensis* and this opinion was followed by others (Ward, 1984). Although the record of *Ae. rossicus* from Japan by Hara (1958) was based on a misidentification (Tanaka *et al.*, 1975) and there is apparently no overlap between the two forms, the arguments given by Peus are not convincing. There are clear and distinct differences between the two species in the shape of the shorter branch of the gonostylus and the structure of the claspettes. In *Ae. rossicus* the shorter branch of the gonostylus is nearly as half as long as the main branch and the claspettes are divided into a longer and a shorter branch. In *Ae. esoensis* the shorter branch of the gonostylus is only one third as long as the main branch and the claspettes are simple. Furthermore the white basal bands on the abdominal terga are always absent in *Ae. rossicus* and generally present in *Ae. esoensis*, at least as patches (Tanaka *et al.*, 1979). Thus until the distribution and status of all the forms involved is clarified, *Ae. rossicus* should be considered as a valid species.

Aedes (Aedes) cinereus Meigen 1818

Female: Medium sized to rather small mosquitoes. The proboscis is dark brown with lighter scales on its ventral surface, and is about as long as the fore femur (Fig. 6.36a). The palps are entirely dark brown. The head is mainly covered with flat dark scales, the vertex with golden narrow curved scales and lateral parts of the occiput with broad yellowish scales. The integument of the scutum is a reddish brown, covered with golden brown narrow scales which are paler on the lateral margins and above the wing bases, giving the scutum a fawn brown colouration. The scutellum has dark setae and pale narrow scales on each lobe. The integument of the pleurites is light brown with patches of broad yellowish white or creamy scales on the propleuron, postspiracular area, mesepisternum and mesepimeron. The scales on the postpronotum are narrow and hair-like, often the lower portion is paler than the upper one. The prealar area is not scaled, but covered with setae. The mesepimeron is bare on the lower half, and lower mesepimeral setae are absent. The anterior part of the fore coxa has a patch of brown scales. The femora and tibiae are dark scaled, with paler scales on their posterior surfaces. There are small, indistinct patches of pale scales on the apices of the femora; the tarsi are entirely dark brown scaled, and pale rings are absent. The wing veins are covered with dark scales. The terga have dark brown scales on the dorsal surface, but without pale transverse bands. The lateral patches of pale scales on each tergum are usually joined, forming longitudinal stripes at the sides of the abdomen, which are not readily visible in dorsal view. The sterna have yellowish white scales. The apex of the abdomen is pointed, with cerci of short or moderate length.

Due to the general colouration adult females of *Ae. cinereus* may, at the first glance, be confused with *Cx. modestus*, but the pointed end of the abdomen and cerci, the claw with subbasal tooth and the lack of pulvilli easily identify it as a member of the genus *Aedes*.

Male: The lobes of tergum IX are as long as they are broad, widely separated, and each lobe bears several slender spine-like setae. The gonocoxite is about twice as long as it is broad, conical, with scales and long setae on its outer surface (Fig. 10.1). The basal lobe is well developed. The basal lobe as well as the part distal to it is covered with dense long setae, and the apical lobe is absent. The gonostylus is inserted well before the apex of the gonocoxite and is unequally divided. The inner and broader branch reaches between 1/3 and 1/2 of the length of the main outer branch, tapers towards the apex and is rounded at its tip. It bears several setae on the lateral margin. The longer main branch is slightly curved and bifurcated

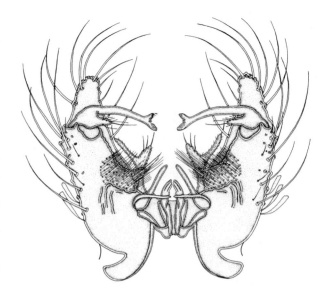

Figure 10.1.   Hypopygium of *Ae. cinereus.*

at the apex giving it a fishtail-like appearance. The outer branch of the fork is usually shorter than the inner branch. Their lengths may vary and often the prongs are nearly the same length, but the outer branch is never longer than the inner branch (Fig. 7.18c, d). The prongs appear to be flattened dorsoventrally and are usually stouter than in *Ae. geminus*. The claspettes are unequally bifurcated with a long and slender branch usually bearing 1–6 long setae and a shorter branch usually bearing 1 apical and 2–3 subapical setae. The paraproct is heavily sclerotised, slender and rod shaped with a narrow apex. The lateral plates of the aedeagus are heavily sclerotised, closed at the base and apex, with the apical half expanded.

Larva: The head is distinctly broader than it is long, and the antennae are slender and nearly as long as the head. The antennal tuft (1-A) is inserted slightly below the middle at about 2/5 of the length of the shaft (Fig. 8.22b). The setae of the labral brush are simple, not denticulated. The postclypeal seta (4-C) is inserted anterior to the frontal setae, and is small and multiple-branched. The frontal setae (5-C to 7-C) are arranged in a

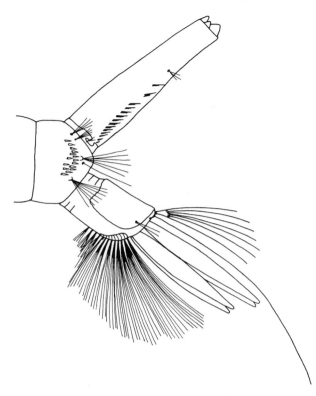

Figure 10.2. Larva of *Ae. cinereus*.

posteriorly curved row (Fig. 8.21b). The inner (5-C) and median (6-C) frontal setae have 5 or more branches, rarely 3–4; the outer frontal seta (7-C) is long and multiple-branched. The comb of abdominal segment VIII usually has 10–16 scales arranged in a partly double row. The individual scale is long with a strong apical spine and small lateral spines. The siphon is slender, index 3.0–4.0 (Fig. 10.2). The pecten consists of about 13–21 weakly sclerotised teeth reaching beyond the middle of the siphon, and the distal pecten teeth are unevenly and more widely spaced than the basal teeth. The siphonal tuft (1-S) is inserted distal to the pecten and usually consists of 3–6 short branches. The saddle is longer than it is wide and extends to the middle of the lateral sides of abdominal segment X or beyond it. The saddle seta (1-X) is double-branched and shorter than the saddle. The ventral brush consists of 8–10 tufts of cratal setae (4-X) on the common base, and 2–4 shorter tufts of precratal setae. The anal papillae are long, at least twice as long as the saddle.

Biology: The larvae can be found in various habitats, but most often they occur at the edges of semipermanent, partly shaded pools of flood plains, in sedge marshes or *Sphagnum* sp. bogs and at the edges of lakes covered by emerged vegetation. The larvae also occur in woodland pools, but they require a higher temperature for larval development than the typical snow-melt mosquitoes. *Ae. cinereus* larvae can be found in these pools when they are subsequently reflooded after rainfall. The larvae usually hatch at a temperature of 12–13°C and the development starts at 14–15°C; the optimum temperature being 24–25°C (Mohrig, 1969). Under these conditions the larvae develop very rapidly and finish their immature stages within 8–10 days. The adults occur later than the typical snow-melt mosquitoes *Oc. rusticus*, *Oc. communis*, *Oc. cantans* or *Oc. punctor*. In central Europe the larvae can be found from April on, in the northern parts of Europe a couple of weeks later, and the adults first occur usually in May and through the summer months until late October. The findings of larvae in *Sphagnum* sp. bogs and other acido-oligotrophic habitats indicate that *Ae. cinereus* seems to be slightly acidophilic. The females feed principally on mammals and bite human hosts readily when available. They attack in numbers at dusk and dawn, but do not bite in an exposed sunlit situation. During the day they rest in the low vegetation, but feed readily after arrival of the prospective host, biting those parts of the host within the vegetative cover (Wesenberg-Lund, 1921). The migration range of *Ae. cinereus* seems to be low; they are practically never met in the open unshaded field. The species has at least two generations per year, in its more northern range only one generation per year may occur; the oviposition takes place in the summer months in suitable dried up depressions prone to subsequent flooding, and hibernation takes place in the egg stage. In many localities *Ae. cinereus* occurs in masses and causes great annoyance to walkers or people seeking recreation in forested areas.

Distribution: *Ae. cinereus* is distributed in the northern Holarctic region and widely spread over Europe. It can be found from Finland to Italy and from Spain to the eastern shores of the Baltic Sea and the North Caucasus. It is distributed in Middle Asia, Kasachstan and Siberia, the Far East and North America.

Aedes (Aedes) geminus Peus 1970

Female: *Ae. geminus* closely resembles *Ae. cinereus* and can be identified with certainty solely through hypopygial characters. Peus (1972) considered both as sibling species and described the adults of *Ae. geminus* in the typical form usually being smaller than the adults of *Ae. cinereus* and the females of the latter usually with a lighter colouration in the typical form than the females of the former. However, the size of an adult is closely related to the nutritional situation during the immature stages and females of *Ae. cinereus* show a variation in their colouration, *e.g.* darker forms may exist, so these characteristics are of no value for identification of the two species.

Male (Fig. 10.3). In *Ae. geminus* the basal lobe is less developed than in *Ae. cinereus* and the covering with long setae seems not to be as densely as it is in the latter species. Whereas in this case it is recommended to compare individuals of the two species to get a clear picture, the immediately visible difference between the two species is the shape of the apical fork of the gonostylus. It is bifurcated in both species, but in *Ae. geminus* the outer branch is longer than the inner one (Fig. 7.18a, b). In *Ae. cinereus* the length of the outer branch never exceeds that of the inner branch. In the typical form the prongs appear to be thinner and more slender in *Ae. geminus* and they are more or less

Figure 10.3.   Hypopygium of *Ae. geminus*.

rounded, whereas in *Ae. cinereus* they are often flattened. The claspette is divided into two branches, the shorter branch bearing 2–3 setae, and in *Ae. cinereus* this branch bears usually 3–4 setae.

Larva: No specific differences, neither in the chaetotaxy nor in other characteristics, can be found in the larvae of *Ae. geminus* and *Ae. cinereus*. Both show individual variability, *e.g.* in the number of the small accessory siphonal setae on the dorsal part of the siphon.

Biology: So far the information available about the biology of *Ae. geminus* is scanty. There is a large overlap in the preferred breeding sites with *Ae. cinereus*, and very often both species can be found together in the same water bodies. Peus (1972) reported *Ae. geminus* to have a lower tolerance against acid habitats, though he could not find the species in mesotrophic and higher acido-oligotrophic swamps, where *Ae. cinereus* was numerous. As *Ae. cinereus*, the species has at least two generations per year and the males form mating swarms of only 10 individuals or less. The females are anthropophilic and can cause great annoyance, when present in large numbers.

Distribution: Because of the close similarity to *Ae. cinereus* it is difficult to assume the whole range of distribution for *Ae. geminus*. Records of *Ae. cinereus* earlier than 1970 or those which are solely based on females or larvae are at best questionable. Lvov (1956) worked on the specific status and distribution of *Ae. esoensis* and *Ae. cinereus* in the eastern Palaearctic region with findings of transitional forms, but never mentioned the characteristic form of the apical fork of the gonostylus which is typical for *Ae. geminus*. It is most likely that *Ae. geminus* does not occur in this region and is distributed mainly in middle and western Europe (Peus, 1972). The species is recorded with certainty from England, northwestern France, Germany, Poland, Czech Republic and southern Sweden. It has also been identified from the southern and eastern shores of the Baltic Sea.

Aedes (Aedes) rossicus Dolbeskin, Gorickaja and Mitrofanova 1930

Female: *Ae. rossicus* is easy to distinguish from the morphologically closely resembling *Ae. cinereus* by comparing the colour of the thorax and the abdomen. In *Ae. cinereus* the colour of the scutum is fawn brown with light brown pleurites, the colour of the abdomen is dark brown, distinctly different to the colour of the thorax. In *Ae. rossicus*, the scutum, the integument of the pleurites and the abdomen are more or less of the same dark brown colour, and the differences in colouration of the thorax and the abdomen are indistinct. In addition, the scales on the pleurites and the sterna are pale yellowish or yellowish white in *Ae. cinereus* and greyish white in *Ae. rossicus*, producing a distinct contrast between the colour of the terga and sterna in the latter species. The proboscis of *Ae. rossicus* is dark scaled on its dorsal surface; the ventral surface has mixed pale and dark scales mainly on the basal half, and the palps

are intermixed with pale and dark scales. The head has narrow pale and broad dark brown scales on the vertex. The lateral parts of the occiput are covered with broad light coloured whitish grey scales, producing a distinct contrast to the scales on the vertex. The eyes are bordered with a narrow row of whitish scales. The scutum has narrow brown scales, often with an inconspicuous pale stripe on the lateral margins, which is more distinct on the posterior part. The upper part of the postpronotum has narrow dark scales, and on the lower part the scales are broad, flat and of a lighter colour. The prealar area has whitish scales and setae; the scales on the mesepisternum do not reach the anterior angle, and the scales on the mesepimeron do not reach the lower margin. The abdomen has dark brown scales on the dorsal surface (the same colour as on the scutum), without pale transverse bands. The lateral pale scales on each tergum are fused, forming longitudinal stripes at the sides of the abdomen, which are hardly visible from above.

Male: The general morphology of the hypopygium is very similar to that of *Ae. cinereus*. A readily visible difference is the shape of the longer branch of the gonostylus, which is never apically bifurcated. It is somewhat flattened distally and rounded at the apex, and the outer margin has fine spatula shaped serrations (Fig. 10.4). The apical part of the gonocoxite, distal to the point of insertion of the gonostylus, tapers more abruptly and seems to be more slender than in *Ae. cinereus*. The claspette is divided into two branches; the shorter branch is broad with 3–4 setae, two of them strong and spine-like, and usually stronger than the setae in *Ae. cinereus*. The longer branch of the claspette is slender with 1–2 setae.

Larva (Fig. 10.5): The larva of *Ae. rossicus* is very similar to that of *Ae. cinereus*. Both species differ in subtle characters, *e.g.* the point of origin of the antennal seta (1-A) and the setation of the prothorax, as indicated in the

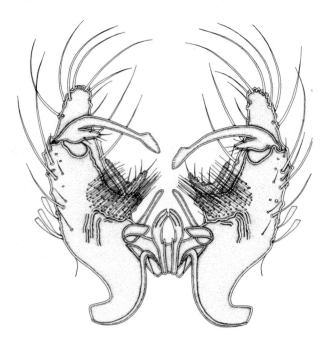

Figure 10.4.   Hypopygium of *Ae. rossicus*.

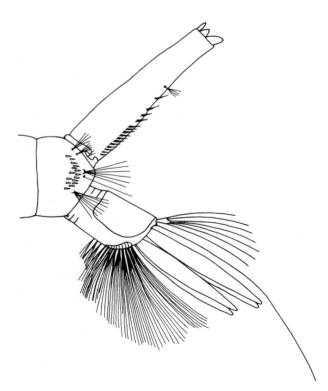

Figure 10.5.   Larva of *Ae. rossicus*.

keys (Fig. 8.22a). Often it is hardly visible and difficult to decide if the antennal tuft is located in the middle of the antenna or slightly below it, sometimes the prothoracic setae may be broken or lost. In these cases one needs to rear the larvae until the adults are hatched and then differentiate between the two species in the adult stage.

Biology: The larvae occur from mid March until the beginning of September in small temporary pools, where they can be found together with *Ae. cinereus* and *Ae. geminus*, but apparently the first generation of *Ae. rossicus* occurs earlier in the year than in the two latter species. Larvae of *Ae. rossicus* can also be found in large numbers in the inundation areas of rivers, often associated with *Ae. vexans* and *Oc. sticticus*. They are rarely found in swampy woodlands in acid breeding waters (Becker and Ludwig, 1981). In the inundated forests of the Upper Rhine valley the females are severe biters, and readily attack humans even during the day. They rarely leave the protected and shaded areas for host seeking, their migration range is limited, thus they are only occasionally found in the open field. Adults occur during the summer months usually from the beginning of May until October. Occasionally single individuals can be found in November when other species of the genera *Aedes* and *Ochlerotatus* have already dis-appeared; apparently they have a higher tolerance against the cold. It is assumed that *Ae. rossicus* produces several generations per year in the temperate zones, in its more northern distribution range it could be less (1 or 2 generations per year). Hibernation takes place in the egg stage.

Distribution: The western European range of its distribution is not so well known as yet. It is supposed that *Ae. rossicus* is distributed in the West as far as the Atlantic Ocean. The species is recorded in Sweden, Norway, France, Germany, Hungary, the Czech Republic, and Yugoslavia and can be found in the Ukraine and north of the Caucasus to the western slopes of the Ural. In the eastern

Palaearctic it can be found in the northern regions of the Far East. The record of *Ae. rossicus* from Japan (Hara, 1958) was based on a misidentification (Tanaka *et al.*, 1975).

10.1.2. Subgenus *Aedimorphus* Theobald

The females are characterised by the numerous erect forked scales on the vertex and occiput. The achrostichal and dorsocentral setae are well developed and numerous. The ante- and postpronotum are usually covered with narrow curved scales, and postspiracular setae are present. The mesepisternum has several upper mesepisternal setae, and the mesepimeron has no setae or scales on its lower part. Sternum VIII has a deep median V-shaped notch apically, and the cerci are long and narrow. In males the gonostylus is usually complex with an expanded apical portion, and typical claspettes divided into a stem and winged filament are absent. The antennae of the larvae are moderately pigmented and slightly curved distally. The postclypeal seta (4-C) is very small and usually multiple-branched. The pecten has 10–23 teeth, the distal 2–4 teeth are spine-like and apically detached. The siphonal tuft (1-S) is located well beyond the middle of the siphon.

The structure of the female genitalia and some male characteristics resemble the species of the subgenus *Stegomyia*, therefore a close relationship between both subgenera is assumed. The subgenus *Aedimorphus* embraces approximately 110 species and subspecies. The majority of them can be found in the Ethopian and Oriental regions. Only one species of *Aedimorphus*, namely *Ae. vexans*, occurs in Europe.

Aedes (Aedimorphus) vexans (Meigen) 1830

Female: Tarsomeres II and III of the fore legs, tarsomeres I to IV of the mid legs, as well as all the tarsomeres of the hind legs have narrow basal pale rings which usually do not exceed more than 1/4 of the length of the tarsomeres (Fig. 6.27a). Compared to other European *Ochlerotatus* species with pale rings on the legs, *e.g. Oc. annulipes, Oc. cantans, Oc. flavescens* or *Oc. excrucians*, the rings are much narrower. The proboscis and palps are dark scaled, and the palps have some white scales apically. The head is covered with narrow curved decumbent pale and dark scales and numerous dark brown erect forked scales which extend anteriorly to the interocular space. The scutal integument is dark brown, and the scutum is covered with narrow curved dark scales and narrow pale scales forming indistinct patches on the anterior submedian and prescutellar dorsocentral areas, as well as the transverse suture and prescutellar dorsocentral areas. The acrostichal and dorsocentral setae are well developed. The postspiracular area has a large patch of narrow curved or moderately broad pale scales. The upper and lower mesepisternal scale patches are present. The mesepimeron has a patch of broad pale scales in the upper part. The tibiae are dark scaled dorsally and light scaled ventrally. The wing veins are covered with moderately broad dark scales and isolated pale scales at the base of the costa (C) and subcosta (Sc). The abdominal terga have white basal bands, and the distal parts are dark scaled. The basal bands on terga III to VI are distinctly narrowed in the middle, forming a bilobed pattern (Fig. 6.27b). Old and worn females which have lost most of their scales can still unequivocally be identified by the distinct V-shaped notch at the apical margin of sternum VIII.

Male: Tergum IX is strongly bilobed with 6–11 setae on each lobe. The gonocoxite is long and moderately broad, with scattered scales on the lateral and ventral surfaces. The basal and apical lobes are absent. The gonostylus gradually expands towards the apex. The spine of the gonostylus is articulated subapically arising from a small tubercle, and is straight (Fig. 10.6). The claspette has a moderately broad basal part, and the apex is slightly expanded and rounded, with a crown of numerous spine-like

Figure 10.6.   Hypopygium of *Ae. vexans*.

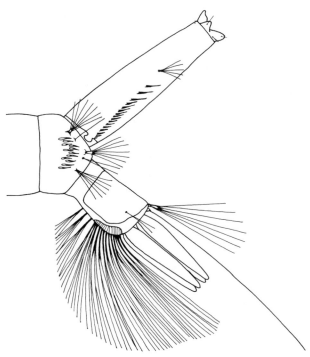

Figure 10.7.   Larva of *Ae. vexans*.

setae, some of them curved apically. The claspette filament is absent. The paraproct has a pointed apex, and the aedeagus is strongly sclerotised, with lateral plates connected at the base.

Larva: The antenna is less than half as long as the head, with numerous spicules scattered over the shaft. The antennal seta (1-A) has 5–10 branches and is inserted below the middle of the antenna. The median setae of the labral brush are serrated apically; a valuable character to separate *Ae. vexans* from the similar larvae of *Ae. rossicus* and *Ae. cinereus*, which both have simple setae. The frontal setae (5-C to 7-C) are arranged in a triangular pattern (Fig. 8.21a), the median frontal seta (6-C) is situated in front of the inner frontal seta (5-C); 5-C has 1–5 branches, 6-C has 1–2 (rarely 3) branches, and 7-C has 6–12 (usually 7–9) branches. The comb with 7–13 scales is arranged in 1–2 irregular rows (Fig. 10.7). Each comb scale has a long stout pointed median spine and small spines at the base. The siphonal index is usually 2.3–3.0. The pecten has 13–18 teeth, the apical 2–3 teeth are larger than the others and detached. The basal teeth have 1–3

lateral denticles. The siphonal tuft (1-S) is situated well beyond the middle of the siphon, with 3–8 short branches, and is about half as long as the width of the siphon at the point of origin. The saddle reaches far down the sides of the anal segment, and the saddle seta (1-X) has 1–2 branches. The ventral brush has 3–4 precratal setae (4-X). The anal papillae are distinctly longer than the saddle.

Biology: *Ae. vexans* is a polycyclic species predominantly breeding in inundation areas such as floodplains of rivers or lakes with fluctuating water levels. The preferred breeding sites are temporary water bodies with neutral to alkaline water, which are present only a few days to weeks after a flood, such as flooded meadows, poplar cultures, willow and reed areas. Usually the larvae hatch in large numbers if the water temperature exceeds 9°C. In central Europe they occur in springtime when the typical snow-melt mosquitoes, such as *Oc. cantans* are already hatched. After flooding the larvae hatch within a few minutes to hours when the flooded water becomes stagnant and the content of oxygen decreases. The hatching behaviour of the larvae is adjusted to the temporary water conditions. After the completion of the embryogenesis, which may last 4–8 days (about 1 week at 20°C), not all the larvae hatch after flooding, but only a proportion of it ("hatching in instalment"). If one population of larvae fail to complete the development due to drying out, a second population can develop during a following flood even if no additional eggs are laid. The hatching rate is particularly high at high water temperatures and after the completion of a diapause, which lasts from September until the beginning of March of the following year in temperate zones. If suitable hatching conditions fail to exist (*e.g.* lack of floods during summer), the eggs are capable of surviving for a long time (at least 5 years).

Ae. vexans as a "summer species" has an optimum temperature of 30°C for its development. At a water temperature of 30°C, the development from hatching of the first-instars to emergence of the adults lasts one week, at a temperature of 15°C it is three weeks. *Ae. vexans* frequently becomes the dominant species during summer months rich in floods and is often the most important nuisance mosquito in temperate climate zones. Often hundreds of larvae per litre of water can be encountered, that is frequently more than 100 million larvae/hectare. Due to a large population pressure after mass emergence the adults frequently migrate long distances from their breeding sites to find a host for the blood-meal and thus may become a serious nuisance, not only close to the breeding sites, but also far away. A migration of up to 15 kilometres (according to the circumstances, the flight capacity is about 1 km/night), occasionally the multiple of that, could be proven. The imigration of females into human settlements, *e.g.* gardens and parks, can cause a considerable nuisance. After the blood-meal, the females lay the eggs 5–8 days at the earliest into damp depressions. A female can lay more than 100 eggs after a single blood-meal; occasionally after repeated blood-meals, multiple egg batches are laid. The preferred hosts are mammals. Both females and males imbibe plant juices in order to cover their energy requirements. However, no eggs are developed without a blood-meal. Under optimum conditions, *Ae. vexans* needs less than three weeks from hatching of one generation to the hatching of the larvae of the next generation (development in water: approx. 6 days; copulation: approx. 2 days; blood-meal: approx. 2 days; egg development: approx. 5 days and embryogenesis about 4 days). It is suspected that only a part of the emigrated population returns to the original breeding sites after the blood-meal, while a considerable part of the population do not return and lay eggs far away from their original breeding sites. Therefore, the migration leads to a natural regulation of the population densities.

Distribution: *Ae. vexans* is distributed almost worldwide and can be found in nearly every country in Europe.

Note on systematics: A subspecies of *Ae. vexans*, ssp. *nipponii* (Theobald), which differs considerably in the scaling of the abdominal terga and the pleurites, has been described in the eastern Palaearctic region.

Medical importance: *Ae. vexans* has many attributes of an ideal vector species. It is widely distributed, can become very abundant, often at the same time when virus activity is at its peak; it feeds readily on humans and domestic animals, and it has been found naturally infected with various arboviruses (Reinert, 1973). Natural infections with western equine encephalitis (WEE) virus, eastern equine encephalitis (EEE) virus and California encephalitis (CE) group viruses have been reported from North America (Wallis *et al.*, 1960; McLintock *et al.*, 1970; Hayes *et al.*, 1971; Sudia *et al.*, 1971). In Europe, *Ae. vexans* is involved in the transmission of Tahyna virus (Aspöck, 1965; Mattingly, 1969; Gligic and Adamovic, 1976; Lundström, 1994).

10.1.3. Subgenus *Fredwardsius* Reinert

The establishment of the monotypic subgenus *Fredwardsius* is based on the type species, *Ae. vittatus* (Reinert, 2000a). The species has long been recognised to not fully conform to any recognised subgenus of *Aedes*. Edwards (1932) included *Ae. vittatus* in the subgenus *Stegomyia* Theobald and placed it in a monotypic group (Group D), apart from other species of the subgenus. Later on, Huang (1977) transferred *Ae. vittatus* from subgenus *Stegomyia* Theobald to subgenus *Aedimorphus* Theobald, mainly based on the structure of the male hypopygium. After a comparision of *Ae. vittatus* with all currently recognised subgenera and genera in the tribe Aedini, Reinert (2000a) found the species to share some characteristics with the subgenera *Stegomyia* and *Aedimorphus*, but it possesses unique and unusual features, that are of subgeneric rank. The combination of these characters separates *Fredwardsius* from all other subgenera and genera within the tribe Aedini., *e.g.* 2–6 well developed lower mesepimeral setae in the female, the greatly expanded distal portion of the gonostylus in the male hypopygium, and the positions and development of head setae 4-C to 7-C in fourth-instar larvae. For a more detailed morphological description of the subgenus, see Reinert (2000a).

Aedes (Fredwardsius) vittatus (Bigot) 1861

Female: The proboscis is as long as the fore femur. Its median part has a band of whitish scales of varying width, which is better developed ventrally than on the dorsal surface. The palps have a few white scales in the middle and the apex is broadly pale scaled. The clypeus has lateral patches of white scales. The antennae are shorter than the proboscis, and the pedicel and first flagellomere are white scaled on the median and lateral parts. The vertex has decumbent curved blackish brown scales, and black erect forked scales. There are two dorsolateral patches of white decumbent scales behind the eyes. The scutum is mainly covered with narrow, curved, blackish brown scales. Three pairs of prominent silvery whitish spots are present, located close to the dorsocentral areas (Fig. 6.24a). The anterior two pairs are usually larger than the posterior pair, which is situated close to the wing roots. All three lobes of the scutellum have broad white scales, and a few black scales may be present at the apex of the middle lobe. Broad white scale patches are present on the ante- and postpronotum, propleuron, postprocoxal membrane and sub- and postspiracular areas. The mesepisternum has upper and lower scale patches. The mesepimeral patch is located in the upper part of the mesepimeron. The femora of all the legs have a white ring close to their apices and white knee spots, and the hind femur is more extensively white scaled at the base. The hind tibia has a distinct median white ring. The tarsi of the fore and mid legs have narrow white basal rings on tarsomeres I to III, and tarsomeres IV and V are entirely dark. The tarsi of the hind legs have broad white basal rings on tarsomeres I to IV, but tarsomere V is entirely white scaled. The wing veins are mainly dark scaled, with a few broad white scales at the base of the costa (C) and a few scattered white scales on the costa (C) and radius (R). The abdominal terga are predominantly covered with black scales. Tergum I has a median longitudinal white band extending to near the apical margin; terga II–VII have narrow white basal bands, and lateral curved white markings separated from the basal bands. Sterna I–VI have lateral and central areas of scattered white scales, and sternum VII is extensively white scaled. Sternum VIII has a deep notch at the middle, and the cerci have dark scales.

Male: Tergum IX is narrowed in the middle, with a distinct setose lobe on each side. The gonocoxite is elongate, and about three times as long as it is wide at its base, without distinct basal and apical lobes (Fig. 10.8). The gonostylus has a very characteristic shape, being greatly

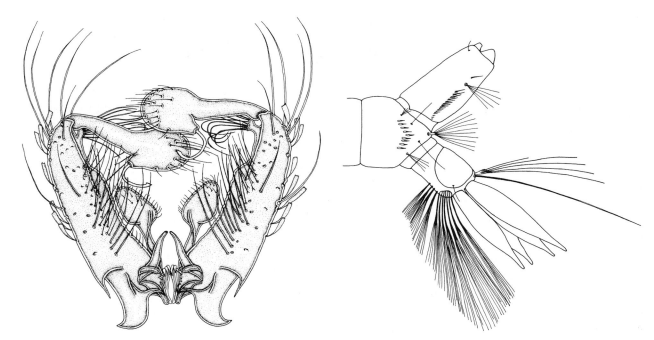

Figure 10.8.   Hypopygium of *Ae. vittatus*.   **Figure 10.9.**   Larva of *Ae. vittatus*.

expanded apically. The apical spine is situated at the base of the expanded apical part, and is long and strongly curved. The claspette is large, with a narrow base, is distinctly swollen from about the middle, and is covered with numerous small setae. The aedeagus is small with several well developed recurved teeth on the lateral and apical parts.

Larva: The antenna is slightly more than half as long as the head, and is covered with very small spicules, although sometimes the antenna is smooth. The antennal seta (1-A) is situated slightly below the middle of the antennal shaft, usually with 2–3 branches (Fig. 8.62a). The postclypeal seta (4-C) has 2–3 branches; the inner and median frontal setae (5-C and 6-C) are simple, and the outer frontal seta (7-C) has 5–7 branches. The comb has 6–9 (usually 8) large comb scales, each scale with a long, pointed median spine and a few thin, short spines close to its base. The siphon is sclerotised along its entire length, except for a very small area at the extreme base, and the

acus is indistinct. The siphon slightly tapers towards its apex (Fig. 10.9). The siphonal index is 2.0 or slightly more. The pecten usually has 15–25 well sclerotised teeth, occupying about 2/3 of the length of the siphon. The distalmost pecten tooth is spine-like, and apically detached (Fig. 8.62b). The siphonal tuft (1-S) is situated below the distalmost pecten tooth, with 3–6 branches, which are about as long as the width of the siphon at the point of origin. The saddle seta (1-X) is short and simple. The upper anal seta (2-X) has 4–6 branches, and the lower anal seta (3-X) is simple and long. The ventral brush has 5–7 tufts of cratal setae (4-X) and 3–4 tufts of precratal setae. The anal papillae are more than twice as long as the saddle, and pointed.

Biology: Females prefer to lay their eggs above the water level of rock pools, occasional utensils, boats, wells or tree holes (Mattingly, 1965). Boorman (1961) found that deep rock holes usually contained larvae of *Ae. vittatus*, particularly those where the water was clear and the bottom covered with a layer of mud and a few dead leaves. *Ae. vittatus* is particularly well suited to such an environment, as its eggs can tolerate desiccation for many years, and the very rapid larval development minimises the likelihood of the immature stages being killed by the pools drying out. In Nigeria, larval and pupal development was completed within six days (Service, 1970a). Larvae of *Ae. vittatus* were found in rock pools in association with those of *Cx. hortensis, Cx. mimeticus, An. claviger* and *Cs. longiareolata*. In France, *Ae. vittatus* is mainly a crepuscular species, but its biting activity continues during most of the night (Rioux, 1958). In the Mediterranean region, the species is found from late spring until the beginning of autumn. Its seasonal incidence depends both on local rainfall and on the water level of rivers. There are records of equal biting intensity both indoors and outdoors. *Ae. vittatus* females sometimes attack people in large numbers (Gutsevich *et al.*, 1974).

Distribution: Mediterranean subregion, Afrotropical and Oriental region. Its northernmost distribution range is probably limited by the 10°C isotherm. Even if the species can survive cold winters in the north, summer temperatures must be sufficiently high to enable the completion of at least one gonotrophic cycle.

Medical importance: Transmission of Yellow Fever virus has been proven experimentally (Gutsevich *et al.*, 1974).

10.1.4. Subgenus *Stegomyia* Theobald

Members of this subgenus are rather small, rarely medium sized mosquitoes. According to Huang (1979) and Gutsevich *et al.* (1974) they are characterised by the following combination of features. The male palps are more than half the length of the proboscis, often slightly longer than the proboscis, and with 5 palpomeres. The female palps are up to 1/4 the length of the proboscis, 4- or sometimes 5-segmented, and when present palpomere V is minute. The vertex is largely covered with broad and flat decumbent scales, erect forked scales are not numerous and restricted to the occiput. The scutum has a characteristic pattern of light scales for each species. The acrostichal and spiracular setae are absent, and postspiracular setae present. The scutellum has broad scales on all the lobes, and the postnotum is bare. The hind tarsus has a basal white ring at least on tarsomere I, and the wings have narrow scales on all the veins. The abdominal terga have white basal bands and often white lateral spots, and the cerci are relatively short. In the male genitalia the basal and apical lobes of the gonocoxite are absent. The gonostylus is usually simple and elongated, or sometimes it is expanded apically or at the base. Typical claspettes, which are divided into a stem and a filament are modified into a structure which is usually lobed, and located at the base of the gonocoxite and covered with numerous setae. The aedeagus is divided into two plates and is strongly toothed apically. In the larvae the antennae are without spicules, the antennal seta (1-A) is usually small and simple. The postclypeal seta 4-C is well developed and branched, and the median frontal setae (6-C) are situated in front of the inner frontal setae (5-C). The comb scales are arranged in a single row, and the base of the siphon has no acus. The ventral brush has 4 or 5 pairs of setae on a common base, without any precratal setae (4-X).

The subgenus is confined to the Old World and mainly occurs in the tropical and subtropical zones throughout this region, except *Ae. aegypti* and *Ae. albopictus* which have been introduced through commerce into the New World. It seems that in *Stegomyia* there are a few widely distributed species and a number of specialised endemic species (Huang, 1979). In Europe the subgenus is represented (or was partly represented in the past) by three species which could be found in the southern countries of the continent. Among the family of Culicidae the subgenus *Stegomyia* is one of the most important subgenera in view of the transmission of pathogens and parasites; it includes several vectors of human filiariasis and a number of severe viral diseases.

Aedes (Stegomyia) aegypti (Linnaeus) 1762

Female: A medium sized dark species with contrasting silvery white ornamentation on the head, scutum, legs and abdomen. *Ae. aegypti* is easily recognised and distinguished from the other members of the subgenus by the white scutal markings which form the typical "lyre-shaped" pattern of the species (Fig. 6.25a). The proboscis is dark scaled; the palps are 1/5 the length of the proboscis with white scales on the apical half, the clypeus with lateral white scales, and the pedicel with large patches of white scales at the sides. The vertex has a median line of broad white scales from the interocular space to the back of the occiput, and white scales also on the sides, separated by patches of dark scales. Erect scales are restricted to the occiput, and are all pale. The scutum is predominantly covered with narrow dark brown scales, with a distinctive pattern of light scales as follows: a small anteacrostichal patch of white scales, a pair of narrow dorsocentral stripes of narrow pale yellowish scales close to the midline extending to the anterior two thirds of the scutum, a short postacrostichal stripe of white scales just in front of the prescutellar area where it forks forming prescutellar dorsocentral stripes which end at the margin of the scutellum, broad lateral presutural stripes which continue over the transverse suture covered with crescent-shaped white scales followed by submedian stripes of narrow white scales reaching the posterior margin of the scutum, a patch of broad white scales on the lateral margin just in front of the wing root and a few narrow curved pale scales over the wing root. The scutellum has broad white scales on all the lobes and a few broad dark scales at the apex of the mid lobe (Fig. 6.23a). The postpronotum has a patch of broad white scales and some dark and pale narrow scales in

the upper part, and the paratergite has broad white scales. The postspiracular area is without scales, but there are patches of broad white scales on the propleuron, and subspiracular and hypostigmal areas. Mesepisternal and mesepimeral patches are present, divided in an upper and lower portion, but not connected. The upper mesepisternal scale patch does not reach the anterior corner of the mesepisternum. The mesepimeron has no lower mesepimeral setae. The coxae have patches of white scales, and all femora have white knee spots; the fore and mid femora with a narrow white longitudinal stripe on the anterior surface. All the tibiae are dark anteriorly; the fore and mid tarsi have a white basal band on tarsomeres I and II, the hind tarsus has a broad basal white band on tarsomeres I to IV, and tarsomere V is all white. The claws of the fore and mid tarsi have a subbasal tooth, and the claws of the hind tarsi are simple. The wing veins are all dark scaled except for a small spot of white scales at the base of the costa. Tergum I has white scales laterally and a median pale patch; terga II to VI have basal white bands and basolateral white spots separate from the bands, and tergum VII has lateral white spots only. Sterna II–IV are largely pale scaled, V and VI predominantly with dark scales, and VII is dark except for a small lateral pale patch.

Male: The palps are as long as the proboscis with white basal bands on palpomeres II–V. The last two segments are slender and upturned with only a few short setae. The posterior margin of tergum IX is deeply concave in the middle, and the lateral lobes are prominent with 3 apical setae. The gonocoxite is slightly more than twice as long as it is wide, with scales restricted to the lateral and ventral surfaces (Fig. 10.10). The gonostylus markedly narrows apically and is curved, with a pointed apical spine (Fig. 7.20c). The claspettes are large, lobe-like, appressed to and occupy most of the mesal surface of the gonocoxite, with numerous setae and several stronger setae, 3 or

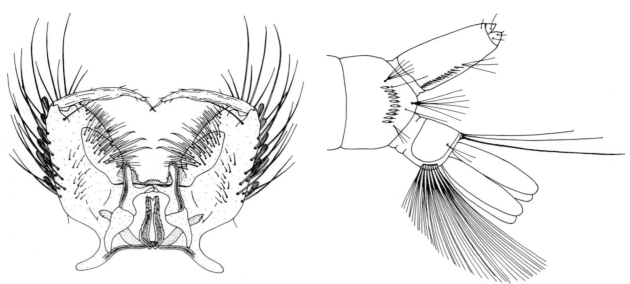

Figure 10.10.   Hypopygium of *Ae. aegypti*.

Figure 10.11.   Larva of *Ae. aegypti*.

4 of which are bent at the tip. The paraproct has an inwardly directed mesal arm, and the aedeagus is strongly serrated.

Larva: The antennae are about half as long as the head and without spicules. The antennal seta (1-A) is small and simple, and inserted slightly beyond the middle of the shaft. The postclypeal (4-C) and median frontal (6-C) setae are displaced far forward toward the anterior margin of the head. 4-C is situated slightly anterior to 6-C, and is well developed with 4–7 short branches. The frontal setae (5-C to 7-C) are long and simple (Fig. 8.60a), and the outer frontal seta (7-C) very rarely have 2 branches. At the base of the meso- and metathoracic setae (9- to 12-M and T) there is a long stout spine which is pointed and hooked at the tip. The comb consists of a single irregular row of 6–12 scales; each comb scale has a long median spine and strong smaller spines at the base forming a "triffid" appearance. The siphon is moderately pigmented; the siphonal index is about 1.8–2.5, and the acus is not developed (Fig. 10.11). The pecten has 8–22 teeth (usually 10–16), evenly spaced or sometimes the distalmost tooth is detached apically, and each tooth has 1–4 lateral denticles. The siphonal tuft (1-S) has 3 or 4 branches, usually inserted close to the distal pecten tooth and just beyond the middle of the siphon. The saddle reaches far down the lateral sides of the anal segment; the saddle seta (I-X) usually has 2 short branches. The ventral brush has 8–10 tufts of cratal setae (4-X), and the precratal setae are absent. The anal papillae are about 2.5–3 times the length of the saddle, sausage-like, and rounded apically.

Biology: In subtropical climates the species is found almost always in the close vicinity of human settlements. The larvae occur in a wide variety of small artificial containers and water recipients of all kinds, both inside and outside of human habitations in gardens and within a circle of 500 m around dwellings, *e.g.* in earthenware pots and water tanks for storing water, uncovered cisterns, rainfilled empty cans or flower pots, broken bottles or discarded motor vehicle tyres. If vegetation surrounds the settlements, the larvae may breed in tree holes, leaf axils, bamboo stumps or coconut shells after heavy rainfall. They can also be found in any artifical and natural water collection in harbours and on ships. The breeding water is mostly clean or has a moderate content of organic matter. The larvae spend a long time under water feeding on the bottom of their breeding sites. The eggs are resistant to dessication and are deposited close to the waterline in the mentioned recipients. At a temperature of 27–30°C the larvae will hatch 2 days after egg deposition, pupation occurs after 8 days and the adults emerge from the pupae 9.5 days after the eggs have been laid. The females feed predominantly during the day in shaded places and only occasionally during the night in lit rooms. Human blood seems to be preferred to that of domestic animals (Carpenter and LaCasse, 1955), the blood-feeding interval is only about 2–4 days. The adults are frequently found resting indoors, *e.g.* in cupboards, closets or behind doors. They do not migrate over long distances, and rarely fly more than several hundred metres from their breeding sites. The species is supposed to have its origin in Africa and subsequently spread over vast areas, mainly the Tropics.

Ae. aegypti is one of the most suitable mosquitoes for laboratory colonies and has been extensively used as a test organism for laboratory research in many fields. At a constant rearing temperature of 22–28°C the species has some outstanding advantages in respect of colonisation: the adults and larvae are easy to handle, rearing is possible in nearly every type of cage and breeding container, and mating takes place even in the smallest space. The females readily feed on a variety of small mammals offered as a blood source and the eggs can be stored for months if necessary, without losing viability. The available literature about *Ae. aegypti*, its biology and medical significance is numerous; a well recognised monograph was published by Christophers in 1960.

Distribution: This cosmotropical species is distributed in the tropical, subtropical and warm temperate regions of both hemispheres.

Its range is mainly limited by the 10°C cold-month isotherms where breeding can continue all year round (Christophers, 1960). Certain populations may extend their summer range considerably north of this line, *e.g.* in North America specimens may be found up to 40° northern latitude in southern Illinois and Indiana, but they are not able to survive during cold winter months; this prevents the establishment of permanent populations. In Europe, prior to 1945, all Mediterranean countries and most major port cities had reported at least occasional introductions of *Ae. aegypti* (Mitchell, 1995). It could be found in Portugal, Spain, France, Italy, Yugoslavia, Greece and Albania, but has now been eradicated or has become rare in many countries where it was previously common and has currently not been reported from the Mediterranean region.

Medical importance: As the major vector of yellow fever virus, *Ae. aegypti* has long been notorious as the "yellow fever mosquito", but it is also an important vector of dengue and several other viral infections.

Synonymy: The species in question has appeared under various other names in the past. The number of names, which are now accepted as synonyms is large and listed in Knight and Stone (1977). Synonyms that are often found in the early literature are *Stegomyia fasciata* of Theobald 1901 (described as *Culex fasciatus* by Fabricius 1805), *Aedes (Stegomyia) calopus* Meigen 1818 and *Aedes argenteus* of Edwards 1921 (described as *Culex argenteus* by Poiret 1787). From about 1935 on, the earliest name, namely *aegypti* given by Linnaeus in 1762, was accepted and is in general use until today.

Note on systematics: A paler variation of the type form, var. *queenslandensis* (Theobald) exists and a subspecies *formosus* (Walker) is characterised by its markedly darker appearance. The latter form is confined to Africa south of the Sahara and has been recorded from the forest or bush away from human settlements, breeding in natural places.

Aedes (Stegomyia) albopictus (Skuse) 1895

Female: The proboscis is dark scaled, about the same length as the fore femur, and the palps are 1/5 of the length of the proboscis with white scales on the apical half. The clypeus is bare and entirely dark. The vertex has broad white scales, and the occiput in the middle is white scaled with dark scales at the sides; erect scales are usually absent. The scutum is mainly covered with narrow dark scales, with a prominent acrostichal stripe of narrow white scales which narrows posteriorly, and extends from the anterior margin of the scutum to the beginning of the prescutellar area where it forks to the end at the margin of the scutellum (Fig. 6.26a). On each side a slender posterior dorsocentral white stripe does not reach the middle of the scutum, but extends about midway to the level of the scutal angle. The supraalar white stripe is incomplete; there is a patch of broad white scales on the lateral margin just before the level of the wing root and a few narrow white scales over the wing root. The scutellum has broad white scales over all the lobes with an apical area of dark scales on the mid lobe. The postpronotum has a large patch of broad white scales and some narrow dark ones in the upper part, and the paratergite has broad white scales. The postspiracular area is without scales, and the subspiracular area has white scales. The mesepisternal patch is divided into large upper and lower patches of white scales. The mesepimeron has upper and lower scale patches connected which form a V-shaped white scale patch, with the open V directed backwards. The coxae have patches of white scales; the fore and mid femora are dark anteriorly and paler posteriorly with apical pale spots. The hind femur has a broad white anterior stripe widening at its base and slightly separated from the apical white scale patch. The tibiae are all dark. The fore and mid tarsi have narrow basal white bands on tarsomeres I and II, the hind

tarsus has broad basal white bands on tarsomeres I to IV, and tarsomere V is all white. The claws are simple without a subbasal tooth. The scales on the wing veins are all dark except for a small spot of white scales at the base of the costa. Tergum I has white scales laterally, terga II to VII have basolateral white spots. In addition, terga III to VI have narrow basal white bands, which widen laterally and do not connect with the spots.

Male: The palps are longer than the proboscis with white basal bands on palpomeres II–V. The last two segments are slender and upturned with only a few short setae. The posterior margin of tergum IX has a conspicuous horn-like median projection and a small setose lobe on each side. The gonocoxite is approximately twice as long as it is wide with a patch of setae on the basomesal area of the dorsal surface (Fig. 10.12). The gonostylus is simple,

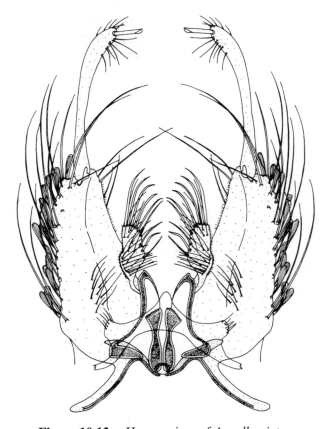

Figure 10.12. Hypopygium of *Ae. albopictus*.

elongated, distinctly swollen apically and has a few thin setae. The spine of the gonostylus is inserted subapically and is blunt ended. The claspettes are large, lobe-like, and occupy most of the mesal surface of the gonocoxite, with numerous long setae and several stronger setae, a few of which are curved apically.

Larva: The head is approximately as long as it is wide. The antennae are about half as long as the head, and without spicules. The antennal seta (1-A) is simple and small and situated close to the middle of the shaft (Fig. 8.62c). The post-clypeal seta (4-C) is located close to the anterior margin of the head. It is well developed with 6–15 branches and has a short stem. The median frontal seta (6-C) is displaced anteriorly, and has 1–2 branches; the inner frontal seta (5-C) is situated posterior to 6-C, and is longer and simple; and the outer frontal seta (7-C) usually has 2–3 branches. The comb has 6 to 13 (usually 8–10) long slender scales in a single row, and each comb scale has a large pointed main spine and fine denticles or fringe at its base. The siphon is short, and tapers distinctly from the middle; the siphonal index is 1.7–2.5 (Fig. 10.13). The number of pecten teeth is between 8 and 14; they are evenly spaced, and each tooth usually has 2 lateral denticles. The siphonal tuft (1-S) has 2–4 branches, and is inserted beyond the distalmost pecten tooth, slightly beyond the middle of the siphon (Fig. 8.62d). The saddle extends to the ventral margin of the anal segment, and the saddle seta (I-X) usually has 2 branches, at least one of them distinctly longer than the saddle. The upper anal seta (2-X) has usually 2 branches, the lower anal seta (3-X) is simple. The ventral brush has 4 pairs of cratal setae (4-X), but the precratal setae are absent. The anal papillae are sausage-like, and about 3 times the length of the saddle.

Biology: The immature stages occur in a variety of small natural and artifical containers, *e.g.* in tree holes, bamboo stumps, coconut shells, rock holes, plant axils or palm fronds and in flower pots, tin cans, water jars, metal

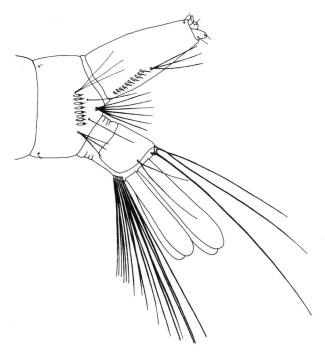

Figure 10.13. Larva of *Ae. albopictus*.

and wooden buckets or drums, broken glass bottles or discarded motor vehicle tyres (Huang, 1972). The eggs are resistant to desiccation, which facilitates their transport in used tyre casings, even over long distances. Continuous breeding throughout the year takes place in tropical and subtropical areas, but in more temperate climatic zones, such as Europe, populations of *Ae. albopictus* are found which show embryonic diapause and overwinter in the egg stage. Several generations per year may occur. Adult females predominantly feed on humans, but may also bite other mammals including rabbits, dogs, cows and squirrels or occasionally avian hosts, *e.g.* Passeriformes or Columbiformes. This feeding behaviour indicates that *Ae. albopictus* is well suited for transmission of a variety of arboviruses that use mammals and birds as their main hosts (Mitchell, 1995). To feed on humans the females readily enter dwellings during dusk and at night, but may also be found biting during the daytime outside houses in shaded situations. *Ae. albopictus* is a mass species in East Asia causing great nuisance wherever it occurs and, although not present before 1990, has become a major pest species in some areas of northern Italy.

Distribution: In the past *Ae. albopictus* was mainly distributed in the Oriental Region and Oceania and thus it got its popular name, the "Asian Tiger Mosquito". In the Palaearctic it occured in Japan and China. In 1985 it was discovered for the first time in the New World (Houston, Texas) and this was the beginning of a rapid spread and discovery of recently introduced populations of *Ae. albopictus* in many parts of the world (Mitchell, 1995). It is now present in over 20 states of the U.S. and in several countries of South America and Africa. Specimens have been found in Australia and New Zealand, but breeding populations have not so far become established there. In Europe, *Ae. albopictus* has probably been present in Albania since at least 1979 (Adhami and Murati, 1987). In the early 1990s it was passively introduced in Italy, due to the international trade of used tyres which provide a suitable habitat for the eggs. The species was first detected in Genoa in September 1990 (Dalla Pozza and Majori, 1992) followed by a rapid spread into other areas of northern and central Italy. In 1999 *Ae. albopictus* was found in western France (Schaffner *et al.*, 2001) and in 2001 in Montenegro (Petrić *et al.*, 2001). The introduction of the species into other countries, especially in southern Europe, may be expected.

Medical importance: *Ae. albopictus* is a vector of dengue viruses and a competent transmitter for numerous other arboviruses as well as *Dirofilaria immitis* (dog heartworm).

Aedes (Stegomyia) cretinus Edwards 1921

Female: The proboscis is completely dark scaled; the palps are about 1/4 the length of the proboscis with whitish scales dorsally on the apical half and dark scales ventrally. The clypeus is bare, and the pedicel has white scales

anteriorly. The vertex has a broad median stripe of broad white scales, and the occiput has a lateral stripe of broad white scales and extensive broad white scaling below; all dark scales are broad and flat. There are narrow white scales at the eye margin, and the erect forked scales are dark. The scutum has narrow dark scales, and an acrostichal stripe of narrow white scales extends from the anterior margin to the beginning of the prescutellar area, where it forks and ends just before the margin of the scutellum (Fig. 6.26b). Dorsocentral white stripes are present on the posterior part of the scutum extending from just posterior to the level of the scutal angle to near the lateral lobes of the scutellum; these stripes are narrow and composed of narrow white scales. The scutum is bordered with a lateral prescutal stripe of narrow white scales which reaches the scutal angle, where after a minute break it continues on with broad white scales and terminates with a few narrow white scales just before the margin of the lateral lobes of the scutellum. The scutellum has broad white scales on all the lobes and a small apical area of dark scales on the mid lobe. The ante- and postpronotum are largely covered with broad white scales. The pleurites have several patches of broad white scales, some of them very densely scaled. The wing veins are dark scaled except for a conspicuous basal spot of pale scales on the costa (C). The fore femur anteriorly has sparse white scales on the basal half and a small white knee spot; posteriorly it is white and the fore tibia is dark. The fore tarsomeres I and II have basal white rings, and III to V are dark. The mid femur anteriorly is dark except for a few white scales at the base and a conspicuous white knee spot, the mid tibia is dark, the mid tarsomeres I and II with basal white rings, III to V dark. The hind femur anteriorly is white almost to its apex and has a conspicuous white knee spot; the hind tibia is dark, and hind tarsomeres I to III have a basal white ring, and hind tarsomere IV has an extreme tip which is dark, but V is white. Abdominal terga II to IV have narrow white basal bands slightly constricted in the middle and not connected to the broad lateral white patches. Sterna II to IV are largely covered with white scales, and V to VII with basal white bands.

Male: The median part of tergum IX is evenly rounded, and the small lateral lobes are strongly sclerotised with 2–4 fine setae. The gonocoxite is oblong, the gonostylus is long and slender, and dilated at the apex where it bears a number of fine setae and a long, slender subapical spine (Fig. 10.14). This spine is markedly longer and more slender than in *Ae. albopictus* (Mattingly, 1954). The claspettes narrow apically, and are covered with setae of varying length, but no specialised setae are present.

Larva: The antenna is smooth, and the antennal seta (1-A) is minute and simple, located at just beyond midway from the base. The outer frontal seta (7-C) is simple. The comb has 9–13 scales arranged in one row with exceedingly minute basal denticles. The siphonal index is about 2.0 (Fig. 10.15). The pecten has 10–13 teeth, some with the main denticle longer and more finely drawn out. The siphonal tuft (1-S) has 3 branches, located about the middle of the siphon with a simple seta situated laterally in the apical third (Fig. 8.61a). The saddle reaches the ventral margin of the anal segment and has a smooth distal edge; the saddle seta (1-X) is simple or 2-branched; the upper anal seta (2-X) has 2–3 branches, and the lower anal seta (3-X) is simple. The anal papillae are longer than the saddle, but at most two thirds of the length of the siphon.

Biology: Little appears to be known about the biology of *Ae. cretinus* (Lane, 1982). Gutsevich *et al.* (1974) reported that larvae were found in tree holes together with those of *An. plumbeus*, *Oc. geniculatus* and *Or. pulchripalpis* and adult females were biting man in shaded areas.

Distribution: Cyprus, Greece, Crete, Georgia and Turkey.

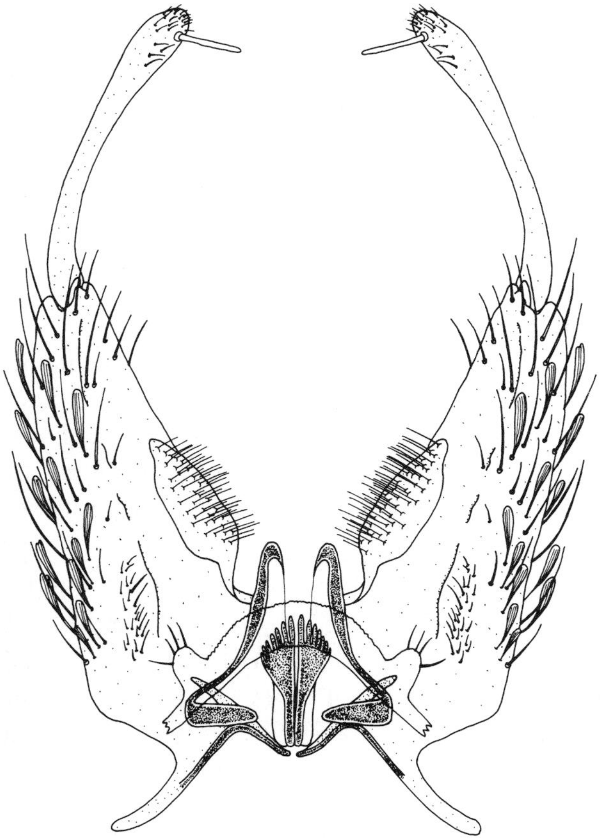

Figure 10.14. Hypopygium of *Ae. cretinus.*

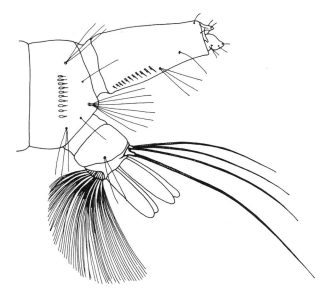

Figure 10.15. Larva of *Ae. cretinus.*

10.2. GENUS *OCHLEROTATUS* LYNCH ARRIBALZAGA

This genus was established by the division of the composite genus *Aedes* into two genera, *Aedes* and *Ochlerotatus*, and restoration of the former subgenus *Ochlerotatus* of genus *Aedes* to generic rank. A further subdivision of *Ochlerotatus* into two sections is proposed (Reinert, 2000c). This classification is primarily based on female and male genitalic characters, *e.g.* the aedeagus in the male hypopygium consisting of two lateral plates that usually bear teeth laterally and/or apically in *Aedes*, and the aedeagus being simple and scoop-like, trough-like, or tube-like in *Ochlerotatus*. Supplemental features supporting the partition of the genera include characteristics of the fourth stage larvae and pupae. For a detailed morphological description and delimitation of *Aedes* and *Ochlerotatus*, see Reinert (2000c). Of the European mosquito species, members of the subgenera *Finlaya*, *Ochlerotatus* and *Rusticoidus* are included in *Ochlerotatus*, which all belong to Section I.

10.2.1. Subgenus *Finlaya* Theobald

The subgenus comprises medium to large species. The length of the palps in the females ranges from very short to 2/3 the length of the proboscis. The head has a mixture of narrow and broad scales, with erect forked scales on the vertex and occiput. The scutum has prominent white and dark scale patterns and variable setation; the dark scales are black or have a metallic tinge. The pleurites have patterns of white scales, and are more extensively scaled in females than in males. The mesepimeral setae are sometimes absent. White knee spots are present, and all tarsomeres are dark. The wing veins are covered with dark scales. The abdominal terga have more or less distinct lateral white patches, sometimes with white basal bands, and the cerci are short and blunt. The length of the male palps varies from half the length of the proboscis to much longer than it. Tergum IX has well developed lobes and more or less spine-like setae. The hypopygium has an elongated gonocoxite, without distinct basal or apical lobes but defined fields of setation. The gonostylus is long and slender, and the claspette is divided into a stem and a filament of different shapes. The paraproct is usually heavily sclerotised, and the aedeagus is pear shaped. The head of the larva is rounded, and the antenna is usually shorter than the head, with a simple antennal seta (1-A). The frontal setae (5-C to 7-C) are simple or may have 2 branches. The abdominal segments are usually covered with stellate setae. The comb scales are large and arranged in a single row. The saddle does not encircle the anal segment, and precratal tufts are present. The ventral and dorsal anal papillae are of different sizes.

The subgenus with its nearly 200 species is one of the largest in the genus *Ochlerotatus*. It is mainly distributed in Asia, Australia and Africa. It is a polymorphic subgenus of which species of North and South America have been revised and placed in other subgenera (Zavortink, 1972). It is doubtful whether the two European species, *Oc. echinus* and *Oc. geniculatus*, really belong to *Finlaya*.

Ochlerotatus (Finlaya) echinus Edwards 1920

Female: Closely related to *Oc. geniculatus* with a similar morphology in all stages. The female of *Oc. echinus* differs from that of *Oc. geniculatus* by patterns of scaling on the head, thorax, abdomen and legs. The proboscis and palps are dark scaled, the head has black setae, the scales on the vertex are dark and laterally with white patches, and there is no white scale border around the eyes. The scutum has two dorsocentral stripes of narrow dark bronze scales divided by a creamy white acrostichal stripe. The supraalar dark scale patch is nearly fused with the dorsocentral stripes. The scutellum has broad whitish scales. The postpronotal scales are whitish, and the mesepisternum and mesepimeron have patches of creamy scales, distinct against the dark integument. The femora have a white ventral stripe and a white knee spot. The tibiae and tarsomeres are entirely black scaled. The wings have rather narrow blackish scales. The abdominal terga are dark scaled with narrow white basal bands, sometimes interrupted in the middle from tergum V on, and on all segments extended laterally into triangular whitish patches. Abdominal segment VIII is broad, and the cerci are short and rounded.

Male (Fig. 10.16): The general shape of the hypopygium is similar to that of *Oc. geniculatus* except for the denser setation of the gonocoxite towards the tip. The claspette stem has a stout seta near the middle and some thin setae at the base, and the claspette filament is hook-like.

Larva: The antenna is more than half as long as the head (longer than in *Oc. geniculatus*), smooth and is not covered with spicules.

The antennal seta (1-A) is simple (Fig. 8.58a). The inner frontal seta (5-C) is simple, the median frontal seta (6-C) is simple or double and the outer frontal seta (7-C) usually has 2–4 branches. Branches of the stellate setae on abdominal segment I are longer than the length of the segment (Fig. 8.59a). The number of comb scales is 11–18, arranged in one row. Each scale is elongated with lateral spines, and a prominent median spine is absent. The siphonal index is 2.5–3.6, and the acus is well developed (Fig. 10.17). The pecten has 15–27 teeth, and occupies at least the basal half of the siphon. Each tooth is very long and spine-like, and occupies at least the basal half of the siphon. The siphonal tuft (1-S) is inserted beyond the middle of the siphon, with 2–4 branches. The anal segment is not entirely encircled by the saddle, and 1–2 precratal tufts (4-X) are present. The anal papillae are broad and long, and the dorsal pair is twice as long as the ventral pair.

Biology: Larvae have been found in the same habitats as *Oc. geniculatus*. In Anatolia and Bulgaria they may also occur in root holes of olive trees. No more is known of the biology of the larvae; they are supposed to feed on microorganisms in the tree holes in the same way as the larvae of *Oc. geniculatus* do. In Portugal larvae and adults were found in August and September (Ribeiro *et al.*, 1988).

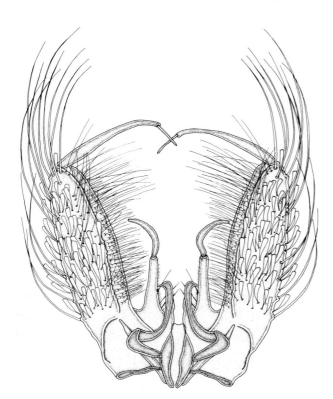

Figure 10.16. Hypopygium of *Oc. echinus*.

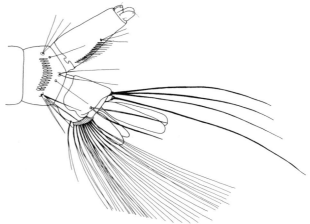

Figure 10.17. Larvae end of *Oc. echinus*.

Distribution: In Europe this species is confined to the Mediterranean region and has been reported in Portugal, Italy, Greece and Bulgaria where it has been found abundantly along the Black Sea coast.

Note on systematics: Edwards (1920) originally placed this species in the subgenus *Ochlerotatus* and transferred it to *Finlaya* later (Edwards, 1932). The same considerations regarding the subgenus affiliation of *Oc. geniculatus* also apply to *Oc. echinus*. As the types of most of the synonymous species of *Oc. geniculatus* are lost, apparent synonymy to *Oc. echinus* is not resolved. From Madeira and the Canary Islands another species within the subgenus *Finlaya, Oc. eatoni* Edwards, has been reported (Knight and Stone, 1977).

Ochlerotatus (Finlaya) geniculatus (Olivier) 1791

Female: Dark scales with a violet tinge especially on the abdomen, the white and blackish pattern of the scutum, the conspicuous white knee spots and the blunt cerci immediately separate the females from all other females of the genus *Ochlerotatus* except the closely related *Oc. echinus*. The proboscis and palps are black scaled, and the vertex is dark with a median light stripe and a narrow band of whitish scales around the eyes. The scutum has two dorsocentral black stripes which sometimes fuse into one anteriorly, or are otherwise completely separated by a pale acrostichal stripe. The submedian and lateral areas of the scutum have creamy or silvery grey scales. Dark anterior and posterior submedian stripes are present, and the scutellum has narrow yellowish scales. The pleurites have patches of broad, whitish scales. The legs are dark; the femora have a white knee spot, and the tibiae and tarsomeres are entirely black scaled. The fore and mid claws have a subbasal tooth. The wings have dark brownish scales. The abdominal terga are black scaled with conspicuous white triangular lateral patches on segments II–VII. Sternum VIII is unusually broad, and the cerci are broad and rounded (Fig. 6.37a).

Male: Tergum IX has somewhat elongated lobes and 4–5 spine-like setae on each lobe. The hypopygium superficially resembles that of the species of the Ochlerotatus Excrucians Complex by lacking a basal lobe, but is separated from it by the absence of an apical lobe (Fig. 10.18). The long and evenly tapered gonocoxite has two areas of dense setation, a basal one with shorter setae, and an apical one with long setae. The claspette stem is short with several setae, and the claspette filament is narrow and somewhat shorter than the stem. The paraproct is heavily sclerotised and bent at the tip, and the aedeagus is pear shaped.

Larva: The larvae are separated from those of all other subgenera of *Aedes* and *Ochlerotatus* (except *Oc. echinus*) by the numerous stellate setae on the thorax and abdomen, the broad and unequally long anal papillae and the single row of large comb scales. The antenna is half as long as the head, is smooth and not covered with spicules

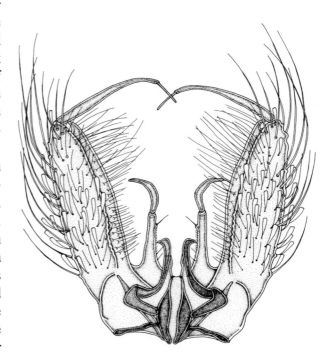

Figure 10.18. Hypopygium of *Oc. geniculatus*.

(Fig. 8.24b). The antennal seta (1-A) is usually simple. The inner frontal seta (5-C) is usually simple, the median frontal seta (6-C) has 1–2 branches and the outer frontal seta (7-C) has 2–4 branches. All setae on the prothorax are double, except for 2-P. Most of the setae on the rest of the thorax and abdomen are of the stellate type. The branches of the stellate setae on abdominal segment I are about the same length as the segment (Fig. 8.59c). The number of comb scales is 8–15 arranged in a single row. Each individual scale is elongated with a swollen base, a strong median spine and small lateral spines at the base. The siphonal index is 2.3–3.2 (Fig. 10.19). The number of pecten teeth is 15–19, and each tooth is long, spine-like, with a few indistinct denticles at the base. The pecten usually occupies less than the basal half of the siphon. The siphonal tuft (1-S) is situated at about the middle of the siphon or slightly below it, with 4–5 branches. The anal segment X is not completely encircled by the saddle, and the latter has a row of long microtrichiae at the distal part. The ventral brush is made up of

7–10 cratal tufts (4-X) and 1–2 precratal tufts. The anal papillae are broad and longer than the saddle, with the ventral pair being shorter.

Biology: The larvae live in tree holes at various heights and in open tree stumps of different deciduous trees as *Quercus* sp., *Fagus* sp., *Alnus* sp., *Betula* sp. and *Juglans* sp. They also occur in mixed forests in old trees and can occasionally be found in ground pools together with *Or. pulcripalpis*, *An. plumbeus* and *Oc. pulcritarsis*, but very rarely in coniferous forests. The species hibernates in the egg stage in northern areas, and in the larval stage in more southern regions. The adults appear during the summer as the development depends on spring and summer rains collected in the tree holes. Females are day and crepuscular biters and readily feed on humans. In southeastern Europe they can occur in masses, viciously attack man in the open, but rarely enter urban areas.

Distribution: Palaearctic region, known in most European countries, its northernmost limit follows that of deciduous or mixed forests. In the Mediterranean region it is reported in northern Portugal, Sardinia, Italy mainland, Greece and extends east to the Caucasus. Also reported from north Africa to Asia Minor.

Note on systematics: When the subgenus *Finlaya* of the old world is revised, this species will most probably turn out not to belong to it. The European species resemble those of the Nearctic subgenus *Protomacleaya*.

Medical importance: Experimentally the species was shown to retain Yellow Fever virus, but no virus transmission has been documented from field investigations (Aspöck, 1996).

10.2.2. Subgenus *Ochlerotatus* Lynch Arribalzaga

The flagellomeres of the female antennae are prolongated distally, the palps are only 1/3 to 1/4 the length of the proboscis, the latter being longer than the fore femur. The scale pattern on the vertex and occiput is variable, often with

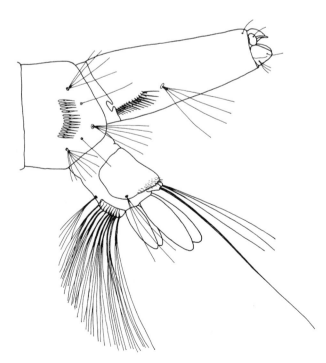

Figure 10.19. Larva of *Oc. geniculatus*.

numerous erect forked scales on the occiput and mixed narrow and broad scales on the vertex and along the eye margin. The thorax in most species has a dark grey to dark brown or blackish integument. The scutum is covered with scales and with rows of acrostichal, dorsocentral and supraalar setae. The scutellum has three lobes and groups of setae and a few narrow scales. The pleurites are extensively scaled with patches of mostly pale to whitish scales. Prespiracular setae are absent. A postprocoxal patch of whitish scales is present in some species. The legs are mostly covered with dark scales, but pale scales may be scattered or grouped to form a knee spot or basal or apical rings, mainly on the tarsomeres. All the tarsal claws have an additional subbasal tooth, and the pulvilli are setous or inconspicuous. The wings are predominantly dark scaled; both the costa and subcosta may have patches of paler scales, and in some species the wing veins are covered with mixed dark and pale scales. The abdomen has elongated cerci and the usually narrowed last segments give the impression of being pointed. The scaling of the abdomen is extensive on both the terga and sterna. It can be rather uniform or display various patterns of banding or mixed colours. The scale patterns on the thorax, legs, wings and abdomen are often used for species identification.

The proboscis of the males is often not longer than the fore femur, and the palps are usually longer than the proboscis, but sometimes as long as the proboscis or shorter. The tarsal claws of the front and mid legs have prolonged main and subbasal teeth, and the hind legs usually have claw shaped teeth as in the females. Tergum IX always has two more or less expressed lateral lobes which usually bear a group of strong or spine-like setae. The gonocoxite in most species has basal and apical lobes, sometimes one or both less expressed, indistinct or absent. The gonostylus is simple with an apical spine. The paraproct has pointed tips, sometimes inwardly curved. Typical claspettes are present, divided into stem and filament. The aedeagus is pear shaped, elongated or rounded.

The antennae of the larvae have a multiple-branched antennal seta (1-A), usually situated at about the middle of the antenna. The lateral palatal brushes are well developed for suspension feeding or brushing. The postclypeal seta (4-C) is inconspicuous, and multiple-branched. The inner frontal seta (5-C) is often situated in front of the median frontal seta (6-C), both pairs being simple to multiple-branched. Prothoracic setae 1-P to 7-P are simple to triple-branched. The number of comb scales is variable from a few to many, arranged in a single or irregular rows. The siphon is well developed, with the siphonal seta (1-S) usually situated at about the middle of the siphon. The pecten has more or less spaced teeth of significant shapes. The saddle partly or fully encircles the anal segment, and the saddle seta (1-X) is usually simple. The cratal and precratal tufts (4-X) are well developed. The anal papillae are of variable shape and size.

Of the nearly 200 species of the subgenus described worldwide, more than half are distributed in the Holarctic region and nearly a quarter in each of the Australian and the Neotropical regions. Only a few species are found in the Oriental and African regions.

In the western Palaearctic and throughout Europe Alphavirus, Flavivirus and three different groups of Bunyavirus were found in a few isolates from *Ochlerotatus* species, such as *Oc. cantans*, *Oc. caspius*, *Oc. communis*, *Oc. flavescens*, *Oc. hexodontus*, *Oc. punctor* and *Oc. sticticus* (Traavik *et al.*, 1985; Lundström, 1994; Aspöck, 1996). Some other parasites have been reported from the *Ochlerotatus* species in Europe, such as the bacterium *Francisella tularensis*. In North America virus vector capacity is documented for several species of the subgenus (Reeves, 1990; Beaty and Marquardt, 1996).

Morphological heterogenity in the subgenus is great, several species groups show distinct characters. As long as no worldwide analysis of the subgenus exists, discrepancies of grouping the species within different regions will prevail. Edwards (1932) first revised and designed species groups and subgroups of *Ochlerotatus* on a worldwide base including Holarctic, South American, Oriental and Australian species. A complete grouping of all European members of the subgenus *Ochlerotatus* does not exist so far. Based on the suggestions for the species grouping given for the Palaearctic region (Martini, 1931), the Fennoscandian region (Natvig, 1948), Germany (Mohrig, 1969) and the former USSR (Gutsevich *et al.*, 1974), the following classification is given, regarding the European species which are included in the keys.

Annulipes Group

Large to medium sized mosquitoes. The tarsi have pale basal rings, which are broad at least on some tarsomeres. Identification of the females is sometimes very difficult and arbitrary. The basal lobe of the gonocoxite is never constricted at the base. The apical lobe is well developed. The claspette filament is winged. The larvae usually have 4–6 precratal setae (4-X). The species included in this group are: *annulipes, behningi, cantans, cyprius, euedes, excrucians, flavescens, mercurator, riparius, surcoufi.*

Caspius Group

The tarsi have pale rings embracing two tarsomeres, the apex of one and the base of the following tarsomere. The apical lobe of the gonocoxite is weakly developed or absent. The species included in this group are: *berlandi, caspius, dorsalis, mariae, pulcritarsis, zammitii.*

Communis Group

Usually medium sized mosquitoes. The tarsi are entirely dark scaled. The claspette filament is differentiated into a well sclerotised ridge and weakly sclerotised transparent wing.

The species included in this group are: *cataphylla, coluzzii, communis, detritus, hungaricus, impiger, nigrinus, leucomelas, pionips, sticticus.*

Intrudens Group

The tarsi are entirely dark scaled, but particular differences can be found in the male hypopygium, which are unique for this group. The basal lobe of the gonocoxite has 3 setae distinctly larger than the rest. One basal seta is well separated from the two distal setae. Dense long setae cover the inner surface of the apical half of the gonocoxite to a various extent. The claspette stem has a thorn shaped process or is distinctly swollen and bent at about the middle. The species included in this group are: *diantaeus, intrudens, pullatus.*

Punctor Group

The tarsi are entirely dark scaled, but members of this group differ from the others by the structure of the tenth abdominal segment of the larvae. The saddle completely surrounds the anal segment or the ventral margins of the saddle are separated by only a very narrow gap. In the male hypopygium the claspette filament is evenly sclerotised without or a narrow transparent wing. The species included in this group are: *hexodontus, nigripes, punctor, punctodes.*

Ochlerotatus (Ochlerotatus) annulipes (Meigen) 1830

Female: The general colouration of the integument is more brownish and the scaling more yellowish than in *Oc. cantans*, but less golden than *Oc. riparius*. The females of these three species resemble each other superficially, however the scale pattern of the scutum is different. The proboscis of *Oc. annulipes* is predominantly creamy white with mixed dark scales. The palps have mixed dark and pale scales, and a sometimes distinct basal light ring. The head has bronze scales and a lateral patch or stripe of creamy white scales. The scutum has a defined median stripe of brown or fawn coloured scales, and the lateral parts are covered with cream coloured or greyish scales. The antepronotum and propleuron have whitish scales. The postpronotum in the upper half has narrow bronze scales, and broad yellowish ones in the lower half. A few post-procoxal scales are usually present. A hypostig-mal patch is absent, but two distinct patches are present on the sub- and postspiracular areas. The mesepisternum has three distinct patches and a few scattered scales at its upper edge. The mesepimeron is covered with creamy white scales in the upper half, and a few scales are present on the lower half and on the mesomeron. The coxae have scattered light scales; the femora are mostly yellowish scaled, the front leg occasionally has a dark spot above the knee, the mid leg is somewhat darker on the dorsal side, and all legs have white kneespots. The tibiae are light scaled, especially the fore tibia, or otherwise speckled with dark scales. There are basal rings of whitish scales on tar-someres I to V, except tarsomere V of the fore legs, which is usually entirely dark scaled. The tarsal rings are variable in breadth, but usually broader than in *Oc. cantans*. The wing veins are covered with intermixed dark and pale scales, the pale scales are usually more yellowish than in *Oc. cantans*. The abdomen has basal white bands on terga I to VII, the last segments rarely have very narrow apical bands. All the terga have some light scales mixed among the darker ones (Fig. 6.34b). The sterna are usually yellowish scaled, with some speckled dark scales.

Male: Tergum IX has two well developed lateral lobes, each with 4–6 strong setae. The basal lobe of the gonocoxite is indistinct or absent, without strong spine-like setae, but numerous thin setae (Fig. 10.20). The apical lobe is well developed. The gonocoxite bears a large amount of long and dense setae at the distal part of its inner ventral surface. This is a unique character within the species of the annulipes group. The gonostylus is curved, and

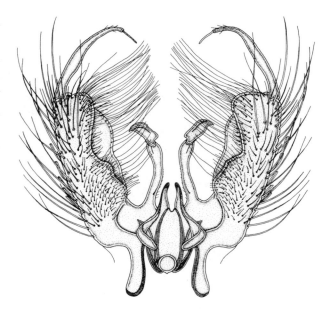

Figure 10.20. Hypopygium of *Oc. annulipes*.

tapers towards the apex. The claspette stem is stout and slightly swollen at the apex, the fila-ment is broad and less than half as long as the stem, and slightly swollen beyond the middle.

Larva: Very similar to that of *Oc. cantans* but with a more tapered distal part of the siphon. The antennae are shorter than the head, and the antennal seta (1-A) has 3–4 branches, positioned at the middle of the antennal shaft. The inner and median frontal setae (5-C and 6-C) have 2–3 branches. The prothoracic for-mula 1-P to 7-P is as follows: 1 (short, simple to 2-branched); 2 (moderately long, simple); 3 (long, simple); 4 (shorter than 2-P, simple); 5 and 6 (simple, long); 7 (long, 3–4 branches). The number of comb scales ranges between 30–40, and each scale has a long median spine (Fig. 10.21). The siphonal index is usually less than 3.0, and the pecten teeth are similar to that of *Oc. cantans*. The siphonal tuft (1-S) has 5–7 branches, situated at about the middle of the siphon. Seta 9-S is prominent, but not stout. The saddle does not encircle the anal segment but covers two thirds of its lateral sides. There are usually six or more precratal tufts (4-X) present, which separate the larvae from those of

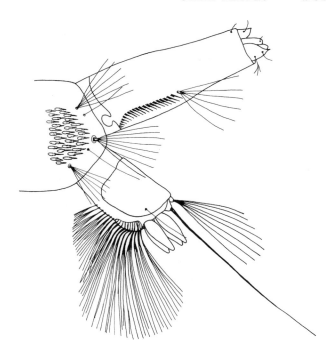

Figure 10.21. Larva of *Oc. annulipes*.

Oc. cantans, that usually have less precratals. The anal papillae are about as long as the saddle or longer.

Biology: *Oc. annulipes* is a monocyclic species occuring in spring. It is widespread throughout Europe, but most abundant in the central parts of the continent, where it can be very dominant locally. It overwinters in the egg stage and the larvae occur at the same time as *Oc. cantans* or slightly later. They breed in open meadow pools, at forest edges and inside decidious forests, preferably in semipermanent pools with leaf detritus, where they are often found together with larvae of *Oc. cantans*. In more open habitats they occur together with larvae of *Oc. flavescens*, *Oc. riparius* and *Oc. excrucians*. The males are found around the breeding sites for several days after emergence. The females are day biters with a crepuscular activity in the areas of high abundance. They are present over several weeks to months during late spring and summer depending on local climate.

Distribution: *Oc. annulipes* is a western Palaearctic species reported from southern Scandinavia to the Mediterranian region. The separation of larvae and females of *Oc. cantans* and *Oc. annulipes* from different areas may be difficult because of the variation of characters such as colouration and setation.

Note on systematics: The resurrection of *Oc. surcoufi* (Theobald) from *Oc. excrucians* (Walker) and the occurrence of the species *Oc. euedes* Dyar in Europe, both with similar larvae and females, further complicates the identification of females caught in traps. The Annulipes Group needs further investigation of a large number of individuals from all species and different geographic areas to analyse the amount of variation and/or covariation in characters.

Medical importance: Tahyna virus has been isolated from *Oc. cantans/annulipes* in Austria (Lundström, 1999).

Ochlerotatus (Ochlerotatus) behningi (Martini) 1926

Female: A medium to large species. *Oc. behningi* has a more pronounced blackish integument and setation than other members of the Annulipes Group. The setae on the pleurites rarely have a golden shine. The proboscis is dark, with intermixed creamy scales along its whole length, and mostly pale scales in the middle. The palps are dark scaled, mixed with a few white scales. The antennae are covered with very dark setae, the pedicel with some white scales. The vertex is covered with creamy scales intermixed with some black ones, forming an indistinct patch, and the occiput has a mixture of narrow golden and dark erect scales and a broad midstripe of yellowish flat scales. The scutum is usually entirely covered with small, narrow golden bronze or rust coloured scales, sometimes indistinct creamy white posterior submedian stripes are present, and the scutellum is covered with dark scales. The antepronotum has a few bronze scales and the propleuron has some whitish scales. The postpronotum has narrow scales, more than two thirds of them are bronze, the rest are creamy coloured. A postprocoxal scale patch is present. The sub- and postspiracular areas have a continuous whitish scaling. The mesepisternum has three distinct patches; an upper patch with creamy scales, a prominent median patch reaching the anterior margin and a small lower patch along the posterior border of the pleurite. One patch covers the upper half of the mesepimeron. The coxae have prominent light scale patches. The femora have mixed whitish and dark scales, with white knee spots. The tibiae as well as tarsomeres I have a somewhat darker pattern. Tarsomeres II to IV have broad whitish basal bands covering up to half of their length, and tarsomere V is dark with a few light scales on the fore and mid legs. The wing veins have dark scales intermixed with a few or many creamy white scales. Some amount of colour variation in scaling might be expected in the females. The abdominal terga are predominantly covered with brownish scales, with scattered creamy white or yellowish scales. Pale scales usually form diffuse median patches, and sometimes short longitudinal bands (Fig. 6.31a). Transverse pale basal or apical bands are absent. The sterna have dominant black scales and a few white basal, narrow bands or lateral patches. The cerci have a few white scales among the dark ones.

Male: Tergum IX has a deep notch and the protruding lateral lobes have numerous thin setae. The gonocoxite is short and stout, with a conical basal lobe which bears many setae of uniform length and width (Fig. 10.22). The apical lobe is well developed. The gonostylus is curved apically, and is narrow with a slender apical spine. A relatively shorter claspette stem differentiates the species from *Oc. cantans*, *Oc. riparius*, *Oc. flavescens* and *Oc. excrucians* males. The claspette filament is narrow, and approximately as long as the stem.

Larva: Similar to that of *Oc. excrucians* and often it is difficult to separate both species. The antennae are short with many distinct spicules, and the antennal seta (1-A) is located below the middle of the antennal shaft. The

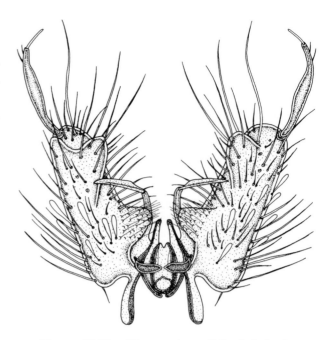

Figure 10.22. Hypopygium of *Oc. behningi*.

postclypeal seta (4-C) has 6–8 branches (Fig. 8.53a). The inner frontal seta (5-C) has 2-4 branches, and the median frontal seta (6-C) has 2–3 branches. A report on material from the Don basin referred to both setae as being double. The prothoracic formula 1-P to 7-P is as follows: 1 (double, short, thin); 2 to 4 (simple); 5 (double); 6 (simple); 7 (triple). The number of comb scales is 18–28 (usually 20–24) arranged in irregular rows (Fig. 10.23). Individual scales have a very protruding median spine and insignificant lateral spines. The siphon is more or less tapered at the apex, and the siphonal index is 3.0–4.0. The pecten teeth are usually evenly spaced, sometimes the two distalmost teeth may be detached apically. Each pecten tooth has a few lateral denticles at its base. The siphonal tuft (1-S) is inserted beyond the middle of the siphon, with 5 branches. The saddle does not encircle the anal segment but covers most of it. Five to six precratal tufts (4-X) are present, and the anal papillae are at least as long as the saddle.

Biology: The species belongs to the early summer species and seems to be monocyclic.

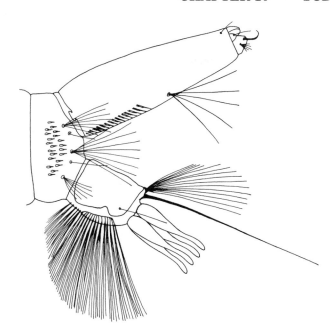

Figure 10.23. Larva of *Oc. behningi.*

Little is known about its general biology. Martini (1931) reported mass occurrence of females as fierce day biters in open spaces from the plains along the Volga river in the Saratov area. The species occurred later than *Ae. vexans.* Females were also found in higher mountainous areas, this could indicate that the females may fly long distances to search for a blood-meal.

Distribution: The distribution is not very well known. It has been recorded in Russia, Ukraine, Slovakia and Poland which seem to confine the species to an eastern European distribution. It has not been reported from the East Palaearctic.

Note on systematics: As no male belongs to the type series (Martini, 1931), the missing features and the amount of variation need further studies. The species certainly needs a closer analysis of female, male and larval characters and comparisons with species of similar colouration.

Ochlerotatus (Ochlerotatus) berlandi (Seguy) 1921

Female: The palps are predominantly dark scaled, with scattered pale scales in the middle, and the tip of the palps is white scaled. The vertex has pale to whitish scales with some diffuse darker scales which are more numerous at the occiput forming a dark triangular patch. The eye margin is covered with long setae. The scutum is mainly covered with pale golden scales, with distinct patches and stripes of dark brown scales. Pale golden scales form a broad median stripe and two lateral stripes. Dark scales form distinct patches on the anterior and posterior submedian areas. Dark posterior dorsocentral stripes extend from the transverse suture to the end of the scutum. Post- and subspiracular scale patches are present. The mesepisternum has a prealar scale patch, and the mesepisternal patch is divided into a larger upper portion and a smaller lower portion. The mesepimeron has two whitish scale patches on its upper half. The femora and tibiae are dark scaled, but scattered pale scales may be present. The tarsi have white apical and basal rings usually present on tarsomeres I and II of the front and mid legs, and on tarsomeres I to III of the hind legs. Tarsomere V of all the legs is entirely pale scaled (Fig. 6.20a). The wing veins are dark scaled; rarely a few isolated pale scales can be present. The abdominal terga have creamy white basal bands which are usually slightly widened laterally, and sometimes expanded into triangular patches. The sterna are black scaled, with more or less developed pale lateral patches, sometimes almost connected in the middle.

Male: The hypopygium is very similar to that of *Oc. pulcritarsis*; no constant differences can be found (Fig. 10.24). *Oc. berlandi* has the lobes of tergum IX well developed and widely separated, and each lobe bears 5–8 spine-like setae. The gonoxocite is about three times as long as it is broad. The basal lobe is weakly developed with one strong, apically recurved, spine-like seta (Fig. 7.38b). The other setae on the basal lobe are thin and of variable length. The apical lobe is indistinct. The gonostylus is somewhat widened in the middle, with a slender

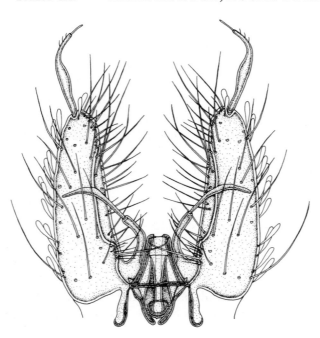

Figure 10.24. Hypopygium of *Oc. berlandi*.

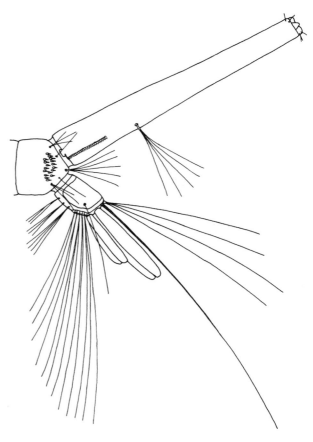

Figure 10.25. Larva of *Oc. berlandi*.

apical spine. The claspette filament is slender and longer than the stem.

Larva: The head is broader than it is long, and the antennae are almost as long as the head. The antennal tuft (1-A) is inserted beyond the middle of the antennal shaft. The postclypeal seta (4-C) is well developed and multiple-branched. The inner frontal seta (5-C) is the most prominent of all the frontal setae, and usually has 9 branches. The median frontal seta (6-C) is situated almost at the same level as 4-C, is less developed than 5-C, and usually has 8 branches. The outer frontal seta (7-C) is well developed, is long and usually has 10 branches. The comb usually has 16–20 comb scales arranged in several irregular rows, each scale with a well developed median spine. The siphon is very long and slender; the siphonal index is 5.5–7.8 (Fig. 10.25). The pecten consists of 19–29 small and blunt teeth. The siphonal tuft (1-S) is long, more than twice the width of the siphon at the point of its origin, and is inserted distinctly below the middle of the siphon, with 3–5 branches. The saddle reaches far beyond the half of the lateral sides of the anal segment. The saddle seta (1-X) is much longer than the saddle, and is simple. The upper anal seta (2-X) has 4–6 branches, the lower anal seta (3-X) is simple, and about as long as the siphon. The ventral brush has 3 precratal setae (4-X). The anal papillae are elongated, sausage shaped, and much longer than the saddle, with the dorsol pair being longer than the ventral pair.

Biology: The species hibernates in the larval or pupal stage and usually has two generations per year. Larvae can be exclusively found in tree holes, very often in those of *Platanus orientalis, Quercus ilex, Q. suber* and *Sophora japonica*, preferably in alkaline waters rich with organic materials. They can usually be found in association with larvae of *Oc. echinus, An. plumbeus* and *Or. pulcripalpis* (Ramos, 1983). Under laboratory conditions, at a temperature of 24°C, larval development lasts for

24 days and the pupal stage for about 7 days. Autogenous egg development was observed. Under natural conditions the larval development lasts for approximately four months (Ramos, 1983). Adult females are mainly zoophilic, but readily bite man either outside or inside human habitations (Ribeiro *et al.*, 1988).

Distribution: *Oc. berlandi* is endemic in the Mediterranean region and has been recorded in Portugal, Spain, France, Italy, Greece, and in Morocco, Algeria, and Tunisia. It was the prevalent treehole species (together with *Oc. geniculatus*) of Sardinia during a five year survey (Marchi and Munstermann, 1987).

Ochlerotatus (Ochlerotatus) cantans (Meigen) 1818

Female: Its greyish integument with dominant dark, blackish brown scaling and fewer scattered white or yellowish white scales on the body and wings distinguishes *Oc. cantans* from *Oc. annulipes*, which has a more yellowish scaling. The white rings on the legs are not as broad as in *Oc. annulipes, Oc. behningi* and *Oc. riparius*. The proboscis has no or few white scales. The palps are dark with a few white scales at the tip. The clypeus has a dark brown integument. The antennae have dark segments with brownish setae, and the pedicel has a few white scales. The vertex is white scaled with a lateral spot of brown scales. The occiput has brownish scales and two median stripes of white scales. The colouration of the scutum is very variable. Typically it is covered with dark brown or bronze brown scales, the lateral parts with greyish white or creamy scales, but sometimes these scales are light brown. A pair of distinct whitish submedian patches are usually present just beyond the scutal angle and sometimes narrow whitish submedian stripes run from the patches to the posterior margin of the scutum (Fig. 6.35a). The scutellum has white and brown scales and light setae. The postpronotum has narrow and yellow upper scales, white and somewhat

broader lower scales, and the postprocoxal membrane is bare. The white patches of the postspiracular and subspiracular areas are fused. The mesepisternum has one upper and two distinct lower patches of white scales, and the mesepimeron has a patch of white scales. The coxae have white scales, and the femora and tibae have mixed dark and pale scales. Tarsomere I of all the legs has more or less mixed scales, tarsomeres II to V have moderately broad white basal rings (Fig. 6.18a), except tarsomere V of the fore legs which is entirely dark scaled. The wings are predominantly dark scaled. Usually a few white scales are scattered on the costa (C) and on some other veins. Abdominal terga I to VIII have white basal bands, sometimes narrow and indistinct. The apical parts of the terga have more or less numerous scattered pale scales (Fig. 6.28b). Sterna I to VIII are whitish with darker lateral patches, and the cerci are predominantly dark scaled and elongated.

Male: The lobes of tergum IX have numerous strong setae. The hypopygium is of a different shape to that of the other members in the Annulipes Group (Fig. 10.26). The basal lobe of the gonocoxite is slender, and distinctly elongated with a large spine-like seta at the base. The apical lobe is present, and covered with short setae. The gonostylus is long with

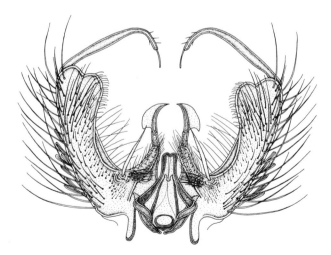

Figure 10.26. Hypopygium of *Oc. cantans*.

a curved tip. The paraproct is long and narrow. The claspette filament is shorter than the stem, prominently winged, and at least as long as it is wide. The shape of the spine-like seta of the basal lobe and the claspette filament seem to be variable characters. The aedeagus is elongated and tapered.

Larva: The antennae are shorter than the head, and the antennal tuft (1-A) is inserted slightly beyond the middle of the antennal shaft. The inner frontal seta (5-C) has 3–5 branches; the median frontal seta (6-C) has 2–3 branches and the outer frontal seta (7-C) has 7–8 branches. The pothoracic formula of setae 1-P to 7-P is as follows: 1 (short, double or triple); 2 (medium long, simple); 3 (very long, simple); 4 (short, simple); 5 and 6 (long, simple); 7 (long, triple). The number of comb scales is 28–40 (usually about 35) arranged in an irregular patch. Each scale has a moderately long median spine. The siphonal index is appoximately 3.0 or less (Fig. 10.27). The pecten teeth are evenly spaced, and each tooth has three or four lateral denticles at the base. The siphonal tuft (1-S) is inserted distal to the last pecten tooth about the middle of the siphon, with 5–12 branches. The saddle does not encircle the anal segment, and covers 3/4 of its lateral sides. The saddle seta (1-X) is simple, and about as long as the saddle. Usually 4–6 precratal tufts (4-X) are present, quite frequently less than in *Oc. annulipes*. The anal papillae are usually as long as or longer than the saddle.

Biology: The larvae develop rather early in spring in southern and central Europe, in northern areas they occur somewhat later but also belong to the early species. Larvae are present until late spring or early summer in varying numbers, depending on the amount of inundation and insolation of the habitat. The species is predominantly monocyclic, but capable of a bicyclic occurrence, the further north, the more obligate monocyclic it is. Under special circumstances another batch of eggs may hatch in late summer. It will mostly originate from unhatched eggs from the previous year. Eggs are hibernating, and larval development lasts from two months to less than four weeks, entirely depending on regional temperature regimes. The larvae occur in open permanent or semipermanent meadow pools and dominate in deciduous or mixed forest pools with scarce aquatic vegetation and a thick layer of leaves on the bottom of the pools. In these habitats they may occur together with larvae of *Oc. annulipes*, *Oc. communis*, *Oc. punctor* and occasionally with late spring species, such as *Ae. cinereus* or *Ae. geminus*. The adults emerge shortly after the trees turn green. It has been reported that females need a period of rest in vegetation before they search for a blood-meal, but local populations differ in this behaviour and some attack soon after emergence. Biting females are encountered most abundantly in lowland regions from late March to June in the Mediterranean region and from late May to early August at the southern borders of the

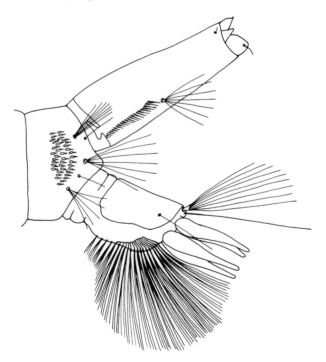

Figure 10.27.   Larva of *Oc. cantans*.

taiga zone. The wide distribution range of *Oc. cantans* over all of Europe comprises plenty of local variations in time of occurrence and biting habits. The life span of the females is between 1 and 2 months. They are frequently found in dense vegetation but also fly over short distances to open spaces such as pastures and river lowlands to feed on cattle and sheep. Their daily biting cycle on humans seems to be bimodal with peaks during dawn and dusk. No autogenous forms have been reported.

Distribution: The species has a west Palaearctic distribution and occurs from the taiga zone in the north to the Mediterranean region in the south.

Medical importance: Flavivirus and Bunyavirus isolates have been reported from Slovakia and Austria (Lundström, 1994, 1999). On material from Slovakia it was shown that the species is susceptible to infection with Tahyna virus.

Ochlerotatus Caspius Complex

Two apparently morphologically identical forms of *Oc. caspius* designated as "species A" and "species B" were detected by electrophoretic analysis of wild populations from Italy (Cianchi *et al.*, 1980). As no morphological and/or biological differences were attributed to those sibling species, they will be treated together under the description of the nominative form. Another species, described as *Aedes duplex*, was recorded from the European part of Russia by Martini in 1926. The two male specimens particularly exhibit differences in their hypopygium with the basal lobe bearing 4 spine-like setae instead of 2 setae. As no further record of any stage of *Ae. duplex* has been made since that time, it is more probable that the two sampled males are only abberant specimens, thus *Ae. duplex* should not be regarded as a distinct species and member of the Ochlerotatus Caspius Complex.

Ochlerotatus (Ochlerotatus) caspius (Pallas) 1771

Female: *Oc. caspius* is very similar to *Oc. dorsalis* in general colouration, but is usually distinguished from the latter by two dorsocentral white stripes which run over the bright fawn coloured scutum. However, the body colouration of *Oc. caspius* is subject to considerable variation. The proboscis and palps are covered with brown and white scales (Fig. 6.30b). The vertex has white and yellowish brown scales intermixed. The scutum has two narrow, white dorsocentral stripes running continuously from its anterior to posterior margin (Fig. 6.22a). The stripes may also be wide and diffuse, and if more yellowish, indistinct against the light brown background, but when the scales are well preserved, the distinction from the scutal pattern of *Oc. dorsalis* is quite easy. Even in species with the scales rubbed off from the central part of the scutum, the anterior and/or posterior parts of the longitudinal stripes are regularly well preserved and visible. The scales on the pleurites are broad and white. Tarsomeres I and II of the fore and mid legs and tarsomeres I to IV of the hind legs have white or cream-coloured basal and apical rings (Fig. 6.19a). The light rings are sometimes indistinct, and hind tarsomere V is entirely white scaled. The wing veins are covered with mixed dark and pale scales (Fig. 6.22b). At the basal quarter of the costa (C), the dark and pale scales are of more or less the same number or the dark scales predominate. The abdominal terga are dark brown scaled, with yellowish scales dorsally and white scales laterally. The terga have basal and apical yellowish bands which are widest in the middle. A longitudinal middorsal yellowish stripe is present, but of varying length (Fig. 6.21a). It is usually present on terga II to IV, otherwise only vaguely expressed by a median widening of the transverse bands. In some specimens the median stripe is present on tergum II only (this pattern resembles that of *Cs. annulata*). The lateral sides of the terga are ornamented with central, triangular, white patches. Tergum VII has mixed dark and pale scales.

Male (Fig. 10.28): The basal lobe gradually arises from the gonocoxite, and is not constricted at the base. Two spine-like setae arise from it, one seta is longer and sharply hooked at the apex; the tip of the hook extends backwards to almost the middle of the spine, and the shorter seta is straight or slightly curved (Fig. 7.35a). The apical lobe of the gonocoxite is inconspicuous, almost bare dorsally. The claspette filament is about as long as the stem, with a narrow unilateral wing.

Larva: Similar to those of *Oc. dorsalis, Oc. detritus, Oc. leucomelas* and *Oc. flavescens*. Average values of some quantitative traits can be used to distinguish between larvae of *Oc. caspius* and *Oc. dorsalis*, but only at population levels (Milankov *et al.*, 1998). The antenna is about half as long as the head, with sparse tiny spicules. The antennal seta (1-A) is inserted slightly below the middle of the antenna, usually with 9 branches which are half as long as the antenna. The postclypeal seta (4-C) has 3–5 short, thin branches. The inner frontal seta (5-C) is inserted well below the median frontal seta (6-C), both are simple, or less frequently 2 branched, but rarely does one of the seta have

3 branches. The outer frontal seta (7-C) has 7–10 branches. The mesothoracic seta 1-M, is simple and moderately long. The comb has 18–28 (usually 20–25) variegated scales arranged in 2–3 irregular rows (Fig. 10.29). At least some of the scales have a distinct median spine (Fig. 8.37b). The siphon slightly tapers in the apical half, and the siphonal index is 1.8–2.6. The pecten has 17–26 (usually 20–22) evenly spaced teeth, the basalmost four teeth are rudimentary. The pecten extends slightly beyond the middle of the siphon. The siphonal tuft (1-S) has 5–10 branches inserted beyond the middle of the siphon. The saddle covers more than half of the lateral sides of the anal segment. The saddle seta (1-X) is about half as long as the saddle, and is simple. The lower anal seta (3-X) is longer than the siphon, and simple. The ventral brush is made up of 2–3 precratal tufts (4-X) and 12–17 cratal tufts (4-X). The anal papillae are short, 0.3–0.9 times as long as the saddle, and lanceolate. The ventral pair is shorter than the dorsal pair.

Biology: *Oc. caspius* is a polycyclic, halophylic species. Sometimes only one

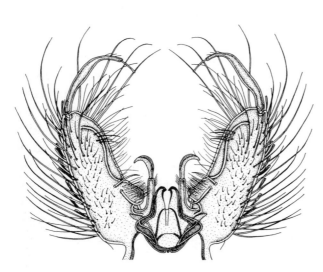

Figure 10.28.   Hypopygium of *Oc. caspius*.

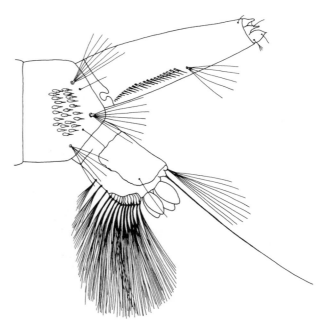

Figure 10.29.   Larva of *Oc. caspius*.

generation per year is produced due to the nature of the breeding site. The species overwinters in the egg stage, the first occurrence of larvae varies with the latitude, but generally takes place at the beginning of the year; this may be February to March in southern parts of Europe. It is regarded as a seaside mosquito that readily breeds in inland salt marshes and freshwaters with 0.5 g NaCl/l (Pires *et al.*, 1982). It is a common species in Atlantic and Mediterranean coastal marches and rockholes. The breeding sites in the coastal area of Portugal are usually restricted to altitudes lower than 50 m (Ribeiro *et al.*, 1989). The larvae develop in open or shaded waters, permanent or temporary water bodies formed by the snowmelt, river floods or coastal marshes subjected to intermittent flooding and rice fields, usually with little vegetation and muddy bottom, often with a high concentration of salt, which may reach values of up to 150 g/l (Bozicic-Lothrop, 1988). The acidity of the breeding sites recorded in Portugal ranged from pH 6.0–7.0 (Pires *et al.*, 1982). The most characteristic freshwater habitats are river valleys, where larvae can breed in large numbers in the floodplains. They can be associated with larvae of many mosquito species, such as *An. atroparvus, An. maculipennis s.s., Ae. vexans, Ae. vittatus, Oc. annulipes, Oc. cantans, Oc. detritus, Oc. intrudens, Oc. mariae, Oc. sticticus, Cx. p. pipiens, Cx. theileri, Cx. impudicus, Cs. annulata* (Bozkov *et al.*, 1969; Ramos *et al.*, 1978; Knoz and Vanhara, 1982; Pires *et al.*, 1982; Marchi and Munstermann, 1987). Although females are strongly exophagic, they enter inhabited areas, houses and cattle sheds if they occur in masses. The females readily bite man and animals both in rural and urban areas (Gutsevich *et al.*, 1974). They often bite during the day and night, but usually most actively search for a blood-meal at dusk. Females are repelled by the lights of standard CDC miniature light traps. The species has a high resistance to heat and drought. Females actively search for blood at temperatures

ranging from 11.5 to 36°C and relative humidity ranging from 47 to 92% (Petric, 1989). They may migrate for moderate distances, up to 10 km. Autogenous development of *Oc. caspius* was detected in Uzbekistan (Chinaev, 1964).

Distribution: It is a Palaearctic species which is more common in southern and dry regions than *Oc. dorsalis* which is of Holarctic origin. *Oc. caspius* is distributed from Europe to Mongolia, north and west China, north Africa, west and middle Asia. In Europe it can be found from England to the central parts of Russia, and from the southwest to the Mediterranean basin. The distributions of the sympatric *Oc. caspius* and *Oc. dorsalis* overlap in most of Europe.

Medical importance: West Nile virus, Tahyna virus and the bacterium *Francisella tularensis*, the causative agent of tularaemia, can be detected in natural populations (Detinova and Smelova, 1973). *Oc. caspius* may play a role in the spread of tularaemia and transmit Tahyna and rabbit myxoma virus in former Czechoslovakia, France and Portugal (Bardos and Danielova, 1959; Joubert, 1975; Pires *et al.*, 1982).

Note on systematics: Most authors consider *Oc. caspius* and *Oc. dorsalis* as separate species according to morphologic, genetic, and mixologic differences (Edwards, 1921; Natvig, 1948; Mohrig, 1969; Lambert *et al.*, 1990; Milankov *et al.*, 1998). Concerning the morphological variation within *Oc. caspius*, Kazantsev (1931) found that the light, "sand-coloured" *Oc. caspius* developed in water with a high salinity, while mosquitoes breeding in fresh water had a more contrasting colouration. Specimens with intermediate morphological characters are common in Europe (Bozicic-Lothrop, 1988) as well as in more eastern regions of the former USSR (Gutsevich, 1977). Three intermediate types of scutal colouration and two intermediate types of male genitalia are described in Europe, but no correlation between water salinity and scutal pattern or development of the basal lobe of the gonocoxite were found (Bozicic-Lothrop, 1988). All the above mentioned differences could suggest the possible presence of subspecies or more than one species in Europe, which was documented by Cianchi *et al.* (1980).

Ochlerotatus (Ochlerotatus) cataphylla (Dyar) 1916

Female: The proboscis is entirely dark scaled, which is the main character to separate the species from *Oc. leucomelas*, which has the proboscis speckled with numerous pale scales. The palps of *Oc. cataphylla* are dark scaled, with numerous scattered white scales. The scutum is usually covered with reddish brown scales, with lateral areas of paler scales. Sometimes a broad median stripe and two posterior submedian stripes of darker scales are present. The pleurites are extensively covered with scales. The upper part of the postpronotum is mostly brown scaled, and the lower part with pale scales. Postpronotal setae are present along the posterior margin of the postpronotum (Fig. 6.41b). The postprocoxal membrane has a patch of pale scales. The subspiracular area is more or less entirely covered with scales, and the hypostigmal and postspiracular scale patches are fused. The upper and posterior part of the mesepisternum are extensively scaled; the scales extending near to the anterior angle. The scales on the mesepimeron end before its lower margin, and mesepimeral setae are present. The femora anteriorly have pale and dark scales. The tibiae and tarsi are mostly dark scaled, but with pale scales especially on the ventral surface, and the tarsi are without white rings. The tarsal claws are small and evenly curved. The wing veins have pale scales on the base of the costa (C) and scattered pale scales along the costa, subcosta (Sc) and R_1; the remaining veins are dark scaled. The terga are dark scaled with broad basal bands of white scales (Fig. 6.38b); sometimes the last terga are mainly covered with white scales.

Male: The lateral lobes of tergum IX have 4–13 (usually 6–8) short spine-like setae. The gonocoxite has well developed basal and apical lobes. All the setae located on the inner side of the gonocoxite above the basal lobe are inwardly directed, and usually overlapping in the middle (Fig. 10.30). The basal lobe is conical with a group of medially directed long setae, one of which is strong, spine-like, and apically curved. The paraproct is sclerotised with inwardly curved tips. The claspette stem is strongly curved, and the claspette filament is half as long as the stem, with a broad unilateral wing. The aedeagus is more or less conical.

Larva: The antenna is less than half as long as the head. The antennal seta (1-A) has 3–5 branches, located at about the middle of the antenna. The postclypeal seta (4-C) has 2–3 short branches. The inner and median frontal setae (5-C and 6-C) are simple; the median frontals are located in front of the inner frontals, and the outer frontal seta (7-C) has 3–6 branches (Fig. 8.35a). The comb is composed of 10–30 (usually 25) scales arranged in 2–3 irregular rows (Fig. 10.31). Each scale has a long terminal spine and smaller spines near

Figure 10.30. Hypopygium of *Oc. cataphylla*.

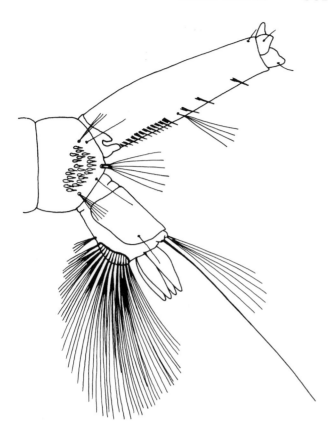

Figure 10.31. Larva of *Oc. cataphylla.*

the base. The siphonal index is about 3.0. The pecten is made up of 13–25 teeth, occupying about 3/4 of the length of the siphon. The 2–4 distalmost teeth are larger and apically detached; all are located beyond the siphonal tuft (1-S). 1-S is situated approximately in the middle of the siphon, with 3–5 branches. The saddle extends about 2/3–3/4 down the sides of the anal segment, and the saddle seta (1-X) is simple and short. The upper anal seta (2-X) has 5–8 branches, and the lower anal seta (3-X) is simple. The ventral brush has 1–2 precratal tufts (4-X). The anal papillae are of variable length, and are tapered.

Biology: *Oc. cataphylla* is a monocyclic species. The most common breeding sites are forest pools in swampy woodlands, *e.g.* alder forests, with neutral to alkaline water. The pools are usually devoid of submerged vegetation but frequently covered with dead leaves at the bottom. In addition, larval populations are recorded from open areas, *e.g.* inundated meadows (Wesenberg-Lund, 1921; Natvig, 1948; Monchadskii, 1951). The larvae hatch immediately after the snow thaw when melt waters or heavy rainfalls flood the depressions. The larvae are usually associated with those of *Oc. rusticus, Oc. cantans* and *Ae. cinereus.* Occasionally the larvae of acidophilic species, *e.g. Oc. punctor* and *Oc. communis* are found in the same breeding sites. In central Europe the adults appear usually in the first half of April, before those of *Oc. cantans* and *Oc. rusticus.* The copulation swarms of male mosquitoes can frequently be observed at a height of about one metre in shaded areas with bushes. Usually the females of *Oc. cataphylla* are a nuisance only in forest areas, where they can bite even during daytime. Repeated blood-meals and ovipositions are possible (Carpenter and Nielsen, 1965).

Distribution: Holarctic, Eurasia and North America; North Europe to South Europe. In Northern Europe the species occurs in the tundra, in central Europe in swampy forests, in south Europe mainly in mountainous areas.

Ochlerotatus Communis Complex

Due to differences in the reproductive physiology, habitat preferences, swarming behaviour, emergence and dispersal patterns, the complex is composed of two or more sibling species (Chapman and Barr, 1964; Ellis and Brust, 1973; Schutz and Eldridge, 1993). Morphometric and electrophoretic comparisons of populations of the Ochlerotatus Communis Complex in North America revealed 4 sibling species, namely *Oc. communis, Oc. nevadensis, Oc. tahoensis* and *Oc. churchillensis.* The females of the latter species are found to be autogenous, whereas the other three species are anautogenous (Brust and Munstermann, 1992). Acording to variable morphological characters in European populations, it is feasible to expect more sibling species of the complex in Europe as well. However, electrophoretical analysis of populations from northern Sweden and Germany has not provided evidence of any differences so far (Weitzel *et al.*, 1998).

Ochlerotatus (Ochlerotatus) communis (De Geer) 1776

Female: The proboscis and palps are dark scaled, the palps rarely have a few white scattered scales. The scutal scale pattern is rather variable, but typically the scutum is covered with yellow to golden scales. A broad median stripe and posterior submedian stripes of dark bronze scales are present. They are separated by narrow stripes of pale scales, which are sometimes fused. The scutellar and supraalar setae are dark brown. A postprocoxal scale patch is absent (Fig. 6.47b), the hypostigmal patch is usually absent, or occasionally with only a few pale scales. The upper mesepisternal patch extends to the level of the anterior angle, and the mesepimeral patch extends to the lower margin of the mesepimeron (Figs 6.45b, 6.46a). Lower mesepimeral setae are present. The femora, tibiae and tarsomeres I of all the legs are mostly dark scaled dorsally and with a few whitish scales ventrally, and the remaining tarsomeres are dark scaled. The tarsal claws are curved with a long subbasal tooth. The wing veins are covered with dark scales, and a few pale scales are scattered at the base of the costa (C) and radius (R). The terga are dark scaled with broad basal bands of white scales.

Male: The lobes of tergum IX are covered with 8–10 short spine-like setae. The hypopygium is similar to that of *Oc. pionips* (Fig. 10.32). The basal lobe of the gonocoxite is rounded, concave at its lower part, with one long, strong spine-like seta. Mesal to it, is a row of closely spaced, inwardly directed long setae. The upper part of the basal lobe has a row of long, prominent, widely spaced, apically strongly curved or sometimes hooked setae. The apical lobe of the gonocoxite is well

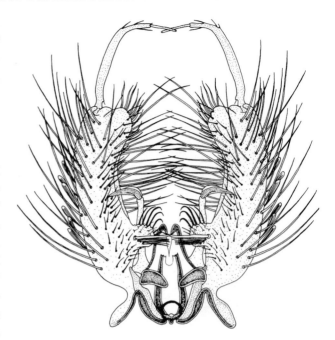

Figure 10.32. Hypopygium of *Oc. communis*.

developed, and rounded apically. The paraproct is strongly sclerotised apically. The claspette stem is long, and the claspette filament is shorter than the stem, and heavily sclerotised with a narrow unilateral wing. The aedeagus is conical, rounded and notched at the apex.

Larva: The antenna is nearly half as long as the head, and the antennal seta (1-A) is situated at about the middle of the antennal shaft, with 6–7 branches. The median frontal seta (6-C) is situated in front of the inner frontal seta (5-C), both pairs are simple and rarely one seta has 2 branches. The outer frontal seta (7-C) has 4–8 branches (Fig. 8.41a). The number of comb scales varies from 40–70, but the average is 60 scales (Fig. 10.33). The scales are arranged in an irregular triangular patch, each individual scale without a prolonged terminal spine, thus appearing to be rounded apically. The siphonal index is 2.3–3.2, usually approximately 2.8. The pecten has 17–26 evenly and closely spaced teeth which do not extend to the middle of the siphon. The siphonal tuft (1-S) has 5–9 branches which are about as long as the width of the siphon at the point of insertion. The

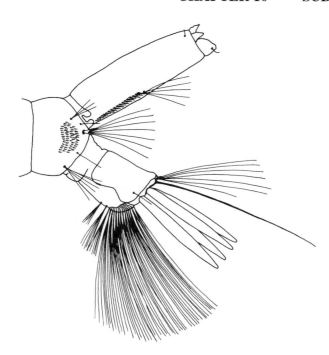

Figure 10.33. Larva of *Oc. communis*.

saddle extends about 3/4 down the sides of the anal segment, and the saddle seta (1-X) is distinctly shorter than the saddle, and is simple. The ventral brush has 2, rarely 3 precratal setae (4-X). The anal papillae are distinctly longer than the saddle.

Biology: *Oc. communis* is usually a monocyclic snow-melt mosquito, which belongs to the most frequent mosquitoes of swampy forests. The preferred breeding sites are acid water bodies which are filled with water during the snow-melt or spring rainfall. Larvae can mainly be found in depressions and ditches without vegetation but with a dense layer of dead leaves at the bottom. Often they are found in strongly acidic waters, *e.g.* with *Sphagnum* sp., with a pH-value of little more than 3.0. They appear only rarely or are absent in waters with neutral reaction, *e.g.* in inundation areas of large rivers. Most of the larvae hatch at temperatures of little more than 0°C, when the breeding sites are still partly covered with ice. In central Europe the larvae hatch from February onwards, adults emerge usually in April. In the laboratory the optimum temperature for the larval development is 25°C. At this temperature the development to the adult stage is completed within 18 days; above 30°C and below 4°C the development is not completed. In central Europe females are troublesome for warmblooded creatures in forest areas from April onwards, particularly during twilight. Females do not migrate long distances from the breeding sites. Usually the population decreases from July on, but Scherpner (1960) found isolated freshly hatched larvae of *Oc. communis* in August.

Distribution: Holarctic, North America and Eurasia. The species is found from northern Europe to the Mediterranean area.

Ochlerotatus (Ochlerotatus) cyprius (Ludlow) 1920

Female: A large species with a light integument and golden to yellowish scaling. Natvig (1948) described *Oc. cyprius* as variable in colour, the golden tinge is sometimes lost and the scales are more or less whitish or yellowish orange in some specimens. It is similar to *Oc. flavescens*, and for separation of the two species, see the description of the latter. The proboscis is yellowish with a few dark scales at the labellum. The palps are 1/4 of the length of the proboscis, with mixed golden and greyish scales. The head is pale scaled, with some dark scales laterally. The antenna is yellowish at the base, with the distal flagellomeres more brownish. The integument of the scutum is light brown with yellowish and golden narrow scales, and sometimes a weak dark median stripe and short dark lateral stripes are present. The postpronotum has narrow golden scales and golden setae. A postprocoxal patch is present. The mesepisternum has two separate cream coloured scale patches, the lower patch reaching the anterior margin. The mesepimeral scale patch nearly reaches the lower margin of the mesepimeron, and lower mesepimeral setae are present (Fig. 6.29b). The fore and mid femora have yellowish scales mixed with dark ones, the hind femur apically has dark scales, and a yellowish knee spot but these are usually indistinct. The tibiae are predominantly yellowish scaled with scattered black scales, which are more numerous towards the apical parts of the mid and hind tibiae. The tarsomeres have broad, yellowish basal rings and dark scaling apically, which is more pronounced at the mid and hind legs. The wing veins are yellow scaled with scattered dark scales. The abdomen dorsally and ventrally has ochre yellowish broad scales mixed with isolated dark scales, and the terga sometimes have more dark scales laterally which never form a continuous transverse band. The cerci are predominantly dark scaled.

Male: Tergum IX has 6–11 spine-like setae on each of the lateral lobes. The gonocoxite is long and slender, and the basal lobe is well developed, with one spine-like seta and several long setae of different widths (Fig. 10.34). The apical lobe is prominent with rather short setae. The gonostylus is long and slender, with an elongated apical spine. The paraproct has a sclerotised and recurved tip. The claspette stem is long and slender, and evenly curved. The claspette filament is more than half as long as

the stem, slightly triangular, and unilaterally winged from the base, without a stalk. The aedeagus is elongated and pear shaped with a narrow opening at the apex.

Larva: The entire body is densely covered with dark spicules (Fig. 8.19a), which identifies the large larvae from all other European *Aedes* and *Ochlerotatus* species. The antennae are shorter than the head, and the antennal seta (1-A) has 1–3 branches. The inner and median frontal setae (5-C and 6-C) are simple or double. The prothoracic setae according to Peus (1937) follow the formula: 1 and 2 (simple); 3 (2-branched); 4 (simple, short); 5 (2-branched); 6 (simple); 7 (3-branched). The number of comb scales is 9–15 arranged in an irregular row (Fig. 10.35). The siphon is long and slender, and moderately tapering. The pecten usually has 19–21 pecten teeth, with several distal teeth detached. Each individual pecten tooth has many lateral denticles. The siphonal tuft (1-S) is situated beyond the distalmost pecten tooth, with 3–4 branches. The anal segment is not encircled by the saddle, the saddle seta (1-X) is about as long as the saddle,

Figure 10.34. Hypopygium of *Oc. cyprius*.

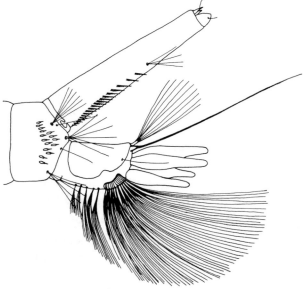

Figure 10.35. Larva of *Oc. cyprius*.

the number of precratal tufts (4-X) is 4–5, and the anal papillae are longer than the saddle.

Biology: The species hibernates in the egg stage. Larvae occur in semipermanent pools along inundated river shores and in snow-melt pools. Peus (1937) found the larvae always in the middle of the pools with an average water depth of 50–80 cm and regarded the species as preferring cold water for development. In Sweden one larva was found in late May in a cold well and adults were caught resting nearby on a flooded meadow surrounded by a deciduous forest (Dahl, 1975). The larvae occur together with those of *Oc. flavescens, Oc. excrucians, Oc. cantans*, and *Oc. riparius*. Adults appear until July and from more eastern areas the species is registered as being a mid spring species (Gutsevitch *et al.*, 1974). The females are aggressive biters, they can attack their hosts in sunlit situations (Peus, 1937). In Siberia *Oc. cyprius* is sometimes a vicious biter (Gutsevitch *et al.*, 1974).

Distribution: It is a rare species in western European regions, reported in Sweden, Finland, Germany, Slovakia, and Poland. It can also be found along the eastern shores of the Baltic Sea; the eastern distribution range stretches from the Ukrainian steppe to Central Kazhakstan.

Ochlerotatus Detritus Complex

Laboratory studies carried out using isoenzyme identification techniques gave evidence for the existence of sibling species within the complex (Pasteur *et al.*, 1977). The then called "species A" and "species B" occurred sympatrically in the Carmargue and showed reproductive isolation. Further ecological and distributional differences between the apparently morphologically identical siblings, *e.g.* frequent autogenity in "species A" which is an exception in "species B", have been recorded (Agoulon *et al.*, 1997; Schaffner, 1998). "Species A" has recently been named *coluzzii*

Rioux, Guilvard and Pasteur (Rioux *et al.*, 1998).

Ochlerotatus (Ochlerotatus) detritus (Haliday) 1833

Female: The species is easily distinguished from all other *Ochlerotatus* members with unringed tarsi by its abdominal colouration. The apical parts of the terga are usually covered with dark and pale scales, with dark scales predominating on the first segments and pale scales on the last segments. The proboscis and palps have dark and pale scales intermixed, and the clypeus is black. The vertex has yellowish white narrow curved scales and blackish brown erect forked scales. The integument of the scutum is brown, and more or less uniformly covered with yellowish brown scales, lighter on the posterior part. Stout black setae are more numerous above the wing bases. The scutellum has patches of yellowish white scales on each lobe. The integument of the pleurites is dark brown with patches of broad, flat yellowish white scales. The mesepimeral scale patch does not reach the lower margin of the mesepimeron, and lower mesepimeral setae are present. The anterior surfaces of the fore and mid femora are conspicuously speckled with pale scales, the tibiae and tarsi have dark scales and more or less numerous scattered pale scales, pale tarsal rings are absent. The wing veins are covered with broad pale and dark scales, sometimes the pale scaling is reduced. The abdominal terga have pale transverse basal bands of more or less uniform width, the apical parts of the terga have dark and pale scales intermixed (Fig. 6.38a). The sterna are mostly pale scaled.

Male: The lateral lobes of tergum IX have 3–8 straight, spine-like setae. The gonocoxite is about three times as long as it is wide (Fig. 10.36). The basal lobe is more or less conical, and covered with fine setae arranged in an irregular row and one conspicuous strong spine-like seta, which is curved apically.

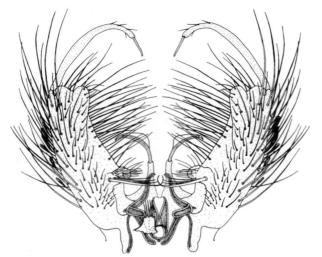

Figure 10.36. Hypopygium of *Oc. detritus*.

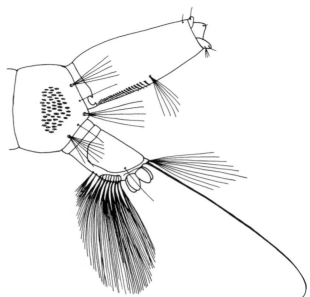

Figure 10.37. Larvae end of *Oc. detritus*.

The apical lobe is well developed carrying straight setae of moderate length. The gonostylus is curved, tapers apically, and the apical spine of the gonostylus is long. The paraproct is well sclerotised in the apical part. The claspettes have a short stem; the distal third of the stem is sometimes constricted. The claspette filament is as long as or longer than the stem, with a long stalk, and slightly bent from the end of the stalk on. The convex side of the filament is evenly widened from about its middle. The aedeagus is small and oval.

Larva: The head is broader than it is long. The antenna is short, about half the length of the head, slightly curved and moderately spiculose. The antennal seta (1-A) is inserted at about the middle of the antennal shaft, with 5–8 branches. The postclypeal seta (4-C) has 2–3 thin and short branches. The inner and median frontal setae (5-C and 6-C) are usually 2–3-branched; 5-C may rarely have 3–5 branches, but is always situated behind 6-C. The outer frontal seta (7-C) has 7–12 branches. The comb has more than 40 scales, usually 45–60, arranged in a triangular patch; each individual scale is blunt ended and fringed with spines of more or less the same length (Fig. 8.37a). The siphon slightly tapers from the middle, and the siphonal index is 2.2–2.8 (Fig. 10.37). The pecten is composed of about 20 (18–27) evenly spaced teeth; rarely the distalmost tooth may be slightly detached, and each pecten tooth has 2–3 ventral denticles. The siphonal tuft (1-S) is inserted at about the middle of the siphon, beyond the distalmost pecten tooth, with 6–10 (usually 8) branches. The saddle reaches more than half way down the sides of the anal segment. The saddle seta (1-X) is simple, and about as long as the saddle. The upper anal seta (2-X) has 8–11 branches, and the lower anal seta (3-X) is simple and longer than the siphon. The ventral brush has 1–3 precratal setae (4-X), but most often only one precratal seta is present. The anal papillae are very short and rounded.

Biology: *Oc. detritus* is a polycyclic species which may have up to three generations per year, according to the latitude of its occurrence. Usually the species hibernates in the egg stage (Gutsevich *et al.*, 1974) and the first larvae occur some weeks later than the majority of the typical snow-melt mosquitoes, when the water temperature of the breeding sites exceeds

10°C (Martini, 1931; Mohrig, 1969). Hibernating larvae were reported from England, lately hatched larvae developed into the fourth-instar during autumn and did not pupate before March of the following year (Service, 1968b). *Oc. detritus* is a typical halophilic species and the larvae occur almost exclusively in breeding habitats with an exceptionally high content of salinity; only occasionally were they found in fresh water. Preferred breeding sites are brackish waters in estuaries and coastal marshes. Larvae can be found in semipermanent ponds in open *Salicornia* sp. and *Tamarix* sp. marshes (depending on fluctuation of the water level), and in stagnant drainage channels or lagunes, with little aquatic vegetation. They are sometimes found together with larvae of *Oc. caspius* and *Oc. dorsalis*, but more often, due to the tolerance of extreme salinity, the species occurs alone in its breeding site. Adults can be found from late March until November in England (Cranston *et al.*, 1987), in central Europe they are recorded from early May until September (Martini, 1931; Mohrig, 1969). The females are persistent biters and readily attack humans, often in large numbers. They may feed during the day, but are predominantly active at dusk. Rioux (1958) considered *Oc. detritus* as a typical exophilic species which enters buildings only occasionally, but Cranston *et al.* (1987) reported frequent entering of houses to feed (endophagy), although no resting indoors. Along with *Oc. caspius*, *Oc. detritus* is the most common coastal species in Europe causing considerable annoyance in villages at the seaside and in their vicinity. Migration distances of about 6 km are recorded from Britain (Marshall, 1938); in southern France the flight range of *Oc. detritus* is estimated to be approximately 20 km (Rioux, 1958). The differences in the biology and behaviour might be attributed to the presence of sibling species of the complex.

Distribution: *Oc. detritus* is a palaearctic species; it occurs along most of the European cost lines, *e.g.* Northern and Baltic Sea as well as Atlantic Ocean and around the Mediterranean basin. Furthermore the species has a scattered distribution in saline inland areas in Europe, North Africa and Southwest Asia. It is possible that the southern range of its distribution is dominated by the sibling species *Oc. coluzzii*.

Ochlerotatus (Ochlerotatus) diantaeus (Howard, Dyar and Knab) 1913

Female: The large dark legged female has a dark brown to greyish integument, golden brownish setae and scales and some blackish brown scaling with a slight metallic shine. The proboscis and palps are blackish brown. The head has pale golden setae between the eyes and golden, narrow curved and erect forked scales on the vertex and occiput. The scutum has narrow pale golden or whitish scales and a broad median or two slightly divided dorsocentral stripes of narrow dark bronze scales and anterolateral dark stripes. The postpronotum has pale yellow or whitish narrow curved scales. The postprocoxal and hypostigmal scale patches are absent. The upper mesepisternal patch of white broad scales does not reach the anterior margin of the mesepisternum, and the lower edge of the patch ends slightly above the level of the anterior angle (Fig. 6.45a). The mesepimeral patch does not reach the lower margin of the mesepimeron. The pleural setae are pale golden. The coxae have white broad scales and yellowish setae. The fore and mid femora are dark scaled anteriorly, white scaled posteriorly, sometimes with basal whitish scaling. The hind femur has more dominant basal white scaling, and a whitish knee spot. The tibiae and tarsomeres are entirely dark scaled. The wing veins are covered with dark scales. The abdominal terga have dark scales, sparsely intermixed with white scales and with large white triangular lateral patches. The lateral patches are connected by distinct white basal bands at least on terga IV to VII, and sometimes narrow white basal bands are present also on terga II and III. The sterna have whitish scales, with indistinct dark triangular scale patches apically. Segment VIII and the cerci are entirely black scaled.

Male: Tergum IX has 6–8 strong, spine-like setae on each lobe. The hypopygium has two characteristic features which separate it from that of the other members of the Intrudens Group (Fig. 10.38). The first is the group of dense setae on the inner apical surface of the gonocoxite and the second is the distinct shape of the claspette filament. The basal lobe of the gonocoxite bears three spine-like setae, two of them located close together at about the middle of the gonocoxite, and the third one is widely displaced towards the base of the gonocoxite. The apical lobe is well developed. Dense, long, inwardly directed setae are located on the apical half of the gonocoxite. The gonostylus is slender and curved with a long apical spine, and usually one additional seta close to the apex. The paraproct has a broad sclerotisation. The claspette stem has a thorn-like process beyond the middle. The claspette filament is broad and somewhat crescent shaped. The aedeagus is pear shaped.

Larva: The larvae differ from all other European *Aedes* and *Ochlerotatus* species by the antennae, which are distinctly longer than the head (Fig. 8.17a). The head is broad, the antennae have numerous spicules. The antennal seta (1-A) has 2–4 long branches. The inner and median frontal setae (5-C and 6-C) have 2–5 branches. The thorax has the prothoracic formula 1-P to 7-P as follows: 1 and 2 (simple); 3 (2-branched); 4 to 6 (simple); 7 (2-branched). The comb has 8–13 scales, each individual scale with a long and strong median spine. The siphon is long and tapering, the siphonal index is 3.2–3.7 (Fig. 10.39). The last two pecten teeth are apically detached, and the siphonal tuft (1-S) is inserted beyond the distalmost pecten tooth, usually with 7–8 branches. The anal segment is not completely encircled by the saddle, but the saddle reaches far down the lateral sides; the saddle seta (1-X) is simple, and shorter than the saddle. The lower anal seta (3-X) is simple and usually longer than the siphon. The ventral brush has 2–3 precratal setae (4-X).

Figure 10.38. Hypopygium of *Oc. diantaeus*.

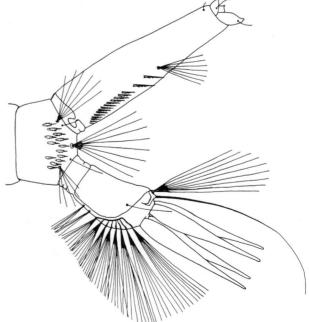

Figure 10.39. Larva of *Oc. diantaeus*.

The anal papillae are very long and slender, and longer than the saddle.

Biology: The species hibernates in the egg stage and is monocyclic. Larvae occur early in spring in temporary water bodies formed after snow-melt. They may be found in boggy localities, in shaded ditches and pools in mixed forests, or in open water pools in more northern and taiga areas. Usually the bottom of the breeding sites is covered with abundant leaf debris. Wood *et al.* (1979) suggest that the exceptionally long antennae might indicate a peculiar feeding behaviour. The larvae tolerate acid conditions and occur together with several common species, such as *Oc. cantans, Oc. cataphylla, Oc. communis, Oc. pionips, Oc. punctor* and *Cs. morsitans.* The adults emerge later than those of *Oc. communis,* due to delayed larval development (Gutsevich *et al.,* 1974). Males can be attracted to mammals where females are present and biting (Jaenson, 1985). Females have been reported to feed on humans in southeastern Russia.

Distribution: A northern Holarctic species, it occurs in central and northern Europe and does extend into the taiga but not into the Arctic zone. It can be found in the southeastern parts of Russia and is also recorded from Canada and northernmost United States.

Ochlerotatus (Ochlerotatus) dorsalis (Meigen) 1830

Female: Most easily distinguished from the similar *Oc. caspius* by its scutal markings and colouration. The scale pattern on the scutum can be variable, but usually is found as described below. The acrostichal and dorsocentral stripes on the scutum of *Oc. dorsalis* are fused to form a dark brown median stripe extending to the pale prescutellar dorsocentral stripes. Posteriorly the stripe is ornamented with a pair of narrow, white lines. There is also a well defined pair of dark posterior submedian stripes. The scales on the other areas of the scu-

tum are ashy white (Fig. 6.22c). Two distinct dorsocentral white stripes, which can be found on the scutum of *Oc. caspius,* are always absent. The proboscis is covered with dark scales, sometimes with scattered pale scales on its middle third. The pleural scales are narrow, and creamy or straw coloured. The tarsomeres have pale rings both basally and apically (Fig. 6.20b). The wing veins are predominantly covered with light scales (Fig. 6.22d). Dark scales are restricted to the apical portion of C, R_1, R_3 and forked portions of M and Cu. The basal quarter of the costa is exclusively white scaled; a character which can be used to separate the species from *Oc. caspius.* The abdominal terga have a whitish grey longitudinal stripe which may not be present on all segments. Indistinct, narrow basal and apical light transverse bands of more or less uniform width are present. Blackish brown or black scales are restricted to two rectangular areas on terga I to V; terga VI and VII are mostly light scaled.

Male: The hypopygium is very similar to that of *Oc. caspius* (Fig. 10.40). The basal lobe of the gonocoxite is constricted at the base with two widely separated spine-like setae. The apex of the longer seta is sometimes not hook

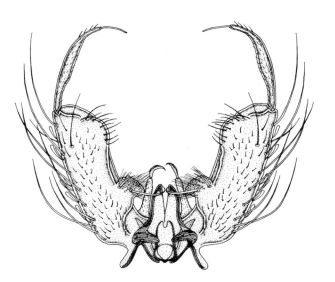

Figure 10.40. Hypopygium of *Oc. dorsalis.*

shaped, otherwise the tip does not extend backwards to more than one third of the spine (Fig. 7.35b). The shorter spine-like seta is straight. The apical lobe is inconspicuous, and usually covered with numerous setae. The claspette filament is shorter than the stem, and strongly curved.

Larva: The differences separating larvae of *Oc. dorsalis* from *Oc. leucomelas* and *Oc. caspius* are not very obvious. The antenna is about half as long as the head, with sparse spicules. The antennal seta (1-A) has 4–7 branches articulated in the middle of the antennal shaft or slightly below it, and is not more than half as long as the antenna. The postclypeal seta (4-C) is short, with 2–5 thin branches. The inner frontal seta (5-C) is situated posterior to the median frontals (6-C), and both pairs are simple, or more frequently with 2 branches (in Nearctic populations 5-C is usually simple, rarely double on one side and 6-C is simple). The outer frontal seta (7-C) has 4–8 branches (usually 5–6). Seta 1-M has 2 long branches, which may be used as a character to separate the species from *Oc. caspius*. The comb has 13–35 (usually 20–25) scales arranged in 2–3 irregular rows (Fig. 10.41). The scales are variable in shape, as in *Oc. caspius*. The siphonal index is 2.5–3.0, rarely less. The pecten has 14–23 evenly or slightly irregularly spaced teeth not reaching the middle of the siphon. The siphonal tuft (1-S) has 3–8 (usually 4–5) branches, and is inserted in the middle of the siphon or slightly below it, usually closer to the base than in *Oc. caspius*. The saddle is reduced to the dorsal half of the anal segment, is strongly pigmented and with a sharply defined lower margin. The saddle seta (1-X) is half as long as the saddle. The anal papillae are rounded; their length greatly varies with the salinity of the breeding water. They are about 1.3 times as long as the saddle in fresh water, but not more than 0.3–0.4 times the length of the saddle in saline water (Gutsevich *et al.*, 1974).

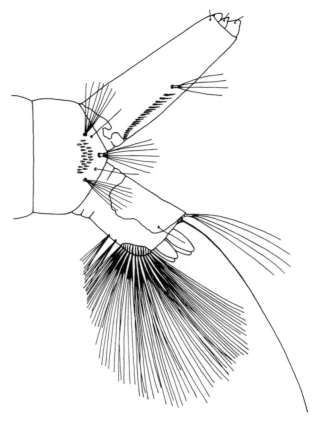

Figure 10.41. Larva of *Oc. dorsalis*.

Biology: *Oc. dorsalis* is a polycyclic and halophilic species that hibernates in the egg stage. There are usually 2–4 generations per year in central Europe depending on the number of floodings. The larvae occur mainly in small, open water bodies and swamps; they are found in permanent or temporary water bodies formed by melted snow, floods, rainfall or groundwater such as inland pastures, roadside and drainage ditches. Females preferably oviposit around saline water bodies but the ground nearby fresh waters can be chosen as well. The larvae can be found in waters with a salt content of up to 12% and a pH value which ranges from 7.0 to 9.3 (Chapman, 1960). They are numerous in Atlantic and Mediterranean coastal marshes but prefer inland saline areas in the Palaearctic region (Bozicic-Lothrop, 1988). Adults can be very numerous in some localities such as river valleys, saline ponds and lakes. The females attack man readily inflicting painful bites (Richards, 1956). They are also characteristic of open habitats in mixed forests and pastures. Bloodsearching activity is recorded from 9 to 30°C and 52–92% of relative humidity in the southern part of central Europe (Petric, 1989). The peak of daily biting activity occurs in the late afternoon and early evening, but females can be quite active during the night, too. They can also search for blood during the day in open areas (Cranston *et al.*, 1987). Females are exophagic but readily enter houses, tents and wards (Waterston, 1918). Autogeny is rare, but was detected in Nevada (Chapman, 1962).

Distribution: *Oc. dorsalis* is a widespread Holarctic salt marsh species in Europe, Central Asia, China, North Russia and North America. In Europe the distribution range extends north to Scandinavia (Natvig, 1948) and south to Greece (Pandazis, 1935).

Medical importance: *Oc. dorsalis* transmits Western Equine Encephalitis virus (WEE), St. Louis Encephalitis virus (SLE), California Encephalitis virus (CE) in the United States (Hammon and Reeves, 1945; Hammon *et al.*, 1952). Japanese Encephalitis virus and the bacterium *Francisella tularensis*, the causative agent of tularemia, were isolated from natural populations (Detinova and Smelova, 1973).

Note on systematics: Three species closely related to *Oc. dorsalis* are the Nearctic *Oc. melanimon* and *Oc. campestris* as well as the Palaearctic *Oc. caspius*. The current status of *Oc. dorsalis* is discussed under the description of *Oc. caspius*.

Ochlerotatus Excrucians Complex

The complex is composed of the Holarctic species *Oc. excrucians* and *Oc. euedes*, some other Nearctic species, *e.g. Oc. aloponotum*, and the Palaearctic, European species *Oc. surcoufi*. The current status of the sibling species within the complex is rather unsatisfying. There exists a great variation between populations from different geographical areas. Apart from differences in the general body colouration of females of *Oc. excrucians*, the shape of the tarsal claw also varies in populations within both North America and Europe (Wood *et al.*, 1979; Dahl, 1984). The variation is expressed by the angle formed between the main and the subbasal tooth, as well as the degree of bending of the former. The "wide type" claw described by Wood *et al.* (1979) as a southern Canadian variant of *Oc. excrucians* is quite similar to some specimens from Italy, France and Germany, designated as *Oc. surcoufi* by Arnaud *et al.* (1976), another "wide type" variant was also found in Sweden (Dahl, 1984). A further confusion is that the European members of the complex, *Oc. excrucians, Oc. surcoufi* and *Oc. euedes*, are not very well studied. However, Arnaud *et al.* (1976) suggested the substitution of all earlier *Oc. excrucians* records in Europe by its European form *Oc. surcoufi*. Since no worldwide comparison of geographic variations and redescriptions of all stages of the species in question has been carried out, this suggestion has not been followed.

Ochlerotatus (Ochlerotatus) euedes (Howard, Dyar and Knab) 1913

Female: For separation from *Oc. excrucians* refer to the description of *Oc. surcoufi*. *Oc. euedes* is usually distinguished from the latter species by the scutal scale pattern. In *Oc. euedes* the dorsocentral stripes of pale scales start at some distance from the anterior margin of the scutum and extend to the transverse suture. The pale lateral stripes continue over the transverse suture and reach the end of the dorsocentral stripes (Fig. 6.35b). In *Oc. surcoufi* the dorsocentral stripes of pale scales usually start at about the level of the transverse suture and reach close to the posterior margin of the scutum. The white lateral stripes are sometimes indistinct, and the transverse suture is covered with well visible stripes of white scales. The abdominal terga of *Oc. euedes* have dark scales and narrow pale basal bands at least on the first segments, otherwise median and lateral patches of white scales and sometimes narrow apical white bands are present. The cerci are dark with scattered light scales.

Male: The hypopygium is similar to those of the other members of the complex (Fig. 10.42). The only difference may be found in the shape of the apical lobe. In *Oc. euedes* it is well developed, usually protruding above the level of the gonostylus articulation, whereas the apical lobe in *Oc. excrucians* and *Oc. surcoufi* usually does not reach the articulation of the gonostylus.

Larva: The large larvae are very similar to those of *Oc. excrucians*. The number of comb scales is 11–19 arranged in 2–3 irregular rows (Fig. 10.43). The siphon is more evenly tapered from the base to the apex, and not as abruptly narrowed as in *Oc. excrucians*. The distalmost pecten tooth is situated well beyond the middle of the siphon, almost reaching 1-S. The siphonal tuft (1-S) is inserted distinctly beyond the middle of the siphon, and the branches of 1-S are shorter than in *Oc. excrucians*.

Biology: The species is monocyclic and hibernates in the egg stage. In Europe, it breeds in the same kind of habitats and occurs at the same time as *Oc. excrucians, Oc. cantans* and *Oc. annulipes*.

Distribution: Reported in the eastern and northern states of the United States and Canada (Darsie and Ward, 1981; Wood *et al.*, 1979). In Europe it has been recorded from southern Scandinavia and, sometimes under the synonym of *Oc. beklemishevi* (Denisova) 1955, from Finland, Poland, Lithuania and European Russia (Snow and Ramsdale, 1999). Very little European material is available.

Medical importance: Alphavirus has been reported from *Oc. euedes* in central Russia (Lundström, 1999).

Note on systematics: *Oc. euedes* (Howard, Dyar and Knab) 1913 was resurrected from synonymy of *Oc. excrucians* and regarded as a separate species. At the same time, *Oc. beklemishevi* (Denisova) 1955, reported from the former USSR and Poland, was put under synonymy of *Oc. euedes* (Knight, 1978; Wood, 1977).

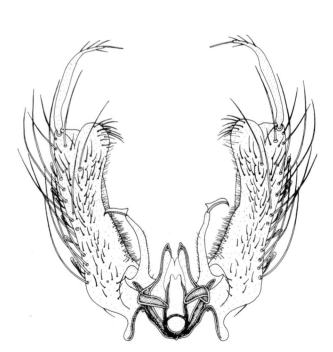

Figure 10.42.   Hypopygium of *Oc. euedes*.

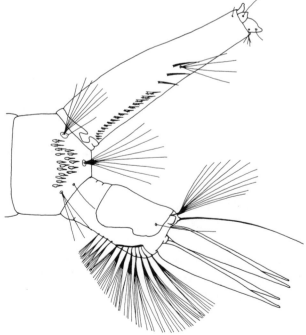

Figure 10.43.   Larva of *Oc. euedes*.

Ochlerotatus (Ochlerotatus) excrucians
(Walker) 1856

Female: A large species, similar in size and body scale patterns to *Oc. euedes* and *Oc. surcoufi*. The best identification character for females of *Oc. excrucians* is the shape of the fore unguis, which is bent in a sharp angle with a very narrow space between the unguis and the subbasal tooth. The integument colour varies from dark in northern specimens to brownish in more southern ones. The colour of scales varies from white to creamy white, and the dark scales from nearly black to very dark brown. The body setae are mostly brownish golden. The proboscis is dark scaled, sometimes with a white scale patch in the middle. The palps have white rings at the basal part of palpomeres I and II. The head has some prominent blackish to bronze erect forked scales. The vertex and occiput have mixed scaling, and a more or less expressed white median stripe and a regular dark spot laterally. The scutum is covered with narrow, bronze scales, which are usually lighter on the lateral parts, sometimes with an indistinct broad median stripe of darker scales. The upper postpronotum has bronze, narrow scales and a patch of broad white ones at the lower part. A postprocoxal scale patch is present, and distinct white subspiracular and postspiracular patches are usually present. In northern specimens the mesepisternum is covered with three distinct white patches. The variability in southern populations is expressed by a larger median patch which may be fused with the lower patch. The mesepimeron has a white patch which occupies half of the sclerite, but never reaches its lower margins. The femora are thoroughly whitish scaled with a few dark scales and dark, irregular bands close to the white knee spots. The tibiae have mixed scaling, which on the fore and mid leg is somewhat lighter, and the amount of white and dark scales may vary. Tarsomere I of all the legs has mixed scales and more or less well defined white basal rings; tarsomeres II to IV have well defined white basal rings, except

for the fore leg, which has an entirely dark scaled tarsomere IV. Tarsomere V of the hind leg has a white basal ring. The fore leg has the main tooth of the unguis sharply bent, almost parallel to the subbasal tooth. However, in northern parts of Europe a "wider" form exists, which nevertheless has the typical bent characteristic of an *Oc. excrucians* unguis (Fig. 6.32a). The wing veins are mainly dark scaled, with pale scales usually present on the costa (C) and subcosta (Sc), and the other veins with a variable amount of pale scales. The terga are dark scaled, with narrow pale basal bands or patches and more or less numerous scattered pale scales in the apical part (Fig. 6.31b). The irregular yellowish basal bands of the terga, which may extend laterally on the last segments, may be a distinct feature of the northernmost populations. The sterna are nearly entirely whitish scaled, with the exception of sterna II to V which have dark apical edges, and the cerci have dark scales.

Male: The hypopygium is of *Oc. excrucians* (Fig. 10.44) very similar to those of *Oc. euedes* and *Oc. surcoufi*. From the latter species it has not been separated in much of the European material. The lobes of tergum IX

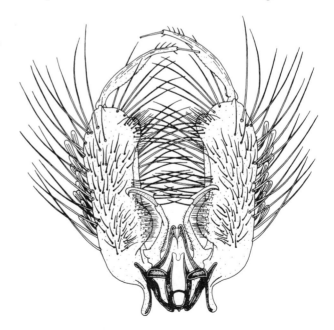

Figure 10.44. Hypopygium of *Oc. excrucians*.

bear 5–7 more or less stout setae. The basal lobe of the gonocoxite is flat, moderately developed or indistinct, almost absent. It is covered with many short setae of similar length and thickness. The apical lobe usually does not reach the articulation of the gonostylus. The gonostylus is slender, curved apically, and somewhat flattened with a moderate long apical spine. The claspette stem is slender, tapers apically, and the claspette filament gradually tapers towards the apex. The aedeagus is elongated.

Larva: Large in size, the antenna is shorter than the head with well developed spicules. The antennal seta (1-A) is inserted at about the middle of the antennal shaft. The postclypeal seta (4-C) has 2–3 short, thin branches. The inner frontal seta (5-C) has 2–3 branches, the median frontal seta (6-C) has 2 branches, and the outer frontal seta (7-C) has 6–7 branches. The prothoracic formula 1-P to 7-P is as follows: 1 (2-branched, long); 2 (simple, medium long); 3 (4–5 very thin, short branches); 4 (medium long, thin); 5 and 6 (long, simple); 7 (3-branched). The abdominal seta 6 of segments I and II (6-I and 6-II) has 2 branches, but is simple on the rest of the segments (Fig. 8.54a). The number of comb scales is 30–38 arranged in a more or less triangular patch; each individual scale has a conspicuous median spine. The siphonal index is 3.4–4.5, and the distal part of the siphon is distinctly tapered (Fig. 10.45). The pecten has 15–24 pecten teeth, with 1–3 teeth apically detached. The siphonal tuft (1-S) is positioned at about the middle of the siphon, beyond the distalmost pecten tooth, with 6 long branches. Seta 9-S is thickened and transformed into a prominent hook. The saddle extends beyond the middle of the lateral part of the anal segment. The ventral brush has 4–6 precratal tufts (4-X). The anal papillae are slender, and longer than the saddle.

Biology: The larvae start to hatch from the overwintering eggs in early spring. More often they hatch later in the year and can be found until the middle of summer in shaded, permanent pools. Only one generation has been recorded from all the different habitats. They occur

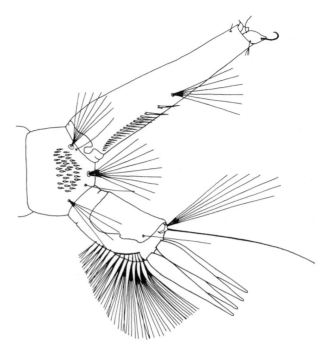

Figure 10.45. Larva of *Oc. excrucians.*

in greater numbers in open semipermanent or permanent pools or pits with vegetation like *Typha* sp. or *Carex* sp. together with larvae of *Oc. cantans* and *Ae. cinereus.* They can also be found in less abundance in a variety of other habitats, especially in mixed forest regions. Larvae have been observed to feed upon *Euglena* sp. and larger protozoans. Larval development can last from a few to many weeks depending on the water temperature. The females are fierce day biters (for that reason they got the name *excrucians*) and can be found until autumn.

Distribution: *Oc. excrucions* is a Holarctic species. Its actual distribution for central and southern Europe is difficult to state, as some of earlier records might contain a mixture of sibling species.

Medical importance: No virus is recorded from Europe from this species, but Swedish specimens showed experimentally vector capacity for Ockelbo virus (Lundström, 1994).

Ochlerotatus (Ochlerotatus) surcoufi (Theobald) 1912

Female: Very similar to *Oc. excrucians.* The most reliable character to distinguish between females of these species is found in the tarsal claws of the fore leg. In *Oc. surcoufi* the main tooth is evenly curved, and the subbasal tooth is diverging, not parallel. The abdominal terga have mixed light and dark scales. Tergum I has a white patch; the following terga have both basal and apical yellowish bands. The apical bands are often reduced to a narrow line, and the basal bands are sometimes laterally irregular. Tergum VII has light scales predominating.

Male: The hypopygium is very similar to that of *Oc. excrucians*, apparently morphologically identical.

Larva: Similar in most characters to *Oc. excrucians*. A striking difference is found in setae 6 on abdominal segments I to VI (6-I to 6-VI) which all have 2 branches (Fig. 8.54b), whereas in *Oc. excrucians* only setae 6-I and 6-II have 2 branches, and setae 6-III to 6-VI are simple. The siphon is long, and tapered towards the apex; the siphonal index is 2.8–3.1 (Fig. 10.46).

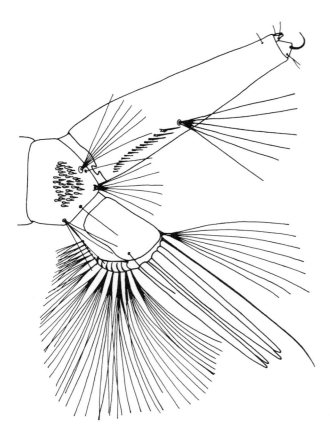

Figure 10.46. Larva of *Oc. surcoufi*.

Seta 9-S is hooked, and not as strong as in *Oc. excrucians*. The ventral brush has 5–6 precratal tufts (4-X).

Biology: In the Pyrenees the species seems to occur in the same habitats as *Oc. excrucians*. The eggs and larvae were found in inundation areas in a Pyrenean valley. Females were caught on man in the same habitat (Arnaud *et al.*, 1976).

Distribution: *Oc. surcoufi* has so far been reported in France (near Paris and the eastern Pyrenees), and on material from Germany and Italy according to Arnaud *et al.* (1976).

Notes on systematics: The species has been resurrected from synonymy with *Oc. excrucians* based on European material by Arnaud *et al.* (1976). In their discussion of the Ochlerotatus Excrucians Complex the authors required biometric and genetic analysis of all sibling species to clarify their status.

Ochlerotatus (Ochlerotatus) flavescens (Müller) 1764

Female: *Oc. flavescens* can be distinguished from other *Aedes* and *Ochlerotatus* species by the abdominal terga which are almost entirely covered with light scales, and only a few scattered dark scales may be present; a character, which is shared only with *Oc. cyprius*. The scutum of *Oc. flavescens* is covered with copper or golden brown scales and the terga with straw coloured scales, which are distinctly paler than the scutal scales. In contrast, *Oc. cyprius* has golden yellowish scales on the scutum and usually ochre yellowish scales on the terga, with the scutum and terga being of more or less the same colour. The proboscis and palps of *Oc. flavescens* have dark and yellowish scales, with the yellowish scales on the proboscis less numerous in the distal half. The vertex and occiput have narrow golden scales and brown erect forked scales, and broad cream coloured scales laterally. The scutellum has narrow yellowish or light brown scales and brown setae on the lobes. The antepronotum has yellow bronze scales. The postpronotum has upper narrow, bronze scales, and lower broader, cream coloured scales. The postprocoxal membrane usually has a patch of pale scales. The hypostigmal, subspiracular and postspiracular scale patches are well developed. The mesepisternum has pale yellow scales; the upper patch reaching its anterior angle. Pale scales on the mesepimeron end slightly before its lower margin, and lower mesepimeral setae are absent (Fig. 6.29a). The femora are brown with yellowish scales intermixed, pale on the posterior surface. The tibiae are mostly yellow scaled; tarsomeres II to IV of the mid and hind legs with broad basal rings of whitish scales (Fig. 6.27c). The wing veins are covered with narrow yellowish scales and scattered dark scales. The terga are covered with straw-coloured scales, sometimes mixed with isolated dark scales (Fig. 6.28a). The sterna have pale yellowish scales.

Male: The lobes of tergum IX are short and narrow, with 5–7 long spine-like setae. The basal lobe of the gonocoxite is slightly flattened with a strong medially directed spine and numerous short setae extending to the middle of the gonocoxite (Fig. 10.47). The apical lobe is prominent, and rounded apically. The gonostylus is slightly expanded in the middle, and the apical spine of the gonostylus is long and slender. The paraproct is strongly sclerotised, and pointed at the apex. The claspette stem is short and straight, slightly tapered distally, with 2–3 long setae on the inner margin near the base. The claspette filament is about as long as the stem, with a cylindrical base and a plate shaped widening on the convex side of the filament, and the apex of the filament is distinctly curved. The aedeagus is long, cylindrical, and notched at the apex.

Larva: One of the largest *Ochlerotatus* larva. The antennae are less than half as long as the head, and entirely covered with distinct spicules. The antennal seta (1-A) is located close to the middle of the antennal shaft, with 5–8 branches reaching near the tip. The inner (5-C) and median (6-C) frontal setae have 2–4 branches, and the outer frontal setae (7-C) have 6–9 branches. The number of comb scales is 17–36 (usually 20–27) arranged in 3 irregular rows (Fig. 10.48). Each individual scale has a strong median spine and several smaller spines at the side, which are about half as long as the median spine. The siphonal index is 3.2–4.0.

Figure 10.47. Hypopygium of *Oc. flavescens*.

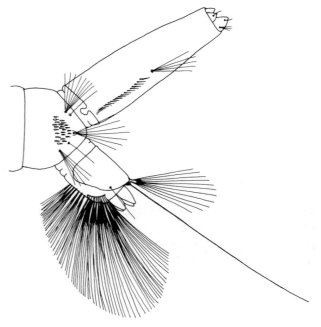

Figure 10.48. Larva of *Oc. flavescens*.

The pecten has 17–28 teeth not reaching to the middle of the siphon, and 1 or 2 distal teeth may or may not be apically detached. The siphonal tuft (1-S) has 4–7 branches, situated in the middle of the siphon, about as long as the width of the siphon at the point of its origin. The saddle extends about three quarters of the way down the sides of the anal segment. The saddle seta (1-X) is simple and approximately as long as the saddle. The upper anal seta (2-X) has more than 7 branches, and the lower anal seta (3-X) is simple and longer than the siphon. The ventral brush has 18–19 tufts of cratal setae (4-X) and 5–7 precratal setae (4-X) extending nearly to the base of the segment. The anal papillae are usually short, and about half as long as the saddle.

Biology: Oc. flavescens usually produces only one generation per year (Mihalyi, 1959; Trpis, 1962). After overwintering in the egg stage the larvae hatch during spring time. They occur predominantly in reeded areas which are not totally shaded. They can frequently be found in brackish marshes along coasts (Gjullin et al., 1961; Mohrig, 1969), but also in partly shaded temporary fresh water ponds in floodplains (Hearle, 1929; Rempel, 1953). The immature stages of this species are tolerant to a wide range of salinity. Usually they occur in neutral or slightly alkaline water, associated with larvae of Oc. cantans, Oc. leucomelas, Ae. rossicus and Oc. sticticus. However, in saline water they can be found together with the larvae of Oc. detritus, Oc. caspius or Oc. dorsalis. The adults prefer open areas when they are searching for a blood-meal. The biting peak of adults occurs during dusk and dawn. Wesenberg-Lund (1921) observed swarming of males near sunset, above nettles. Although Oc. flavescens is widely distributed it usually does not occur in great numbers.

Distribution: It is a Holarctic species, reported from Eurasia and North America. In Europe the species is widely distributed except in some mediterranean countries.

Ochlerotatus (Ochlerotatus) hexodontus (Dyar) 1916

Female: A medium sized species, similar to that of Oc. punctor, but can be distinguished from the latter by the pale bands on abdominal sterna III to VI which are of uniform width or only slightly constricted in the middle and the large patch of pale scales at the base of the costa (C). The proboscis and palps are dark scaled, the palps rarely with a few white scales. The head has narrow dark brownish scales and erect forked scales dorsally. The pedicel usually has some whitish scales. The scutum is more or less uniformly covered with rusty brown or yellowish brown scales, usually without a darker median pattern, and occasionally with a pair of indistinct dark submedian stripes (Fig. 6.49b). The supraalar and scutellar setae are yellowish or yellow brown. The prosternum is scaled with yellow white scales, the upper 3/4 of the postpronotum with rust brown scales. The postprocoxal membrane has pale scales. A hypostigmal scale patch is absent, the subspiracular patch is divided into an upper and lower portion, and the postspiracular scale patch is well developed. The mesepisternum has a pale upper patch extending to the anterior angle, narrowly separated from the prealar patch, and the lower mesepisternal patch is confined to the posterior margin. The mesepimeron has yellowish scales extending to near the lower margin, and 1–3 mesepimeral setae are present. The femora are dark scaled, speckled with pale scales, and pale on the posterior surface. The tibiae and tarsi are dark brown with scattered pale scales especially towards the apices of the tarsomeres, but pale basal rings on the tarsomeres are absent. The wing veins are covered with dark scales, and pale scales are present at the base of the costa (C), and subcosta (Sc) often forming a large patch. The abdominal terga are dark scaled with basal bands of greyish white scales of uniform width, sometimes slightly constricted in the middle (Fig. 6.48b). The sterna are covered with greyish pale scales.

Male: The colouration and scaling are much as for the females except that the prosternum is not scaled. The hypopygium of *Oc. hexodontus* (Fig. 10.49) is similar to that of *Oc. punctor*, see the description of the latter species.

Larva: Very similar to that of *Oc. punctor*, but may be separated from the latter by the size and number of comb scales. In *Oc. hexodontus* the number of comb scales is usually 6–9, the individual scales are large, and usually longer than the distalmost pecten tooth. In *Oc. punctor* the number of comb scales is more than 10, and individual scales are shorter than the distal three pecten teeth. In Nearctic populations there also exist differences in the thoracal setation (Wood *et al.*, 1979), which is probably also true for European populations. In *Oc. hexodontus* seta 5-P is branched and setae 1-M, 1T and 3-T are usually short and multiple-branched. In *Oc. punctor* seta 5-P is usually simple, and 1-M, 1-T and 3-T are longer and usually simple or double. *Oc. hexodontus* has antennae much shorter than the head, and the antennal seta (1-A) is small, and inserted slightly below the middle of the antennal shaft, with multiple branches not reaching to the tip. The inner (5-C) and median (6-C) frontal setae are usually simple, rarely with

2 branches, and the outer frontal setae (7-C) have 3–6 branches. The siphon is straight, tapers in the apical half, and the siphonal index is about 3.0 (Fig. 10.50). The pecten has 12–22 evenly spaced teeth situated in the basal half of the siphon. The siphonal tuft (1-S) is located beyond the distalmost pecten tooth, usually with 4–5 branches. The anal segment is entirely encircled by the saddle, and the saddle seta (1-X) is simple and about as long as the saddle. The ventral brush has 16–18 cratal setae (4-X) and two, rarely one, precratal setae. The anal papillae are lanceolate, and usually twice as long as the saddle or longer.

Biology: *Oc. hexodontus* is a monocyclic species which belongs to the dominating mosquitoes in areas with long severe winters and short summers. The species occurs in tundra areas close to the tree border where the immature stages develop numerously in oligotrophic snow-melt ponds. In high mountain regions, the larvae can be found in small snow-melt pools with little or no vegetation. The larvae hatch at water temperatures of just above 0°C. The larvae are frequently found associated with those of other snow-melt mosquitoes, such as *Oc. communis*, *Oc. punctor*, *Oc. pionips* and *Oc. pullatus*. The adults of *Oc. hexodontus* can also occur in

Figure 10.49.   Hypopygium of *Oc. hexodontus*.

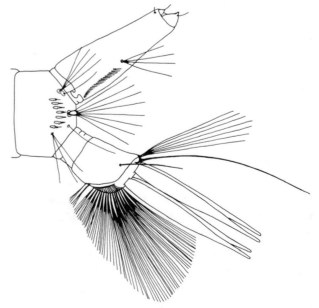

Figure 10.50.   Larva of *Oc. hexodontus*.

mountainous regions close to the tree border and furthermore, in larger numbers if the number of the other species has already decreased. The females are very aggressive biters and may approach their hosts even during strong winds.

Distribution: *Oc. hexodontus* is a Holarctic species found in North America and Eurasia, from northern Scandinavia to eastern Sibiria, subarctic tundra and taiga, and above the tree line in mountainous regions.

Ochlerotatus (Ochlerotatus) hungaricus (Mihalyi) 1955

Female: A small species, with blackish brown scaled proboscis and palps. The occiput has narrow whitish scales dorsally and broad whitish scales and scattered dark scales laterally, which may form a small patch. The scutum is covered with greyish white scales and a median stripe of dark brown scales, which is divided into two prescutellar dorsocentral stripes not reaching the posterior margin of the scutum. The scutellum has pale narrow scales and light and dark brown setae on the lobes. The postprocoxal membrane is without scales, and a hypostigmal scale patch is absent. The upper mesepisternal scale patch reaches the anterior angle of the mesepisternum, and the mesepimeral scale patch does not reach the lower margin of the mesepimeron. The femora of the fore legs are predominantly pale scaled in the basal half, and the tibiae of the hind legs have dark scales on the anterior surface. The tarsomeres are dark scaled, and pale basal rings are absent. The wing veins are covered with dark scales, and a few isolated pale scales may be present at the base of the costa (C). The abdominal terga have blackish brown scales and pale basal bands which are slightly narrower in the middle and connected with pale lateral triangular patches. The sterna have whitish scales and broad apical dark bands.

Male: The gonocoxite is about three times as long as it is wide at the base, and is covered with scales and setae of different length (Fig. 10.51). The basal lobe of the gonocoxite is well developed, and conical, with 2 strong

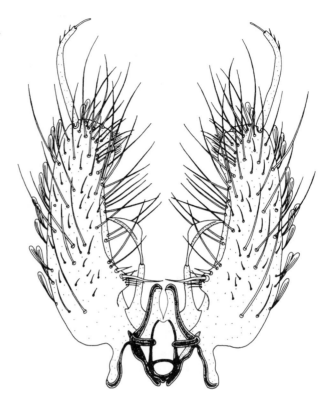

Figure 10.51. Hypopygium of *Oc. hungaricus.*

spine-like setae of different length and several short setae. The longer spine is slightly curved in the middle, the other spine is short and straight (Fig. 7.34b). The apical lobe is weakly developed, and indistinct, with numerous short setae. The gonostylus gradually tapers towards the apex, and the apical spine is moderately long and slender. The paraproct is inwardly curved and pointed. The claspette stem is short and straight; the claspette filament has a long stalk and a broad unilateral wing at the convex side. The aedeagus is tubular shaped.

Larva: The head is very dark, blackish brown. The antenna is slightly more than half as long as the head. The antennal seta (1-A) is located distinctly below the middle of the antennal shaft, at about 1/3 of the length from the base, with 5–8 branches. The inner frontal seta (5-C) is simple, the median frontal seta (6-C) has 1–2 branches, and the outer frontal seta (7-C) has 6–8 branches. The prothoracic formula (1-P to 7-P) is as follows: 1 (long,

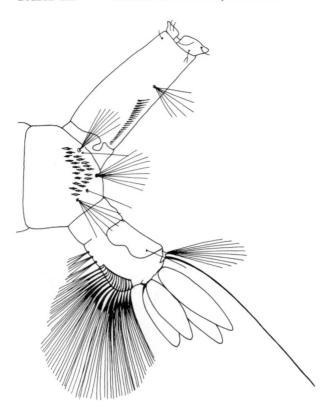

Figure 10.52. Larva of *Oc. hungaricus*.

simple); 2 (medium long, simple); 3 (short, 2–4 branched); 4 (short, simple); 5 and 6 (long, simple); 7 (long, 2-branched). The comb has 16–24 (average 20) scales, arranged in 2 irregular rows (Fig. 10.52). Each individual scale has a long median spine and several shorter spines at the sides. The siphon is cylindrical, only slightly tapering towards the apex, and the siphonal index is 2.0–2.4. The pecten has 19–24 evenly spaced teeth, reaching beyond the middle of the siphon. Each pecten tooth has several lateral denticles at its ventral margin. The siphonal tuft (1-S) has 3–6 branches, situated well beyond the middle of the siphon and well separated from the distalmost pecten tooth. The saddle extends approximately half way down the sides of the anal segment, and the saddle seta (1-X) is simple and shorter than the saddle. The upper anal seta (2-X) has 8–11 branches, the lower anal seta (3-X) is simple and longer than the siphon. The ventral brush has 13–16 cratal setae

(4-X) and 3 precratal setae. The anal papillae are lanceolate, distinctly longer than the saddle, and the dorsal pair is longer than the ventral pair.

Biology: Because the distribution range of the species seems to be limited, little is known about its biology. Larvae are typically found in the floodplains of larger river systems. Mihalyi (1961) collected the larvae in shaded floodwater pools under trees along the Danube river. The species has been found from April to September, thus more than one generation per year is assumed. The females are severe biters and readily feed on humans in shaded situations even during the day.

Distribution: *Oc. hungaricus* has so far been reported from Austria, Slovakia and Hungary (Snow and Ramsdale, 1999).

Ochlerotatus (Ochlerotatus) impiger (Walker) 1848

Female: This small species may sometimes be mistaken for small specimens of *Oc. punctor*. Its integument is brownish grey with conspicuous long and black setae on the scutum. The proboscis and palps are entirely black scaled. The vertex has a spot of dark brown scales. The occiput is covered with erect black setae and white scales. The antenna is black, and the pedicel has some scattered white scales. The scutum has a median stripe of bronze narrow scales, which is narrower anteriorly and a lateral stripe of white narrow scales. The scutellum has white narrow scales. The postpronotum has a few bronze scales, otherwise broad white scales, and postpronotal setae are scattered on the entire postpronotum (Fig. 6.41a). A postprocoxal patch is present, the subspiracular and postspiracular scale patches are present, but a hypostigmal patch is absent. The number of postspiracular setae is ten or fewer (Fig. 6.42b). The mesepisternum has three more or less distinct white scale patches, the middle one not reaching the anterior margin. The mesepimeron has a large white patch of scales. All the coxae have white scales, the femora and tibiae of all the legs are dark with scattered white scales, which are less numerous than in *Oc. nigripes*, the tibiae have conspicuous black setae, and all the tarsomeres are dark scaled. The tarsal claws are sharply bent in the middle (Fig. 6.42a). The wing veins are usually entirely dark scaled, a few pale scales are sometimes present at the base of the costa (C) and radius (R). The scaling of the legs and wings seems to vary between Scandinavian and other circumpolar populations. The abdominal terga have broad basal white bands.

Male: The lobes of tergum IX are rounded, with long setae. The gonocoxite has short setae predominating on the inner side (Fig. 10.53) The basal lobe of the gonocoxite is well developed, and conical, with one spine-like seta. The apical lobe is small and indistinct. The gonostylus is slender, and somewhat broadened in the middle. The paraproct is broadly sclerotised. The claspette filament is about as long as the stem, with a unilateral wing, and the aedeagus is elongated.

Larva: The small fourth-instar larva can easily be mistaken for earlier instars of *Oc. hexodontus* or *Oc. punctor* where the saddle is not fully developed and does not encircle the anal segment. The antennae are very short. The inner and median frontal setae (5-C and 6-C) are always simple, and the outer frontal seta

(7-C) has 3 branches. The prothoracic formula 1-P to 7-P is as follows: 1 (simple or 2-branched, short); 2 (simple, little longer); 3 (simple, long); 4 (simple, short); 5 (simple or 2-branched, long); 6 (simple, long); 7 (3-branched, long). The number of comb scales is 10–14, each scale has a long median spine and several short spines at the base. The siphonal index is 2.8–3.0. The pecten teeth are evenly spaced, each tooth with one long, lateral denticle (Fig. 10.54). The siphonal tuft (1-S) is situated slightly below the middle of the siphon, with 4–6 long branches. The saddle covers approximately half of the anal segment. The ventral brush has 2 tufts of precratal setae (4-X), and the anal papillae are longer than the saddle (Dahl, 1997).

Biology: The species is monocyclic. The larvae appear very early in the year and can be found in ponds at the borders of birch-willow shrubs in subarctic areas. They develop fast in snow-melt ponds, preferably with some rotten

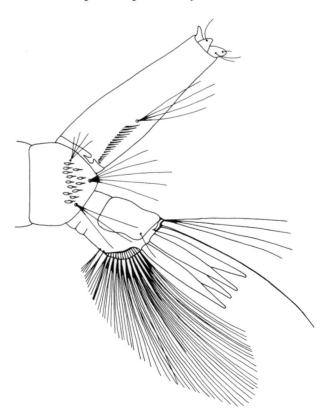

Figure 10.53. Hypopygium of *Oc. impiger*.

Figure 10.54. Larva of *Oc. impiger*.

plant material on the bottom. The small females are most obnoxious biters during the day in the subarctic and arctic areas of Europe. Females start to bite very soon after emergence. However, they do alight quickly from their hosts and are thus not easy to catch on humans. The biology of the species, together with that of *Oc. nigripes*, has been more thoroughly studied in arctic Canada (Wood *et al.*, 1979).

Distribution: *Oc. impiger* is a circumpolar species and in North America it is found, together with *Oc. nigripes*, under extreme arctic conditions. It seems to have a more southern limit than the latter species. In Europe the distribution of *Oc. impiger* is restricted to the tundra of Scandinavia, but its whole distribution range is not well known, it has been recorded in southern Norway (Natvig, 1948) and in the tundra of Russia (Gutsevich *et al.*, 1974).

Ochlerotatus (Ochlerotatus) intrudens (Dyar) 1919

Female: A dark greyish species without metallic reflexions in the scales. The presence of a hypostigmal patch of pale scales separates the females from those of *Oc. communis*, *Oc. diantaeus* and *Oc. nigrinus*. However, the patch is composed of a few scales, which may often be rubbed off in old specimens. The proboscis is black scaled, the palps have scattered white scales, and the head has some pale golden setae between the eyes and yellowish white narrow scales on the vertex. The scutum has two dorsocentral stripes of narrow curved bronze scales, divided by a narrow pale acrostichal stripe, which is sometimes indistinct. The anterior submedian and whole lateral areas are covered with pale scales, and two distinct posterior submedian dark stripes are present. The scutellum has pale yellowish scales and setae. The postpronotum has yellowish-brown narrow curved scales and some whitish scales at the posterior corner. The postprocoxal patch is absent, and a hypostigmal patch is usually present. The sub- and postspiracular areas have patches of white scales. The upper mesepisternal

patch of scales is small, and divided into two or more portions, not reaching the anterior margin of the mesepisternum. The prealar patch of scales is displaced apically, and the lower third of the mesepimeron is without scales. The lower mesepimeral setae are usually absent (Fig. 6.44b). The coxae have white scaling, the rest of the legs are almost entirely dark scaled, the fore femur sometimes with some scattered white scales. The wing veins are almost entirely dark scaled, and the base of the costa (C) and radius (R) sometimes have a few white scales. The abdominal terga are dark scaled, with distinct basal white bands.

Male: The hypopygium is similar to those of *Oc. diantaeus* and *Oc. pullatus*, but differs in the shape of the claspette and the setation on the apical half of the gonocoxite (Fig. 10.55). The lobes of tergum IX are prominent, with a row of 4–7 spine-like setae. The basal lobe of the gonocoxite has three large setae which are all spine-like, one of them well separated from the others. The apical lobe of the gonocoxite is well developed, and a small subapical zone of the gonocoxite is covered with long dense setae directed more distally. The gonostylus is curved, and tapers towards the apex. The claspette stem has a thorn shaped process at about its middle. The claspette filament is strongly bent. The aedeagus

Figure 10.55. Hypopygium of *Oc. intrudens*.

is pear shaped, relatively short and with a wide opening.

Larva: The head is broader than it is long, and the antenna is shorter than the head, and covered with spicules. The antennal seta (1-A) is short and inserted slightly below the middle of the antennal shaft. The inner frontal seta (5-C) has 3–5, usually 4 branches and the median frontal seta (6-C) has 3–5, usually 3 branches (Fig. 8.35c). The number of comb scales is 12–17, each individual scale is elongated and pointed with lateral spines along the basal part. The siphon is long and slightly tapered (Fig. 10.56). Usually 2–3 distalmost pecten teeth are large, spine-like, without lateral denticles and more widely spaced. The distalmost tooth may be located slightly beyond the point of 1-S insertion. The siphonal tuft (1-S) is inserted slightly beyond the middle of the siphon. The anal segment is not entirely encircled by the saddle, the number of precratal tufts (4-X) is 1–2, and the anal papillae are slender and longer than the saddle.

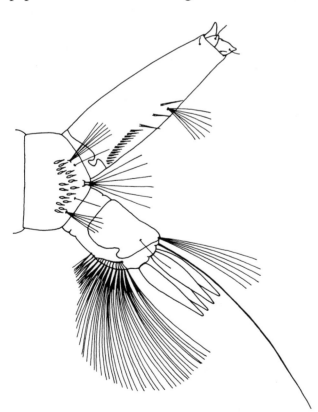

Figure 10.56. Larva of *Oc. intrudens*.

Biology: The species is monocyclic at least in the northern parts of its distribution range. Whether it can have two generations in southern areas is not known. Eggs are overwintering and larvae are found from early spring on until the beginning of summer in many types of temporary pools, which can be located in forests, without submerged vegetation but covered with dead leaves at the bottom, or in more open areas in grassy or snow-melt water pools as well as along floodplains. The larvae are often found together with those of *Oc. communis*, *Oc. diantaeus* and *Oc. punctor*, but they emerge before any of these other species (Wood *et al.*, 1979; Barr, 1958). The first adults occur from early spring on to midsummer depending on the latitude. The females are locally dominant, fierce biters which also enter houses.

Distribution: Holarctic species, recorded from central to northern, and eastern to southeastern Europe, it occurs from the steppe in southeastern Europe to the taiga in the north. The distribution range does not seem to extend into western Europe.

Ochlerotatus (Ochlerotatus) leucomelas (Meigen) 1804

Female: *Oc. leucomelas* is similar to *Oc. cataphylla* in sharing the character of intermixed pale and dark scales on the wing veins, but differs from the latter in the proboscis which is speckled with numerous pale scales, whereas in *Oc. cataphylla* the proboscis is uniformly dark scaled. The palps are dark brown with scattered pale scales. The vertex and occiput have narrow yellowish white scales and dark erect forked scales, and pale decumbent scales at the sides. The scutum has golden bronze scales, the lateral and prescutellar areas usually have straw coloured scales. The scutellum is dark brown with narrow yellowish white scales. The postpronotum has broad white scales, more yellowish in the upper part; and the scales of the pleurites and legs are as for *Oc. cataphylla*. The postprocoxal membrane has a patch of pale scales. The hypostigmal, subspiracular and postspiracular scale patches are fused. The scales of the mesepisternum do not extend to its anterior angle and the mesepimeral scale patch ends slightly before the lower margin of the sclerite; lower mesepimeral setae are present. The tibiae and tarsi are mostly dark scaled, with pale scales especially on the ventral surface. The tarsi do not have pale rings. The wings are covered with dark scales and numerous scattered pale scales on all veins. The terga have broad basal bands of white scales, the distal parts are mainly dark scaled, and sometimes the last terga are predominantly covered with white scales. The sterna are almost entirely whitish scaled, with some dark scales at the base.

Male: The lobes of tergum IX usually have more than 10 short spine-like setae which are directed slightly outwards; in *Oc. cataphylla* these setae are usually less numerous and directed distally. The gonocoxite is long and slender, with numerous long setae on the inner surface, and the basal setae do not overlap in the middle (Fig. 10.57). The basal lobe is conical with a medially directed, apically strongly recurved long spine and several long setae. The apical lobe of the gonocoxite is well developed and rounded at the apex. The paraproct is strongly sclerotised in the lateral parts, with an inwardly pointed apex. The claspette stem is long and slender, and strongly curved. The claspette filament is very long, with a small plate shaped unilateral widening beginning at the base of the filament and ending distinctly before the apex. The aedeagus is cylindrical and notched at the apex.

Larva: The head is broader than it is long. The antennae are nearly half as long as the head, with weakly developed spicules, usually ventrally located in rows. The antennal seta (1-A) is located in the middle of the antennal shaft with 3–6 branches which are half as long as the antenna. The postclypeal seta (4-C) has 4 short branches, the median frontal seta (6-C) is situated in front of the inner frontal seta (5-C), and both pairs are simple, 5-C rarely with 2 branches, and the outer frontal seta (7-C) has 3–6 branches. The comb has 18–30 scales (average 24) arranged in 2–3 irregular rows (Fig. 10.58). Each individual scale varies in shape, the dorsal scales are without a long median spine, the ventral scales have a prominent median spine, and both types have several spines at the sides. The siphon tapers in the apical third, and the siphonal index is 2.5–3.1. The pecten has 13–18 (usually 15) more or less evenly spaced teeth located in the basal third of the siphon. The siphonal tuft (1-S) is situated below the middle of

Figure 10.57. Hypopygium of *Oc. leucomelas*.

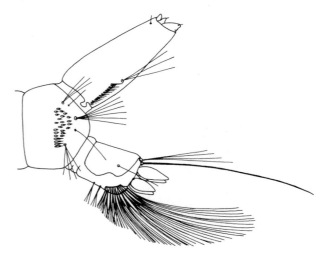

Figure 10.58. Larva of *Oc. leucomelas*.

the siphon with 3–8 branches. The saddle extends down to 2/3 of the sides of the anal segment, and the saddle seta (1-X) is nearly as long as the saddle. The lower anal seta (3-X) is simple, the upper anal seta (2-X) has 4–9 branches, and is half as long as 3-X. The ventral brush has 15–18 tufts of cratal setae (4-X) and 1–3 precratal setae (4-X). The shape of the cratal setae is characteristic, the branching of the tufts starts far from the base, thus the stem of each tuft is very long. The anal papillae are not longer than the saddle and are tapered apically, with the ventral pair shorter than the dorsal pair.

Biology: *Oc. leucomelas* is a monocyclic snow-melt mosquito. The larvae occur early in the year after the snow-melt but seldom in extensive numbers. They are usually found at the edge of forests, in flooded meadows or reeded areas together with larvae of *Oc. cantans, Oc. communis, Oc. rusticus, Ae. rossicus* and *Oc. sticticus*. In shadowed areas the number of larvae usually decreases. They also occur in slightly saline water associated with *Oc. detritus, e.g.* in flooded meadows along coasts, thus the larvae are obviously tolerant to variations in the salinity of the breeding sites. The pH can be slightly acid to alkaline (Natvig, 1948). In central Europe the adults usually emerge in May and they decrease in numbers by early July.

Distribution: *Oc. leucomelas* is a European species. It is mainly distributed in northern, central and eastern Europe and has a limited distribution in the southern European countries, where it is solely reported in Spain.

Ochlerotatus Mariae Complex

Three sibling species, *Oc. mariae, Oc. zammitii* and *Oc. phoeniciae*, have been recognised so far in the complex (Bullini and Coluzzi, 1982; Coluzzi *et al.*, 1976; Coluzzi and Bullini, 1971; Coluzzi and Sabatini, 1968). Apart from slight morphological differences between the species, a various degree of hybrid sterility was demonstrated. The sterility involves both hybrid sexes between *Oc. phoeniciae* and the other two

members of the complex, while only F_1 hybrid males were found to be sterile in the cross between *Oc. mariae* and *Oc. zammitii*. The larvae of all species have an apparently identical adaptation to breeding in rock pools along the Mediterranean coasts. The distribution range is western Mediterranean for *Oc. mariae* and eastern Mediterranean for *Oc. zammitii* and *Oc. phoeniciae*. The latter has been recorded from the coasts of Cyprus, southeastern Turkey, Lebanon and Israel (Coluzzi *et al.*, 1974). No overlap between these distribution ranges has so far been recorded. According to the updated checklist of European mosquitoes by Ramsdale and Snow (1999), *Oc. mariae* and *Oc. zammitii* correspond to the aforementioned geographical "isolation" within the Mediterranean basin, except for the Greek record of *Oc. mariae*, which most probably is *Oc. zammitii*.

Ochlerotatus (Ochlerotatus) mariae (Sergent and Sergent) 1903

Female: This species differs from *Oc. caspius* in the ornamentation of the scutum, the black scaling of the legs and in the absence of a longitudinal light stripe on the abdomen. The proboscis is long and slender, with dark brown scales, sometimes mixed with whitish scales in the middle, giving the impression of being ringed. The palps are one sixth of the length of the proboscis, and covered predominantly with dark brown scales, with a white tip and some scattered whitish and cream coloured scales. The scutum is largely covered with rust brown to golden scales, usually with some whitish scales, which may sometimes form an indistinct, variable scutal pattern. The scutal setae are predominantly blackish brown. The scutellum has three groups of white sickle shaped scales and golden brown or dark setae. The pleurites have a brown integument and small patches of white scales. A postprocoxal patch of white scales is present. The ventral surface of the femora are white scaled, and the anterior surface and tibiae are predominantly dark scaled with speckled white scales. The tarsi are conspicuous by their white apical and basal rings on some tarsomeres. Tarsomeres V of the fore and mid legs are whitish with some scattered dark scales, and tarsomere V of the hind leg is completely white scaled. The wing veins are covered with dark and pale scales, the pale scales are more numerous on the costa (C), subcosta (Sc) and radius (R). The colouration of the abdominal segments is variable (Figs 6.21b, c, d). The general appearance is dark, and the terga have narrow basal bands of pale scales which are usually widened laterally into triangular spots. The sterna are covered with whitish and cream coloured scales and dark lateral spots in the apical half, and the cerci are slightly projected.

Male: The lobes of tergum IX have 4–6 long setae. The basal lobe of the gonocoxite is moderately convex, densely covered with setae, some of which may be slightly thicker and longer than the others, but strong spine-like setae are absent. The apical lobe of the gonocoxite is indistinct (Fig. 10.59). The claspette has a short and straight stem, and the filament is nearly as long as the stem, narrow and slightly curved, and without a transparent wing. The aedeagus is more or less tubular.

Larva: The antenna is shorter than the head, slightly curved, with weakly developed spicules. The antennal seta (1-A) is situated in the middle of the antennal shaft, with 6–9 branches. The postclypeal seta (4-C) is very thin and short, and branched. The inner and median frontal setae (5-C and 6-C) are simple, the median frontals are situated in front of the inner frontals, and the outer frontal seta (7-C) usually has 7 branches. The comb is composed of 16–25 scales arranged in 2–3 irregular rows, each individual comb scale with a distinct median spine and a varying number of smaller spines of different size at the base (Fig. 10.60). The siphon is short, slightly tapered, and the siphonal index is 1.4–2.0. The pecten consists of 15 or more thin teeth which are longer distally, reaching to the middle of the siphon. Each pecten tooth usually has four or more lateral denticles at the base. The siphonal tuft (1-S) is situated slightly beyond the middle of the siphon, with 6–7 branches, which are as long as the width of the siphon at the point

Figure 10.59. Hypopygium of *Oc. mariae*.

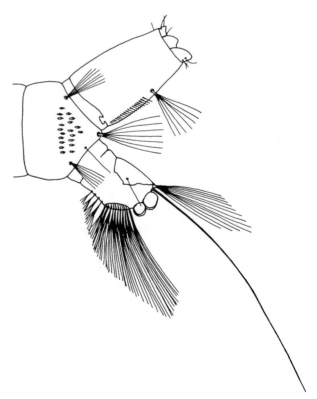

Figure 10.60. Larva of *Oc. mariae*.

of its origin. The saddle is weakly developed, extending slightly to the sides of the anal segment. The saddle seta (1-X) is simple. The upper anal seta (2-X) has 12–14 branches, the lower anal seta (3-X) is simple, and more than twice as long as the siphon. The ventral brush has 11–13 cratal tufts (4-X) and 4–5 precratal tufts. The anal papillae are very short and spherical.

Biology: The larvae can exclusively be found in rock pools on the sea shore, often in the surf zone. The usual concentration of salt in such pools is 2–4%, but the larvae are able to tolerate a much higher concentration, up to 20% (Rioux, 1958). In the Mediterranean region *Oc. mariae* has several generations per year and its larvae are commonly found from March to October. Full embryonic diapause is observed when eggs are incubated at relatively low temperatures (less than 16°C) and a short photoperiod. A photoperiod also induces a remarkable change in oviposition behaviour of *Oc. mariae* females. Coluzzi *et al.* (1975) demonstrated that the adult females readily oviposit when originating from larvae reared at a long day photoperiod, while they are very reluctant to oviposit in the same situation when reared at a short day photoperiod. This species can frequently cause nuisance in rocky coastal Mediterranean areas.

Distribution: *Oc. mariae* which belongs to the so called "Tyrrhenean" type of the complex inhabits western Mediterranean coasts, ranging from the Algarve to the Italian western coast, including the western coast of Sicily. It can also be found in Tunisia, Algeria and the Balearic Islands (Coluzzi *et al.*, 1975; Ribeiro *et al.*, 1988).

Medical importance: According to Ribeiro *et al.* (1988) *Oc. mariae* is not known as a vector of any disease, but Gutsevich *et al.* (1974) reported the species to transmit the parasite of bird malaria, *Plasmodium relictum*.

Ochlerotatus (Ochlerotatus) zammitii (Theobald) 1903

Very similar to *Oc. mariae* in all stages. It seems to be somewhat more robust in appearance with a more distinct colouration pattern. While the scutum of *Oc. mariae* is without special ornamentation, *Oc. zammitii* sometimes has a scutum with two light creamy white longitudinal stripes. The hypopygium of the male is identical to that of *Oc. mariae*, but the larvae differ from those of the latter species by the more numerous spicules on the antenna and the pecten teeth usually having fewer than 4 lateral denticles at the base (Darsie and Samanidou-Voyadjoglou, 1997; Seguy, 1924).

Distribution: Adriatic coasts, eastern coast of Sicily, the whole Island of Malta, and Ionean and Aegean coasts (Coluzzi *et al.*, 1974; Labuda, 1969; Regner, 1969).

Ochlerotatus (Ochlerotatus) mercurator (Dyar) 1920

Female: The proboscis and palps are dark scaled, and sometimes the palps have a few pale scales at the tip. The pedicel has mixed pale and dark scales. The vertex is yellowish scaled, and the occiput has a pair of dark lateral spots (Fig. 6.30a). The antepronotum and lower part of the postpronotum are pale scaled, and the upper part of the postpronotum has brown scales. The scutum has a broad median stripe of dark reddish brown scales, sometimes divided by a narrow acrostichal stripe of pale scales. The posterior submedian areas are brown scaled, and all other areas of the scutum are covered with yellowish scales. A postproxocal scale patch is present, the sub- and postspiracular patches are present, but a hypostigmal patch is absent. The mesepisternum has a prealar patch and upper and lower mesepisternal patches of pale scales, the upper patch not reaching the anterior angle of the mesepisternum. Almost all of the mesepimeron is covered with pale scales. The femora and tibiae of the fore and mid legs have mixed white and dark scales on the anterior surface, and white scales predominate on the posterior surface. The hind femur is light scaled, and the hind tibia has a longitudinal light stripe. The tarsi are predominantly dark scaled with pale basal rings. Tarsomeres I of all the legs have a diffuse basal ring and scattered pale scales almost reaching to the apex. Tarsomeres II and III of all the legs have a more or less distinct basal ring, which is broadest on hind tarsomere III, where it embraces about half of the length of the tarsomere. Tarsomeres IV and V of the fore legs and tarsomere V of the mid legs are usually entirely dark scaled. Tarsomere V of the hind legs sometimes has a few white scales at the base. The tarsal claws are relatively large, and bent near the middle at some distance from the base of the sub-basal tooth. The wing veins are entirely dark scaled, sometimes some isolated pale scales are present at the base of the costa (C). The terga are predominantly dark scaled, terga I and II with basal white bands reduced to the median part. The rest of the terga have fully developed transverse basal white bands, which are widened laterally into triangular patches, and are most distinct on terga VI and VII. Narrow apical bands are sometimes present on the last segments. The cerci are long and distinctly projecting.

Male: The lobes of tergum IX have 6–12 spine-like setae. The gonocoxite is elongated, with well developed basal and apical lobes

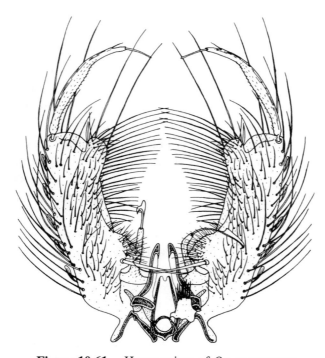

Figure 10.61.   Hypopygium of *Oc. mercurator*.

(Fig. 10.61). The inner surface of the gonocoxite is covered with long inwardly directed setae, and at least several setae located just above the basal lobe do not overlap in the middle. The basal lobe is conical, and densely covered with thin setae and one medially directed very long spine-like seta which is slightly curved at the apex. The gonostylus is somewhat broadened in the middle, with a slender apical spine. The claspette filament is longer than the stem, with a long stalk, abruptly broadened into a unilateral wing beyond its middle. The paraproct is strongly sclerotised in its apical part, and the aedeagus is elongated.

Larva: The head is wider than it is long. The antenna is about half as long as the head, and is covered with spicules. The antennal seta (1-A) is inserted slightly below the middle of the anntenal shaft, with 8–13 (usually 9–10) branches. The inner frontal seta (5-C) has 3–6 branches, usually four or more are present on one side, and the median frontal seta (6-C) has 1–4 branches. The prothoracic formula 1-P to 7-P is as follows: 1 (long, simple, rarely 2 branches); 2 (short, simple); 3 (short, 2–3 branches) (Fig. 8.47a); 4 (short, simple); 5 (long, 2–3 branches); 6 (long, simple); 7 (long, 3 branches). The comb is composed of 23–36 (usually about 30) comb scales arranged in several irregular rows (Fig. 10.62). Each scale has indistinct median and lateral spines which become smaller toward the base. The siphon tapers towards the apex, and the siphonal index is 3.3–3.5. The pecten consists of 20–29 (usually 24) closely spaced teeth occupying from about 1/3 to slightly less than half of the siphon length. Each pecten tooth has 2–6 lateral denticles. The siphonal tuft (1-S) is distinctly longer than the width of the siphon at the point of its origin, with 4–7 (usually 5) branches. The saddle covers more than half of the lateral sides of the abdominal segment. The saddle seta (1-X) is distinctly shorter than the saddle, and is simple, or rarely with 2 branches. The upper anal seta (2-X) has 8–13 branches. The ventral brush has

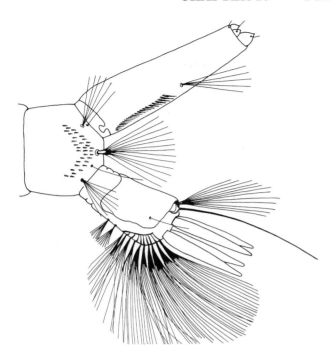

Figure 10.62. Larva of *Oc. mercurator*.

erect forked scales at the sides. Flagellomere I of the antennae is entirely dark scaled and thickened, and flagellomeres I to III are distinctly shorter than the others. The scale pattern of the scutum is similar to that of *Oc. sticticus*, and the hypostigmal and postprocoxal scale patches are absent. The upper mesepisternal scale patch extends to the anterior angle of the mesepisternum, and the pale scales on the mesepimeron end distinctly above its lower margin. The wing veins are covered with dark scales, with scattered pale scales at the base of the costa (C), the entire length of the subcosta (Sc) and the media (M), proximal to the cross veins. The abdominal terga are dark scaled, with broad basal pale bands of more or less uniform width (Fig. 6.50b). On some terga the bands may be slightly, if at all, constricted.

Male (Fig. 10.63): The differences between the hypopygia of *Oc. nigrinus* and *Oc. sticticus* are subtle and usually obvious when directly compared to each other.

2–6 precratal setae (4-X). The anal papillae are slightly pigmented, tapered, and distinctly longer than the saddle.

Biology: The species may have two generations per year. The larvae can be found from May onwards in semipermanent water bodies, such as ditches and small ground pools, or less frequently in permanent waters in open marshes or in old river arms. Adults were encountered from the end of June until the beginning of September. Gutsevich and Dubitsky (1987) stated that nowhere is the species very numerous.

Distribution: Northern Holarctic region, in Europe found in the northern part of European Russia as well as in the mountains of Crimea and Caucasus.

Ochlerotatus (Ochlerotatus) nigrinus (Eckstein) 1918

Female: Closely related to *Oc. sticticus* and very similar to it in all stages. The proboscis and palps are dark scaled. The occiput has dark

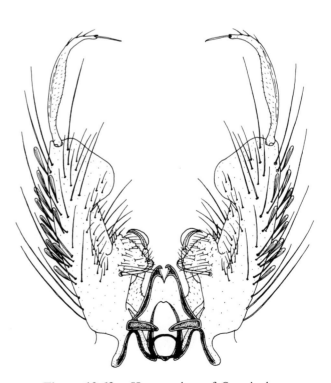

Figure 10.63. Hypopygium of *Oc. nigrinus*.

In *Oc. nigrinus* the basal lobe of the gonocoxite is not as much constricted at the base as it is in *Oc. sticticus* and its upper part is broad and rounded, not slender and crescent shaped (Fig. 7.48b). The apical lobe arises abruptly from the gonocoxite, is widest at its base and rounded in its middle part, and not apically rounded as in *Oc. sticticus*. In all other characteristics the male genitalia of the two species are nearly identical.

Larva: Similar to that of *Oc. sticticus*, but can be separated by the inner (5-C) and median (6-C) frontal setae which are usually simple, very rarely 2-branched (Fig. 8.49c). The antenna is slightly less than half as long as the head, and the antennal seta (1-A) is situated at about the middle of the antennal shaft, with 3–5 branches not reaching the tip. The number of comb scales is usually 10–12, rarely more than 20 (Fig. 10.64). Each scale is more elongated and the median spine slightly longer than in *Oc. sticticus*. The siphon is short and straight, and the siphonal index is about 2.5. The pecten teeth are more or less evenly spaced, reaching beyond the middle of the siphon. The siphonal

tuft (1-S) is situated slightly beyond the distalmost pecten tooth, with 4–7 branches. The saddle extends far down the sides of the anal segment, and the saddle seta (1-X) is simple and shorter than the saddle. The ventral brush usually has 4, rarely 3 precratal setae (4-X). The anal papillae are longer than the saddle, and of varying length.

Biology: *Oc. nigrinus* prefers to breed in open terrains, preferably flooded meadows in river depressions, mostly associated with *Ae. vexans*, however, it is not as widely distributed and by far not so numerous as *Ae. vexans* (Eckstein, 1918, 1920; Peus, 1933).

Distribution: Southern Scandinavia, Finland, Denmark, Germany, France, Poland, Estonia, northern Urals to West Siberia.

Ochlerotatus (Ochlerotatus) nigripes (Zetterstedt) 1838

Female: A medium to large species, easy to identify by its dense blackish body setation, especially on the scutum, its blackish brown integument, and the entirely white scaling of the pleurites. The proboscis is entirely dark scaled, and the palps are dark with some scattered white scales. The head is covered with dark brownish scales, and the occiput with some golden brown erect scales, mixed with grey and dark scales. The antenna is dark greyish, and the pedicel has mixed pale and dark scales. The scutum and scutellum have prominent and dense black or very dark brown setae. The scutum is covered with narrow dark brown scales, occasionally with spots of light scales laterally and on the prescutellar dorsocentral area. All the light scale patches on the pleurites are white. The postpronotum has narrow dark brown scales, and a lower patch only either with narrow or broad white scales, and setae scattered on the entire postpronotum. A postprocoxal patch is usually present, but a hypostigmal patch is absent. The subspiracular and postspiracular areas have white scale patches. The number of postspiracular setae is

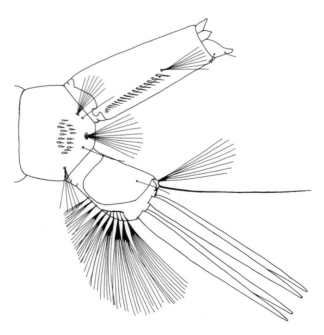

Figure 10.64.   Larva of *Oc. nigrinus*.

14 or more (Fig. 6.42d). The mesepisternum has dense upper and lower patches of narrow white scales. The mesepimeron has a broad upper whitish scale patch. The wing veins are covered with dark scales, and pale scales are usually present at the base of the costa (C) and radius (R), sometimes also at the bases of other veins. All the coxae have white scale patches and long dark setae. The femora, tibiae and tarsomeres I of all the legs are dark scaled with scattered paler brownish scales. Tarsomeres II to V are dark scaled, with a few pale scales sometimes present on tarsomeres IV and V. The abdominal terga have white basal bands. The sterna are entirely covered with white scales, and the cerci are covered with dark scales and are very setous.

Male: The lateral lobes of tergum IX have numerous, heavily sclerotised setae; a character which is shared with *Oc. impiger*. The gonocoxite has long setae predominating on its inner side (Fig. 10.65). The basal lobe of the gonocoxite is well developed, with numerous setae of different length, but of more or less the same thickness. The apical lobe is small and indistinct. The gonostylus is broad, and the apical spine of the gonostylus is slender. The paraproct has inwardly curved tips. The claspette stem tapers towards the apex, and the claspette filament is more than half as long as the stem, with a unilateral wing. The aedeagus is oval shaped with a rather wide opening.

Larva: The antenna is short and less than half as long as the head. The antennal seta (1-A) is situated beyond the middle of the antennal shaft, and has 1–3 short branches. The inner and median frontal setae (5-C and 6-C) are usually simple, sometimes with 2 branches. The prothoracic setae are heavily aciculated; the formula 1-P to 7-P is as follows: 1 (1–2 branches, very short); 2 (simple, medium long); 3 (3-branched); 4 (1–3 branches,

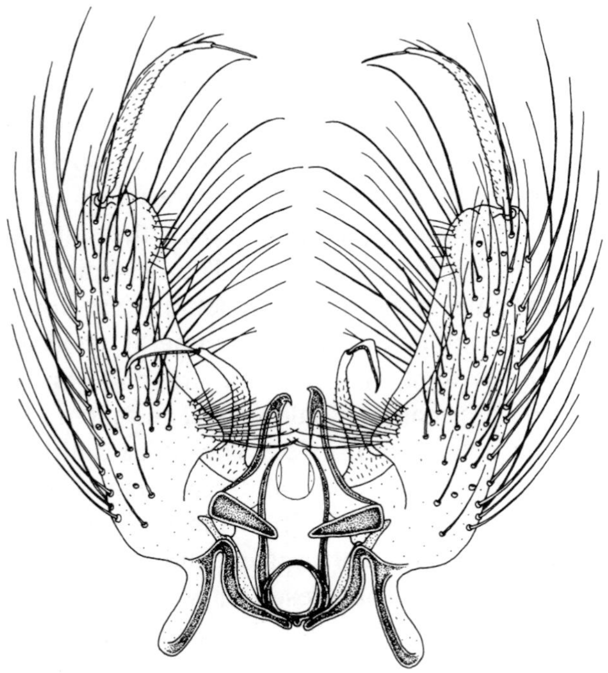

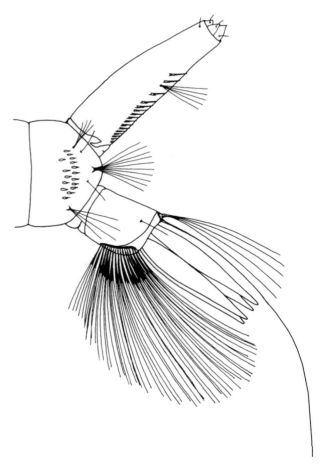

Figure 10.65. Hypopygium of *Oc. nigripes*.

Figure 10.66. Larva of *Oc. nigripes*.

medium long); 5 and 6 (2–3 branches); 7 (multiple branched). The number of comb scales is 8–19 and each individual scale has a long median spine and insignificant small lateral spines. The siphonal index is 2.6–3.1 (Fig. 10.66). The pecten has 1–3 distal teeth apically detached, with at least one tooth located beyond the siphonal tuft (1-S). Each pecten tooth is elongated with only one long lateral denticle. The siphonal tuft (1-S) is as long as or longer than the width of the siphon at the point of its insertion, with 2–6 branches. The saddle completely encircles the anal segment, but sometimes a narrow gap may be present at the ventral surface. The saddle seta (1-X) is short, and less than half as long as the saddle. The anal papillae are at least 1.5 times longer than the saddle.

Biology: It is a monocyclic species. In the Canadian tundra (beyond 80°N) females were observed to oviposit only in direct sunlight during the warmest part of the day. Eggs were laid several cm above the water level at the edges of ponds. Such an oviposition behaviour is connected with the search for south facing sites where the snow melts earliest in the next season (Corbet and Danks, 1975). In Europe, the larvae have been found in high mountains in Dalarna, Sweden, but mostly occur in the northernmost tundra in willow and birch shrub marches in the arctic zone in temporary water bodies formed by snow-melt. They are encountered in deep and less oligotrophic pools, as well as in shallow puddles surrounded by *Carex* sp. at higher treeless elevations in the scandinavian mountains. The larvae are easily disturbed and may spend a long time at the bottom of their breeding sites. The larval development lasts several weeks, adults emerge a little later than those of *Oc. impiger*. Copulation takes places during swarming (Corbet, 1965). Females were observed attempting to feed on various birds and mammals (Wood *et al.*, 1979), facultatively autogeny is reported by Corbet (1964). The females are not such agressive biters as *Oc. impiger* females.

Distribution: *Oc. nigripes* is one of the highest arctic, circumpolar species. It mainly occurs north of the tree line in the entire northern hemisphere. Together with *Oc. impiger* its distribution range extends further to the north than any other mosquito species.

Ochlerotatus (Ochlerotatus) pionips (Dyar) 1919

Female: Similar to *Oc. communis* but can be separated from the latter by the presence of the postprocoxal scale patch. A medium sized species, usually a little larger than *Oc. communis*. The proboscis and palps are entirely dark scaled. The head has narrow yellowish brown scales dorsally and broad appressed yellowish scales laterally and numerous pale erect forked scales. The scutum is covered with golden bronze or yellowish grey scales, with a dark brown median stripe, occasionally divided by a narrow acrostichal stripe of yellowish scales, and a posterior submedian stripe of dark brown scales (Fig. 6.49a). The scutellum has narrow curved yellowish scales and light brown to brown setae on the lobes. The postpronotum has narrow brown scales dorsally, becoming pale ventrally. The postprocoxal membrane has a patch of pale scales. A hypostigmal patch is absent, but the subspiracular and postspiracular scale patches are well developed. The upper and lower mesepisternal scale patches are fused with the prealar patch, reaching the anterior angle of the mesepisternum. The mesepimeron has a pale scale patch reaching the lower margin, and 1–4 lower mesepimeral setae are usually present. The femora are dark brown scaled with scattered pale scales, and the posterior surface is pale scaled. The tibiae and tarsi have dark scales, and pale basal rings on the tarsomeres are absent. The wing veins are covered with narrow dark scales, and rarely a few pale scales may be present at the base of the costa (C) and radius (R). The abdominal terga are dark scaled, each with a pale basal band of more or less uniform width or slightly widening at the sides. The sterna are covered with whitish scales, and more or less distinct narrow dark apical bands are present.

Male: The hypopygium is very similar to that of *Oc. communis* and sometimes they are difficult to separate (Fig. 10.67). The upper part of the basal lobe bears a row of long, prominent setae, which are straight or only slightly curved, but never strongly curved or hooked as in *Oc. communis* (Fig. 7.44). No other distinct differences exist in the hypopygia of the two species. In case of the identification being uncertain by the hypopygium alone and the whole specimen is available, the length of the palps and the postprocoxal scale patch should be taken into consideration. In males of *Oc. pionips* the palps are slightly shorter than the proboscis and the postprocoxal scale patch is present. In males of *Oc. communis* the palps are longer than the proboscis and the postprocoxal scale patch is absent.

Larva: The head is wider than it is long. The antenna is about 2/3 as long as the head, slender and slightly curved with spicules. The antennal seta (1-A) is situated slightly below the middle of the antennal shaft, with 7–13 (usually 9) branches which do not reach the tip. The postclypeal seta (4-C) is small, with 3–5 short branches. The inner (5-C) and median (6-C) frontal setae have 3–5 (rarely 6) branches (Fig. 8.42a), and in *Oc. communis* both pairs are simple, rarely with 2 branches. The outer frontal seta (7-C) has 5–9 branches. The comb of abdominal segment VIII has more than 60 scales arranged in an irregular patch (Fig. 10.68). Each individual scale is without a prolonged median spine, thus appears to be rounded apically, and the lateral margin is fringed with short spines, decreasing in size towards the base (Fig. 8.42b). The siphon is straight, tapers towards the apex, and the siphonal index is 2.5–3.0. The pecten has 18–24 evenly spaced teeth, confined to the basal half of the siphon. The siphonal tuft (1-S) is located beyond the distalmost pecten tooth at about the middle of the siphon, with 4–9 branches, which are slightly longer than the width of the siphon at the point of its insertion. The saddle extends far down the sides of the anal segment, and the saddle seta (1-X) is simple and a little shorter than the saddle. The upper anal seta (2-X) has 9–13 branches, the lower anal seta (3-X) is

Figure 10.67. Hypopygium of *Oc. pionips*.

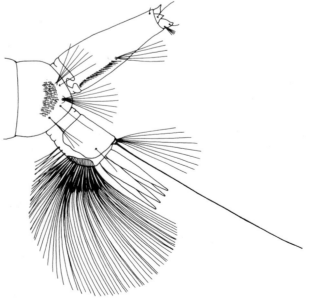

Figure 10.68. Larva of *Oc. pionips*.

simple, and is much longer than the siphon. The ventral brush has 17–21 tufts of cratal setae (4-X) and 2–3 precratal setae. The anal papillae are lanceolate, pointed, and longer than the saddle.

Biology: *Oc. pionips*, like *Oc. communis*, is a typical monocyclic snow-melt mosquito, whose larvae hatch during the snow-melt in early spring. Its breeding sites are snow-melt ponds in boggy forests up to an altitude of more than 1000 m, where larvae often are associated with those of *Oc. communis*, *Oc. hexodontus* and *Oc. punctor*. Gjullin *et al.* (1961) observed that *Oc. pionips* develops slower than the associated *Aedes/Ochlerotatus* species. The adults seem to appear later and are often less numerous than the other snow-melt mosquitoes.

Distribution: Holarctic species, occurs in North America and northern Eurasia.

Ochlerotatus (Ochlerotatus) pulcritarsis (Rondani) 1872

Female: Very similar to the females of *Oc. berlandi*. A slight difference exists in the scutal colouration pattern. Whereas the scutum of *Oc. berlandi* is distinctly contrasted by dark and pale golden scales, *Oc. pulcritarsis* exhibits a weaker pattern of pale and dark scales on the scutum, and looks rather uniformly golden brownish coloured. However, the median and lateral stripes may be somewhat lighter than the submedian spots.

Male (Fig. 10.69): The hypopygium is very similar to that of *Oc. berlandi*. The spinelike seta on the basal lobe of the gonocoxite is usually less curved at the apex, and rarely hooked.

Larva: The head is square shaped, and somewhat broader than it is long. The antenna is almost as long as the head, and the antennal shaft is smooth and without spicules. The antennal seta (1-A) has 3–4 branches, inserted at about the middle of the antennal shaft (Fig. 8.58c). The postclypeal seta (4-C) is small and multiple-branched. The frontal setae (5-C to

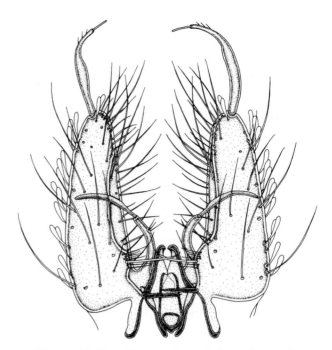

Figure 10.69. Hypopygium of *Oc. pulcritarsis*.

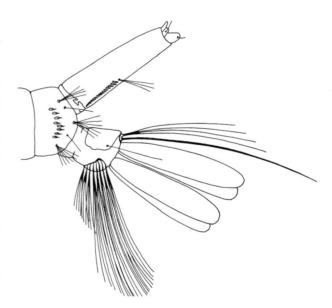

Figure 10.70. Larva of *Oc. pulcritarsis*.

7-C) are well developed and multiple-branched. The comb consist of 6–10 (usually 8) scales arranged in one row (Fig. 10.70). Each scale is large with a well developed median spine, and small spines on the basal half of the scale. The siphon is dark, almost black, slightly tapering in the apical half, and the siphonal index is 4.0–5.0. The pecten consists of 18–22 evenly spaced teeth. Each individual tooth is blunt ended with 3–6 heavily sclerotised lateral denticles. The siphonal tuft (1-S) has 3–4 branches, is about as long as the width of the siphon at the point of origin, and is situated below the middle of the siphon. The saddle covers about half of the anal segment. The saddle seta (1-X) is longer than the saddle and is simple. The upper anal seta (2-X) has 4–5 branches of variable length. The lower anal seta (3-X) is longer than the siphon, and simple. The anal papillae are very long, sausage shaped, and several times longer than the saddle.

Biology: Hibernation takes place in the egg stage, and the species usually has two generations per year, but sometimes only one generation occurs. The larvae can be found in tree holes, stumps and among roots of deciduous trees, such as *Quercus* sp., *Platanus* sp. and *Ulmus* sp. the latter being preferred, as well as in olive tree holes (Shannon and Hadjinicolaou, 1937). The water temperature of the breeding sites never exceeds 21°C even in southern European climatic conditions. Similar observations were made along the Black Sea shore in Bulgaria (Bozkov *et al.*, 1969). Larval development may last up to 2 months. Adult females are anthropophilic; they bite outside houses during daytime (Rioux, 1958). Adult mosquitoes of both sexes were found in stables and houses in a village where no trees were present. This may indicate that the species is facultatively zoophilic and could breed in artificial containers or have a considerable migrating capacity (Shannon and Hadjinicolaou, 1937).

Distribution: *Oc. pulcritarsis* is principally a species of the Mediterranean region. Its northern distribution range reaches as far as the Czech Republic. It is also found in central and southeastern Asia.

Note on systematics: One subspecies, ssp. *asiaticus* Edwards is reported in Uzbekistan, Turkmenia and Pakistan.

Ochlerotatus (Ochlerotatus) pullatus (Coquillett) 1904

Female: A medium sized species. The proboscis is dark scaled, and the palps are predominantly dark scaled with a few scattered pale scales at the joints of the palpomeres. The clypeus is blackish brown, and the pedicel is dark brown with a few pale scales. The vertex and occiput are clothed with pale yellowish narrow scales and erect forked scales of the same colour on the dorsal part and usually broad appressed yellowish white scales laterally. The integument of the scutum is black, and the scutum is covered with yellowish brown narrow curved scales, disconnected by several bare longitudinal stripes and areas to which they contrast, mainly due to the exposure of the dark integument. The transverse suture, prescutellar area and lateral ends of the scutum are also devoid of scales. The setae of the scutum are usually golden brownish, sometimes blackish brown, and are more numerous on the posterior part. The scutellum has narrow pale scales and yellowish brown setae on the lobes. The pleurites have patches of broad yellowish white scales, a hypostigmal patch of scales is present, and the postprocoxal membrane is bare. The mesepisternal patch of scales is divided into an upper and lower portion, and the upper portion does not reach the anterior margin of the mesepisternum. The mesepimeral patch of scales reaches near the lower margin of the mesepimeron, and 1–5 lower mesepimeral setae are present (Fig. 6.44a). The femora have dark brown and pale scales intermixed, but are darker apically with a pale knee spot. The tibiae and tarsomeres I are dark brown, and speckled with pale scales, especially on the ventral surface. The remaining tarsomeres are entirely dark scaled. The wing veins are covered with narrow dark scales, and with patches of pale scales at the base of the costa (C), radius (R) and anal vein (A). Abdominal tergum I has a broad patch of white scales, terga II to VII are blackish brown scaled with a basal transverse band of white scales which are sometimes slightly widened laterally, and the abdominal sterna are mostly white scaled. The cerci are exceptionally long and conspicuous.

Male: The lobes of tergum IX have 5–7 short stout setae. The gonocoxite is long and slender, about three times as long as it is wide, with its inner surface densely covered with long setae (Fig. 10.71). The basal lobe of the gonocoxite is prominent and has 3 setae which are distinctly larger than the rest; one long stout apically curved spine-like seta is inserted dorsally and two sinuous, flattened and slightly lanceolate setae arising from the ventral surface of the basal lobe (Fig. 7.31b). The apical lobe is well developed, thumb-like, with numerous setae at the tip. The gonostylus is about half as long as the gonocoxite, curved, slightly expanded before the middle, bearing a few small setae close to the apex, and the apical spine of the gonostylus is long and slender. The paraproct is heavily sclerotised apically. The claspette stem is long, swollen and strongly bent in the middle, without a thorn shaped process; the basal half is stout, and the distal half is slender. The claspette filament

is shorter than the stem, prominently winged on the convex side, and the aedeagus is longer than it is wide.

Larva: The head is slightly wider than it is long. The antenna is about half as long as the head, slightly curved and covered with spicules. The antennal seta (1-A) is inserted at the basal third of the antennal shaft, with about 5 branches, which do not reach to the tip of the antenna. The postclypeal seta (4-C) is small, with 4–5 branches, the frontal setae are multiple-branched, the inner (5-C) and median (6-C) setae have at least 3 branches (mostly 4–6), and the outer frontal seta (7-C) has 8–13 branches (Fig. 8.42c). The prothoracic setae 2-P and 3-P are nearly as long and strong as 1-P (Fig. 8.47b). The comb has 40–60 scales arranged in a large triangular patch; the lateral scales are more or less pointed, and fringed with smaller lateral spines and a longer median spine. The pecten has 15–25 teeth, which are evenly spaced and situated close together (Fig. 10.72). The siphon uniformly tapers towards the

Figure 10.71.   Hypopygium of *Oc. pullatus*.

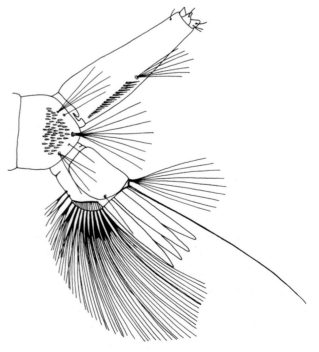

Figure 10.72.   Larva of *Oc. pullatus*.

tip from near the middle, and the siphonal index is 3.0–3.5. The siphonal tuft (1-S) is situated more or less at the middle of the siphon, beyond the distalmost pecten tooth, with 5–8 branches. The anal segment has a saddle which extends more than half way down the side, and the saddle seta (1-X) is simple and shorter than the saddle. The upper anal seta (2-X) has 6–10 branches, and the lower anal seta (3-X) is long and simple. The ventral brush has about 12–15 tufts of cratal setae (4-X) and 1–3 precratal tufts. The anal papillae are about twice as long as the saddle, and pointed.

Biology: *Oc. pullatus* produces one generation per year; the larvae hatch from overwintering eggs from early spring until late summer, depending on the elevation of their occurrence. They can be found in the tundra mostly in small clear snow-melt pools and in mountainous regions in a variety of breeding sides, *e.g.* in puddles and pools without vegetation created by overflow of mountain streams or after heavy rain fall, small clear lakes with a rocky bottom or boggy holes (Kaiser *et al.*, 2001). The larvae may be associated with those of other mountainous *Aedes/Ochlerotatus* species, *e.g. Oc. communis* and *Oc. punctor*. Their development lasts longer than that of their associates and consequently the adults occur later, mainly in the summer months. The females readily attack their hosts during any time of the day in forested areas. Male swarming takes place after sunset in openings of the forest. The species is usually found in small numbers, but adults can be abundant in some localities, generally far remote from human habitations (Carpenter and LaCasse, 1955).

Distribution: *Oc. pullatus* is a northern holarctic species with a disjunct distribution range. In central and southern Europe it is restricted to mountainous regions, including the Pyrenees, Alps, Dinari Mountains, Tatras, Carpathians, Balkans, and Rhodopi Mountains and can be found up to very high elevations

(2000 m and higher). In the northern parts of its distribution it occurs in the arctic tundra, lowlands and plains in Eurasia as well as in North America.

Ochlerotatus (Ochlerotatus) punctodes (Dyar) 1922

Female: The adult females cannot be distinguished from those of *Oc. punctor*. Differences occur mainly in the larval stage (Knight, 1951).

Male (Fig. 10.73): Separable from *Oc. punctor* by the shape of the basal lobe of the gonocoxite and the claspette filament. The basal lobe has an irregular triangular shape, and the basal part of the lobe abruptly arises from the gonocoxite, whereas in *Oc. punctor* the basal part of the lobe arises more gradually from the gonocoxite. The claspette filament is relatively long and narrow, and of almost the same length as the stem and weakly sclerotised, but in *Oc. punctor* the filament is shorter than the stem and strongly sclerotised (Fig. 7.42b). Carpenter and LaCasse (1955) stated that the spine of the basal lobe is more weakly developed in *Oc. punctodes* and sometimes is difficult to separate from the long setae at the base

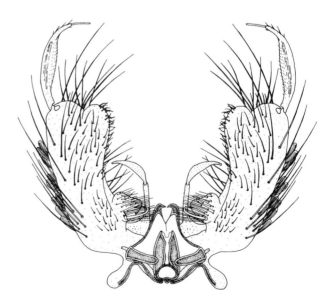

Figure 10.73. Hypopygium of *Oc. punctodes*.

of the lobe. In *Oc. punctor* the spine of the basal lobe is strongly developed and distinctly different to the long setae at the base of the lobe.

Larva: The inner (5-C) and median (6-C) frontal setae are usually simple (Fig. 8.32b), whereas in *Oc. punctor* both pairs are usually 2-branched. The saddle does not completely encircle the anal segment, and extends to near the mid-ventral line (Fig. 10.74). The anal papillae are usually much shorter than in larvae of *Oc. punctor*, varying from much shorter to slightly longer than the saddle.

Biology: *Oc. punctodes* belongs to the typical salt marsh fauna in subarctic regions (Frohne, 1953). The larvae occur predominantly in the more saline waters, while the larvae of the closely related *Oc. punctor* prefer fresh water habitats. In these habitats the larvae of *Oc. punctodes* can be found together with those

of *Oc. punctor, Oc. communis, Oc. impiger, Oc. excrucians* and *Oc. flavescens*. Frohne (1953) found newly hatched larvae of *Oc. punctodes* in Alaska at the end of April and a few larvae were still present in August.

Distribution: *Oc. punctodes* is a subarctic species. It is reported from Alaska and in Europe from Norway, Sweden, Finland and the European region of Russia.

Ochlerotatus (Ochlerotatus) punctor (Kirby) 1837

Female: A medium sized species. The proboscis and palps are dark scaled. The occiput has golden yellowish narrow scales and yellow erect forked scales dorsally, with broad appressed creamy white scales laterally. The scutum is covered with yellowish brown scales, usually with a median stripe of dark brown scales, rarely divided by an acrostichal stripe of yellowish scales, and the posterior submedian areas with dark brown scales. The scutellum has yellowish brown scales and light brown setae on the lobes. The prosternum has no scales on the anterior surface, or occasionally has scattered pale scales, but is not as extensively covered with scales as in *Oc. hexodontus*. The postpronotum has narrow yellowish brown scales anteriorly and broader paler scales at the posterior margin. The postprocoxal membrane has pale scales (absent in *Oc. communis*). A hypostigmal patch is absent, the subspiracular patch is divided into an upper and lower portion, and the postspiracular patch is well developed. The upper and lower mesepisternal scale patches are fused, and extend to the anterior angle of the mesepisternum, narrowly separate from the prealar patch (Fig. 6.47a). The scales on the mesepimeron extend to its lower margin, and 1–5 lower mesepimeral setae are present. The femora, tibiae and tarsomeres are mostly dark scaled. The claws of the fore legs are elongated, and gradually curving distal to the subbasal tooth. The wing veins are usually entirely dark scaled, with rarely a few pale scales present

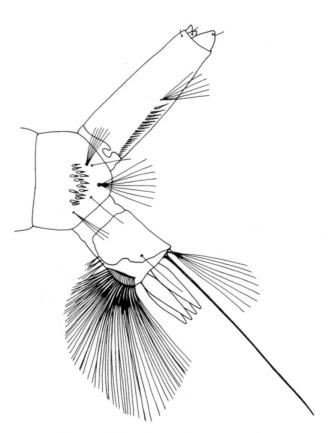

Figure 10.74.   Larva of *Oc. punctodes*.

(less than in *Oc. hexodontus*, *Oc. communis* and *Oc. pionips*) at the base of the costa (C). The abdominal terga are dark scaled with basal bands of white scales, which are distinctly confined in the middle or even interrupted on the more anterior terga (Fig. 6.48a). The sterna are covered with greyish white scales, and the apices of some of the sterna are dark scaled medially.

Male: The lobes of tergum IX are sclerotised, each bearing several spine-like setae. The gonocoxite is about three times as long as it is wide, the basal lobe of the gonocoxite is well developed, and more or less triangular shaped (Fig. 10.75). The basal part has a row of long setae, a long apically recurved spine, and the apical part is densely covered with short setae. The apical lobe is broadly rounded with short curved setae extending downwards to near the middle of the gonocoxite. The gonostylus is slightly expanded in the middle with several small setae before the apex, and the apical spine of the gonostylus is slender. The paraproct is strongly sclerotised, with the apex inwardly pointed. The claspette stem is short and slightly curved near the middle. The claspette filament is shorter than the stem, lanceolate, widened in the middle, strongly sclerotised and not transparent, with a curved apex. The aedeagus is cylindrical, and notched at the apex.

Larva: Very similar to that of *Oc. hexodontus*. For separation of the two species see the description of the latter. The antennae are less than half as long as the head, and covered with numerous spicules. The antennal seta (1-A) is situated in the middle of the antennal shaft or slightly below it, with 4–7 branches not reaching the tip. The postclypeal seta (4-C) has 2–4 short branches. The inner (5-C) and median (6-C) frontal setae have 1–3 branches, usually 2-branched, and the outer frontal seta (7-C) has 2–8 branches (Fig. 8.32a). The comb has 10–25 small scales, usually arranged in 2–3 irregular rows or a triangular patch, and each scale has a prominent median spine and several smaller spines in the basal part. The siphon tapers in the apical half, and the siphonal index is about 3.0 (Fig. 10.76). The pecten has 14–26 evenly

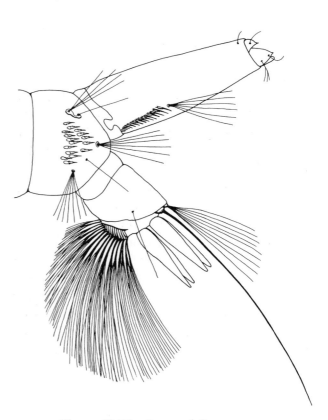

Figure 10.75. Hypopygium of *Oc. punctor*.

Figure 10.76. Larva of *Oc. punctor*.

spaced teeth confined to the basal half of the siphon. The siphonal tuft (1-S) is located beyond the distalmost pecten tooth, with 3–9 branches, and is about as long as the width of the siphon at the point of origin. The saddle completely encircles the anal segment, and the saddle seta (1-X) is usually as long as or a little longer than the saddle. The upper anal seta (2-X) has 5–9 branches, and the lower anal seta (3-X) is simple and long. The ventral brush has 16–19 tufts of cratal setae (4-X) and 1–2 precratal setae. The anal papillae taper, are of variable length but are always distinctly longer than the saddle.

Biology: *Oc. punctor* is a snow-melt mosquito, which has a preference for swampy forests with boggy waters. The larvae hatch during the snow-melt, when the water temperature is only a little above 0°C. While Horsfall (1955) and Monchadskii (1951) found larvae only in springtime, they occur in southern Germany also in the summer after strong rainfall sometimes together with the larvae of *Ae. cinereus* and *Cs. alaskaensis* (Peus, 1929; Vogel, 1933, 1940; Becker and Ludwig, 1981). Some larvae may overwinter together with those of *Oc. rusticus* and *Cs. morsitans*. Larvae of this acidophilic species can be found in great numbers in boggy waters with *Sphagnum* sp., where the pH-value can be less than 4.0. The optimum temperature for the development of *Oc. punctor* is 25°C, however the lowest mortality rate is found at temperatures of 15°C. At 25°C larval and pupal development lasts 10–17 days, at 20°C 15–22 days, at 15°C 20–26 days and at 10°C 33–41 days. At 30°C the larvae die in the first and second larval stages. In early spring the larvae of *Oc. punctor* occur together with those of *Oc. communis*, and a little earlier than the larvae of *Oc. diantaeus* and *Ae. cinereus*. In central Europe, the adults occur in the second half of April, mostly later than those of *Oc. communis*, but earlier than *Oc. cantans*. The adults prefer sheltered terrain and seldom

migrate out of the forest. Their peak biting activity is at dusk; on sultry days and in strongly shaded situations they can be troublesome even during daytime.

Distribution: *Oc. punctor* is a Holarctic species and can commonly be found in North America and Eurasia.

Ochlerotatus (Ochlerotatus) riparius (Dyar and Knab) 1907

Female: The species has a brownish integument. The proboscis is predominantly white scaled and the palps are dark scaled with scattered white scales or a narrow white band. The vertex and occiput are covered with bronze golden narrow scales and a small lateral, white patch. The scutum has narrow, bronze and golden gleaming scales, and a broad median stripe of darker scales is usually present. Small pale anterior submedian patches close to the transverse suture may be present, and the pre-scutellar bare space is usually surrounded by pale scales (Fig. 6.33a). The scutellum has bronze scales and a diffuse pale patch on the median lobe. The postpronotum is covered with narrow bronze scales in the upper half and with narrow sickle shaped pale scales in the lower half. Small white scale patches on the antepronotum and propleuron are present, and the postprocoxal membrane has a small white patch. A small hypostigmal patch of scales is sometimes present. The postspiracular area and paratergite have pale scale patches. The mesepisternum has three distinct patches along the posterior margin, and the mesepimeral patch of pale scales covers a little more than half of the mesepimeron. The coxae have white scale patches, the fore femur and all the tibiae have mixed scales dorsally, with white scales ventrally, and the femora of the mid and hind legs are a little darker. Tarsomeres I of all the legs are predominantly white scaled with a diffuse basal ring (sometimes absent), tarsomeres II to IV have a basal white ring of different width, tarsomeres V are entirely dark scaled, and the tarsal claws are evenly curved (Fig. 6.32b). The wing veins are covered with dark greyish scales, which on the costa (C) and subcosta (Sc) are mixed with white scales. The abdominal terga have distinct white or pale basal bands, which are sometimes diffuse or interrupted in the middle forming indistinct triangular patches laterally. All terga have scattered white scales apically which usually form apical bands at least on segments VI to VIII (Fig. 6.34a). The sterna are covered with broad white scales.

Male: The lateral lobes of tergum IX have 4–7 setae. The shape of the hypopygium is similar to that of *Oc. cantans*. The basal lobe of the gonocoxite is slightly longer than it is broad at its base, and not that elongated as in the latter species, with one long apically curved spine-like seta (Fig. 10.77). The apical lobe is well developed, and covered with short, thin setae. The gonostylus is somewhat broadened in its middle part, with a slender apical spine. The claspette stem is slender, and the claspette filament is moderately broad, and unilaterally winged. The aedeagus is elongated and pear shaped.

Larva: The antenna is shorter than the head and covered with numerous spicules. The antennal tuft (1-A) is situated at about the middle of the antenna, with moderately long branches not reaching the tip. The inner and median frontal setae (5-C and 6-C) usually have 2 branches (Fig. 8.49a); 5-C is sometimes 3-branched. The prothoracic formula (1-P to 7-P) is as follows: 1 (long, simple); 2 and 3 (short, simple); 4 (medium, simple); 5 and 6 (long, simple); 7 (long, double or triple). The number of comb scales is 6–12 arranged in one row (Fig. 10.78). Each individual scale is elongated with a long median spine. The siphonal index ranges between 3.5 and 4.0. The 2–3 last pecten teeth are apically slightly detached. Each pecten tooth has one larger and 2–3 smaller lateral denticles. The siphonal tuft (1-S) is situated beyond the middle of the siphon and beyond the distalmost pecten tooth, with 4–5 long branches. Seta 9-S is rather long and curved, but is not

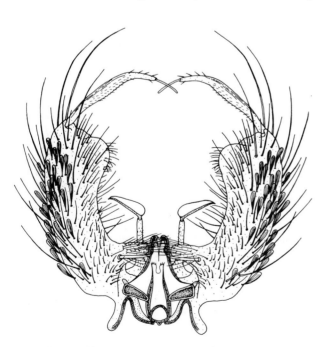

Figure 10.77. Hypopygium of *Oc. riparius*.

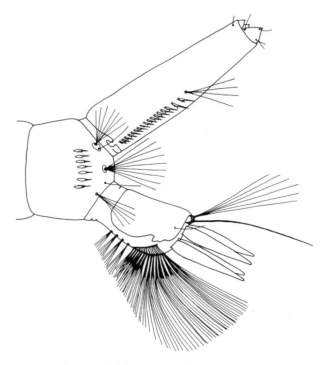

Figure 10.78. Larva of *Oc. riparius*.

transformed into a strong hook. The saddle does not encircle the anal segment, but covers most of its lateral sides. The saddle seta (1-X) is simple and shorter than the saddle. The ventral brush has 4–6 precratal tufts (4-X). The anal papillae are about as long as the saddle.

Biology: The species is uncommon or rare over most of its range (Wood *et al.*, 1979), thus its biology is not well known. It hibernates in the egg stage and is probably monocyclic, although females can be found throughout the season. The larvae and adults occur at the same time as *Oc. excrucians* and *Ae. cinereus*. Breeding habitats are pools at mixed forest edges, or pools in deciduous forests, usually rich in leaf debris on the bottom. Larvae can also be found in peat bogs in the open (Gutsevich *et al.*, 1974).

Distribution: Nearctic and North Palaearctic species, which has mostly been reported in northern and central Europe. It is usually rare, sometimes locally abundant, but is rather widespread in temperate areas.

Ochlerotatus (Ochlerotatus) sticticus (Meigen) 1838

Female: A medium sized species, with the proboscis and palps dark scaled. The pedicel has pale scales on its median part, and flagellomere I is yellowish at the base. Flagellomeres I to III are of the same length as the others. The vertex is predominantly covered with pale yellowish scales, and the occiput with pale erect forked scales. The scutum has a dark, median longitudinal stripe sometimes separated by a narrow acrostichal stripe of pale scales. The lateral parts of the scutum are pale yellowish scaled, and the posterior submedian stripe is reddish to dark brown. The upper part of the postpronotum has dark scales, and the lower third is mostly pale scaled. The postprocoxal membrane is without scales, and a hypostigmal patch is absent. The subspiracular and postspiracular scale patches are well developed. The mesepisternum has greyish white scales, and the upper mesepisternal patch extends to

near the anterior angle, narrowly separated from the prealar patch. The mesepimeral patch of scales ends distinctly above the lower margin of the mesepimeron, and lower mesepimeral setae are absent. The tibiae and tarsi are dark scaled dorsally and mostly pale scaled ventrally, and tarsomeres V of all the legs are mostly dark scaled. The tarsal claws are moderately and evenly curved, each with a small subbasal tooth. The wing veins are usually entirely dark scaled, rarely some isolated pale scales may be present at the base of the costa (C). The abdominal terga are dark scaled, terga II to IV with pale basal bands distinctly constricted in the middle, and on the following terga the basal bands are interrupted forming triangular pale patches at the lateral sides (Fig. 6.50a).

Male (Fig. 10.79): The basal lobe of the gonocoxite is constricted at its base, the apical part is slender and not attached to the gonocoxite, and is more or less crescent shaped (Fig. 7.48a). The lobe is densely covered with short setae and a rather large and prominent

Figure 10.79. Hypopygium of *Oc. sticticus*.

spine, which is curved apically and located at the constricted base. The apical lobe is well developed, gradually arising from the gonocoxite, apically rounded, and covered with short setae. The gonostylus is slightly expanded in the middle, with several small setae close to the apex. The apical spine of the gonostylus is long and slender. The paraproct is strongly sclerotised, inwardly curved and pointed at its apex. The claspette stem is short and straight, and the filament is short, about half as long as the stem, with a small bilateral wing. The wing abruptly expands close to the base of the filament, and more or less gradually tapers towards the apex. The aedeagus is pear shaped.

Larva: The antenna is nearly half as long as the head, covered with less numerous, but prominent, coarse spicules. The antennal seta (1-A) is inserted slightly below the middle of the antennal shaft, with 4–5 branches not reaching the tip. The postclypeal seta (4-C) is small, and situated between the median frontal setae (6-C), with 1–4 short branches. The inner frontal seta (5-C) has 2–4 branches, the median frontal seta (6-C) is usually 2-branched, and the outer frontal seta (7-C) usually has 5 branches. The number of comb scales is 18 to 27 arranged in 2–3 irregular rows (Fig. 10.80). Each individual scale has a median spine 1.5 times the length of the subapical spines. The siphon is straight, gradually tapering towards the apex, and the siphonal index is 2.5–3.0. The pecten teeth are more or less evenly spaced extending beyond the middle of the siphon, and the siphonal tuft (1-S) is inserted beyond the pecten, with 4–6 branches not exceeding the width of the siphon at the point of insertion. The saddle extends far down the sides of the anal segment, and the saddle seta (1-X) is simple and shorter than the saddle. The ventral brush has 1–2 precratal setae (4-X). The anal papillae are long and pointed, and often 2.0–2.5 times longer than the saddle.

Biology: *Oc. sticticus* is a polycyclic species. The larvae occur mainly in temporary

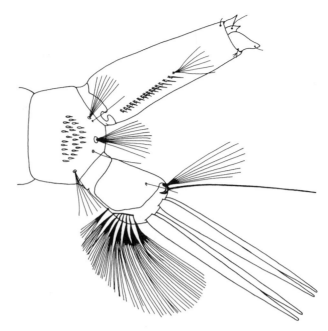

Figure 10.80. Larva of *Oc. sticticus.*

water bodies after floods and are regularly associated with those of *Ae. vexans*; it is often the second most frequent mosquito after *Ae. vexans*. The larvae can certainly hatch at lower water temperatures (<8°C) than the larvae of *Ae. vexans*. In central Europe they are associated with the larvae of *Ae. rossicus, Ae. cinereus* and *Oc. cantans* after floods during spring time, when *Ae. vexans* is not ready to hatch in masses. However, the peak of development occurs during floods in summer. In contrast to *Ae. vexans, Oc. sticticus* has an unequivocal preference for shaded breeding waters in the flood plains, which are often covered by trees, the pH-value of the waters usually ranges from neutral to alkaline. The optimum temperature for the development of *Oc. sticticus* is about 25°C. The development from hatching of first-instar larvae to the emergence of the adults lasts 6–8 days at 25°C, 10–14 days at 20°C, 18–19 days at 15°C and 37 days at 10°C. The females can migrate considerable distances when searching for a blood-meal, distances of more than 20 kilometres have been observed (Hearle, 1926).

Adults of *Oc. sticticus* stay predominantly in covered terrains, *e.g.* in flood plains of river systems covered with trees where the females frequently become a nuisance. The peak of biting activity of females is in twilight, however, they can also bite during the day in shaded situations. The females may lay up to 150 eggs in shady, damp depressions which will become flooded by rising water levels.

Distribution: The holarctic species is widespread in Europe and occurs from northern Europe to the Mediterranean area and to Siberia in the east. It has also been reported from North America.

10.2.3. Subgenus *Rusticoidus* Shevchenko and Prudkina

The adults are large sized mosquitoes with numerous erect forked scales on the occiput and lateral parts of the vertex. The scutum is covered with narrow scales including most of the prescutellar area, and the scutellum with curved narrow scales and numerous long setae on all lobes. The postpronotum has only broad flat scales. Both the anteprocoxal and postprocoxal membrane have a patch of broad pale scales. The pleurites are extensively covered with pale scales and numerous setae. The abdominal terga have dark scales but with extensive pale scaled areas. In the male genitalia tergum IX has narrow lobes on each side with several short setae. The basal lobe of the gonocoxite bears several long lanceolate setae, and the claspette filament is short and more or less onion or triangular shaped. The fourth-instar larvae of European *Rusticoidus* species are easily distinguished from all other *Aedes* and *Ochlerotatus* species by having several pairs of setae inserted dorsally at the siphon and a variable branched small seta located laterally, usually close to the distal pecten teeth in addition to the siphonal tuft (1-S).

The members of the subgenus were previously placed in the former subgenus *Ochlerotatus* of *Aedes* as a separated globus (*Feltianus*) or group (*rusticus*-group) (Martini, 1931; Edwards, 1932) and this classification was followed by others (Natvig, 1948; Mohrig, 1969). In 1973, Shevchenko and Prudkina established the new subgenus *Rusticoidus* based on the structures of the male genitalia and the larval siphon of *Oc. refiki* as a monotypic subgenus, but confusion has existed since then as to which species should be included in the subgenus. According to Reinert (1999a) the following European species are now included in the subgenus *Rusticoidus*: *Oc. krymmontanus*, *Oc. lepidonotus*, *Oc. quasirusticus*, *Oc. rusticus*, *Oc. refiki* and *Oc. subdiversus*. Additionally, two North American species, *Oc. bicristatus* (Thurman and Winkler) and *Oc. provocans* (Walker), were transferred from the former subgenus *Ochlerotatus* to *Rusticoidus* (Reinert, 2000b).

Oc. krymmontanus was described from the southern slopes of the Crimean Mountains (Alekseev, 1989). Due to the lack of material for examination and the uncertain specific status, the species is not included here.

Ochlerotatus (*Rusticoidus*) *lepidonotus* (Edwards) 1920

Female: Can be separated from other *Rusticoidus* species by the presence of whitish scales on the postnotum. The proboscis and palps are black scaled, the palps with scattered pale scales. The vertex and occiput have narrow yellowish scales and dark erect forked scales, and the lateral parts of the head have whitish scales. The pedicel is black with dense white scales, and the clypeus is blackish brown. The integument of the scutum is blackish brown, and the scutum is covered with narrow yellowish scales. The scutellum has pale yellowish scales and setae on the lobes. The postnotum is dark brown, with a group of narrow whitish scales. The pleurites are extensively covered with pale scales, and the postpronotum with light yellowish or brown scales. The femora and tibiae have light yellowish scales, except for black scales at the apices. The tarsi are dark brown scaled with scattered white scales on entire tarsomere I and the basal part of tarsomere II. The wing veins are covered with intermixed yellowish white and dark scales, with dark scales predominantly located on the costa (C), radius 1 (R_1) and anal vein (A). The abdominal terga are entirely covered with greyish white scales, with a more or less prominent median part with black scales on terga II to V. The sterna are entirely whitish scaled.

Male (Fig. 10.81): The base of the gonocoxite has one lobe arising gradually from the gonocoxite, more or less conical in shape, with a group of lanceolate, flattened setae, but additional smaller lobes are absent. The apical lobe is less developed than in other members of *Rusticoidus*, and is hardly visible. The gonostylus is strongly curved apically, with several small setae near the apex, and the apical spine of the gonostylus is long and straight. The paraproct is inwardly curved at the apex. The claspette stem is relatively short and thick, and more or less straight. The claspette filament is short, and of an irregular triangular shape. The

aedeagus has a broadly rounded apex and small lateral denticles.

Larva: The head is wider than it is long. The antenna is less than half as long as the head, and is entirely covered with spicules. The antennal seta (1-A) is located slightly below the middle of the antennal shaft, with 5–6 branches. The postclypeal seta (4-C) is short with 2–3 thin branches. The inner frontal seta (5-C) has 3–4 branches, the median frontal seta (6-C) has 1–3 branches and the outer frontal seta (7-C) has 8–10 branches. The number of comb scales is 6–11, arranged in an irregular row (Fig. 10.82). Each scale is large with a prominent median spine, and sometimes 1 or 2 lateral spines of almost the same size as the median spine. The siphon tapers from the basal third, the siphonal index is 2.7–4.1, and the dorsal surface of the siphon has 2 pairs of setae. A single additional seta which is longer than the width of the siphon at the point of its insertion, is located on the lateral side of the siphon close to the distalmost pecten tooth. The pecten has 9–21 teeth (average 13–15), sometimes 1–2 distal most teeth are apically detached, and

Figure 10.81. Hypopygium of *Oc. lepidonotus*.

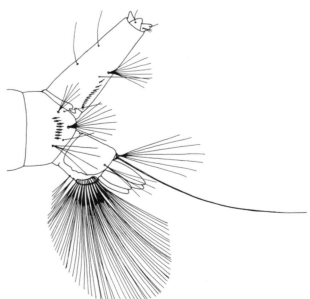

Figure 10.82. Larva of *Oc. lepidonotus*.

the pecten does not extend beyond the basal half of the siphon. The siphonal tuft (1-S) is inserted slightly below the middle of the siphon, beyond the distalmost pecten tooth, with 8–13 branches. The saddle extends far down the sides of the anal segment, and the saddle seta (1-X) has 3–5 branches, and is longer than the saddle. The upper anal seta (2-X) has 7–9 branches, and the lower anal seta (3-X) is simple and much longer than the siphon. The ventral brush has 14–16 tufts of cratal setae (4-X) and 2 precratal setae. The anal papillae are usually shorter than the saddle, lanceolate, and the dorsal pair is longer than the ventral pair.

Biology: *Oc. lepidonotus* is a rare species and recorded so far from south eastern Europe only, thus little information is available about its biology. It is apparently a monocyclic species, which hibernates in the egg stage (Medschid, 1928). Larvae occur by early spring after heavy rainfalls or snow-melt. They may be associated with larvae of *Oc. detritus* when the breeding sites are slightly saline (Gutsevich *et al.*, 1974). Adults are predominantly found at the end of April and during May. Martini (1931) reported numerous females biting horses at dawn in Asia Minor.

Distribution: Greece and Turkey.

Ochlerotatus (Rusticoidus) quasirusticus (Torres Canamares) 1951

Female: Very similar to *Oc. rusticus* in all stages, and the females particularly, closely resemble each other. The scutal pattern is less distinct than in *Oc. rusticus*, the scutum is covered with narrow yellowish brown scales, disconnected by several bare longitudinal stripes at the acrostichal, anterior dorsocentral and posterior submedian areas to which they contrast, mainly due to the exposure of the dark integument. The subcosta (Sc) is without pale scales, but other distinct differences do not exist.

Male: The lobes of tergum IX are situated close together, with 5–8 short spine-like setae. The gonocoxite is more or less straight, not as

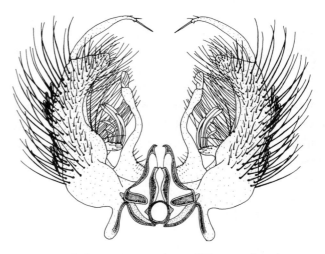

Figure 10.83. Hypopygium of *Oc. quasirusticus*.

rounded as in *Oc. rusticus* (Fig. 10.83). The basal lobe of the gonocoxite is well elongated, the distal margin has 6–8 strongly curved lanceolate setae, and the apical portion of the basal lobe has 2–3 long and strong setae (Fig. 7.28b). The gonostylus is distinctly bent in the apical third, and the apical spine of the gonostylus is long, slender and nearly straight, but never s-shaped as in *Oc. rusticus*.

Larva: The antennae have large and strong spicules, more developed than in *Oc. rusticus*. Prothoracic seta 3-P has 3 branches. 1–2 distalmost pecten teeth are apically detached, and situated beyond the middle of the siphon (Fig. 10.84). The siphonal seta (1-S) is attached within the pecten. The dorsal additional pairs of setae are longer than the width of the siphon at its base, as opposed to being shorter than the width of the siphon at its base in *Oc. rusticus*. The lateral additional seta is usually simple, about as long as the width of the siphon at the point of origin, and longer than in *Oc. rusticus*. Seta 9-S is strongly developed, large and hooked. The anal papillae are distinctly longer than the saddle, with the dorsal pair nearly twice as long as the saddle.

Biology: Very similar to that of *Oc. rusticus*. Some larvae hatch in late autumn and hibernate in the larval stage, others hatch in

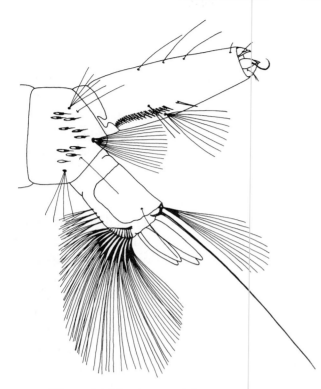

Figure 10.84. Larva of *Oc. quasirusticus*.

early spring of the following year (Encinas Grandes, 1982). They are predominantly found in permanent water bodies within the forest or in ditches and flooded meadows. Often they are associated with larvae of *Oc. refiki*, *Oc. excrucians*, *Cx. hortensis*, *Cs. fumipennis* and *An. atroparvus* (Torres Canamares, 1951). Adults can be found from May to the end of June and they are often encountered close to their breeding sites, indicating that their migration ability is limited. The females are mainly zoophilic but also feed on humans, predominantly in shaded situations.

Distribution: *Oc. quasirusticus* is so far reported only from Spain, where it belongs to one of the most abundant *Ochlerotatus* species in certain regions (Encinas Grandes, 1982). Due to the great similarity to *Oc. rusticus* it was probably overseen or misidentified in the mosquito collections of other countries in the past, thus its whole distribution range is not clarified yet. The species is believed to occur also in north Africa (Encinas Grandes, 1982).

Ochlerotatus (Rusticoidus) refiki (Medschid) 1928

Female: *Oc. refiki* closely resembles *Oc. rusticus*, but differs from the latter in the colouration pattern of the terga. The proboscis and palps are predominantly dark scaled, with scattered pale scales of varying number, mainly at the base of the proboscis. The vertex and occiput have yellowish white narrow scales and erect forked scales of the same colour. The pedicel is dark with a row of white scales at the base, and the clypeus is dark brown. The integument of the scutum is blackish brown, and covered with yellowish scales and a median stripe of dark scales, sometimes divided by a narrow yellowish acrostichal stripe. The scutellum has whitish narrow scales on the lobes. The postprocoxal membrane has pale scales, and the postpronotum has broad flattened scales, dark brown or black in the upper part and white in the lower part. The hypostigmal and spiracular scale patches are fused. White scales on the mesepisternum extend more or less to the anterior angle and to the lower margin, and the mesepimeral patch extends to the lower margin of the mesepimeron. The femora are predominantly pale scaled, the tibiae and tarsomeres I mostly have dark scales, and tarsomeres II to V are entirely dark scaled. The wing veins are covered with dark scales, with scattered pale scales at the base of the costa (C), subcosta (Sc) and radius (R). The colour of the abdomen greatly varies. The terga are often covered with dark scales and indistinct pale basal bands not widened in the middle and scattered pale scales at the distal part of the terga, sometimes with a narrow band of pale scales at the apical margin. Sometimes pale scales predominate on all terga, with only a few scattered dark scales (Fig. 6.40b). In contrast to *Oc. rusticus* the pale scales on the terga never show the tendency to form a longitudinal stripe in the middle.

Male: The lobe of tergum IX has 5–10 short spine-like setae. The basal lobe is situated close to the middle of the gonocoxite; the ventral division with an elongated basal part is covered with 15–16 slightly curved lanceolate setae arranged in several rows (Fig. 10.85). The dorsal division of the basal lobe is small, and situated apically to the ventral part with 2 long and strong setae directed towards the apex. The gonostylus is bent apically, and the apical spine of the gonostylus is slender and more or less straight. The paraproct is strongly sclerotised, inwardly curved at the apex, and pointed. The claspette stem is long, curved at the basal part and swollen at the apex. The claspette filament is elongated and pointed at the apex, and transversely striated. The aedeagus is moderately rounded at the apex with small lateral denticles.

Larva: The head is wider than it is long. The antenna is less than half as long as the head, is slightly curved and covered with spicules. The antennal seta (1-A) is located at about the middle of the antennal shaft, with 5–6

short branches. The postclypeal seta (4-C) is thin, with 3 branches. The median frontal seta (6-C) is usually simple, rarely with 2–3 branches, and is situated before the inner frontal seta (5-C), which has 2–5 (usually 3) branches. The outer frontal seta (7-C) usually has 6–9 branches. The number of comb scales is 6–11 arranged in a row, and each individual scale has a strong median spine and several shorter spines at the base (Fig. 10.86). The siphon is straight, gradually tapers towards the apex, and the siphonal index is 3.0–4.0. The dorsal surface of the siphon has 3 pairs of additional setae, and another seta with 2–5 short branches is located on the lateral side of the siphon close to the distalmost pecten tooth. The pecten has 12–21 teeth occupying the basal third of the siphon, each tooth has a prominent main spine and 3–4 lateral denticles. Sometimes 1–2 distalmost pecten teeth are apically detached. The siphonal tuft (1-S) has 6–9 branches, and is located at about the middle of the siphon or slightly below it, but always beyond the distalmost pecten tooth. The saddle extends down to near the lower margin of the anal segment, and the saddle seta (1-X) has 1–3 branches, and is longer

Figure 10.85. Hypopygium of *Oc. refiki*.

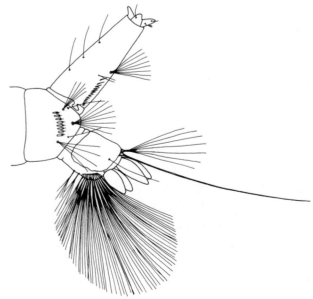

Figure 10.86. Larva of *Oc. refiki*.

than the saddle. The upper anal seta (2-X) has 7–9 branches, and the lower anal seta (3-X) is simple and longer than the siphon. The ventral brush has 13–15 tufts of cratal setae (4-X) and 2–3 precratal setae. The anal papillae are about as long as the saddle, and lanceolate.

Biology: The biology of *Oc. refiki* is similar to that of *Oc. rusticus*. It is a rare species. The early larval stages of this monocyclic snow-melt mosquito can be found after heavy rainfall in late autumn. They can survive even when the breeding sites are covered with ice during winter time. However, the majority of the population overwinters in the egg stage and hatching takes place during the snow-melt early in the year. Typical breeding sites are semipermanent water bodies in swampy woodlands, *e.g.* with *Alnus glutinosa*; occasionally larvae can be found in flooded meadows (Vogel, 1933). Larvae of *Oc. refiki* are often associated with those of *Oc. rusticus, Oc. cantans*, and *Oc. cataphylla*. The water of the breeding sites usually has a neutral to alkaline pH-value. In central Europe the larvae pupate in April and the first adults occur at the end of April or early May. They prefer shaded areas where they may bite humans and mammals even during daytime, however, the biting activity is usually highest at dusk. The adults do not migrate much and prefer to stay close to their breeding sites in shaded areas.

Distribution: In Scandinavia, *Oc. refiki* has been reported from Sweden (Dahl, 1975). It is widely distributed in the other parts of Europe, *e.g.* France, Spain, Italy, Switzerland, Germany, Czech Republic, Yugoslavia, Slovakia, Hungary, and Romania. Outside Europe the species can be found in Asia Minor.

Ochlerotatus (Rusticoidus) rusticus (Rossi) 1790

Female: A large mosquito, with a dark scaled proboscis and palps, and a few scattered pale scales at the base of the proboscis. The vertex and occiput have narrow yellowish white scales, and the eyes are bordered with narrow white scales. The pedicel is dark brown with a circle of whitish scales, and the clypeus is blackish brown. The integument of the scutum is blackish brown, covered with golden bronze scales and a median stripe of dark scales, usually divided by a narrow acrostichal stripe, but sometimes there are two more dark stripes in the posterior submedian areas. The lateral parts of the scutum have cream coloured scales. The scutellum is dark brown with narrow yellowish white scales and pale setae on the lobes. The postprocoxal membrane has pale scales. The pleurites are extensively covered with yellowish or white scales. The postpronotum has broad flattened scales, blackish brown in the upper part and whitish in the lower part. The hypostigmal and spiracular scale patches are fused. The mesepisternum has scales extending to the anterior angle and lower margin, the scales of the mesepimeron extend more or less to the lower margin, and lower mesepimeral setae are present. The femora have pale yellowish scales on the ventral surface and dark scales on the dorsal surface, the tibiae and tarsomeres I have pale and dark scales intermixed, and tarsomeres II to V are almost entirely dark scaled. The wing veins are predominantly covered with dark scales, with scattered pale scales at the base of the costa (C) and on the subcosta (Sc). They are most numerous at the apex of the subcosta. Abdominal tergum I has two patches of yellowish white scales and pale setae; the other terga are dark scaled with pale basal bands which are usually widened middorsally and show a tendency to form a longitudinal stripe in the middle, at least on the apical terga (Fig. 6.40a). Dark parts of the terga often have scattered pale scales. The sterna are predominantly whitish scaled.

Male: The apical margin of tergum VIII is densely covered with long, inwardly curved setae. The lobe of tergum IX has 5–7 short spine-like setae. The gonocoxite has dense long setation on the entire inner surface

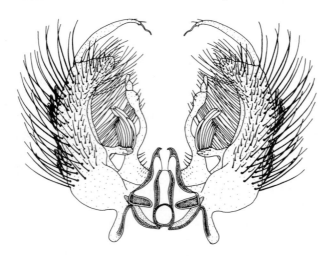

Figure 10.87.   Hypopygium of *Oc. rusticus*.

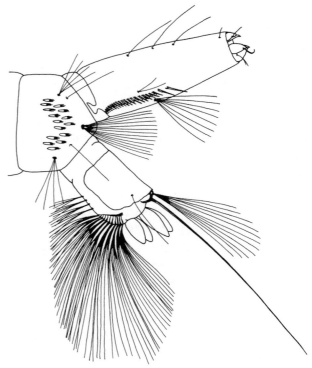

Figure 10.88.   Larva of *Oc. rusticus*

(Fig. 10.87). The basal lobe of the gonocoxite has a constricted stem-like base and a group of lanceolate, flattened setae arranged more or less in a row. The apical lobe is well developed, protruding beyond the level of the insertion point of the gonostylus, with numerous short setae. The gonostylus is slightly curved at the apex with several small subapical setae, and the apical spine of the gonostylus is twisted and distinctly S-shaped. The paraproct is strongly sclerotised, and inwardly curved at the apex. The claspette stem is long, distinctly curved in the middle, and slightly swollen at the apex. The claspette filament is short, more or less onion-like, without a plate shaped widening, and is transversely striated. The aedeagus is moderately rounded at the apex with small lateral denticles.

Larva: The head is wider than it is long. The antennae are approximately half as long as the head, slightly curved and adorned with numerous spicules (Fig. 8.24a). The antennal seta (1-A) is located at about the middle of the antennal shaft or slightly below it, with 5–6 branches. The median frontal seta (6-C) has 2, rarely 3, branches, situated before the inner frontal seta (5-C), which has 3, rarely 2, branches, and the outer frontal seta (7-C) usually has

8 branches. The number of comb scales is 10–18, arranged in two irregular rows, each individual scale with a strong median spine, 1–2 shorter lateral spines and small spines at the base. The siphon is straight, tapering in the apical half, and the siphonal index is approximately 3.0–3.5 (Fig. 10.88). The dorsal surface of the siphon has 3, rarely 4, pairs of additional setae, and another seta with 1–2 thin branches is located on the lateral side of the siphon beyond the median pecten teeth. The pecten has 15–25 teeth not extending to the apical third of the siphon; the basal and median teeth have 2–3 lateral denticles, 1–3 distal pecten teeth are detached and spine-like. The siphonal tuft (1-S) is located at about the middle of the siphon within the distal pecten teeth, with 6–8 branches. The saddle extends about three quarters of the way down the sides of the anal segment, and the saddle seta (1-X) is simple, and nearly as long as the saddle. The upper anal seta (2-X) has

more than 6 branches, and is half as long as the lower anal seta (3-X) which is simple and longer than the siphon. The ventral brush has 11–16 tufts of cratal setae (4-X) and 3–4 precratal setae. The anal papillae are about half as long as the saddle, and the dorsal pair is longer than the ventral pair.

Biology: *Oc. rusticus* is a monocyclic snow-melt mosquito which predominantly occurs in swampy woodlands with a high level of ground water, or more rarely it can be found in floodplains. Larvae are able to hatch during heavy rainfalls in autumn when the water level rises. The diapause of these larvae is terminated by the decreasing temperatures in autumn. Usually the larvae which hatch in autumn hibernate in the second and third larval instar; they can even survive under a closed coverage of ice. The high content of dissolved oxygen in cold water or bubbles of oxygen under the ice which are produced by assimilating plants enable the larvae to cover their demand of oxygen and to survive; usually they are attached to these oxygen bubbles. However, during a severe winter the mortality rate can be very high. For this reason *Oc. rusticus* usually does not occur in areas where the isotherm of January is less than −1°C (Kirchberg and Petri, 1955). Often the larvae of *Oc. rusticus* are associated with those of *Cs. morsitans* or with overwintering larvae of *An. claviger*. A second larval population of *Oc. rusticus* and *Cs. morsitans* hatches from hibernating eggs in early spring shortly after the snow-melt. Thus, first and fourth-instars of these two species can be found together with numerous first-instar larvae of *Oc. communis* and *Oc. punctor* which usually hatch in early spring from hibernating eggs. Typical breeding sites are ditches or deeper depressions with vegetation, *e.g. Carex* sp. or *Phragmites* sp. The larvae of *Oc. rusticus* are seldom found in shallow water bodies because of the high risk that the entire water body may freeze. Larvae are preferably found in breeding sites with a pH-value of 5.0–8.0; they are rare or absent in water bodies with pH values less than 5.0. The optimum temperature for the larval development in the laboratory is 15–20°C. The time of development is about 66 days at a constant temperature of 10°C, 28–29 days at 15°C and 23–25 days at 20°C. Although the larvae of *Oc. rusticus* belong to the first occurring species of snow-melt mosquitoes, they pupate and emerge after the species which hatch in early spring from hibernating eggs such as *Oc. communis* and *Oc. punctor*. In central Europe adults usually emerge at the end of April. Females are vicious biters in shaded situations. The adults prefer to stay in forested areas and do not migrate long distances, usually not more than two kilometres (Schäfer *et al.*, 1997).

Distribution: *Ae. rusticus* is widely ditributed throughout Europe and can also be found in north Africa and Asia Minor.

Ochlerotatus (Rusticoidus) subdiversus (Martini) 1926

Female: The proboscis and palps are dark scaled, often with some scattered pale scales. The vertex has whitish narrow scales, and the occiput has pale and dark scales. The integument of the scutum is blackish brown, the scutum has a median stripe of yellowish bronze scales, lighter in the median and darker in the lateral part, but sometimes the stripe is indistinct. The scutellum has narrow whitish scales and pale setae on the lobes. The postnotum is blackish brown without a group of scales. The pleurites have dense patches of silvery to yellowish scales, and the upper part of the postpronotum has straight or slightly curved bronze scales. Postprocoxal and hypostigmal patches are present. The femora and tibiae have light and dark scales intermixed, and the tarsi are dark scaled with numerous pale scales on tarsomeres I and the basal part of tarsomeres II. The wing veins are predominantly covered with narrow dark scales, and pale scales are present mainly at the base and the anterior part of the wing. The abdominal terga have greyish white scales and a varying number of dark scales, forming indistinct spots. Pale transverse bands are absent. There is great variation in the colouration of the species. Gutsevich *et al.* (1974) described a light form with numerous pale scales on the abdominal terga, proboscis and palps of females. Transitional forms have also been found.

Male: The apical margin of tergum VIII is densely covered with long, thin setae. The lobes of tergum IX are situated close together, with 7–10 short, thick setae. The base of the gonocoxite has two or more lobes, but only one lobe bears a group of lanceolate, flattened setae, the other smaller lobes have long hair-like setae (Fig. 10.89). The apical lobe is weakly developed with short, slightly curved setae. The gonostylus is curved, with several small setae near the apex, and the apical spine of the gonostylus is more or less straight. The paraproct is well sclerotised, and inwardly curved at the apex. The claspette stem is long and slightly curved in the basal part. The claspette filament is short, and more or less triangular shaped with a pointed apex. The aedeagus is not as broadly

rounded at the apex as in *Oc. lepidonotus*, and has small lateral denticles.

Larva: The head is wider than it is long. The antenna is about half as long as the head, and densely covered with spicules. The antennal seta (1-A) is located at about the middle of the antennal shaft, with 3 branches. The postclypeal seta (4-C) is short, with 2 branches. The inner frontal seta (5-C) has 2 branches, the median frontal seta (6-C) has 2–3 branches and the outer frontal seta (7-C) has 5 branches; 6-C is situated in front of 5-C. The number of comb scales is usually 14–15 arranged in two irregular rows, and each scale is large with a prominent median spine and small spines at the base. The siphon is straight, tapering in the apical third, and the siphonal index is 3.0–3.3 (Fig. 10.90). The dorsal surface of the siphon

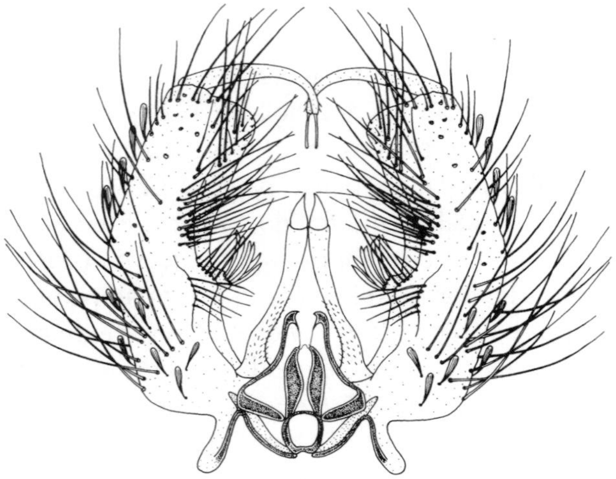

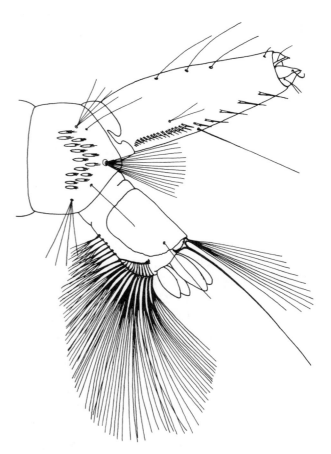

Figure 10.89.   Hypopygium of *Oc. subdiversus.*

Figure 10.90.   Larva of *Oc. subdiversus.*

has 3–4 pairs of additional long setae, and another thin seta with 2 branches is located on the lateral side of the siphon close to the median pecten teeth. The pecten has 3–4 distalmost teeth atypical, spine-like, widely spaced, almost reaching the apex of the siphon. The siphonal tuft (1-S) is simple, and nearly twice as long as the width of the siphon at the point of its origin. It is attached at about the middle of the siphon within the pecten. The saddle extends down 3/4 of the sides of the anal segment, and the saddle seta (1-X) is simple and as long as the saddle. The upper anal seta (2-X) usually has 11 branches, and the lower anal seta (3-X) is simple and about as long as the siphon. The ventral brush has 13–15 tufts of cratal setae (4-X) and 6–7 precratal setae. The anal papillae are shorter than the saddle, lanceolate, and the dorsal pair is longer than the ventral pair.

Biology: *Oc. subdiversus* is a monocyclic species. Larvae occur in early spring after heavy rainfalls or snow-melt. Adults may be very numerous in some localities and massive attacks of biting females on man and live stock have been observed (Gutsevich *et al.*, 1974).

Distribution: European part of Russia to Kazakhstan and southern Siberia.

10.3. GENUS *CULEX* LINNAEUS

Members of this genus are usually small to medium sized species with sparse pleural scaling. The scutellum is distinctly trilobed, prespiracular setae are absent, all claws are without a subbasal tooth. The abdomen is blunt ended with short, oval cerci. The last three characters separate *Culex* species from nearly all *Aedes* and *Ochlerotatus* species. The antenna is as long as or shorter than the proboscis, and is covered with numerous setae. The proboscis is usually dark, sometimes with scattered pale scales, and the length of the female palps is about 1/3 that of the proboscis, and most often shorter. The eyes meet in the middorsal line or are slightly separated, the vertex has numerous erect scales, and the occiput has narrow and broad scales. The scutum is covered with narrow scales and a full pattern of setae, but acrostichal setae are sometimes absent. The scaling on the pleurites is reduced and variable. The dorsal margin of the mesomeron is not in line with the hind coxa. Tarsomere I of the hind legs is as long as or longer than the hind tibia (except in subgenus *Barraudius*), pulvilli are present in most species. The wing has narrow scales on all veins and a long radial fork. All abdominal segments are nearly equally broad, usually banded or with lateral triangular patches of pale scales. The males are smaller than the females, but scaling and setation are quite similar. The antenna has numerous flagellate whorls, and the palps are long with five palpomeres exceeding the length of the proboscis. The morphology of the *Culex* hypopygium is very complicated and belongs to one of the most complex structures in the family of Culicidae. Usually the gonocoxite is devoid of scales (except in subgenus *Barraudius*) and bears only one lobe, which is displaced mesally, hence its name subapical lobe. It may be divided and ornamented with several setae of different length and shape. Typical claspettes are absent, and the aedeagal apparatus is of a complicated structure with several features of unknown homology. These surround the phallosome, which is dorsally covered by the membranous proctiger and the sclerotised and fused or unfused parts of the paraproct, which distally bear a more or less crown-like area of spines or denticles. The head of the larva is much broader than it is long. The antenna is spiculate in most species and longer than the head in many species. The antennal tuft (1-A) usually has multiple branches situated in the distal half of the antennal shaft. The mouth parts are adapted to filter feeding with well developed anteromedian and lateral palatal brushes with an exception in the predacious larvae of the tropical subgenus *Lutzia*, which have the labrum and the brushes modified for capturing prey. The frontal seta 5-C is located distally to 6-C, and all setae are usually multiple-branched. Setae 1-P to 3-P are located on a common sclerotised plate. Abdominal seta 4-I is plumose, but never palmate, and setae 6-I to 6-VII have a variable number of long branches. The number of comb scales is usually between 40 and 60, and each individual scale is relatively small. The siphonal index is variable, and the siphon is usually slender and moderately to extremely long. At least 4–6 siphonal setae (1-S) are inserted along the ventral surface of the siphon, either paired or arranged in a straight or zig-zag row. The pecten teeth in some species are widely spaced distally. The saddle usually completely encircles the anal segment, and the saddle seta (1-X) is often branched. The precratal tufts (4-X) are usually reduced in number or absent, and the anal papillae are variable in shape and size.

The genus *Culex* with more than 750 described species from 25 subgenera worldwide comprises only a few non tropical species. In Europe, species of the subgenera *Barraudius, Culex, Maillotia* and *Neoculex* can be found and most of them have a Mediterranean and/or central European distribution.

Several tropical *Culex* species from Asia and Africa are well known for transmission of lymphatic filariasis and various viral diseases.

10.3.1. Subgenus *Barraudius* Edwards

Members of the subgenus are small brownish species. The proboscis of the female is shorter than the fore femur, the margin of the eyes is usually ornamented with narrow scales, and the vertex has erect and broad light scales. The scutum has uniform brownish scales, and the scutellum is usually light scaled. The hind tarsomere I is distinctly shorter than the hind tibia. The abdominal terga are dark scaled without transverse pale bands, but basolateral pale

patches of scales may be present, and the sterna have light scales. The palps of the male are longer than the proboscis, without long setae, but are covered with a few short spines. The gonocoxite is covered with small scales on the outer surface. The subapical lobe of the gonocoxite arises slightly beyond the middle, not as far apically situated as in the majority of other *Culex* species, with a number of spine-like or hair-like setae; any transparent, broad scale-like setae are absent. The gonostylus is slender, the paraproct apically has a group of small spines forming a paraproct crown, and the aedeagus has dorsal and ventral arms. The head of the larva is broader than it is long, the antenna is as long as the head or slightly longer and more spiculate towards the tip, and the antennal seta (1-A) is multiple-branched. The comb scales are numerous, small and elongated. The pecten occupies about half of the siphon length, each tooth has several lateral denticles. The siphonal tufts (1-S) are arranged in a zigzag row on the ventral side of the siphon, and the main tracheal trunks are broad. The saddle entirely surrounds the anal segment; precratal setae (4-X) are absent, and the anal papillae are short.

The small subgenus *Barraudius* embraces only four species so far, *Cx. richeti* Brunhes and Venhard is only known from Nigeria and *Cx. inatomii* Kamimura and Wada reported from Japan. The two other members of *Barraudius, Cx. modestus* and *Cx. pusillus*, are partly distributed in the European region.

Culex (Barraudius) modestus Ficalbi 1889

Female: The proboscis is dark brown, paler on its ventral surface from the base to the middle, and slightly swollen at the apex. The palps, clypeus and flagellum of the antenna are dark brown. The vertex has dark brown setae which are directed anteriorly between the eyes. The head is covered with brown or yellowish narrow scales, with some broader pale scales on each side, and the occiput has dark brown erect and forked scales. The integument of the scutum is brown, and covered with chestnut-brown scales, lighter on the scutellum and in front of it. Fine blackish setae are scattered mainly along the margin of the scutum, along the dorsocentral stripes and above the wing roots. The setae are more conspicuous and longer on the scutellum. The pleurites are pale brown, with small patches of pale scales on the mesepisternum and upper mesepimeron, 3–6 postpronotal setae, and one lower mesepimeral seta is present. The legs are mainly dark brown, the fore and mid femora have pale scales on the posterior surface, the hind femur is pale except for the dorsal surface which is brown scaled, and the pale knee spot is distinct. The tibiae are dark brown dorsally, with pale scales on the ventral surface. All tarsomeres are dark scaled, and the hind tarsomere I is shorter than the hind tibia (Fig. 6.51a). The wings are entirely dark scaled, and the cross veins are well separated. The terga are dark brown scaled, transverse pale bands are absent, but lateral pale patches usually form a continuous pale border on either side of the abdomen. The sterna are uniformly covered with pale yellowish scales. The abdomen is blunt ended, which separates the species from the similarly coloured females of *Ae. cinereus*.

Male: The palps are almost devoid of setae, and are longer than the proboscis. The long palps separate the males from the similarly coloured males of *Ae. cinereus* which have palps which are considerably shorter than the proboscis. The gonocoxite is approximately twice as long as it is wide with more or less dense scales on its outer surface (Fig. 10.91). The lobe of the gonocoxite is situated slightly beyond the middle of the gonocoxite and is divided into two distinct tubercles. The proximal one bears 2–3 spines of different size, one or two of them may be curved apically, and the more distal tubercle carries one spine and two strong setae. Broad transparent scale-like setae are absent. The gonostylus is long and slender (longer than in the similar *Cx. pusillus*), usually more than half as long as the gonocoxite,

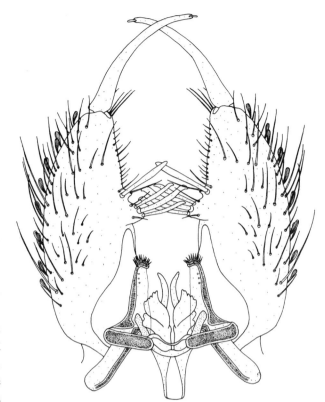

Figure 10.91.   Hypopygium of *Cx. modestus*.

evenly tapering apically and curved in the apical half. The apex of the paraproct has one row of spines forming a paraproct crown. The ventral arm of the aedeagus is short, only slightly curved or nearly straight at the apex and not extending beyond the paraproct crown. The dorsal arm of the aedeagus is conspicuously bent upwards.

Larva: The antenna is moderately spiculate, slightly longer than the head, curved, darkly pigmented at the base and distinctly narrowing from the insertion point of the antennal seta (1-A) to the apex. Seta 1-A is inserted beyond the middle of the antennal shaft, and is half as long as the antenna, with 15–25 branches. The inner frontal seta (5-C) has 3–5 branches, the median frontals (6-C) have 3–4 branches and the outer frontal seta (7-C) has 7–8 branches. The comb consists of a patch of 50 or more fringed scales which are more or less rounded apically. The siphon is straight, the main tracheal

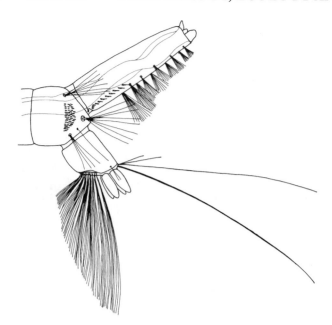

Figure 10.92.   Larva of *Cx. modestus.*

trunks are broad, and the siphonal index is about 4.0–5.0 (Fig. 10.92). The pecten has about 12 relatively widely spaced teeth situated in the basal half of the siphon, and most of the teeth have 4 or 5 lateral denticles. Setae 1-S has 10–12 tufts arranged in a more or less ventral zig-zag row. The basalmost tuft arises proximal to the distalmost pecten tooth, and the distalmost tuft is located close to the apex of the siphon. Each tuft is slightly shorter, or rarely slightly longer than the width of the siphon at the point of its insertion. The saddle is as long as it is wide or slightly longer and completely encircles the anal segment. The saddle seta (1-X) is small with 2–3 branches. The upper anal seta (2-X) has 3–4 branches, one branch longer than the others, and the lower anal seta (3-X) is long and simple. The ventral brush has 10–13 cratal setae (4-X). The anal papillae are shorter than the saddle, slender and tapering.

Biology: The larvae show a preference for shallow sunlit habitats and are frequently found on meadows, in irrigation channels, inundation areas of rivers or rice fields. Other common breeding waters are ground pools, ponds, swamps and marshes with rich vegetation, the water may be fresh or slightly saline. In southern Europe they are mainly found in salt water marshes (Ribeiro *et al.*, 1988) and rice fields. The larvae occur from late spring until late autumn and they are often found together with those of *Anopheles* species. In central Europe the seasonal maximum of the adult population is registered from the beginning of July to late September. Usually the females do not enter buildings, but readily bite man outside, often during the day at sun and wind exposed places. They may cause a considerable nuisance in some regions, especially in late summer when the floodwater *Aedes* and *Ochlerotatus* species have already vanished.

Distribution: *Cx. modestus* is widely distributed in the Palaearctic region from England to southern Siberia. It is recorded from middle and southwest Asia, north India and north Africa. In Europe it is a common species in the southern and central countries.

Medical importance: The species has repeatedly been reported as an arbovirus vector of two different Bunyaviremia, Tahyna and Lednice (Lundström, 1994) and is also regarded as a potential vector of West Nile arboviruses (Ribeiro *et al.*, 1988). In addition, it has been found naturally infected with tularemia (Gutsevich *et al.*, 1974).

Culex (Barraudius) pusillus Macquart 1850

Female: It can be distinguished from the closely related *Cx. modestus* by its commonly darker colouration and the separated basolateral pure white patches of scales on the abdominal terga, which do not form a continuous longitudinal stripe usually present in *Cx. modestus*. The proboscis, palps and clypeus are dark brown, and the proboscis is apically swollen and shorter than the fore femur. The pedicel is dark brown with a tuft of brown scales, and the flagellum is dark brown. The occiput is brown with golden narrow scales and dark forked scales, and the sides are scattered with whitish scales. All decumbent scales of the vertex are narrow and golden, with a few dark setae, directing anteriorly between the eyes. The integument of the scutum is brown, and covered with brown narrow scales and blackish brown setae. The scutellum is brown, occasionally with a pale greenish shine, with brown coarse setae. The pleurites are brown with large patches of pale scales, but prealar scales are absent. The femora have brown scales, usually slightly paler on the ventral surface, with an indistinct knee spot. The tibiae and tarsomeres are dark brown scaled, and the hind tarsomere I is shorter than the hind tibia. The wings are entirely dark scaled, and the cross veins are well separated. The abdominal terga are dark brown with separated basolateral spots of white scales, and the sterna are uniformly covered with whitish grey scales.

Male: The gonocoxite has scales on the outer surface (Fig. 10.93). The lobe of the gonocoxite is situated slightly beyond the middle of the gonocoxite, with numerous setae and 2–3 strong spines of different size which are recurved apically, one of them may be more or less flattened. A simple strong seta is situated more distally on a small separated elevation. The gonostylus is straight or slightly curved, relatively short, less than half as long as the gonocoxite, with a broad base and tapering apically. Four to five small setae are located on the outer surface of the gonostyle close to its apex. The apex of the paraproct is densely covered with small spines forming a paraproct crown. The ventral arm of the aedeagus is long and curved at the apex and usually extends beyond the paraproct crown. To distinguish *Cx. pusillus* males from those of the similar *Cx. modestus*, the shape of the gonostylus and the length of the ventral arm of the aedeagus should be taken into consideration as indicated in the key. These two characters are usually sufficient for a positive identification.

Larva: The body is nearly transparent, pale yellowish green, and the head and siphon are slightly pigmented. The antennae are weakly spiculate, approximately as long as the head, straight, darkly pigmented at the base and

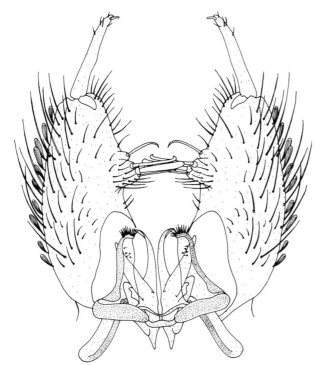

Figure 10.93. Hypopygium of *Cx. pusillus*.

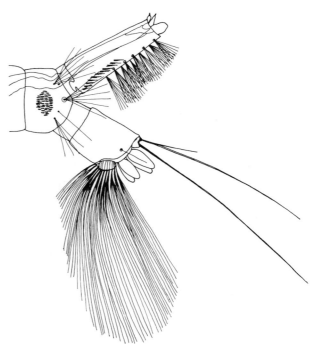

Figure 10.94. Larva of *Cx. pusillus*.

distinctly narrowing from the insertion point of the antennal seta (1-A) to the apex. Seta 1-A is inserted slightly below the middle of the antennal shaft, and is half as long as the antenna or slightly more, with 15–27 branches. The frontal setae are long, the inner and median frontals (5-C and 6-C) have 2–3 branches, and the outer frontal seta (7-C) has 7 branches. The comb consists of a patch of 50 or more long fringed pointed scales (Fig. 10.94). The siphon is short, more or less cylindrical, the main tracheal trunks are broad, and the siphonal index is about 3.0. The pecten has 11–13 teeth, the distalmost teeth extending slightly beyond the middle of the siphon, and each individual tooth is long and slender with 1–3 lateral denticles. Setae 1-S have 8–10 tufts arranged in a more or less zig-zag row along the ventral surface. 2–3 basalmost tufts are situated between the distal pecten teeth. The tufts with about six or more branches, are each at least as long as the width of the siphon at the point of its insertion. The anal segment is slightly shorter than the siphon, the saddle is approximately as long as it is wide, and completely encircles the anal segment. The saddle seta (1-X) is short with 1–2 branches. The upper anal seta (2-X) has 2–3 branches; one branch is longer than the others, and the lower anal seta (3-X) is long and simple. The ventral brush has about 12 cratal setae (4-X). The anal papillae are half as long as the saddle, and the dorsal pair is slightly longer than the ventral pair.

Biology: *Cx. pusillus* is apparently a halophilic species. The larvae are mainly to be found in coastal breeding places, *e.g.* saline marshes and swamps, lagoons or stagnant pools with or without vegetation. Occasionally they occur in saline inland waters like salt lakes or oases, but rarely in fresh water. Common associates in the breeding waters are larvae of *Oc. caspius, An. multicolor* and *An. pulcherrimus. Cx. pusillus* is not a very common species, therefore little is known about its biology and the adult behaviour. They are frequently found between September and November (Martini, 1931). Females have never been observed entering dwellings or stables and probably do not feed on humans (Senevet and Andarelli, 1959).

Distribution: Eastern Mediterranean, Middle East and southwest Asia, north Africa. In Europe, *Cx. pusillus* is solely reported from eastern and southern Greece (Samanidou-Voyadjoglou and Darsie, 1993). This may represent its northernmost record.

10.3.2. Subgenus *Culex* Linnaeus

The body of the females is small to medium sized, and the proboscis in some species has a pale ring in the middle, but otherwise is entirely dark scaled. The vertex has numerous erect scales, and the occiput mainly has narrow scales with a few broad scales. The setae and scale patterns of the scutum and pleurites are as for the genus *Culex*. Lower mesepimeral setae are usually absent, but in some species 2–3 setae may be found. The coxae have few scales, and the wing scaling is as for the genus *Culex*. The abdominal terga may or may not have basal pale bands. The sterna are usually scaled and some species have laterosternal patches of pale scales. The proboscis of the males is usually lacking a pale ring. The gonocoxite is without scales, the gonostylus is flattened and bent, and an apical spine is present. The structure of the aedeagus is very complex with differently shaped plates of outer and inner divisions. The paraproct bears distally a crown-like area composed of heavy spines. The antenna of the larva is spiculate, and is usually not longer than the head. Other morphological characteristics and setation as for the genus *Culex*.

The subgenus *Culex* comprises the most species of the genus, but the main portion of it is of tropical distribution. In Europe two species occur in northern parts, the rest in central and southern Europe. *Cx. p. pipiens* occurs throughout the continent, whereas the southern and northern limits of *Cx. torrentium* are not fully understood.

Other species have a more southern distribution or are limited to the Mediterranean and adjacent areas.

Culex (Culex) brumpti Galliard 1931

Female: According to Galliard (1931) *Cx. brumpti* is a relatively large mosquito. The proboscis is dark with an indistinct pale median ring, the scutum is covered with brownish scales, sometimes slightly paler on its posterior part, and the pleurites have five patches of pale scales. The legs have pale spots at the femoro-tibial and tibio-tarsal joints, and the tarsi are all dark, without pale rings. The wings are entirely dark scaled. The abdominal terga have basolateral triangular pale spots.

Male (Fig. 10.95): The subapical lobe of the gonocoxite has a group of three long spine-like setae, two broad, transparent scale-like setae and two shorter narrow setae, one of them curved at the apex. At the base of the lobe there is an additional long, narrow hair-like seta. The gonostylus is expanded beyond the middle and then tapers apically. The apex of the paraproct has several rows of stout spines. The dorsal arm of the aedeagus is robust basally and directed outwards at an angle in the form of a sharp tooth, and the ventral arm of the aedeagus

is slender and curved, and extended apically in a fan shaped process (Fig. 7.62a).

Larva: Relatively large in size, with a curved antenna, abruptly narrowed distal from the insertion point of the antennal seta (1-A). 1-A is situated at 2/3 the length of the antennal shaft, with multiple branches. The inner (5-C) and median (6-C) frontal setae have 2–3 branches, the outer frontal seta (7-C) has 6–8 branches. The comb is arranged in a patch of 22–44 fringed scales (Fig. 10.96). The siphon is long and slender, the siphonal index is about 6.0–7.0. The siphonal setae (1-S) consist of 4–6 pairs of tufts, with 2–3 branches. The tufts are very short, much shorter than the siphon at the place of insertion, and all tufts arise beyond the distalmost pecten tooth. The pecten has about 8–11 teeth occupying the basal quarter to third of the siphon. The main tracheal trunks are

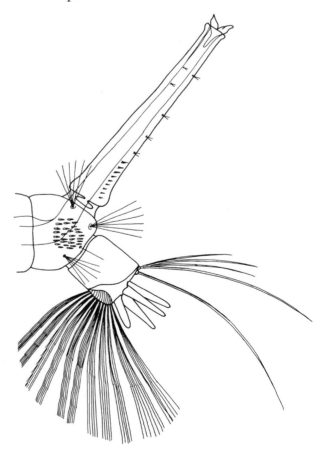

Figure 10.95. Hypopygium of *Cx. brumpti.*

Figure 10.96. Larva of *Cx. brumpti.*

broad. The anal segment is completely encircled by the saddle; the upper anal seta (2-X) has 3–4 branches of varying length, and the lower anal seta (3-X) is simple and nearly as long as the siphon. The ventral brush has about 11 tufts of cratal setae (4-X), and precratal setae are absent. The anal papillae are pointed and nearly as long as the saddle.

Biology: Because of its limited distribution range, little information is available about the biology of *Cx. brumpti*. In Corsica, the larvae were found in pools with aquatic vegetation, and in river beds (Galliard, 1931; Aitken, 1954). Both authors sampled the larvae in August. Nothing is known about the feeding behaviour of adult females.

Distribution: Corsica, Sardinia.

Culex (Culex) laticinctus Edwards 1913

Female: The proboscis is dark, but slightly paler on the basal half of the ventral surface. The palps are usually entirely dark scaled, but palpomere III may have a few apical pale scales. The clypeus and flagellum are dark brown, and the pedicel and flagellomere I have a few pale scales. The erect forked scales of the vertex are light yellowish brown, and the curved scales are white to pale yellowish. The integument of the scutum is brown, and the scutal scales are light yellowish-brown, with paler scales located anteriorly, laterally and on the prescutellar area. The scutal setae are prominent, brown, and darker than the scales. The scutellum has narrow pale curved scales and darker setae on all three lobes. The pleural integument is yellowish-brown, with yellowish to golden-brown setae and patches of narrow pale scales. Postspiracular and prealar scales are absent; the prealar area has 7–14 setae, and usually 2–3 lower mesepimeral setae are present. The fore coxae have pale scales on the anterior surface, and the mid and hind coxae have a longitudinal area of pale scales. The fore and mid femora have dark scales anteriorly and pale scales on the posterior surface, and inconspicuous pale knee

spots are present. The hind femur is mainly pale scaled with a dorsal stripe of dark scales widening to encircle the distal end. The tibiae are dark scaled anteriorly, whitish posteriorly, and the apex of the hind tibia has a prominent pale spot. The tarsi are entirely dark scaled. The wings are dark scaled with a short line of pale scales on the costa located near the humeral cross vein. Tergum I has a median posterior patch of pale scales, terga II to VII have broad pale basal bands, covering between 1/2 and 2/3 of each segment, and tergum VIII has lateral patches of pale scales (Fig. 6.54a). Sterna I to VII have cream coloured scales, and sternum VIII is usually entirely white scaled.

Male: The tergum IX has small lobes bearing 4–9 small setae. The gonocoxite is stout, with dense patches of long and short setae around the apex and near the subapical lobe (Fig. 10.97). The lobe is distinctly divided but not prominent; the proximal portion has three stout setae, slightly flattened and bent distally, and the distal portion has two setae similar in shape, and a crescent shaped, scale-like seta with a pointed apex. The gonostylus is relatively short, sharply bent in the middle and then tapering towards the apex, with a crest of small sharp ridges close to the tip (Fig. 7.63a). The

Figure 10.97. Hypopygium of *Cx. laticinctus*.

apex of the paraproct has several rows of spines, and the ventral arm of the paraproct is well developed and curved.

Larva: The head is wider than it is long, and the antenna is about 2/3 the length of the head, curved and strongly spiculate, with a darker distal part. Seta 1-A has about 25 branches (19–30). Seta 1-C is long and rather stout, and the postclypeal seta (4-C) is short and simple. The inner frontal seta (5-C) usually has 4–5 branches, the median frontal seta (6-C) usually has 4 branches, both reaching just beyond the anterior margin of the head, and the outer frontal seta (7-C) has about 7 branches, longer than 5-C and 6-C. The comb has more than 30 evenly fringed scales (Fig. 10.98). The siphon is straight, evenly tapering towards the apex, and the siphonal index averages 3.5 (2.8–4.6). The siphonal setae (1-S) consist of about seven pairs of tufts arranged in a more or less ventral zigzag row, with the penultimate pair of tufts displaced laterally. Each tuft has 6–9 branches about as long as the width

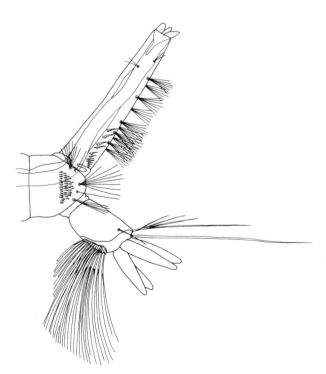

Figure 10.98. Larva of *Cx. laticinctus*.

of the siphon at its point of origin, and 2–3 basalmost pairs of tufts arising within the pecten. The lateral and most distal tufts are smaller and have fewer branches than the others, and are shorter than the width of the siphon at the place of insertion. The pecten has 10–16 long curved teeth each with 3 or 4 basal denticles, extending to near the middle of the siphon. The saddle completely encircles the anal segment, and is longer than it is broad; the saddle seta (1-X) is small, usually simple, sometimes with 2 branches, the upper anal seta (2-X) has 4–5 branches of varying length, and the lower anal seta (3-X) is simple. The ventral brush has about seven pairs of multiple-branched cratal setae (4-X), and precratal setae are absent. The anal papillae are pointed, and about as long as the saddle.

Biology: *Cx. laticinctus* seems to have been more common in the past than it is today (Harbach, 1988). It was frequently collected in cisterns and concrete basins or artificial pools, tanks and barrels in gardens (Aitken, 1954). Nowadays it is more often found in stream pools, rock pools, swamps or ditches. The larvae usually occur in fresh water, but are occasionally found in brackish water. They are often collected with larvae of other *Culex* spp., *Anopheles* spp., *Cs. annulata*, *Cs. longiareolata* and *Ur. unguiculata. Cx. laticinctus* is mainly to be found in the summer months, although a few specimens may occur during the rest of the year. Adult females have never been observed entering houses (Senevet and Andarelli, 1959), and it is not known whether they bite man (Harbach, 1988). Ribeiro *et al.* (1988) noted that it is apparently a zoophilic species with little or no medical importance.

Distribution: The range of *Cx. laticinctus* extends from the Canary Islands eastwards through the countries around the Mediterranean Sea, Somalia, Ethopia, Sudan, the Arabian Peninsula, to the Middle East and southwestern Asian countries. In Europe it is reported in Portugal, Spain, Romania, Italy, Greece (Crete) and Yugoslavia.

Culex (Culex) mimeticus Noe 1899

Female: It is the only European *Culex* species with spots of pale scales on the wing veins and pale rings on the tarsomeres. The head has narrow pale yellowish scales dorsally, brownish laterally. The lateral patches of broad white scales are quite prominent. Three basal palpomeres are covered with blackish brown scales, and the apex of the terminal palpomere is usually white scaled. The basal and apical part of the proboscis is entirely dark scaled, with a broad median white ring. The scutum is covered with yellowish to golden scales in the middle, and greyish or whitish scales laterally. The pleural sclerites have patches of white scales. The femora and tibiae of the fore and mid legs are usually entirely blackish brown scaled anteriorly, sometimes speckled with some pale scales, and the posterior surface is white scaled. The anterior surface of the hind tibia usually has a distinct longitudinal stripe of pale scales. Tarsomeres I to IV of all the legs have pale rings extending to both sides of the articulation, and the basal rings are broader. The wing veins are predominantly covered with blackish brown scales, but numerous pale scales form a characteristic spotted pattern (Fig. 6.52a). The anterior margin of the wing has three white spots; the first spot at the middle of C, entails C and Sc; the second at half distance toward the apex, entails C, Sc and R_1; and the third spot near the apex of the wing, usually entails C, R_1 and R_2. Pale scales are also present at the furcations of R_{2+3} and M, the broad midportion of R_4, the midportion of Cu_1 and the basal portion of A. Cu_2 is almost entirely covered with blackish brown scales; only a very short apical part may be pale scaled. The abdominal terga are dark brown to black with transverse pale basal bands (usually 1/3 as wide as the tergum) and large lateral pale spots. Terga VI and VII have narrow apical bands. The sterna are white scaled, except for a narrow apical portion.

Male: Similar to the female, but the contrast between the dark and light scales on the wing veins seems to be weaker. The palps are longer than the proboscis, with blackish brown scales and usually three dorsal patches of pale scales, one broad median patch on palpomere III, and two narrow basal patches on palpomeres IV and V, and the apex of palpomere V is covered with pale scales. The proboscis has a median pale ring. The antennae are brown scaled, with pale rings. The gonocoxite has relatively short setae predominating on the inner surface (Fig. 10.99). The subapical lobe has a broad, leaf shaped appendage and usually 4–6 large flattened setae. The gonostylus tapers apically. The ventral arm of the aedeagus has 1–2 denticles, and the dorsal arm usually has three finger-like processes (Fig. 7.65a). The apical crown of the paraproct is large, and made up of numerous spines. The ventral arm of the paraproct is thin, and the apex is slightly curved ventrally.

Larva: Similar to the larvae of the subgenus *Neoculex*. The main tracheal trunks are narrow, less than half as wide as the siphon. The head is relatively broad, often dark. The antenna is as long as the head, usually light with a darker

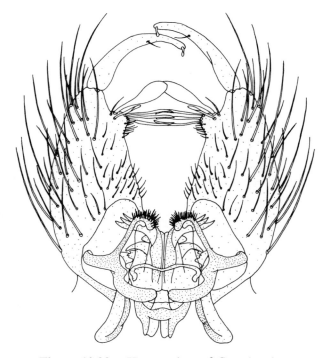

Figure 10.99.   Hypopygium of *Cx. mimeticus.*

apex, with the basal 2/3 covered with tiny spicules. The antennal seta (1-S) is multiple-branched, and inserted just above the spiculate region of the antennal shaft. The position of the subapical setae (2-A and 3-A) as well as the width of the tracheal trunks are valuable discriminative characters between *Cx. mimeticus* and *Cx. theileri*. In the former species the subapical setae are inserted between 1/3 and 1/2 of the distance between the apical setae (4-A to 6-A) and the antennal setae (1-A) (Fig. 8.71a); in the latter they are located adjacent to the apical setae. The labral seta (1-C) of *Cx. mimeticus* is relatively thickened in the middle, and often denticulated in the apical part. The postclypeal seta (4-C) has 2–4 branches, the inner frontal seta (5-C) has 3–5 branches, the median frontal seta (6-C) has 2–3 branches, and the outer frontal seta (7-C) is 5–7 branched. The comb has 20–35 scales (usually 25–30), each scale with a distinct, strong median spine and lateral rows of small spines (Fig. 8.70a). The siphon is straight,

gradually tapering apically, and the siphonal index is 4.5–7.0, usually about 6.0 (Fig. 10.100). The number of pecten teeth is 12–18 (usually 13–15), most often evenly spread over the basal third of the siphon. 5–6 pairs of siphonal setae (1-S) are inserted beyond the pecten. The basalmost 3–4 pairs of tufts have 4–5 branches, they are at least twice as long as the width of the siphon at the point of their insertion. Two distal pairs of tufts have 2–3 thin branches, their length equal to the siphon width at the point of their insertion, and a penultimate tuft is articulated laterally. The saddle entirely encircles the anal segment, and the saddle seta (1-X) has 2 branches. The upper anal seta (2-X) has 2 branches, with the dorsal branch more than half as long as the ventral branch. The lower anal seta (3-X) is simple. The anal papillae are 1.5–2 times longer than the saddle.

Biology: *Cx. mimeticus* is a polycyclic, orophilic (associated with mountainous regions) species. It is quite common in some parts of the Mediterranean region. The larvae occur in small, shallow pools in dried up streams and torrent beds containing growth of *Spirogyra* sp. (Aitken, 1954), rock pools and in shallow margins and backwaters of rapidly flowing mountain streams overgrown with aquatic vegetation. The breeding water is usually crystal clear with a pH ranging between 5.2 and 6.0 (Ribeiro *et al.*, 1977). The larvae may be associated with those of *An. atroparvus, An. claviger, An. marteri, An. petragnani, An. superpictus, An. cinereus hispaniola, Cx. pipiens, Cx. theileri, Cx. perexiguus, Cx. hortensis, Cx. impudicus, Cx. territans* and *Cs. longiareolata* (Ribeiro *et al.*, 1977; Gutsevich *et al.*, 1974). They are reported from an altitude of 3055 m in Tibet (Feng, 1938). The species is moderately orophilic in Portugal, occuring from 150 m to 1100 m (Ribeiro *et al.*, 1989). The females are zoophylic, but occasionally enter houses and bite man (Sicart, 1951).

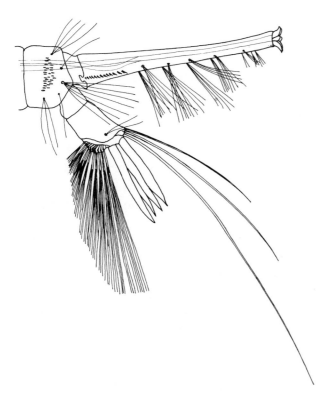

Figure 10.100. Larva of *Cx. mimeticus.*

Distribution: Oriental region, southern parts of the Palaearctic (Mediterranean, Middle East, Iran, Nepal, Tibet, India, Vietnam, China, Japan). In Europe it is recorded in the southern parts of the continent in France, Italy, Yugoslavia, Macedonia, Bulgaria, Greece, Cyprus and Russia.

Culex (Culex) perexiguus Theobald 1903

The species is closely related to *Cx. univittatus* and exhibits a very similar external morphology in all life stages. Thus the following description of *Cx. perexiguus*, based on Harbach (1988), will highlight the differences between the two species as there is a little chance that *Cx. univittatus* might occur in some parts of the Mediterranean region.

Female: A small sized mosquito. The proboscis is dark, with whitish scales ventrally except at the base. The pedicel and flagellomeres have few pale scales. The head has golden brown scales, and the vertex is pale scaled with forked brown scales laterally. The integument is brownish, the scutal scales are narrow and golden brown, and pale scales form a pair of faint submedian spots. The postpronotum has golden brown scales which become paler posteriorly. The upper and lower posterior border of the mesepisternum and anterior part of the mesepimeron have patches of white scales. The legs are dark scaled on the anterior surface, and the hind tibia has a more or less distinct longitudinal anterior pale stripe (Fig. 6.55a). The tarsomeres are dark scaled dorsally, with pale scales ventrally. The wings are dark scaled, with a short line of pale scales on the costa (C). The abdominal terga have slightly convex white basal bands connected with the basolateral patches. The sterna are whitish scaled, sometimes with dark scales scattered or in patches.

Male: Tergum IX has two slightly elevated lobes with long, somewhat spaced setae. In *Cx. univittatus* the lobes seem to be almost flat and the coverage of setae is denser. The gonocoxite has numerous long and short setae

Figure 10.101. Hypopygium of *Cx. perexiguus.*

(Fig. 10.101). The subapical lobe is slightly divided; the proximal portion has 3 long and strong setae, and the distal portion has three shorter hair-like setae and one broad spatulate seta. In *Cx. univittatus* the latter seta is narrower. The gonostylus is expanded beyond the middle, and has two setae and a short apical spine. The aedeagus has a stout ventral arm and concave apex, without any spines, which is shorter than the dorsal arm (Fig. 7.62b). In *Cx. univittatus* the ventral arm is as long as the dorsal arm, and they seem to be in line in a lateral view of the aedeagus (preparation necessary). The ventral arm of the paraproct is long and recurved at the apex.

Larvae: The head is broader than it is long, and the antennal seta (1-A) has 19–27 branches. The frontal setae 5-C and 6-C usually have 2–3 branches. The number of comb scales is 35–55; each individual scale is elongated and rounded apically. The siphon is usually long and slender, and the siphonal index may vary (Fig. 10.102). The number of pecten teeth is 8–15; the larger basal teeth are widely spaced, each with 3–4 lateral denticles. The siphonal seta

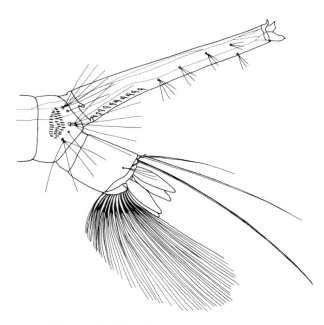

Figure 10.102. Larva of *Cx. perexiguus.*

(1-S) consists of five pairs of lateral tufts usually positioned beyond the pecten. In *Cx. univittatus* the number of pairs is six and they are inserted more ventrolaterally, with the two basalmost tufts (1a-S and 1b-S) attached within or close to the pecten. All tufts in both species have short branches not exceeding the width of the siphon at the point of insertion. The saddle entirely encircles the anal segment, and the anal papillae are short.

Biology: The species is common during summer and autumn. The larvae can be found in many kinds of stagnant water collections, *e.g.* clean to moderately polluted swamps, ponds, streams, pools, wells, usually with emergent vegetation and occasionally in man-made containers. The adult females probably prefer to feed on birds (Harbach, 1988). This has also been reported for *Cx. univittatus* from Turkmenia (Gutsevich *et al.*, 1974). Martini (1931) reported the species (as *Cx. univittatus*) as biting humans inside houses at night.

Distribution: The species is recorded in Portugal, Spain, Italy including Sicily, Macedonia, Bulgaria, Greece and Turkey (Snow and Ramsdale, 1999). *Cx. perexiguus* is found in Asia Minor, southwestern Asia towards India and in northern Africa (Harbach, 1988).

Medical importance: From Israel and Egypt *Cx. perexiguus* has been reported as a vector of West Nile virus (Harbach, 1988). Despite the statement that females mainly feed on birds, Gutsevich *et al.* (1974) reported *Cx. univittatus* to be a vector of West Nile fever. In South Africa both Sindbis and West Nile virus are transmitted by *Cx. univittatus* (Jupp, 1996).

Note on systematics: The species name which has hitherto been used for European material is *Cx. univittatus*, but Harbach (1999) stated that *Cx. univittatus* is restricted to the temperate highlands in the East African Subregion of the Afrotropical Region and identified a few specimens from Greece, Italy and Turkey as *Cx. perexiguus* based on characteristics of the male genitalia and larvae. He suggested regarding the species that occurs in southern Europe as *Cx. perexiguus* rather than *Cx. univittatus*.

Culex Pipiens Complex

The complex consists of several species, subspecies, forms, races, physiological variants or biotypes according to various authors. At present it includes the names *Cx. pipiens pipiens* Linnaeus, *Cx. p. pipiens* biotype *molestus* Forskal, *Cx. p. quinquefasciatus* Say, *Cx. p. pallens* Coquillett, *Cx. p. restuans* Theobald and *Cx. torrentium* Martini in the Holarctic as well as two Australian subspecies, *Cx. p. australicus* Dobrotworsky and Drummond and *Cx. p. globocoxitus* Dobrotworsky.

The status of the three first names has been taxonomically stabilised by designation of neotypes (Harbach *et al.*, 1984; Harbach *et al.*, 1985; Sirivanakarn and White, 1978). It is now generally accepted that the former *Cx. pipiens molestus* (Harbach *et al.*, 1984) is not separated from the subspecies *Cx. pipiens pipiens* and is designated as a biotype as no genetical differences have been found (Bourguet *et al.*, 1998). However new data based on protein electrophoresis revealed a significant genetic distance between the two forms (Becker *et al.*, 1999).

Females of the complex are very difficult to separate in field material. In several reared populations it took eight variables and a discriminant analysis to discern between *pipiens, molestus* and *quinquefasciatus* females and overlapping was considerable (Kruppa, 1988). Thus there is no reliable character yet for discrimination between *pipiens* and *molestus*.

The former *Cx. quinquefasciatus* Say and *Cx. quinquefasciatus pallens* Coquillet are currently regarded as subspecies of *Cx. pipiens* (Miller *et al.*, 1996). They freely hybridise but show a difference in the male hypopygial morphology (Kruppa, 1988). *Cx. pipiens pipiens* and *Cx. torrentium* are two separate sibling species (Harbach, 1985; Dahl, 1988; Harbach, 1988; Miller *et al.*, 1996) defined by genetic characteristics and different morphology in some life stages.

Culex (Culex) pipiens pipiens Linnaeus 1758

Female: A medium sized mosquito, with a yellowish brown to dark brown integument. The antennae are dark, and the pedicel and flagellomere I have a few tiny white scales. The palps are mainly black scaled, and the proboscis has cream coloured scales ventrally. The head has dark forked scales and some paler scales laterally. The scutum has delicate golden brown scales, which are lighter laterally. The scutellum has narrow, pale yellow scales and dark setae. The postpronotum has golden brown scales. The pleurites have yellowish or white scale patches on the mesepisternum. Postspiracular and prealar scales are absent, or rarely a few scales may be present. The absence of scales in *Cx. p. pipiens* provides an almost reliable character for separation from *Cx. torrentium* females which have a few prealar scales when undamaged. The coxae have a small patch of dark scales, the femora have a yellowish apical border but otherwise are dark scaled, and the hind femur has mostly whitish scales. The tibiae and tarsi are dark scaled, the hind tibia lacks a longitudinal pale stripe (Fig. 6.55b). The scaling of the wings is dark, and the subcosta (Sc) intersects the costa (C) beyond the furcation of R_{2+3} (Fig. 6.52b). The abdominal terga are predominantly dark scaled; tergum II has a small basomedian whitish spot, and terga III to VII have whitish to yellowish, narrow basal bands which expand laterally (Fig. 6.54b). The sterna are yellowish scaled.

Male: The subapical lobe of the gonocoxite has seven large simple and one broad scale-like setae (Fig. 10.103). The gonostylus is broad with several small and two spiniform setae apically. The aedeagus has several complicated plates and lobes of which the ventral and dorsal arms are of diagnostic character and are used for the so called DV/D formula (Sirivarnakan and White, 1978; Ishii, 1991). The shape of the ventral arm is slender, blade-like with a sharply pointed apex. The dorsal arm is tubular, stout and distinctly truncated at the apex (Fig. 7.64b). The ventral arm of the paraproct varies in length from being an inconspicuous knob to

Figure 10.103. Hypopygium of *Cx. p. pipiens*.

a conspicuous extension which is never recurved. The paraproct has numerous stout setae forming a brush-like crown. Separation from *Cx. torrentium* is possible by the blunt apex of the dorsal arm and a different shape of the ventral arm of the paraproct.

Larva: The head is wider than it is long, and the antenna is shorter than the head. The postclypeal seta (4-C) is short and simple, and all the frontal setae are long. The inner frontals (5-C) have 5–6 branches, the median frontal seta (6-C) is 4–5 branched, and the outer frontals (7-C) have 6 branches. The prothoracic setae 1-P to 3-P are long and simple, 4-P has 2 branches and is somewhat shorter than the other setae, 5-P and 6-P are simple and long; 7-P is 2-branched and long. The metathoracic seta 1-T is shorter than half the length of 2-T. The number of comb scales is about 40; each individual scale is short and widened at the apex, and evenly fringed. The siphon is slender, evenly tapered towards the apex, and the

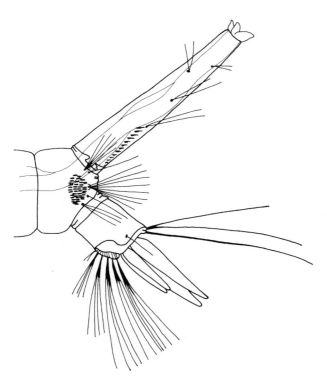

Figure 10.104. Larva of *Cx. p. pipiens*.

siphonal index is between 4.8 and 5.0 (Fig. 10.104). The number of pecten teeth is 13–17, evenly spaced up to seta 1a-S. Each pecten tooth has a long pointed tip and three lateral denticles. The siphonal tuft (1-S) consists of four widely spaced double-branched pairs of setae which arise distal to the pecten in an irregular row. The anal papillae are elongate, and the dorsal pair has twice the length of the saddle. The larvae of the species complex are very similar. Those of *Cx. p. pipiens* and *Cx. torrentium* can only be separated from each other when specimens are very well preserved; differences in abdominal setation as indicated in the keys (Fig. 8.77c). The larvae of *Cx. quinquefasciatus* have a significantly shorter siphon (the siphonal index is about 4.0).

Biology: Overwintering females lay their eggs on the water surface in batches as egg rafts of usually 150–240 eggs. Females which have taken a small blood-meal before overwintering have a poorer chance of surviving (Mitchell and Briegel, 1989). The larvae hatch within one or two days and complete their development to adults in about one to a few weeks, depending on the temperature. They are able to inhabit nearly every kind of water collection. The first larvae often occur together with those of *Anopheles* species and can be found in semi-permanent waters, larger pools with vegetation, rice fields, along river edges in still zones and in inundation areas, occasionally also in tree holes. The larvae frequently occur in man-made water bodies such as flooded cellars, construction sites, water barrels and tin cans, metal tanks, ornamentational ponds and containers in gardens and in churchyards. They even tolerate a small amount of salinity and can occur in rock pools. The species can develop up to several generations per year depending on climatic conditions. The females are anautogenous, ornithophilic, eurygamous, and overwinter in diapause. Occasionally it has been observed to feed on mammals in the field or on mice in the laboratory. In southern France anautogenous populations with stenogamous males of *Cx. p. pipiens* have been reported from the field. Its biotype *molestus* occurs more frequently in human environments.

Distribution: *Cx. p. pipiens* is widespread in the Holarctic region and found throughout Europe. Its distribution seems to be more northern in the easternmost European parts than in Scandinavia. Its northern borders throughout Europe need renewed analysis, as it has not always been separated from *Cx. torrentium*. It has been introduced into Australia and probably also into South America and eastern and South Africa. In the southeasternmost areas of the northern hemisphere *Cx. p. pipiens* occurs sympatrically with *Cx. p. quinquefasciatus*.

Note on systematics: The species is highly variable in its pattern of integument and scaling colour, hence it has numerous synonyms worldwide. From European material nine valid synonyms exist (Knight and Stone, 1977; Knight, 1978; Ward, 1992, 1994). Two further invalid subspecies descriptions on European material and now treated as synonyms are the former *Cx. p. torridus* Iglisch and *Cx. p. erectus* Iglish (Dahl, 1988; Harbach, 1988). There are different views as to the specific status of *Cx. pipiens* and *Cx. quinquefasciatus*. They are still regarded as one species based on their mode of hybridisation and molecular studies which have not confined a species specific status (Bourguet *et al.*, 1998). However, the males can be separated by morphological characters (Sirivarnakarn and White, 1978; Kruppa, 1988).

Medical importance: *Cx. p. pipiens* seems to play a minor role as an arbovirus vector in Europe. Nevertheless, one report of high incidence of West Nile virus from Romania (Nicolescu, 1998) and a record of low incidence of Ockelbo virus from wild caught females (Lundström, 1994) demonstrate the capacity of the species for arboviruses. Worldwide *Cx. p. quinquefasciatus* has been involved in the

transmission of lymphatic filariasis, especially in southeast Asia and Africa, and in the transmission of human or equine encephalitis in north and south America.

Culex pipiens pipiens biotype *molestus* Forskal 1775

The biotype cannot be separated from *Cx. p. pipiens* on single morphological characters in adults. Kruppa (1988) found a statistical difference in the larval mentum between *pipiens*, *quinquefasciatus* and *molestus* in reared strains. No reliable genetic marker was found to separate *molestus* from *pipiens* (Bourguet *et al.*, 1998). However, four biological criteria for the biotype *molestus* have been established: autogeny, stenogamy, anthropophyly and not overwintering in diapause.

Female and male: Colouration, setae and scaling of head, thorax, abdomen, wings and legs as in *Cx. p. pipiens* (Harbach *et al.*, 1984). The males are stenogamous, meaning copulation with females in very limited spaces without swarming is possible. This has led to a search for male morphological characters to distinguish between *pipiens* and *molestus* and recently two have been reported (Oljenicek and Zoulova, 1994). The first is the length of palpomere IV, which gave a reliable difference with the palpal length of *molestus* being 1.5 times that of *pipiens*, a result in contrast to earlier observations (Harbach, 1988). The second feature is the average length of the antennal seta compared to the average length of the antenna, expressed in an antennal index. This index is 3.5 in *pipiens* and ranges between 4.1 and 4.4 in *molestus*. These morphometric characters were measured on reared material.

Larva: The morphological characters are so variable that only identification based on eight characters with statistical analysis gave some discrimination between *pallens, molestus* and *pipiens*. More reliable is the discrimination between all three mentioned taxa against

Cx. p. quinquefasciatus (Kruppa, 1988). However, it is claimed by many authors that a difference in the siphonal length can be found with a shorter siphon in *molestus* than in *pipiens* (Olejnicek and Zoulova, 1994). Previously this has been correlated with highly polluted larval habitats characterised by a high content of ammonium (Gabinaud *et al.*, 1985).

Biology: A well accepted biological character of the biotype *molestus* is the autogeny of the females, which might even occur in several forms (obligate or facultative) as among other culicid species (O'Meara, 1985; Clements, 1992). Without taking any blood-meal the females lay much fewer eggs and can bite readily after the first batch (Harbach *et al.*, 1984). Females can reproduce throughout the year without a diapause; they are found in dark humid places in man-made habitats. The males are stenogamous; they readily mate in confined spaces without swarming. The most common larval habitats are dark and moist cellars of large buildings in towns, underground sewage constructions and man-made water containers in dark, humid places.

Distribution: Females of the complex, registered as *molestus* because of biting man both indoors and outdoors have been reported from many of the largest cities throughout Europe. There are records from very northern cities in the European parts of Russia. No recent summary of the occurrence of *molestus* throughout Europe is available.

Note on systematics: *Cx. molestus* was described as a species by Forskål 1775 on Egyptian material. Harbach *et al.* (1984) selected a neotype and at the same time stated that *molestus* was not a valid species. All claims to identify *molestus* on morphological characters failed. This might depend on an extensive gene flow between the populations. A certain hybridisation between the forms has been postulated for natural conditions and was tested in laboratory experiments (Chevilon *et al.*, 1995).

Culex (Culex) torrentium **Martini 1925**

The species is very similar to *Cx. p. pipiens* in all stages, except for larval seta 1-T (Harbach *et al.*, 1985), the prealar scale patch (Service, 1968a), the pointed and twisted apex of the dorsal arm of the aedeagus and the curved ventral arm of the paraproct in the male hypopygium, and the pattern of the egg chorion (Dahl, 1988). Wild caught females are larger in average body size than those of *Cx. p. pipiens*. This character is very subtle and needs much experience to recognise it. Martini (1931) highlights some differences in scaling which seem to give some guidance for specimens from central and east Europe.

Female: The proboscis is dark with a whitish midpart and pale scaling ventrally, and the palps are dark brown scaled. The antenna is dark with a whitish tinge. The head has yellowish flat scales on the occiput and vertex and around the eyes; the setae are golden to dark brown. The scaling on the scutum is brownish, apically lighter, and on the pleurites is brownish with some whitish scales. A patch of prealar scales is present in most newly emerged specimens. The coxae have patches of light scales, the femora are dark, ventrally whitish, with a white knee spot. The tibiae are dark scaled, ventrally whitish, and the hind tibia has no longitudinal pale stripe. All the tarsomeres are dark brown with some light ventral scaling. The wing veins are covered with brownish, elongated scales. The abdomen is dark brown with pale yellowish basal bands on all segments which are not expanded laterally.

Male: Although there seem to be some constant features in the scaling of the proboscis and legs of the males, the most reliable characters for the separation of *Cx. p. pipiens* from *Cx. torrentium* are found in the hypopygium (Fig. 10.105). The dorsal arm of the aedeagus is pointed and twisted at the apex and not blunt as in *Cx. p. pipiens*. In *Cx. torrentium* the ventral arm of the paraproct

Figure 10.105.   Hypopygium of *Cx. torrentium*.

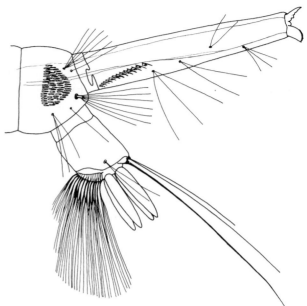

Figure 10.106.   Larva of *Cx. torrentium*.

is always long and recurved, without much variability in shape. In *Cx. p. pipiens* it varies from being vestigious to conspicuous, but is never recurved. The shape of the cercal sclerite of the paraproct is broader and shorter in *Cx. torrentium* than in *Cx. p. pipiens*.

Larva (Fig. 10.106): The larvae differ from those of *Cx. p. pipiens* and its biotype *molestus* in some very subtle characters. Harbach *et al.* (1985) found some differences in the thoracic and abdominal setation as indicated in the keys (Fig. 8.77a).

Biology: The larvae occur throughout the warmer season, often together with those of *Cx. p. pipiens* in both unpolluted and polluted habitats, such as edges of slow running streams, in vegetation at borders of lakes, semi permanent pools, marshy areas, man-made containers and reservoirs of sewage plants. The development seems to be slower than in *Cx. p. pipiens* which might result in only one generation per year in northern areas. Females are ornithophilic and have never been reported to bite man, not even in laboratory colonies. Both sexes are most active in nectar feeding between 22.00 hours and 03.00 hours (Andersson and Jaenson, 1987). However,

in captivity the females live longer and lay their eggs more scattered through time and in smaller batches. The pattern of the eggs exochorion is more widely spaced in *Cx. torrentium* than in *Cx. p. pipiens*; this character needs verification from other geographic areas (Dahl, 1988).

Distribution: The species seems to have a wide distribution in the temperate Palaearctic region and neither its southernmost or northernmost borders are well established.

Medical importance: In northern Europe Ockelbo virus (Alphavirus) has been isolated from *Cx. pipiens/Cx. torrentium*. In the laboratory the vector competence of *Cx. torrentium* was much higher than that of *Cx. pipiens* (Lundström, 1994).

Culex (Culex) theileri Theobald 1903

Female: A large mosquito, with a brown general colouration. The most striking feature is the presence of a longitudinal pale stripe on the anterior surface of all the femora and tibiae, particularly on the fore and mid legs; the palps are predominantly brown scaled. Palpomeres III and IV have pale scales dorsally, they are sometimes largely pale scaled. The proboscis is brown scaled with more or less numerous pale scales predominating in the middle, but not forming a distinct ring. The head has white, yellowish or golden brown narrow decumbent scales and brown erect scales. The scutum has golden brown scales medially, yellowish white laterally, and is whitish at the prescutellar area. The postspiracular area has a broad patch of pale scales. The upper and lower mesepisternal patches of scales are fused forming a stripe at the posterior margin of the mesepisternum. The femora and tibiae of the fore and mid legs have an anterior longitudinal white stripe along their whole length (Fig. 6.53a), and sometimes the fore femur has a linear row of white spots instead. The tarsomeres of the fore and mid legs are entirely dark. Tarsomere I of the hind leg has a distinct white stripe on its anterior surface, but the remaining tarsomeres are completely brown scaled. The basal half of the costa (C) has pale scales. Sometimes the subcosta (Sc) and radius (R-R$_1$) also carry some pale scales (Aitken, 1954). The other wing veins are entirely dark scaled. There are pale scales on abdominal tergum I and sometimes median scale patches on terga II and III. Terga IV to VII have broad basal pale bands which are usually triangularly produced posteriorly, sometimes with median yellowish dots, which may form a pale median longitudinal stripe. Tergum VIII is completely covered with yellowish scales.

Male: The palps are longer than the proboscis at least for the length of palpomere V, brown scaled, with more or less numerous pale scales which sometimes form rings. The lobe of tergum IX has 10–12 setae. The subapical lobe of the gonocoxite has four flattened blade-like setae, a thin hair-like seta and a leaf shaped seta which is relatively broad and elliptical (Fig. 10.107). The gonostylus is sickle shaped with a slender subapical spine. The ventral arm of the aedeagus has 2–4 strong lateral teeth, and the dorsal arm is simple and pointed apically.

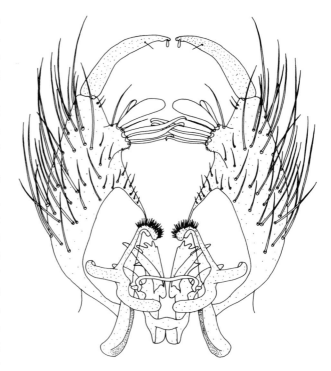

Figure 10.107. Hypopygium of *Cx. theileri*.

Larva: The antennal shaft is strong, three quarters of the length of the head, with a darker base and apical part. The antennal seta (1-A) is multiple-branched (up to 30 branches), and the subapical setae 2-A and 3-A are inserted close to the apex (Fig. 8.71c), and not separated from the apical setae as in *Cx. mimeticus*. The labral seta (1-C) is dark, strong, spiniform, and gradually tapering. The postclypeal seta (4-C) is 1–3-branched. The inner frontal seta (5-C) is 2–5-branched, the median frontal seta (6-C) has 2–4 branches, and the outer frontal seta (7-C) is 6–10 branched. The number of comb scales ranges between 12–44, usually with 20–30 scales (Fig. 10.108). Each individual scale has a strong median spine, with or without lateral rows of smaller spines. The siphon slightly tapers apically, and the siphonal index ranges from 4.0 to 8.0, usually 5.0–6.0. The main tracheal trunks are broad, at least half as wide as the siphon, with an oval cross section. The pecten is restricted to the basal third of the

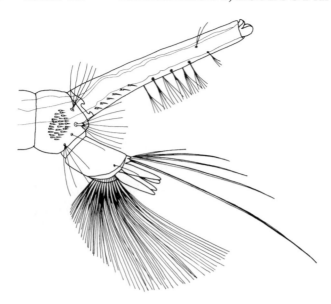

Figure 10.108.   Larva of *Cx. theileri.*

siphon, with 5–15 widely spaced teeth. Five, sometimes six pairs of siphonal setae 1-S, are inserted beyond the pecten (an odd number of setae could be present as well). The three basal pairs are subequal in length, 4–6-branched, situated in a zigzag row on the ventral surface of the siphon, and are about as long as the width of the siphon at the point of insertion. The two distal pairs are weaker and shorter, inserted laterally and subventrally, with 2–5 branches. The saddle seta (1-X) has 1–3 branches, the upper anal seta (2-X) has 3–4 branches, and the lower anal seta (3-X) is simple and longer than the siphon. The ventral brush has 14 well developed cratal setae (4-X), without precratal setae. The anal papillae are about as long as the saddle, slender, pointed, and both pairs are of similar length.

Biology: A polycyclic species recorded from a broad range of elevations. It is reported to be common at an altitude of 1000–3000 m in the Himalayan areas (Sirivanakarn, 1976). The larvae occur in spring in flooded meadows, stagnant or slowly moving streams, ditches, rock pools, drains, swamps and rice fields but also

frequently artificial containers and strongly polluted water (Aitken, 1954; Ramos *et al.*, 1977; Sirivanakarn, 1976). They usually breed in fresh or slightly saline water (2 g NaCl/l), but tolerate a salinity up to 16.6 g NaCl/l and pH 5.5–9.5 (Gutsevich *et al.*, 1974; Pires *et al.*, 1982; Ramos *et al.*, 1977). The larvae can often be found in association with those of *An. algeriensis, An. atroparvus, An. claviger, An. labranchiae, An. maculipennis s.s., An. superpictus, Ae. vexans, Ae. vittatus, Oc. caspius, Oc. detritus, Cx. modestus, Cx. laticinctus, Cx. pipiens, Cx. perexiguus, Cx. hortensis, Cx. impudicus, Cs. longioreolata, Cs. annulata, Ur. unguiculata* (Coluzzi, 1961; Bozkov *et al.*, 1969; Pires *et al.*, 1982; Ramos *et al.*, 1977). The females are zoophilic, but sometimes feed on man and bite mainly in the open, occasionally in large numbers, also entering houses and other buildings (Gutsevich *et al.*, 1974; Ramos *et al.*, 1977).

Distribution: Disjunct through the Ethiopian (south, east and north Africa), Palaearctic (Mediterranean, Ukrainian steppes, Crimea), Middle East and the east Oriental region (India, Burma, China). In Europe reported from Portugal, Spain, France, Italy, Yugoslavia, Greece, Hungary, Bulgaria and Ukraine.

Medical importance: In South Africa, Sindbis and West Nile viruses were found in wild populations (McIntosh, 1975). *Cx. theileri* is known to be a carrier of Rift Valley Fever virus and canine *Dirofilaria* in North Africa and Portugal (Smith, 1973; Ribeiro *et al.*, 1983, 1988).

10.3.3. Subgenus *Maillotia* Theobald

The palps of the females are much shorter than the proboscis. The head has mixed white and dark to golden scales. The scutum and pleurites have many whitish and dark, somewhat narrow scales, sometimes in more distinct patterns. Prealar scales and usually postspiracular scales are present. A white knee spot is present, and the tarsi are uniformly dark scaled. The abdomen may or may not have pale basal bands or lateral patches. The wing veins are covered with long, narrow dark scales. Mohrig (1969) mentioned an important wing character, which separates species of the subgenus *Maillotia* from those of the subgenera *Culex* and *Neoculex*. In *Maillotia* the ending of Sc into C corresponds roughly with the branching of R_{2+3}. The palps of the males are longer than the proboscis. The gonocoxite has a mesally displaced subapical lobe bearing broad, heavy spines and a flat apical distention which reaches beyond the joint of the gonocoxite and gonostylus. The latter is bent and broad with several setae and one apical spine. The aedeagus is insignificant, the paraproct is broad and crowned with denticles and stout spines. The head of the larvae is usually broader than it is long. Seta 3-P is nearly as long as setae 1-P and 2-P. The comb scales are arranged in an irregular patch; each individual scale is usually elongated and fringed with numerous thin spines. The siphon is very long and slender. The pecten has widely spaced teeth, and the siphonal tufts (1-S) are arranged in a more or less regular ventral row covering at least two thirds of the siphonal length, and some distal tufts are laterally displaced. The cratal setae (4-X) are situated close to the anal papillae.

The small subgenus consists of about ten species only. The majority of this is distributed in the Ethiopian region including Madagascar or in westernmost Asia. One species, namely *Cx. hortensis*, is regularly found in the European region with a distribution in the Mediterranean area and central Europe. Another species of *Maillotia*, *Cx. deserticola*, which was transferred from subgenus *Neoculex* (Harbach, 1985), has a very limited distribution in Europe. There is one doubtful record from Corsica (Schaffner, 1998) and one confirmed record from the Zaragosa Province in Spain (Ramos *et al.*, 1998). The species was found in an area well known as a faunistical island regarding other taxa of insects (Eritja *et al.*, 2000). Because of its isolated occurrence in only one location, *Cx. deserticola* is not included in the keys, and no detailed morphological description is given.

Culex (Maillotia) hortensis Ficalbi 1889

Female: The general appearance of the female is similar to *Cx. territans* with greyish unspotted wings but usually paler scaling on the scutum and thorax. The relative broad pale apical bands on the abdominal terga and their distinct median widening at least on some terga differentiate the females from both *Cx. territans* and other *Culex* species in Europe. The scaling of the palps is variable, often with pale scales at the apex forming a ring on the last palpomere, but sometimes entirely dark scaled. The proboscis is usually entirely dark scaled, sometimes with scattered pale scales on the ventral surface. The occiput has light scales, the eyes are bordered with whitish scales, and the vertex is whitish with some dark scales. The scutum has dark setae and is mostly brownish scaled, very often whitish, narrow scales form lateral stripes, and the scutellum always has whitish narrow scales. The mesepisternum and mesepimeron have a few pale scale patches. The coxae have white scales on the ventral surface, and the femora are white scaled; the hind femur has dark scales dorsally. White knee spots are present, and the tibiae and tarsomeres are dark with a few white scales on the ventral surface. The apex of the hind tibia has a white spot which is sometimes difficult to detect. The wing veins are entirely dark scaled, except for the basal 1/5 of the costa which has pale scales. The end of Sc is nearly aligned with the furcations of R_{2+3} and M (Fig. 6.56a). Terga I to III have broad pale apical bands with a median widening; the rest of the terga have narrow pale apical bands.

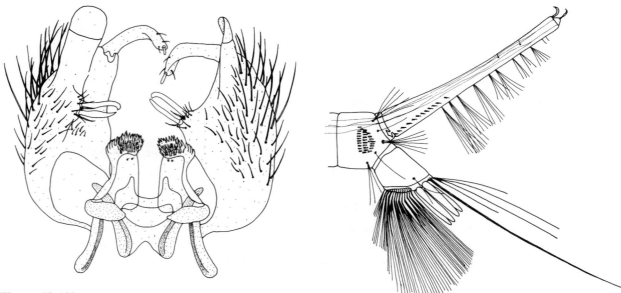

Figure 10.109.   Hypopygium of *Cx. hortensis*.

Figure 10.110.   Larva of *Cx. hortensis*.

Male: The palps are almost completely devoid of setae. The lobes of tergum IX are inconspicuous. The species is easily differentiated from all other European members of the genus *Culex* by a broad, flattened, sclerotised process at the apex of the gonocoxite (Fig. 10.109), which extends distinctly beyond the base of the gonostylus. The gnostylus is bent, and the aedeagus has a dorsal and ventral bridge. The paraproct has denticles and several rows of spines.

Larva: The head is much wider than it is long, and the antennae are long and slender, apically with extremely long spines. The antennal seta (1-A) has about 10 branches, and the antennal shaft is covered with several short spicules around the insertion point of 1-A. The inner and median frontal setae (5-C and 6-C) have 2 branches, and the outer frontal seta (7-C) has at least 5 branches. The prothoracic seta 3-P is nearly as long as 1-P and 2-P (Fig. 8.66a). The comb scales are arranged in an irregular triangular spot, and each individual scale may be of two shapes, either long, narrow and pointed or shorter and rounded apically. The siphon is long and slender, and the siphonal index is between 6.5 and 8.0, with at least 4–5 pairs of long siphonal tufts (1-S) situated in a more or less ventral row (Fig. 10.110). The number of pecten teeth is 12, and they are widely spaced towards the middle of the siphon. The ventral brush consists of 12–14 tufts of cratal setae (4-X).

Biology: Little is known about the phenology and general biology of *Cx. hortensis* as the species is rather uncommon and seems to occur in greater numbers only sporadically. The larvae usually occur in clear water with a certain amount of algae and other vegetation, but also in rice fields, small ponds, unused wells or garden pots. Hibernation takes place in the female stage, daytime resting sites are dark places, *e.g.* wooden stables. The females usually do not feed on humans.

Distribution: In Europe, *Cx. hortensis* is frequently found in the Mediterranean region. It occurs on the Canary Islands and is distributed through Spain, France, Italy and Greece up to central Europe, where it is rare. It can also be found in north Africa, middle Asia and India.

10.3.4. Subgenus *Neoculex* Dyar

Small to medium sized species, the vertex has erect and narrow scales. The palps are not longer than one quarter of the length of the proboscis, with few scales. The scutum has stout acrostichal and dorsocentral setae, and is covered with narrow uniform scales and bare areas in between. The pleurites are sparsely scaled, and the coxae are without scales. The tarsi are dark scaled. Pale transverse bands of the terga, if present, are located at the apex of each tergum. The palps of the males are as long as or longer than the proboscis, and the claws of the hind legs have a subbasal tooth. The gonocoxite is covered with more or less dense setae, but typical scales are absent. The subapical lobe has spine-like or hair-like setae, broad transparent scale-like setae are absent, and the gonostylus is simple. The apex of the paraproct is usually adorned with one row of large denticles. The aedeagus is simple, formed by one pair of sclerites connected by 1 or 2 transverse tergal bridges. The thorax and abdomen of the larvae have minute spicules, visible under high magnification (200×) only. The numerous comb scales are slender and elongated, and bordered with thin spines. The siphon is long and slender, sometimes slightly widening at the apex, the siphonal index is usually more than seven, and the main tracheal trunks are narrow. The pecten teeth are slender, covering between 1/5 and 1/3 the length of the siphon. The siphonal tufts are situated in an irregular row at the ventral surface of the siphon.

The subgenus *Neoculex* consists of less than 30 species, mainly distributed in Africa and the Australian region. Three distinct species are reported from the European region. A remarkable fact is that no member of the subgenus *Neoculex* is known to take its blood-meal from humans. In nature they were observed to exclusively feed on frogs, small mammals and birds (Mohrig, 1969). The egg laying behaviour seems to be typical for the subgenus (Mattingly, 1970). The egg rafts are usually not laid directly upon the water surface, but above the water line. The hatching larvae reach the water by wriggling down after emergence.

Culex (Neoculex) impudicus Ficalbi 1890

Female: A small black mosquito, which differs from the similar *Cx. territans* by the apical pale bands on the abdominal terga which are narrowed in the middle, sometimes interrupted and incomplete, forming triangular lateral patches. The pale spot at the apex of the hind tibia, typical for *Cx. hortensis*, is weakly defined or completely absent. The proboscis is dark brown scaled. The occiput has greyish brown setae, a row of white flat scales around the eye margin, and a pair of long brown setae projecting forward between the eyes. The scutum is covered with greyish brown or blackish narrow scales. The acrostichal and dorsocentral setae are brown, and long blackish setae are situated at the margins of the scutum and at its posterior part. The integument of the pleurites is brown or ashy grey with patches of pale scales on the antepronotum, propleuron, mesepisternum and mesepimeron, and one lower mesepimeral seta is present. The coxae have patches of pale scales, and the femora are dark scaled on the dorsal surface, with pale scales ventrally, and an indistinct knee spot. The hind tibia is dark at the apex or with only a few pale scales, and all the tarsi have dark scales. The wings are entirely dark scaled. The terga are dark scaled with median narrowed or interrupted apical bands. The sterna are usually pale scaled, or occasionally two basal patches of dark scales may be present on each side of some sterna.

Male: The palps are well covered with setae. The gonocoxite is prominently exposed (hence the specific name), compact in appearance, slightly longer than broad at its base and covered with numerous long and conspicuous

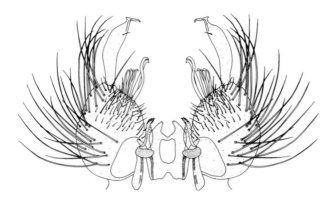

Figure 10.111.   Hypopygium of *Cx. impudicus.*

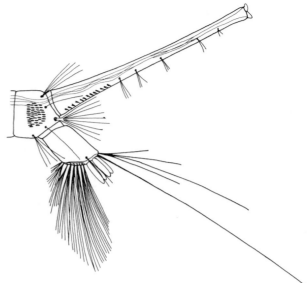

Figure 10.112.   Larva of *Cx. impudicus.*

setae on its outer surface (Fig. 10.111). The densely setose gonocoxite is unique for this species and makes identification of males of *Cx. impudicus* an easy matter. The subapical lobe is situated beyond the middle of the gonocoxite with two long pointed setae which are recurved at their apices; one of which has a more or less sigmoid form. Additionally, the lobe bears several smaller, pointed and apically recurved setae and numerous short, thin, straight setae. The gonostylus is constricted subapically, then extends into a "T"-shaped apical part; a feature unique among European *Culex* members. The apex of the paraproct is inwardly curved with a row of several mesally directed denticles and several spines. The aedeagus is made up of two simple plates connected by two transverse bridges at the base and close to the apex, with 3–4 small denticles at their apices.

Larva: The larvae are extremely difficult to separate from those of *Cx. territans*. Both species closely resemble each other in chaetotaxy and other characters. As indicated in the keys, *Cx. impudicus* usually has anal papillae bluntly ended and about half as long as the saddle, subequal in length (Fig. 10.112). On the other hand, in *Cx. territans* larvae the anal papillae are pointed, usually as long as the saddle, and the ventral pair may be slightly shorter than the dorsal. Unfortunately, the length of the

papillae may vary between specimens, and other distinct characters to distinguish between the two species do not exist.

Biology: The larvae can be found in fresh and stagnant water, generally with considerable vegetation and in well shaded places (Aitken, 1954). They occur in pools, small ponds, rock pools, ditches, marshes or along the edges of streams. Occasionally they are found in rice fields (Rioux, 1958). The larvae are often associated with those of *An. maculipennis s.l., An. claviger* and *Cs. annulata.* In north Africa they frequently occur together with larvae of *An. algeriensis, Cx. hortensis, Cx. pipiens* or *Cs. longiareolata* (Sicart, 1940; Senevet and Andarelli, 1959). Hibernation seems to take place in the adult stage. The larvae are most abundant between May and August, the population maximum of the adults is usually reached in late summer. Like the other European members of the subgenus *Neoculex,* the females of *Cx. impudicus* preferably feed on amphibians and are not known to feed on humans.

Distribution: *Cx. impudicus* is mainly distributed in the Mediterranean region. It is a common species in southern and central Portugal (Ribeiro *et al.*, 1988) and reported from Spain, France, Italy, Corsica, Sardinia, Sicily and Greece. In north Africa it can be found in Algeria and Tunisia and it is recorded from Iran (Knight and Stone, 1977).

Culex (Neoculex) martinii Medschid 1930

Female: A small mosquito, which differs from the other members of the subgenus *Neoculex* and those of the subgenus *Culex* by the abdominal terga, which are dorsally covered with uniform reddish brown scales, and pale apical or basal transverse bands are absent. The general abdominal colouration of *Cx. martinii* is similar to *Cx. modestus*, but it is distinguished from the latter species by the longer hind tarsomere I and by the wing venation. In *Cx. modestus* the subcosta (Sc) ends distinctly distal to the level of the cross vein r-m, whereas in *Cx. martinii* Sc ends more or less at the level of r-m. The proboscis is dark scaled, slightly swollen at the apex, the palps are very short with dark scales, and the clypeus is brownish. The vertex and occiput have small brownish scales, with patches of broader pale scales at the sides. The integument of the scutum is yellowish, and covered with small, golden brown scales and long brownish setae. The integument of the pleurites is brownish-yellowish with dark setae. The femora have brownish scales, slightly lighter on the ventral surface, and a knee spot is absent. The tibiae and tarsi are entirely dark scaled, and the tarsi are without pale rings. The wings have dark scales on all veins. The abdominal terga are reddish brown, without pale transverse bands, and the sterna are entirely yellowish pale scaled.

Male: The palps are dark, and longer than the proboscis by about the length of palpomere V, and not swollen apically. The gonocoxite is short and stout (Fig. 10.113), slightly more than twice as long as it is wide, and the subapical lobe is situated well beyond the middle of the

Figure 10.113. Hypopygium of *Cx. martinii.*

gonocoxite, with two moderately long, stout spines and a row of several stronger setae located distal to the spines. The gonostylus is short and broad, distinctly widened beyond the middle, with two small setae situated close together at about two thirds of its length. The apical spine of the gonostylus is relatively long, and blunt ended. The apex of the paraproct is widened with a convex row of apical denticles and several small setae inserted subapically. The aedeagus is divided into two simple plates connected by apical and basal transverse bridges.

Larva: The head is distinctly broader than it is long. The antenna is about 3/4 the length of the head, slightly curved in the basal part and entirely covered with spicules. The antennal seta (1-A) has 22–26 branches. The inner frontal seta (5-C) is 2-branched, the median frontals (6-C) are exceptionally long, reaching close to the apex of the antenna, with 1–2 branches, and the outer frontal seta (7-C) has about 5 branches. The comb is composed of a patch of 35–40 elongated scales; each scale is apically rounded and fringed with thin spines.

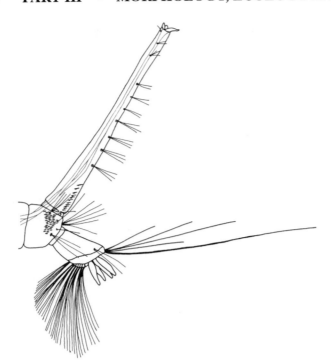

Figure 10.114.   Larva of *Cx. martinii.*

The siphon is long and slender, evenly tapered towards the apex, the siphonal index is 7.5–11.0, and the main tracheal trunks are narrow (Fig. 10.114). The pecten has 11–16 teeth occupying the basal 1/5 of the siphon, and each pecten tooth has 3–4 short lateral denticles. Four to six pairs of siphonal tufts (1-S) are situated in an irregular row at the ventral surface of the siphon, 1–2 apical tufts are laterally displaced, and usually the basalmost tuft (1a-S) is inserted beyond the last pecten tooth. Each tuft has 2–5 branches, slightly longer than the width of the siphon at the place of origin, and the apical tufts are shorter. The anal segment is elongated, and entirely surrounded by the saddle. The saddle seta (1-X) has 2 branches, the upper anal seta (2-X) has 4 branches, and the lower anal seta (3-X) is simple. The ventral brush is composed of 11–12 tufts of cratal seta (4-X), and the anal papillae are about half as long as the saddle.

Biology: *Cx. martinii* is not a very common species, thus the information about its biology is very limited. In Germany, a considerable

number of larvae were found once in a boggy alder forest in October (Mohrig, 1969). It is not known if this is the preferred breeding site of the species or whether it breeds in a wide variety of habitats. Nothing is known about the number of generations per year in Europe, the hibernation principles, the time of appearance of the adults, their period of main flight activity or host seeking behaviour. It seems likely that the species does not feed on humans but prefers to bite amphibians and/or birds, like the other European members of the subgenus *Neoculex*.

Distribution: *Cx. martinii* is mainly distributed in the eastern Mediterranean region, Asia Minor and middle Asia. It is reported from Italy, Yugoslavia, Croatia, Hungary, Turkey, Germany and in north Africa from Morocco.

Culex (Neoculex) territans Walker 1856

Female: The proboscis and palps are dark scaled, and the proboscis is slightly swollen at the apex. The clypeus and pedicel are dark brown; the first flagellomere is slightly swollen, and the flagellum is dark brown with black setae. The occiput has narrow curved pale to golden scales and brown erect forked scales, and broad whitish scales laterally. The integument of the scutum is usually light brown, sometimes darker, and covered with narrow light brown scales; paler scales on the anterior and lateral margins of the scutum and the prescutellar space. Dark setae at the margin of the scutum are relatively long. The scutellum is brown with greyish narrow scales and long black setae. The antepronotum is densely covered with pale scales. The pleurites are dark brown or greyish with patches of broad whitish scales. The femora are dark scaled on the anterior surface, and pale posteriorly, with small pale knee spots. The tibiae are dark anteriorly and pale on the ventral surface, and the apex of the hind tibia is without a pale spot. The tarsi are entirely dark scaled, although a pale stripe may be present on tarsomere I. The wings are entirely covered with narrow dark scales, and the cross veins are well separated. The end of Sc is distinctly displaced towards the wing base compared to the furcations of R_{2+3} and M (Fig. 6.56b). Tergum I has a median patch of pale scales, and the remaining terga are dark brown scaled with evenly narrow apical bands of pale scales usually joining a small pale triangular patch on each side. Tergum VIII is almost entirely dark. The sterna have pale scales often with a greenish tinge.

Male: The last two palpomeres have long setae. The subapical lobe of the gonocoxite has two basalmost setae which are long and broad, flattened, slightly sinuous and with recurved apices (Fig. 10.115). A few other setae are located distally, similar in appearance, but smaller, and additionally one or two long and hair-like setae are present. The gonostylus usually tapers evenly from the base towards the apex. The apical spine of the gonostylus is somewhat expanded at the tip. The paraproct is without ventral arms, and the paraproct crown has an inwardly curved row of denticles. The aedeagus consists of two simple plates which are connected by transverse bridges at the base and close to the apex, the plates are ornamented with small denticles at their apices.

Larva: The head is distinctly broader than it is long, the antennae are about as long as the head, inwardly curved, densely covered with spicules from the base to the insertion point of the antennal seta (1-A), and distinctly narrowed from the insertion point to the apex. Seta 1-A has 25–32 branches, and is situated at about 2/3 the length of the antennal shaft. The frontal setae are long, the inner (5-C) usually has 2 branches, the median (6-C) is usually simple and situated nearly in front of the inner setae, and the outer frontal seta (7-C) has 8–9 branches. The comb has 50 or more, apically rounded scales which are completely fringed (Fig. 8.70b). The siphon is long and slender, with a siphonal index of 6.0–7.0, tapering to near the apex, where it slightly but distinctly expands, and the main tracheal trunks are narrow (Fig. 10.116). The pecten has 12–16 teeth occupying the basal 1/4 to 1/3 of the siphon, each pecten tooth with one or two rather stout lateral denticles on their ventral margin. Seta 1-S consists of 4–6 pairs of tufts, usually commencing at the point where the pecten ends, and rarely the basalmost tuft may arise slightly within the pecten. Each tuft has 2–4 branches of variable length, usually twice as long as the width of the siphon at the point of its origin. The distalmost tuft is shorter and arises slightly out of line laterally. The saddle completely encircles the anal segment and is spiculate on the dorso-apical part; the saddle seta (1-X) is usually double, the upper anal seta (2-X) has about 4 branches, and the lower anal seta (3-X) is very long and simple. The ventral brush has about 7 pairs of multiple-branched cratal setae (4-X). The anal papillae are pointed, and are usually as long as the saddle; the ventral pair may be slightly shorter than the dorsal pair.

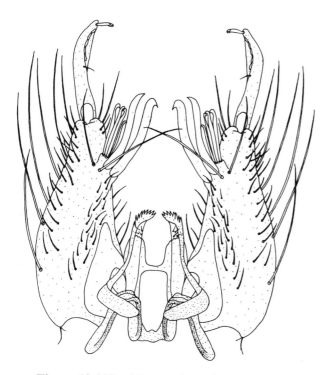

Figure 10.115. Hypopygium of *Cx. territans*.

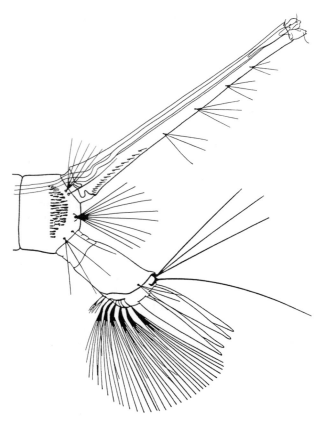

Figure 10.116.   Larva of *Cx. territans*.

in early spring and the first larvae can be found from the end of April, or beginning of May until September; the population maximum is reached in late summer. The females are not known to bite man, but predominantly feed on amphibians, especially *Rana* sp., reptiles and birds.

Distribution: *Cx. territans* is widely distributed throughout Europe. Its range stretches into central Asia and northern Africa. In the Nearctic region it is known from the United States including Alaska and Canada.

Note on systematics: *Cx. territans* was formerly synonymised with *Cx. apicalis* Adams, which was originally described from Arizona, United States. After a revision of the subgenus *Neoculex* in the United States, Bohart (1948) gave evidence that they are two distinct species with a distribution of *Cx. apicalis* in the southern parts of the United States and that of *Cx. territans* in the northern states and Alaska. The European species were first identified as *Cx. apicalis*, but it was shown by Mattingly (1953) and others that they were identical with *Cx. territans*. Thus, all European records in the literature of *Cx. apicalis* prior to 1950 actually referred to *Cx. territans*.

Biology: In the northern regions, *Cx. territans* has probably only one generation per year, but in the southern parts of its distribution range, it is polycyclic, as typical for other *Culex* species. Preferred larval habitats are permanent bodies of water such as ponds, swamps, pools along streams, edges of lakes or drainage channels with a slow water flow, often associated with dense vegetation. The larvae prefer cooler water in shaded situations (Mohrig, 1969) and are often found together with those of *An. maculipennis s.l.* and *An. claviger*. Gutsevich *et al.*, (1974) reported about larval habitats being in strong sunshine in temperate latitudes and completely shaded in southern regions. The larvae are rarely found in heavily polluted water. In the Nearctic they also occur in artificial containers and other small bodies of water (Wood *et al.*, 1979). Adult females reappear from hibernation

10.4. GENUS *CULISETA* FELT

The genus was formerly known under the name *Theobaldia* Neveu-Lemaire 1902 but later on it was realised that this name had already been used for a genus of molluscs since 1885. According to the rules of nomenclature the name *Theobaldia* was consequently replaced by the name *Culiseta* Felt.

The genus embraces mainly medium-sized to large, dark mosquitoes. The females have a straight proboscis and short palps. Prespiracular setae are present, usually of pale colour, but postspiracular setae are absent. The prealar area has setae, but is usually found without scales. The base of the radius has a few setae which are more numerous on the ventral surface of the wing. The abdomen is blunt ended, the cerci are short and rounded apically, the claws are simple, without a subbasal tooth, and pulvilli are absent. In males the length of the palps may vary in individuals of the different subgenera. The gonocoxite is rather long, a basal lobe is present, but an apical lobe may be present or absent. The gonostylus is simple, the apical spine is not longer than its maximum width (except in *Cs. glaphyroptera*), and claspettes are absent. The larvae are large to very large with the head being wider than long. The comb scales are numerous and blunt ended, and the siphonal tuft (1-S) is always present, inserted near the base of the siphon, and the pecten is present. Abdominal segment X is completely surrounded by the saddle (except in *Cs. longiareolata*), which is pierced by one or more tufts of precratal setae (4-X).

The larvae of the genus are generally found in semipermanent and permanent pools, rarely in other locations. As regards the feeding behaviour of the adult females, some species are known to feed exclusively on birds, but others, especially those of the subgenus *Culiseta*, readily attack man and other mammals and are known to be severe biters.

The distribution of the genus *Culiseta* is almost worldwide, but largely confined to the more temperate zones of the Holarctic region. It is a relatively small genus including approximately 40 valid species and subspecies, which are spread over seven subgenera. Throughout the European region, 10 species of three subgenera, *Allotheobaldia, Culicella* and *Culiseta*, are recorded.

In the European members of the genus *Culiseta*, both adult and larval morphological characters can be used to separate the three subgenera. Such an extent of congruity in larval and adult subgeneric characters cannot be found in any other European genus of the Culicidae. Furthermore, the subgenera *Culiseta* and *Culicella* exhibit striking differences regarding their larval and adult behaviour, which will be described in more detail in the subgeneric sections.

10.4.1. Subgenus *Allotheobaldia* Broelemann

The females have a conspicuous colouration pattern on the scutum, and the palps of the males are shorter than the proboscis, and distinctly thickened at the apex. The gonocoxite is without an apical lobe, and tergum IX has two prominent lateral lobes. The head of the larva is small with mouth parts adapted to feed on the substrate. The antennae are short, with a weakly developed antennal seta (1-A). The siphon is short and not sclerotised at the base, and the saddle is weakly developed and plate shaped. The females deposit the eggs in boat-shaped rafts on the water surface, similar to the species of the genus *Culex*.

The subgenus *Allotheobaldia* is represented by only one species, *Cs. longiareolata*, which is distributed in the southern Palaearctic region.

Culiseta (Allotheobaldia) longiareolata (Macquart) 1838

Female: *Cs. longiareolata* can easily be distinguished from all other European species of the genus *Culiseta* by its distinct longitudinal pale stripes on the scutum, which resemble a lyre in shape and the femora and tibiae with pale scales aggregated into conspicuous spots or stripes. The proboscis is blackish brown, the palps are dark brown with pale scales, the latter predominating on the dorsal part. The tip of the palps is almost entirely pale. The antennae are blackish brown, and the pedicel and first two flagellomeres have white scales. The head has dense white scaling along the margins of the eyes, broad white scales also in the median line of the vertex and on the lateral parts of the occiput. The scutum has light brown, narrow scales. A narrow acrostichal stripe of pale scales extends from the anterior margin to the scutellum; in addition there are narrow dorsocentral and lateral stripes, which are connected over the transverse suture. The scutellum and pleurae have patches of white scales except on the upper part of the postpronotum, where the scales are of a creamy yellowish colour. The legs are blackish brown with pale spots and longitudinal stripes on the femora, tibiae and tarsomeres I. All the tarsi have pale basal bands on tarsomeres I to III, and tarsomere V is usually entirely dark. The wing veins are covered with dark scales except the costa, which is covered with pale scales along its entire anterior surface. Dark scales are aggregated at the base of R_s, cross veins (r-m and m-cu) and the furcations of M and Cu giving an appearance of spots. The cross veins are well separated. The scales of the terga vary in colour, but usually form broad white basal bands, while in the apical part of the terga a mixture of yellowish creamy and brown scales is more frequently found. Tergum VIII is usually entirely white scaled.

Male (Fig. 10.117): The main character to separate *Cs. longiareolata* from all other European members of *Culiseta* is the conspicuous tergum IX, which is expanded laterally into two long and slender, sclerotised lobes bearing tiny spine-like setae at their apices (Fig. 7.69a). The gonostylus is broadened apically, bluntly ending with two short, pointed subapical spines. The aedeagus is thick and strongly sclerotised.

Larva: The antennae are short, and the antennal seta (1-A) is articulated in the apical third of the antenna, is short and usually 2-branched, rarely with 1 or 3 branches. The head has simple inner (5-C) and median (6-C) frontal setae, rarely 2-branched, and outer (7-C) frontal setae with 3–4 branches (Fig. 8.79a). The number of comb scales shows great variation (40–75), and seta 3-VIII is strongly developed with multiple branches (Fig. 10.118). The siphon is more or less conical in shape, with an index between 1.5 and 2.0. The pecten has 7–13 short teeth arranged in an irregular row and

Figure 10.117. Hypopygium of *Cs. longiareolata*.

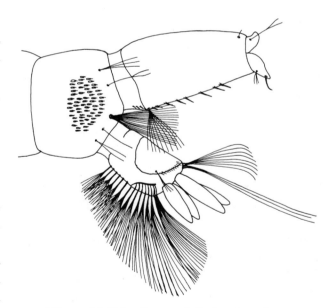

Figure 10.118. Larva of *Cs. longiareolata*.

occupying up to 80% of the length of the siphon. The basalmost teeth are smaller and inserted in the unsclerotised basal part of the siphon. The siphonal tuft (1-S) is shorter than seta 3-VIII, with 10–15 branches arising close to the margin of the sclerotised part of the siphon. The abdominal segment X has a plate shaped saddle extending along about half of its lateral sides. The saddle has dense short spines at its posterior margin, the saddle seta (1-X) is short, not more than half the length of the saddle. The length of the anal papillae varies according to the salinity of the breeding water, but is 0.5–1.5 times as long as the anal segment.

Biology: The larvae can be found in rock holes and in any kind of artificial container, *e.g.* wooden and metal barrels, or tanks built of concrete and wells. Rarely do they occur in natural water bodies like pools, ditches and drain canals. The larvae are able to tolerate a slight salinity and a high degree of pollution and are often likely to be found together with those of *Cx. p. pipiens* and *Cx. mimeticus*. At temperatures of 20–25°C the larval development lasts about 20–22 days. The larvae spend most of their time at the surface of the breeding site and rarely descend to the bottom. The pupae of *Cs. longiareolata* are able to lie passively on the bottom of their breeding sites for a length of time (Peus, 1954). Hibernation takes place in the larval stage. In the temperate climatic zones adults can be found from February to November. The females of *Cs. longiareolata* do not enter dwellings and rarely bite man outside but are regarded as vectors of blood parasites in birds. In populations from Turkmenia autogenous egg deposition has been observed (Rioux *et al.*, 1975).

Distribution: In Europe, *Cs. longiareolata* is widely distributed in the Mediterranean region from Spain and Portugal in the west to the European part of Turkey in the east, in France as far north as Paris, and in Switzerland and southern England. It has also been recorded from the Canary Islands, Madeira and the Azores. The species can be found in southern Ukraine and the lower Volga area as far as the northern slopes of Caucasus. Outside Europe the distribution stretches from middle and southwest Asia, to India and Pakistan and middle Africa.

10.4.2. Subgenus *Culicella* Felt

The wings are usually without spots, except in *Cs. ochroptera* which may have an indistinct dark spot at the base of R_{4+5}. The cross veins (r-m and m-cu) are well separated; the distance between them being at least the length of m-cu (Fig. 6.57a). The indistinct tarsal pale ringing shows the tendency to include both the basal and apical parts of the tarsomeres. The palps of the males are as long as or longer than the proboscis, the gonocoxite is without an apical lobe, and the aedeagus is weakly sclerotised. The antennae of the larvae are longer than the head, with antennal seta (1-A) arising from a point near to the apex, well developed and multiple-branched in the form of a broad fan. The mouth parts are adapted for suspension feeding. The siphon is long and slender, with an index of more than 5. The pecten consists of a few small inconspicuous teeth at the siphon's base, except for *Cs. fumipennis*, which has in addition to the pecten teeth several stout, spine-like setae on the ventrolateral surface of the siphon. The external morphology of the larvae is adapted to their behaviour as suspension feeders. Usually they are found breathing at the water surface, but they can spend hours submerged. They are sometimes attached to active respiratory parts of plants from which they may take air bubbles. Others lay on the ground of the breeding sites with their backs towards the bottom, producing a flow of water with their mouth parts and filtering microorganisms. The species of the subgenus hibernate in the larval or egg stage. The eggs are deposited singly on the ground above the residual water level, as species of the genera *Aedes* and *Ochlerotatus* usually do.

In the European region the subgenus *Culicella* consists of four species.

Culiseta (Culicella) fumipennis (Stephens) 1825

Female: *Cs. fumipennis* is a large mosquito with a dark brown scutum and unspotted wings. The abdominal terga are dark brown scaled with basal, yellowish white bands of uniform width. The legs are largely dark with narrow pale rings at the bases of the tarsomeres. It closely resembles *Cs. litorea* and *Cs. morsitans* but differs from the latter in having a mainly dark proboscis with pale scales laterally and ventrally in its middle third and the dark scales on most of the abdominal sterna are usually aggregated to form an inverted "V" (Fig. 6.58a, b). In *Cs. morsitans* the proboscis is usually entirely dark scaled and the dark and pale scales on the abdominal sterna are intermixed. However, these two characters show some variation, *e.g.* the proboscis of *Cs. morsitans* with a few scattered pale scales or sometimes the abdominal scale colouration patterns of *Cs. fumipennis* are very indistinct. The most reliable diagnostic features are the narrow pale basal rings on all tarsomeres of the hind legs of *Cs. fumipennis*. In *Cs. morsitans* tarsomeres IV and V of the hind legs are entirely dark. Other distinct characters to separate females of these two species do not exist. The differences between *Cs. fumipennis* and *Cs. litorea* may be found on the fore legs. In *Cs. fumipennis* the pale rings include the apical and basal parts between tarsomeres III–IV and IV–V (Fig. 6.59a), and in *Cs. litorea* the apical parts of tarsomeres III and IV are entirely dark scaled (Fig. 6.59b).

Male (Fig. 10.119): The gonocoxite is slender and conical. The basal lobe has 3–4 strong setae. The aedeagus is weakly sclerotised, with lateral sclerites converging apically. The male genitalia of *Cs. fumipennis* and *Cs. morsitans* can be separated with certainty only when directly compared to each other. In *Cs. fumipennis* the gonostylus is somewhat more slender and abruptly constricted shortly beyond the base; the ratio of its median width to its entire length is about 1:16. In *Cs. morsitans* the gonostylus is stout, and more gradually tapered towards the apex and the ratio of the median width to the entire length is about 1:12.

Figure 10.119.   Hypopygium of *Cs. fumipennis*.

The median lobe of tergum VIII bears up to 3 stout spine-like setae in *Cs. fumipennis*, and 3–7 strongly sclerotised longer setae in *Cs. morsitans*. These two characters combined should enable the separation of the two species with a certain confidence.

Larva: In the larval stage *Cs. fumipennis* is easy to distinguish from all other European members of the subgenus *Culicella* by its siphon (Fig. 10.120). It bears, in addition to the pecten teeth, large isolated spine-like setae irregularly scattered on its ventrolateral surface. The most distal one extends well beyond the middle of the siphon. The antennae are longer than the head, the antennal seta (1-A) is large and multiple-branched. The inner (5-C) frontal seta is 2–4 branched, the median (6-C) is 2-branched and the outer (7-C) has 5–6 branches. The comb consists of 120–160 long and slender scales fringed with small, fine spines. Seta 3-VIII has 4–6 long branches. The siphonal tuft (1-S) consists of 4–5 branches and is distinctly

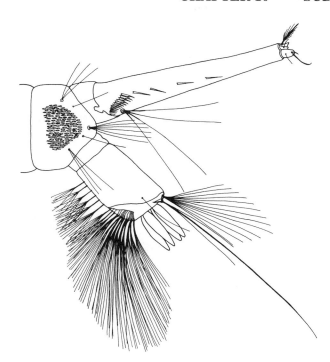

Figure 10.120. Larva of *Cs. fumipennis*.

longer than half of the siphon. A pair of conspicuous long, multiple-branched tufts is situated dorsally near the tip of the siphon. Seta 9-S on the posterolateral valves is strongly developed, and hook shaped. The abdominal segment X is long, completely surrounded by the saddle with six tufts of precratal setae (4-X) piercing through it, and the saddle seta (1-X) is simple. The anal papillae are lanceolate, and about half as long as the saddle.

Biology: In Sardinia, larval catches were recorded from January through April (Marchi and Munstermann, 1987), but in the central and more northern parts of Europe, the larvae of *Cs. fumipennis* first occur in spring during the months of April and May. The preferred breeding sites are open, unshaded water bodies such as shallow temporary pools with rich vegetation or covered with duckweed (*Lemna* spp.). The larvae also occur among the grassy margins of permanent ponds or swamps. They are often associated with larvae of *Cs. morsitans*, *Cx. territans* and *Cx. hortensis*, occasionally

with those of *An. claviger*. They feed on microorganisms and spend most of their time submerged. Very rarely the larvae can be found in water with a high salinity (Martini, 1931). Little is known about the feeding habits of the adult females. They were never observed to enter any kind of dwellings or bite man and domestic animals. It is likely that they don't feed on mammals, but take their blood-meals from birds or reptiles, like *Cs. morsitans*. Although adult females were captured in the middle of the year, in July and August, it is not known if *Cs. fumipennis* has more than one generation per year.

Distribution: A holarctic species, widely distributed throughout Europe, from southern Scandinavia to the east Balticum and southwards to the Ukraine and northern Caucasus. It is recorded from nearly every country in central and southern Europe and occurs around the Mediterranean basin to northern Africa.

Note on systematics: *Cs. setivalva* was formerly regarded as a valid species (Knight and Stone, 1977) but was put under synonymy of *Cs. fumipennis* (Danilov, 1984; Ward, 1992).

Culiseta (Culicella) litorea (Shute) 1928

Female: *Cs. litorea* was first described as a variety of *Cs. morsitans* (Shute, 1928), but subsequently rised to species rank, mainly due to striking differences in the male hypopygium (Marshall and Staley, 1933). The females, like those of *Cs. fumipennis*, have largely pale scaled sterna with a pattern of dark scales forming an inverted "V", but the pattern is not always visible, especially in pinned specimens. In this case *Cs. litorea* might be easily confused with *Cs. morsitans*. The separation between *Cs. litorea* and *Cs. fumipennis* is far more difficult. According to Marshall (1938) the pale rings on the last two tarsal joints of the fore and mid legs are less distinct in *Cs. litorea* than in *Cs. fumipennis*, and on the hind legs of *Cs. litorea* the rings are either inconspicuous or absent. Because none of these differences are absolute, it might be impossible to identify *Cs. litorea* females with certainty, especially if the determination is based on a single or a few specimens.

Male: The male genitalia are very distinctive and the only reliable means to separate *Cs. litorea* from its close relatives. The gonocoxite is more or less twice as long as it is wide, parallel sided and blunt ended (Fig. 10.121). The basal lobe is elongated and bears two stout setae, one of which reaches as far as or beyond the apex of the gonocoxite. In *Cs. fumipennis* and *Cs. morsitans* the gonocoxite is more slender and conical, tapering towards the apex and none of the conspicuous stout setae of the basal lobe reaches the apex of the gonocoxite. Furthermore, the male palps of *Cs. litorea* are usually shorter than the palps of the two other species. Marshall and Staley (1933) found the ratio of the length of the palps to the length of the proboscis to be about 1.14–1.22 in *Cs. litorea* and about 1.33–1.40 in *Cs. morsitans*, but Service (1970b) reported an overlap of this character in the two species, which is therefore not always valuable for a positive identification. Marshall (1938) mentioned a further diagnostic character. The base of the gonostylus appears to be distinctly more bulbous in *Cs. litorea* than in the two other species.

Larva (Fig. 10.122): The larvae of *Cs. litorea* are readily distinguished from those of *Cs. fumipennis* by having no isolated spine-like setae in addition to the pecten teeth, but the separation from *Cs. morsitans* is less reliable. The only character which might allow identification of fourth-instar larvae of the two species can be found in the length of the siphonal tuft (1-S) compared to the length of the siphon. In British populations Marshall (1938) found in *Cs. morsitans* the length of 1-S usually less than, and in *Cs. litorea* more than, 0.4 times the length of the siphon. However, Aitken (1954) found larvae of *Cs. litorea* from Sardinia with the length of S-1 being less than 0.4 times

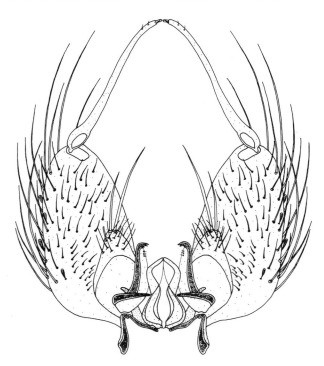

Figure 10.121.   Hypopygium of *Cs. litorea*.

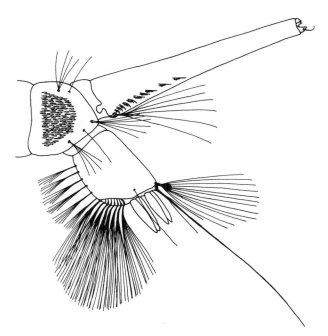

Figure 10.122.   Larva of *Cs. litorea*.

the length of the siphon, and Rioux (1958) was unable to separate larvae of the two species from the Mediterranean area by this character. There is a considerable overlap in the above mentioned ratios between the two species, which invalidates reliable larval identification (Service, 1970b).

Biology: The biology of *Cs. litorea* closely resembles that of *Cs. morsitans* in many aspects. Larval habitats are pools, small ponds or ditches, but unlike *Cs. morsitans*, which can commonly be found in both open (sunlit) and densely shaded situations, the larvae of *Cs. litorea* are always restricted to habitats of the first type. It is mainly a coastal species that can tolerate slightly brackish water, but larvae are not restricted to salt water habitats, where they can often be found together with those of *Oc. detritus*. Frequently they can be found in semipermanent fresh water habitats or channels with a rich growth of vertical vegetation (Rioux, 1958). Aitken (1954) found larvae in a large coastal fresh water marsh, containing abundant growth of cattails (*Typha* sp.). *Cs. litorea* feeds principally on avian blood; occasionally a few females were found to bite reptiles or mammals, including man (Cranston *et al.*, 1987). Hibernation takes place in the larval stage and they are able, like *Cs. morsitans*, to survive under a coverage of ice during the winter months (Rioux, 1958).

Distribution: *Cs. litorea* is known from Ireland, England, France, Spain and Italy (Sardinia) in Europe and from Algeria. It has a more restricted distribution than *Cs. morsitans*.

Culiseta (Culicella) morsitans (Theobald) 1901

Female: The general appearance of *Cs. morsitans* is in many respects very similar to *Cs. litorea* and *Cs. fumipennis* in all stages. The few characters which might facilitate distinction between the three relatives are given in the descriptions of the two latter species. *Cs. morsitans* is a rather large species. The proboscis is uniformly dark scaled, occasionally with a few pale scales in its middle third, and slightly thickened in its apical part (Fig. 6.58c).

The clypeus is dark brown, the palps are about one quarter of the length of the proboscis, and are mainly dark scaled with a few pale scales at the tip. The pedicel is brown, and the flagellum of the antennae is blackish brown with black setae. The vertex has long dark setae, and the occiput has narrow yellowish-white scales and dark erect forked scales dorsally and broad yellowish-white scales laterally. The integument of the scutum is dark brown, with narrow brown and yellowish-golden scales. There is golden scaling mainly on the acrostichal and dorsocentral stripes, on either side of the prescutellar area, on the anterior part of the supraalar area and on the anterior submedian area. The scutellum is dark brown with patches of narrow yellowish-white scales and long dark setae on the lobes. The mesepisternum, mesepimeron and lower part of the postpronotum have small patches of whitish scales. The prespiracular setae are yellowish, numerous, but postspiracular setae are absent. The femora and tibiae are dark brown, with pale scales at the ventral surface. Pale knee spots are well developed, and all tibiae have pale rings at their apices. The tarsi are dark with faint pale rings involving both ends of the joints. Rings at the joints of tarsomeres III–IV and IV–V are inconspicuous or absent on the fore and mid legs and always absent on the hind legs. The wing veins are entirely dark scaled, without spots, and the scales are narrow and dark brown. The cross veins r-m and m-cu are widely separated and without scales. The terga are dark brown with narrow yellowish white basal bands. The sterna are clothed with mainly pale scales; less numerous dark scales are irregularly scattered, not forming any pattern (Fig. 6.58d).

Male (Fig. 10.123): The hypopygium is very similar to that of *Cs. fumipennis*; for differential characters see the description of the latter. The posterior margin of tergum VIII has a broadly rounded lobe bearing a group of 3–7 stout, spine-like setae. The lobe of tergum IX is slightly elevated with long hair-like setae

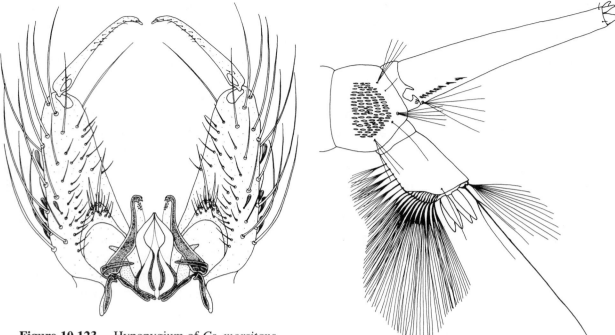

Figure 10.123.   Hypopygium of *Cs. morsitans*.

Figure 10.124.   Larva of *Cs. morsitans*.

(Fig. 7.69c). The gonocoxite is conical, about 2.5 times as long as it is wide at its base, tapering apically, and the basal lobe of the gonocoxite is well developed with 2–4 conspicuously stout setae. The gonostylus is slightly bulbous at the base, gradually tapering towards the apex. The apical spine of the gonostylus is short and stout. The paraprocts are strongly sclerotised, with two, rarely three apical teeth. The aedeagus is weakly sclerotised and pointed at the apex.

Larva: Very similar to *Cs. litorea*; for separation of the two species in the larval stage see the description of the latter. The head is exceptionally large in relation to the body, and more than 1.5 times wider than it is long. The antennae are as long as or slightly longer than the head, spiculate, and curved with a darkly pigmented, tapering apex. The antennal seta (1-A) is large, forming a fan-like tuft with 18–25 branches, inserted at the upper third or quarter of the antennal shaft and reaching well beyond its tip. The postclypeal seta (4-C) is small and simple, and situated anteriorly to the frontal setae. The inner (5-C) and median (6-C)

frontal setae have 2–3 branches; 6-C is very long (Fig. 8.85c). The outer frontal seta (7-C) has 6–8 branches. The prothoracic setae (1-P to 7-P) are very long, usually with 1–2 branches. The lateral abdominal setae on the first two segments (6-I and 6-II) have 3–4 branches, those on segments III to VI (6-III to 6-VI) are simple. The comb has more than 90 scales closely arranged in a large triangular patch (Fig. 10.124). The individual scales are long and narrow in the middle, slightly widened at the base, with the apical part rounded and fringed with small spines laterally and apically. The siphon is straight, long and slender, and slightly tapered towards the apex. The siphonal index ranges from 5.0–7.0, and the pecten consists of 6–11 teeth which are confined to the basal fifth or quarter of the siphon. The smaller basal teeth usually arise from the membrane proximal to the siphon, and the distal teeth are detached. The siphonal tuft (1-S) has 4–5 branches, and is distinctly longer than the basal width of the siphon. The anal segment is long and narrow, and completely ringed by the saddle. The saddle seta (1-X) is simple, and slightly shorter than the saddle. The ventral brush consists of 12–15 cratal setae (4-X) and 5–6 precratal setae which pierce the saddle. The anal papillae are variable in length, lanceolate, and the dorsal pair might be slightly shorter than the ventral pair.

Biology: *Cs. morsitans* is a monocyclic species. The eggs are deposited during early summer in the moist substrate above the residual water level. Hatching occurs in autumn when heavy rainfall leads to a rise of the water level in the breeding site. Usually the larvae grow to the second or third-instar in the same year. Depending on the weather conditions, fourth-instar larvae can sometimes be found prior to November, but pupation never occurs until the next spring (Marshall, 1938). During winter time the larval development is postponed. The larvae often descend to the bottom of the breeding site, where they lie in an inverted position with their head setae and tip of the siphon in contact with the ground. They are able to survive for considerable periods under a coverage of ice, but entire freezing of the breeding site leads to a high mortality. Although hibernation usually takes places in the larval stage, newly hatched first-instar larvae can be observed in early spring. This is in accordance with the situation in the Nearctic region, where *Cs. morsitans* overwinters in the egg stage in most of Canada and first-instars appear from April on (Wood *et al.*, 1979). In Europe the larvae can be found from autumn to spring/early summer in a variety of breeding sites, including pools, small ponds or ditches and even in slowly flowing waters. Predominantly they occur in swampy woodlands and temporary water bodies in forests or at their edges in both, open and shaded situations. They are able to tolerate a considerable amount of salinity and also develop in slightly brackish water. In spring time the larvae can often be found together with those of *Oc. rusticus* and first-instars of *Oc. punctor* and *Oc. communis*, and later in the year they occur together with larvae of *An. claviger*. In Britain they have been found with larvae of *Cs. fumipennis* and *Cs. litorea* (Cranston *et al.*, 1987). The adults emerge from April on and can be observed until October. The females take their blood-meal mainly from birds and occasionally from reptiles and small mammals. It is assumed that in central Europe the females attack humans very rarely, if ever, but Horsfall (1955) reported that *Cs. morsitans* was a serious pest in eastern Europe and the former USSR. The daytime resting sites of the adults are hollows of trees and uninhabited buildings or cellars, but they are rarely found in leafy shrubbery or in grassy vegetation (Service, 1971b).

Distribution: *Cs. morsitans* is widely distributed throughout the Palaearctic region. It is very common in almost every European country and its range stretches from southern Scandinavia into north Africa and from the Northern Sea and Atlantic Ocean eastwards to west Siberia and southwest Asia. The form *dyary* (Coquillett) that was formerly regarded as a distinct species, but put under synonymy of *morsitans* (Wood *et al.*, 1979; Ward, 1984), can be found in the Nearctic region.

Medical importance: *Cs. morsitans* was found to be a carrier of Ockelbo virus in Sweden (Francy *et al.*, 1989).

Culiseta (Culicella) ochroptera (Peus) 1935

Female: Females of *Cs. ochroptera* may be confused with *Cs. morsitans* on a first glance, but they are usually smaller in size and more gracile than the latter. The most striking differences are the generally more brownish colouration of *Cs. ochroptera* (general colouration of *Cs. morsitans* more greyish) and the tibiae of the fore legs which are predominantly yellowish scaled, whereas the fore tibiae of *Cs. morsitans* are mainly dark scaled. The proboscis of *Cs. ochroptera* is usually densely covered with pale scales, with a dark apex.

The palps are dark brown, sometimes with pale scales at the apices of palpomeres III and IV. The antennae are dark brown, and the base of the first flagellomere is whitish-yellow. The head has narrow pale scales and broad erect black scales. Along the margin of the eyes there are long, curved, dark setae. The anterior part of the scutum is uniformly golden brown or rusty brown without pale scales. The supraalar and prescutellar area has narrow, whitish brass coloured scales, which usually form indistinct, narrow, longitudinal stripes from the middle of the scutum. The scutellum is brown with patches of whitish brass coloured scales on the lobes. The posterior margin of the scutellum has dark, slightly curved setae. The postnotum is ochre brown coloured. The postpronotum has uniformly brown scales. The femora are dark on the anterior surface, and pale yellow on the posterior surface with a white patch of scales at their apices. The tibiae of the fore legs are pale yellowish scaled except for a narrow longitudinal dark band on the anterior surface. The tibiae of the mid and hind legs are pale on the anterior and posterior surface and dark on the ventral and dorsal surface. The tarsi of the mid and hind legs have narrow pale basal rings on tarsomeres I to III and sometimes a few pale scales at their apices, and tarsomeres IV and V are entirely dark. The pale rings are often indistinct. The costa has a varying number of pale or ochre coloured scales along the entire anterior margin, but other scales on the wing veins are dark. At the base of R_{4+5} the dark scales may be aggregated to form a small but distinct dark spot. The colouration of the abdominal terga may be variable. Typically the terga are brown scaled with yellowish scales forming indistinct narrow basal and apical bands, and tergum VIII is completely pale scaled. Sometimes the apical bands or both the apical and basal bands are absent, and isolated pale scales are scattered at the bases of the terga. The sterna have a diffuse pattern of intermixed pale and dark scales.

Male: The median lobe of tergum VIII has 6–8 strong setae arranged in an irregular row, and the lobes of tergum IX usually have 7–10 slightly curved setae. The gonocoxite is elongated, and is at least 2.5 times as long as it is wide at its base (Fig. 10.125). The basal lobe of the gonocoxite has 5–8 strong setae, which are slightly inwardly curved. The paraproct is strongly sclerotised, with 1–2 teeth at the apex. The aedeagus is more or less oval, and weakly sclerotised.

Larva: The larvae of *Cs. ochroptera* resemble superficially those of *Cs. morsitans*, but they are readily distinguishable from the latter by having the inner frontal seta (5-C) with 5–9 branches (Fig. 8.85a) and the comb scales at the posterior margin of the comb with strongly sclerotised longitudinal mid ridges (Fig. 10.126). In *Cs. morsitans* the inner frontal seta is 2–3 branched and the comb scales lack sclerotised mid ridges. *Cs. ochroptera* has an antenna which is slightly longer than the head, dark pigmented at the base and in the tapering

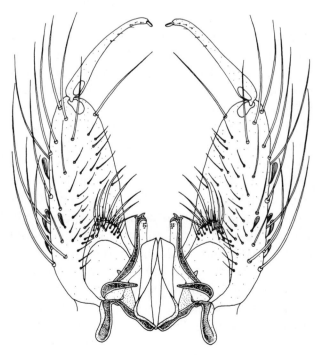

Figure 10.125.   Hypopygium of *Cs. ochroptera*.

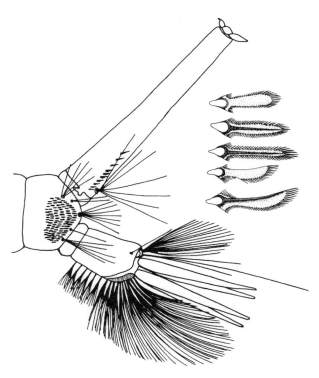

Figure 10.126. Larva of *Cs. ochroptera* and enlarged comb scales.

apical part. The postclypeal seta (4-C) has 2–4 small branches, the median frontal seta (6-C) is always 2-branched, and the outer frontal seta (7-C) has 7–13 branches. The comb has 60–95 scales of different shape, some with a dark longitudinal mid ridge. The siphonal index is 5.0–7.0, and the siphon slightly tapers apically. The pecten has 7–10 teeth; the 2–3 distal teeth are simple spines without lateral denticles and more widely spaced. The siphonal tuft (1-S) has 5–10 branches. The anal segment is longer than it is broad, and the saddle completely surrounds the segment. The upper anal seta (2-X) has 12–23 branches, and the lower anal seta (3-X) is 2–3 branched. The ventral brush has 10–22 tufts of cratal setae (4-X) and 5–8, usually 6, precratal setae. Often only one precratal seta is not situated on the saddle. The anal papillae are narrow and lanceolate, and 1.5–2.0 times longer than the saddle.

Biology: *Cs. ochroptera* is a very rare mosquito and therefore the knowledge about its biology is scanty. In central and eastern Europe the larvae can be found in peat bogs, where they sometimes appear associated with larvae of *Cs. alaskaensis*. In the east of their distribution range they can also occur in large shallow marshes, in forest ponds and ditches or at the muddy shores of lakes. It seems likely that *Cs. ochroptera* has at least two generations per year and that hibernation apparently takes place in the larval or the egg stage, but Gutsevich *et al.* (1974) reported hibernation of adult females in the eastern Ukraine. The adult females rarely bite humans; they seem to feed mainly on birds and amphibians.

Distribution: Forest zones of the Palaearctic region, from central Europe through west Siberia to northeast China. Its southernmost distribution range reaches Romania. It is also distributed from the Fennoscandian region (Finland) southeast to the Caucasus.

10.4.3. Subgenus *Culiseta* Felt

The cross veins (r-m and m-cu) are in one line or slightly separated, and the distance between them is not longer than the length of m-cu (Fig. 6.57b). If the tarsi are pale ringed, the rings are confined to the basal parts of each tarsomere. The palps of the males are as long as or longer than the proboscis, and the last two segments are distinctly swollen. The aedeagus is strongly sclerotised, and pointed towards the apex and often curved or hooked. The larvae have antennae shorter than the head, and the antennal seta (1-A) is weakly developed. The siphon is short, and the index is less than 4.0. The pecten is short and often prolonged towards the end of the siphon by a row of long, thin setae. The larvae are mainly bottom feeders, they spend most of their time submerged at the bottom of their breeding sites feeding on the substrate. The females of the subgenus hibernate as adults. They deposit the eggs in boat shaped egg rafts on the surface of the water like the species of the genus *Culex*. In the Palaearctic region the subgenus is represented by five species and one subspecies.

Culiseta (Culiseta) alaskaensis (Ludlow) 1906

Female: *Cs. alaskaensis* is a large mosquito resembling both *Cs. annulata* and *Cs. subochrea* in having a median longitudinal pale band on tergum II and conspicuous spots on the wings, resulting from aggregations of dark scales in certain areas. But unlike *Cs. alaskaensis*, the other two species are ornamented with a subapical white ring on the femora and a median white ring on tarsomeres I of all legs. In addition, the pale basal rings on tarsomeres II–IV are much less distinct than in the other two species. The proboscis and palps are mainly dark scaled with a few scattered pale scales in the basal half of the proboscis and throughout the palps. The antennae are dark brown, and the pedicel and first flagellomere have yellowish-white scales on the inner surface. The occiput has narrow curved whitish scales and dark erect forked scales on its dorsal part and broad whitish scales on its lateral part. The integument of the scutum is dark brown, with dark and whitish scales. Two indistinct longitudinal narrow stripes or patches of pale scales may be visible, and the lateral parts of the scutum are usually lighter than the median part. The scutellum has narrow whitish scales and dark setae on the lobes. The pleurites have sparse patches of narrow curved pale scales, and the prespiracular setae are yellowish. The femora anteriorly are dark brown with pale scales intermixed, and the posterior surface and apices are white scaled, with no subapical pale ring. The tibiae are dark brown with scattered pale scales, and the tarsi are dark with pale basal rings on tarsomeres II–III of the fore and mid legs and on tarsomeres II–IV of the hind legs. Tarsomere I of the hind leg is without a

median light ring (Fig. 6.62b). The wing veins have narrow dark scales, which are aggregated in some areas forming conspicuous spots. The costa (C), subcosta (Sc) and radius 1 (R_1) have scattered pale scales throughout their length. The base of the subcosta bears a dense tuft of yellowish setae on the ventral side of the wing. The terga are blackish brown with rather broad basal white bands, which are widened laterally, especially on the last segments. The sterna are white scaled, with a few dark scales intermixed.

Male (Fig. 10.127): The two hypopygial characters which distinguish *Cs. alaskaensis* from *Cs. annulata* and *Cs. subochrea* are firstly a group (2–10) of short dark stout setae located in the middle of the posterior margin of tergum VIII, and secondly the slightly convex apical lobe of the gonocoxite which is covered with short thin setae. Both characters are absent in *Cs. annulata* and *Cs. subochrea*. The lobe of

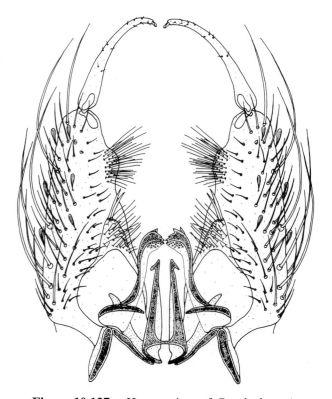

Figure 10.127.   Hypopygium of *Cs. alaskaensis*.

tergum IX has a row of long curved setae. The gonocoxite is covered with many long setae on the outer surface, and the basal lobe has 2, or rarely 3, strong setae. The gonostylus is curved and tapered apically, and the apical spine is short and bifurcated. The aedeagus is sclerotised, long, conical, tapered and hooked at the apex.

Larva: The antennae are less than half as long as the head, and the antennal seta (1-A) is multiple-branched and inserted near the middle of the antenna. The postclypeal seta (4-C) is short, thin and 3-branched. The inner frontal seta (5-C) has 5–7 branches, the median (6-C) is 2–3 branched and the outer (7-C) has 8–11 branches (Fig. 8.80a). The abdominal segment VIII has 35–50 comb scales arranged in a triangular patch (Fig. 10.128). The siphon is short and broad, slightly tapered apically, with an index of 2.5–3.0. The pecten has 6–8 spine-like teeth on the basal fifth of the siphon followed by an even row of 16–18 hair-like setae extending to near the apical quarter of the siphon. The abdominal segment X is completely surrounded by the saddle, and the saddle seta (1-X) is

inconspicuous, and much shorter than the saddle. The ventral brush is well developed with 3–4 precratal setae (4-X), at least two of them piercing the saddle. The anal papillae are of variable length, at least as long as the saddle, and pointed.

Biology: Larvae of *Cs. alaskaensis* can be found from late spring on, in a variety of habitats. They favour small open pools formed by melting snow which do not dry up in summer. These pools usually have a considerable amount of fallen leaves at the bottom and little aquatic vegetation, in the tundra they inhabit swamps. The larvae are often associated with those of *Oc. excrucians*, *Oc. flavescens* and *Oc. cantans*. In the northern parts of its distribution range *Cs. alaskaensis* is apparently a monocyclic species with one generation per year. In the more temperate southern regions several generations per year can be expected (Mohrig, 1969). The species apparently hibernates as adult females in tree cavities, caves and cellars, often together with *Cx. p. pipiens*. The females leave their winter habitats usually earlier than other mosquito species. *Cs. alaskaensis* frequently feeds on humans. In the tundra zones it is well known as the large snow-melt mosquito that readily attacks man and reindeer in early spring.

Distribution: *Cs. alaskaensis* is a holarctic species that is typical for the boreal and tundra zones of Fennoscandia, Siberia and Alaska. In the Palaearctic it is distributed from Britain and Norway in the west to the far East. In central Europe its southern distribution stretches towards the northern slopes of the Alps. In this part of its distribution range, *Cs. alaskaensis* is usually restricted to the higher mountains. Outside Europe it can be found in mountainous regions of Iran, Pakistan and northern India.

Note on systematics: A subspecies of *Cs. alaskaensis*, ssp. *indica* Edwards 1920, has been described with a distribution range from the Caucasus to Middle Asia and southern India. Its general colouration is much lighter than the nominate form, the scutum is covered with

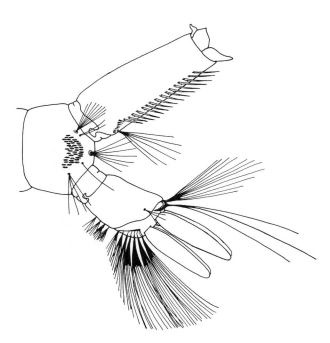

Figure 10.128. Larva of *Cs. alaskaensis*.

golden yellowish scales and the pale basal bands on the terga are often broader. The male genitalia and larvae do not distinctly differ from the nominate form.

Culiseta (Culiseta) annulata (Schrank) 1776

Female: *Cs. annulata* is a large, dark brown mosquito with whitish markings on the abdomen and the legs. It can be distinguished from *Cs. alaskaensis* by the presence of subapical white rings on the femora and conspicuous white rings in the middle of tarsomeres I. *Cs. annulata* is closely related to *Cs. subochrea*, the characters separating the two species are given in the description of the latter. The proboscis of *Cs. annulata* is speckled with pale and dark scales, which are darker in the apical part, and the labellum is dark brown. The clypeus is dark brown, the palps are dark with scattered pale scales, which are especially abundant at the apices and with a conspicuous pale spot at the joints of palpomeres II and III. The antennae are dark brown, and the pedicel has a few whitish scales on the inner surface. The head has pale narrow scales and dark erect forked scales on the occiput, and the eyes are bordered with yellowish white scales and dark, stout setae. The scutum has narrow dark brown and pale scales. The posterior submedian area has two pale spots, and the prescutellar area has whitish scales. The scutellum is brown with whitish scales and black setae, and the postnotum is brown or dark brown. The pleurites have patches of broad, whitish scales, and the postpronotum is predominantly pale scaled. Hypostigmal, subspiracular and postspiracular patches are present. The mesepimeral patch of scales almost reaches the lower margin of the mesepimeron. Prespiracular setae and lower mesepimeral setae are present. The legs have dark brown scales and conspicuous white rings. The femora are predominantly dark scaled with scattered pale scales, distinct white subapical rings and pale knee spots, and the tibiae have pale and dark scales intermixed basally and are white scaled apically. Tarsomere

I has a noticeable white ring in the middle and white rings also at the bases of tarsomeres II to IV, and tarsomeres V of all the legs are entirely dark scaled (Fig. 6.62a). The wings are largely covered with dark scales, which are aggregated to form distinct dark spots at the base of R_s, at the cross veins and the furcations of R_{2+3} and M. Some scattered pale scales are mainly found in the basal part of the costa (C), subcosta (Sc) and radius (R). The cubitus (Cu) is entirely dark scaled. The cross veins (r-m and m-cu) usually form a straight line (Fig. 6.63a). The abdominal terga have whitish basal bands, and the apical parts are uniformly dark scaled. Tergum II has a narrow basal band and a characteristic longitudinal median white band. Tergum VIII is predominantly pale scaled. The sterna have yellowish white scales.

Male: The posterior margin of tergum VIII is usually without stout setae, or occasionally a few setae may be present. The lobe of tergum IX has 8–12 hair-like setae. The gonocoxite is conical, gradually tapered towards the tip (Fig. 10.129). The basal lobe of the gonocoxite is well developed, with 2 (rarely 3) strong setae conspicuously stouter than the rest. The apical lobe is usually absent or indistinct. The gonostylus is long and slender, with a short

Figure 10.129.   Hypopygium of *Cs. annulata*.

apical spine. The paraproct is strongly sclerotised, recurved, with apical teeth, and the sclerites of the aedeagus are separated and also strongly sclerotised.

Larva: The head is broader than it is long, and the antennae are less than half as long as the head and straight. The antennal seta (1-A) is inserted just beyond the middle of the antennal shaft, with 10–15 branches, not reaching to the apex of the antenna. The distance between the postclypeal setae (4-C) is about the same as the distance between the inner frontal setae (5-C) (Fig. 8.82a). 5-C has 4–8 branches, the median frontal seta (6-C) has 1–3 branches, and the outer seta (7-C) has 6–14 branches (Fig. 8.78a). The comb usually has 35–50 scales, rarely more (Fig. 10.130). The individual scales are slightly narrowed in the middle part, blunt ended and uniformly fringed with small spines. The siphon distinctly tapers apically, with a siphonal index of 3.2–4.0. The siphonal tuft (1-S) is inserted close to the base of the siphon, usually with 9–10 branches, and is

about as long as the width of the siphon at the base. The pecten has 11–18 spine-like teeth, followed by a row of 11–21 thin, hair-like setae occupying approximately 2/3 the length of the siphon. The saddle completely surrounds the anal segment, and its ventral surface is only half as long as its dorsal surface. The saddle seta (1-X) is much shorter than the saddle, usually with 3 branches. The upper anal seta (2-X) has 13–19 branches, and the lower anal seta (3-X) has 3 branches. The ventral brush usually has 16–18 cratal setae (4-X) and 2–3 precratal setae, one or two of which perforate the saddle. The anal papillae are lanceolate, and usually as long as the saddle.

Biology: In central Europe the larvae usually occur from early spring onwards. The population increases in the summer months and reaches its maximum in September. Depending on the latitude of its occurence, *Cs. annulata* may have one to three generations per year. The eggs are laid in rafts, which are composed of appoximately 200 eggs. The species breeds in a wide variety of permanent and semipermanent habitats including natural and artificial water recipients, both in open and shaded situations. Larvae can be found in stagnant pools, ponds, ditches, water troughs and other artificial containers such as barrels collecting rain water. Dense populations could be found in tanks containing manure water, so it seems probable that a high content of nitrogen provides an additional attraction for the females to lay their eggs (Mohrig, 1969). The larvae are able to tolerate a high content of salinity and can also be found in brackish water (Marshall, 1938). Aitken (1954) found larvae breeding in a tree hole in Corsica. In artificial containers they are often found in association with those of *Cx. p. pipiens*. In natural breeding sites the larvae occur together with those of *Cs. subochrea* and *Cs. morsitans* (Natvig, 1948). Hatching of *Cs. annulata* larvae takes place 3–5 days after egg deposition, the larval development depends on the temperature. Martini (1931)

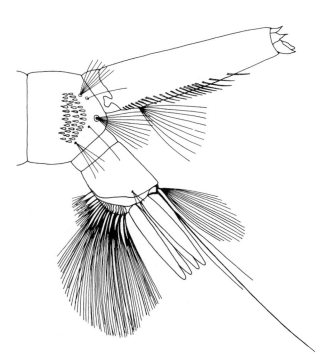

Figure 10.130. Larva of *Cs. annulata*.

estimated the overall time from egg deposition to the emergence of the adult being 18 days at a temperature of 20–23°C and 16 days at a temperature of 24–27°C; above 31°C no larvae survived. Usually the species hibernates in the adult stage and the first females occur in early spring when they leave their winter shelters. At this time they readily attack man and mammals during the day, but in the summer months they show a more nocturnal biting activity and frequently enter houses or stables to feed on humans or domestic animals. Occasionally they may also take their blood-meal from birds. In former Yugoslavia the females could be sampled more efficiently in CO_2 baited traps than on man (Petric, 1989). The females of *Cs. annulata* can often be found inside houses even during the day from early autumn on, when they search for their winter habitats. They hibernate in cellars, attics of dwellings or in sheds of domestic animals, where they can be extremely annoying during winter time, when the hibernation is interrupted by rising temperatures or humidity. Winter habitats can also be found far away from human settlements in tree cavities, stacks of wood or other natural shelters. When the winter is mild, or in the southern range of its distribution, hibernation can also take place in the larval stage.

Distribution: *Cs. annulata* is widely distributed throughout Europe, but it is more common in the north than in the south, where it seems to be largely replaced by *Cs. longioreolata* (Edwards, 1921). The distribution range of *Cs. annulata* extends into north Africa, Asia Minor and southwest Asia.

Medical importance: The species is known to be a potential vector of Tahyna virus (Ribeiro *et al.*, 1988) and transmitter of some plasmodia of birds (Gutsevich *et al.*, 1974).

Culiseta (Culiseta) bergrothi (Edwards) 1921

Female: The integument is dark brownish with some lighter pleural parts. The colour of the light scales is variable, from white to creamy white. The antennae have some white scales on the pedicel, and the eyes are bordered with white scales. The proboscis is covered with black scales, and the palps are dark with scattered pale scales. The vertex has white flat scales, and the occiput has upright narrow golden scales and brown setae. The scutum is covered with narrow bronze scales and a more or less distinct varying pattern of white scales and bronze brown setae. White broader scales cover the front, sides and back of the scutum in an irregular pattern and form a narrow, incomplete acrostichal stripe. Dorsocentral stripes of narrow white scales and a postacrostichal patch of white scales are usually present forming a scutal pattern similar to that of *Cs. subochrea*. However, submedian stripes are not present. The scutellum is covered with whitish scales and light setae. The antepronotal and propleural setae are long and dominating, the postpronotum has narrow bronze and some lower, broader whitish scales. The mesepisternal patch is large with elongated white scales in its upper part, and the mesepimeron is covered with scales in its upper third. Usually not more than 15 prespiracular setae and not more than 10 lower mesepisternal setae are present (Fig. 6.61a). The coxae of the fore and hind legs have long, conspicuous setae. The femora have white scales on the ventral surface and a mixture of white and dark scales on the dorsal surface. The tibiae are dark with either light dorsal stripes or mixed scales, especially on the front legs, and the knee spots are white. Tarsomeres I–V of all the legs are dark, with occasionally some white scales on tarsomere I; the claws are simple. The wing veins are covered with elongated, dark scales, and spots are always present, although often inconspicuous. The spots result from aggregations of dark scales at the furcations of R_{2+3} and M and the cross veins (r-m and m-cu). The lining up of r-m and m-cu can vary from nearly a straight line to a slight displacement of m-cu towards the wing root, but never distal to r-m. The abdomen is dark with pale basal bands on the terga, which are narrower in the last segments. The sterna are pale with some scattered dark scales.

Male: Somewhat smaller than the males of other northern European *Culiseta* species. They share the feature of having an extremely long tarsomere I of the fore leg (exceeding the length of tarsomeres II–V of about one fifth) with males of *Cs. fumipennis*. The median lobe of tergum VIII has 4–18, usually more than 10, strong spine-like setae (Fig. 7.71b). The gonocoxite has long and numerous setae laterally and apically (Fig. 10.131). The basal lobe is weakly developed, with two strong spine-like setae of almost equal length. The apical lobe is indistinct, and covered with dense setae. The gonostylus is long, more than half as long as the gonocoxite, and the apical spine is blunt ended. The paraproct is strongly sclerotised, with a curved apex, bearing 3–4 teeth. The aedeagus is oval shaped, slightly sclerotised in the lateral parts with sigmoid, pointed tips.

Larva: The head is considerably broader than it is long, and the antennae are about half as long as the head. The antennal seta (1-A) has 4–6 branches situated slightly proximal to the middle of the antennal shaft (Fig. 8.83c). The frontal setae 5-C and 6-C are nearly in a row, with multiple branches. The prothoracic setae are all very long. Numerous comb scales are arranged in a triangular spot (Fig. 10.132). Each scale has a long, narrow stem, which is rounded apically and fringed with small spines. The siphon is slightly tapered apically, with a siphonal index of 3.0–3.5. The pecten has 12–15 spine-like teeth, and each tooth has 2–3 slender lateral denticles, followed by a row of longer hair-like setae which reach distally about 2/3 the length of the siphon. The anal segment is entirely encircled by the saddle, and the ventral brush has 3–4 precratal setae

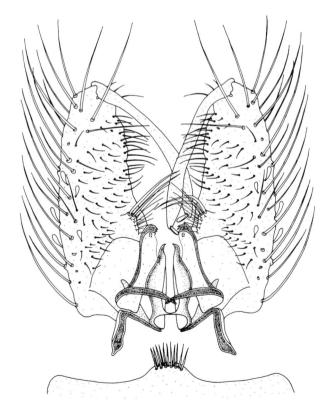

Figure 10.131. Hypopygium of *Cs. bergrothi*.

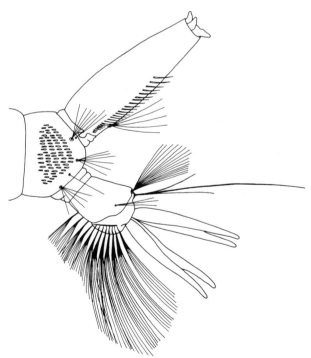

Figure 10.132. Larva of *Cs. bergrothi*.

(4-X), usually two of them piercing the saddle. The anal papillae are very long and slender.

Biology: In the northern parts of its distribution range the species breeds in open pools and has only one generation per year. The larvae inhabit swamps and pools in the southern tundra and shaded forest pools in coniferous and mixed forests, they may also tolerate polluted water. In the southern parts of its distribution range several generations per year may occur. Hibernation takes place in the female stage (Gutsevich *et al.*, 1974) and larvae can be found together with other early spring species, such as *Oc. communis*. In Scandinavia the larvae have been recorded until the summer months. The females are most abundant from June to August. They predominantly bite cattle and other mammals and seldom enter houses.

Distribution: This northern species is found throughout the tundra and northern taiga zone of the Palaearctic, as well as in high mountainous areas both in Scandinavia (Natvig, 1948) and in Japan (Tanaka *et al.*, 1979).

Culiseta (Culiseta) glaphyroptera (Schiner) 1864

Female: *Cs. glaphyroptera* is very similar to *Cs. bergrothi*, the other member of the subgenus *Culiseta* with entirely dark tarsomeres. However, it is easy to distinguish males and females of these two species (Natvig, 1948). In *Cs. bergrothi* the eyes are distinctly bordered with light scales and the palps are dark with scattered pale scales. On the other hand, in *Cs. glaphyroptera* the eyes are not bordered with light scales and the palps are entirely dark. The head has whitish scales laterally, the vertex has a tuft of short light scales between the eyes, and the occiput in the posterior part has upright dark brown scales. The proboscis and palps are dark brown. The integument of the scutum is dark brown with golden brown sickle-shaped scales, which form a narrow acrostichal stripe on the anterior part of the scutum and two dorsocentral and lateral, more indistinct, stripes.

The scutellum is dark with scales of the same colour as on the scutum. The pleurites of the thorax have patches of whitish flat scales. The lobes of the antepronotum are blackish brown with pale narrow scales. The number of prespiracular setae usually ranges from 16 to 22 and that of the lower mesepisternal setae from 12 to 18 (Fig. 6.61b). The legs are dark scaled, and the posterior surface of the femora and tibiae has light scales, which sometimes form a longitudinal line; the knee spot is distinct and yellowish white. The tarsomeres are entirely covered with dark brown scales, and pale rings are absent (Fig. 6.60a). The wing veins are dark scaled, and the cross veins are slightly separated. Wing spots are absent or only weakly developed. When present, the spots result from aggregations of dark scales at the furcations of R_{2+3} and M and the cross veins (r-m and m-cu). The terga are dark brown with indistinct bands of light brown to pale scales in the basal third of each tergum. The sterna are more or less uniformly covered with yellowish white scales, or sometimes narrow bands of brownish scales are present at their posterior margins.

Male: The lobe of tergum IX has numerous long, thin setae. The gonocoxite is relatively short, and the basal lobe is well developed with two long and prominent setae which are strongly bent in the middle (Fig. 10.133). The apical lobe has a characteristic tuft of long scale-like setae. The gonostylus is long and expanded subapically, and the apical spine of the gonostylus, unlike in other species of the subgenus *Culiseta*, is long. The aedeagus is cylindrical in its proximal part, tapered beyond the middle, and the tips are strongly sclerotised.

Larva: The head is broader than it is long, and the antennae are 2/3 the length of the head, and slightly curved inwardly. The antennal seta (1-A) is inserted close to the middle of the antenna and is about half as long as the antennal shaft, usually with 10 branches (Fig. 8.83a). The postclypeal setae (4-C) are situated

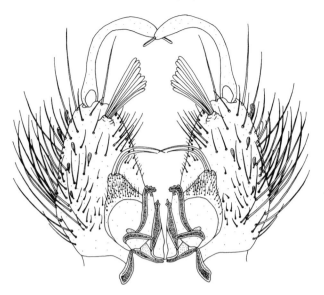

Figure 10.133. Hypopygium of *Cs. glaphyroptera*.

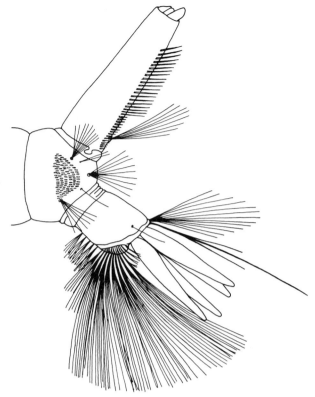

Figure 10.134. Larva of *Cs. glaphyroptera*.

before the inner frontals and are close together. The frontal setae are well developed; the inner frontals (5-C) have 7–9 branches reaching to the labrum, the median (6-C) have 5–7 branches and the outer (7-C) are 8–12-branched. The comb consists of approximately 70 scales, which are arranged in a crescent shaped patch (Fig. 10.134). The comb scales are elongated with a broad base and rounded apex. The siphonal index is 3.4–3.5. The pecten has 15–21 spine-like teeth followed by a row of 19–22 longer hair-like setae reaching to the apical quarter of the siphon. The siphonal tuft (1-S) is inserted close to the base of the siphon, with 8 branches which are distinctly longer than the width of the siphon at the point of its origin. The anal segment is completely surrounded by the saddle, which is ventrally half as long as it is dorsally. The saddle seta (1-X) is short with 1–2 branches. The upper anal seta (2-X) usually has 14 branches, forming a fan, and the lower anal seta (3-X) has 3–4 branches. The ventral brush has 13–16 tufts of cratal setae (4-X) on the common base, with five precratal setae, three of which pierce the saddle. The anal papillae are 1.5–2.5 times longer than the saddle, and pointed.

Biology: *Cs. glaphyroptera* can only be found in mountainous regions. The larvae are mostly restricted to partly shaded cool breeding sites. They usually occur in the beds of small mountain rivers or streams and in tubs of larger rocks, where the water remains after the first flood in spring. They prefer water with fallen leaves and a rich amount of detritus rather than a clean stony bottom with a poor development of algae and other organisms. Occasionally the larvae can be found in ground pools close to small brooks and sources together with *Cx. territans*. Apfelbeck (1928) reported common findings of *Cs. glaphyroptera* larvae in tree holes of mountain forests in Yugoslavia, where they were found in association with typical tree holes breeders such as *An. plumbeus* and

Oc. geniculatus during the summer months. Later in the year, at the end of November, the larvae disappeared, but adults could be found in sheltered situations, such as in caves. *Cs. glaphyroptera* hibernates in the adult stage. Little is known about the feeding behaviour of the females. It seems likely that they feed on birds or small mammals which inhabit the forest.

Distribution: The species is widely distributed throughout the mountains of central and southeastern Europe. In addition it has been found in the mountainous regions of western Ukraine and Crimea. It was thought that *Cs. glaphyroptera* must also be present in the northern parts of Europe, but Natvig (1948) examined material from Scandinavia and clearly demonstrated that all specimens formerly referred to as *Cs. glaphyroptera* were in fact *Cs. bergrothi*. As a consequence, *Cs. glaphyroptera* was erased from the list of Fennoscandian mosquitoes (Dahl, 1997).

Culiseta (Culiseta) subochrea
(Edwards) 1921

Female: *Cs. subochrea* is very similar to the closely related *Cs. annulata* in all stages and only the differences between the two species are mentioned here. The scutal integument and most of the scutal scales are yellowish in *Cs. subochrea*, whereas in *Cs. annulata* the integument is more brownish and the pale scutal scales are creamy white. The legs of *Cs. subochrea* are more speckled than those of *Cs. annulata* due to a greater number of pale scales scattered on the femora, tibiae and tarsomeres I, hence in the general appearance the contrast of pale and dark colour of the legs is not so conspicuous. The subapical white rings on the femora are less distinct and the yellowish white basal rings on the tarsi are much broader than in *Cs. annulata*. The spots on the wings formed by aggregated dark scales are less obvious in *Cs. subochrea* and in addition to some pale scales on the costa (C), subcosta (Sc) and radius

(R), more or less numerous pale scales are scattered along the cubitus (Cu). In *Cs. annulata* vein Cu is entirely dark scaled. In *Cs. subochrea* the cross veins r-m and m-cu are usually slightly separated, the distance between them not being longer than the vein m-cu (Fig. 6.63b). In *Cs. annulata* the cross veins usually form a straight line. The terga of *Cs. subochrea* have indistinct pale bands formed by yellowish scales, the dark areas in the apical half of the terga are scattered with more or less numerous yellowish scales. In *Cs. annulata* the terga have more distinct whitish basal bands and the apical half of the terga lack pale scales, they are uniformly dark scaled.

Male (Fig. 10.135): In *Cs. subochrea* the basal lobe of the gonocoxite carries 3–5 strong setae conspicuously stouter than the rest and the median lobe of tergum VIII usually bears several stout setae. In *Cs. annulata* the number of strong setae on the basal lobe is 2, rarely 3, and the median lobe of tergum VIII is usually devoid of stout setae. Furthermore, Peus (1930) referred to the number of thin setae on the lobe of tergum IX, which ranges from 13–24 in *Cs. subochrea* and from 8–12 in *Cs. annulata*. Unfortunately, these characters are variable and overlap to a certain degree, thus they cannot be precisely used in all cases, especially if

Figure 10.135.   Hypopygium of *Cs. subochrea*.

the identification is based on a single hypopygium with three strong setae on the basal lobe. In this case it is most likely that the specimen belongs to *Cs. subochrea* when the median lobe of tergum VIII bears several stout setae and to *Cs. annulata*, when the stout setae are absent.

Larva: In *Cs. subochrea* the distance between the postclypeal setae (4-C) is less than the distance between the inner frontal setae (5-C) (Fig. 8.82b), whereas in *Cs. annulata* both pairs of setae are usually about the same distance apart. Again, the two ranges of variation overlap considerably and it is not always possible to separate larvae of the two species according to this character. No other structural differences between the larvae of the two species seem to exist, but Ribeiro *et al.* (1977) pointed out that in material from Portugal, the length of the siphonal tuft (1-S) is distinctly shorter than the width of the siphon at the base in *Cs. subochrea*, and the length of 1-S is about the same as the width of the siphon at its base in *Cs. annulata*. So far this observation has not been verified with material from other locations.

Pupa: A character which seems to be generally constant and enables one to distinguish between *Cs. subochrea* and *Cs. annulata* with certainty is found in the pupal stage. The minute denticles which fringe the pupal paddle are long and pointed in the former species and considerably shorter and blunt ended in the latter one.

Biology: Due to the rare findings of *Cs. subochrea* in western and central Europe, the biological data are scanty, but the biology seems to be similar to that of *Cs. annulata*. Hibernation usually takes place in the adult female stage preferably in farm buildings and cellars, but in its southern distribution range the larvae may also hibernate and the dormant phase of the females does not last very long. Several generations per year may occur. As in *Cs. annulata*, the larvae can be found both in fresh water habitats, like ditches, ponds or garden tanks, and habitats with a varying degree of salinity up to more than 1/3 that of sea water (Marshall, 1938). Larvae were found in rice fields with clean water, associated with those of *An. atroparvus* and *Cx. pipiens* (Ribeiro *et al.*, 1988), but they exhibit a remarkable preference for saline breeding sites (Gutsevich *et al.*, 1974; Mohrig, 1969; Rioux, 1958). The females bite man and domestic animals and they are more exophagic than those of *Cs. annulata*. They were reported to feed on man during the day and far away from dwellings (Rioux, 1958). The mating behaviour is oriented to stenogamy. An autogenous egg laying of one female reared in the laboratory was observed (Marshall, 1938).

Distribution: *Cs. subochrea*, a Palaearctic species, can be found in nearly all European countries, reaching from southern Fennoscandia to the Mediterreanean region including western and central Europe, but it is not a very common mosquito. Its range stretches into north Africa and the near East and it is widely distributed in middle Asia.

Note on systematics: The taxonomic status of *Cs. subochrea* is still controversial. It was originally regarded as being merely a variety (desert form) of *Cs. annulata*, but then rose to species rank due to its distribution not being confined to desert areas and the conspicuous colouration of the adults (Edwards, 1921). Likewise, Peus (1930) referred to several hypopygial, pupal and larval characters differentiating the two species and this was confirmed by Marshall (1938). Later it was brought back to subspecies status of *Cs. annulata* by Maslov (1967), in accordance with the opinion of other Russian authors (Shtakelberg, 1937; Monchadskii, 1951). This change was based on the comparison of characters and their variability of both forms in almost all of the distribution range. The authors stated an overlap in distribution of the subspecies and its nominative form in western and central Europe (Gutsevich *et al.*, 1974). Based on material from Portugal,

Cs. subochrea was reranked as a valid species by Ribeiro *et al.* (1977) and this was followed by others (Ward, 1984). This opinion is supported by the fact that larvae of both forms could be found sympatrically in the same breeding sites in many localities of western and central Europe (Rioux, 1958; Aitken, 1954; Peus, 1951). The specific characters were always kept and intermediate forms are not known, indicating that the gene flow between both forms is inhibited. Further research, based on morphological and genetical studies, demonstrated that *Cs. subochrea* should be regarded as a distinct species (Cranston *et al.*, 1987).

10.5. GENUS *COQUILLETTIDIA* DYAR

The proboscis of the females is moderately long (about 1.5 times longer than the thorax), and uniformly broad. The palps are short, 1/4 the length of the proboscis or shorter. The vertex is covered with numerous erect forked scales. The acrostichal, dorsocentral and lateral setae of the scutum are well developed. The scales on the scutum are usually narrow, and decumbent. Prespiracular setae are absent in all subgenera, and postspiracular setae are absent in the subgenus *Coquillettidia*. The upper mesepisternal setae are well developed, and upper mesepimeral setae are usually present. The mesepisternal and mesepimeral patches are small, with decumbent pale scales. The legs usually have pale rings. Tarsomere I of the hind legs is shorter than the hind tibia. The claw lacks a subbasal tooth, and typical pad-like pulvilli are absent. The wing veins are covered with a mixture of pale and dark, narrow and broad scales. The abdomen is truncated, and segment VIII is short and broad. The cerci are short and blunt. The palps of the males are about as long as the proboscis, and the apical segment is generally covered with numerous long setae. The fore and mid legs have a pair of claws, unequal in shape and size. Abdominal segment VIII has a distinctly sclerotised tergum, and segment IX is bilobed with sparse long setae. The basal lobe of the gonocoxite is small but distinct, and tapered apically. It bears one or more setae as long or longer than the lobe. The gonostylus is usually short, enlarged apically with a short apical spine. Claspettes are absent, and the apex of the paraproct is pointed or denticulated. The head of the larva is much wider than it is long, and slightly sclerotised. The antennae are extremely long, at least 1.5 times as long as the head. The part beyond the articulation of the antennal seta (1-A) is slender and whip-like. Abdominal segment I has the lateral tracheal branches extended into a pair of heart shaped air-sacs which lie partly in the metathorax. The siphon is short, conical, and strongly sclerotised. The lobes of the spiracular openings are dark, sclerotised, and folded together. They form a tapered saw-like piercing apparatus for penetrating aquatic plant tissue in order to obtain oxygen. A similar modification of the siphon is characteristic for larvae of some other none-European genera or subgenera, *e.g. Culex (Lutzia)*, *Mimomyia (Mimomyia)* and *Hodgesia* (Gillett, 1972). Abdominal segment X is long and slender, and the saddle completely encircles the segment.

Larvae of all members of the genus obtain the oxygen from the submerged parts of aquatic plants by penetrating the plant tissue with their highly specialised siphon. The pupae are also attached to plant tissues in order to take oxygen from the aerenchyma by their sclerotised, pointed respiratory trumpets. The cuticle of the trumpets is softened subapically at a line of weakness, and brakes off prior to the time of adult emergence. The females lay the eggs on the water surface in rafts of variable shape.

The species are grouped in three subgenera, *Austromansonia* (only one representative from New Zealand), *Rhynchotaenia* (13 species mostly recorded from the Neotropical, some from the southern Nearctic region) and *Coquillettidia*. The last subgenus contains 43 species, more than half of them are distributed in the Ethiopian region, the rest in the Oriental and Australian regions, one in the Nearctic region. Only two species of the subgenus occur in the Palaearctic region, these are *Cq. buxtoni* and *Cq. richiardii*.

10.5.1. Subgenus *Coquillettidia Dyar*

Coquillettidia (Coquillettidia) buxtoni (Edwards) 1923

Female: Readily distinguished from *Cq. richiardii* by the presence of exclusively or mostly dark coloured scales on the proboscis, palps, wing veins and tarsomeres. The scales are blackish brown pigmented with structural metallic purple. The head has pale decumbent scales and brown upright scales. The scutum has golden brown scales, which are paler around the wing bases and at the posterior margin. The femora and tibiae are dark scaled with scattered ochre coloured scales. The latter form indistinct longitudinal stripes on the ventral surface and a spot at the apex. The dark scales on the wing veins are narrower than in *Cq. richiardii* but still broader than in members of the genus *Culex*. The terga are mainly dark coloured with a violet hue; whitish scales are restricted to triangular basolateral patches, sometimes connected by narrow basal bands, and the sterna have pale basal bands.

Male: The palps are longer than the proboscis. The hypopygium is quite similar to that of *Cq. richiardii*. The lobes of tergum IX possess 4–5 thin setae. The main difference between the two species is found in the shape of the gonostylus. In *Cq. buxtoni* the basal half of the gonostylus is stem-like, not constricted in the middle. The apical half is considerably bulky and then gradually narrows towards the apex. The outer margin of the gonostylus bears four small and tiny setae apically (Fig. 10.136b).

Larva: Closely resembles that of *Cq. richiardii*. The head is about 1.5 times wider

than it is long. The postclypeal seta (4-C) is multiple-branched. The inner frontal seta (5-C) is short and 8-branched. The median frontal seta (6-C) has 5–7 and the outer frontal seta (7-C) has 9 branches. The comb has 16–22 scales arranged in an irregular row, dorsally in a partly doubled row. The individual scales have a pointed terminal spine. Seta 1-VIII has 5–7, usually 6 branches (Fig. 8.87b), setae 2-VIII and 3-VIII are 2-branched, and 4-VIII and 5-VIII are usually 3–4 branched. Setae 2-VIII, 4-VIII and 5-VIII are at most half as long as seta 3-VIII. The abdominal segment X resembles that of *Cq. richiardii* but the saddle is covered with numbers of rows containing 2–8 spicules on a common base (Fig. 10.137b). The saddle seta (1-X) is 4 branched.

Biology: Due to its limited distribution range and rare findings the data on the biology of the species are scanty. The larvae were found attached to the roots of *Acorus* sp. and *Typha* sp. (Coluzzi and Contini, 1962). Eggs are laid in boat shaped rafts (Guille, 1975). The females bite man in open areas (Gutsevich *et al.*, 1974).

Distribution: Mediterranean subregion of the Palaearctic. In Europe the species is present in Spain, France and Italy, and also reported from Romania and Ukraine (Snow and Ramsdale, 1999).

Coquillettidia (Coquillettidia) richiardii (Ficalbi) 1889

Female: The scales on the wing veins are much broader than those of any other European species (Fig. 6.6b). The apex of the proboscis is slightly broader and distinctly darker than the preceeding portion, and sometimes the pale scales form a median ring. The base of the proboscis has intermixed yellowish and brown scales, sometimes with the dark scales predominating. A pale ring is present in the middle of tarsomere I of all the legs, a pattern similar to that of *Cs. annulata* and *Cs. subochrea*. The palps are short, not exceeding 1/4 the length of the proboscis, and are covered with mixed yellowish and brown scales. The vertex has yellowish golden narrow, curved, decumbent scales and dark upright forked scales. The scutum is brown coloured, with narrow, curved brown and golden scales. The mesepisternal and mesepimeral patches have broad, whitish scales. The femora and tibiae are basally sprinkled with yellowish and brown scales, and apically pale scaled. Tarsomere I of all the legs has a pale ring in the middle, which is sometimes indistinct or absent. If so, the legs are mainly covered with pale scales. Broad pale basal rings are usually present on tarsomeres I–III of the fore legs and all tarsomeres of the mid and hind legs. The pale rings are particularly distinct on the hind tarsomeres. The wing veins are covered with broad, intermixed yellowish and brown scales. The terga are brown scaled, with scattered pale scales more numerous at their bases. Basolateral triangular patches of yellowish scales are present, and the scales may form inconspicuous basal bands which are constricted in the middle similar to those of *Oc. punctor*. The sterna are pale scaled.

Male: The palps are nearly as long as the proboscis. The lobes of tergum IX have 8–10 setae. The gonocoxite is short and stout. The basal lobe is heavily sclerotised with a strong rod-like spine. The gonostylus is widened basally just above the articulation to the gonocoxite, distinctly constricted and flexed in the middle, enlarged again at the beginning of the distal third and then tapered apically (Fig. 10.136a). The outer side of the gonostylus has 6–7 tiny setae and two more setae on its inner side, close beyond the middle. The apical spine of the gonostylus is short. The apex of the paraproct is strongly sclerotised and denticulated.

Larva: The head is wider than it is long. The antennae are very long, 1.5–2 times longer than the head. The long terminal filament is hardly visible on a white background. The

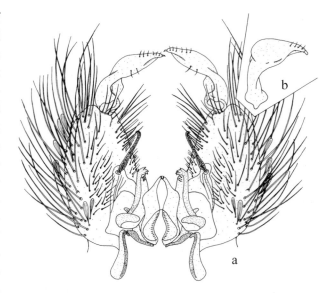

Figure 10.136. Hypopygium of (a) *Cq. richiardii* and gonostylus of (b) *Cq. buxtoni*.

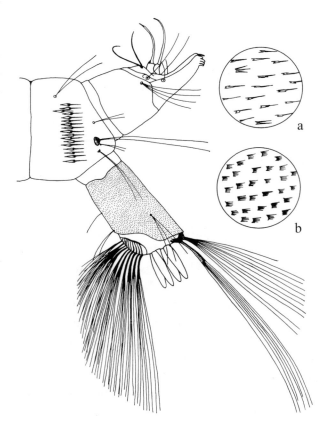

Figure 10.137. Larva of *Cq. richiardii* and spicules on the saddle of (a) *Cq. richiardii*; (b) *Cq. buxtoni*.

antennal seta (1-A) has 15–20 branches. The postclypeal seta (4-C) usually has 5–6 branches, situated anterior to the frontal setae. The inner frontal seta (5-C) is short, matching the postclypeal seta. The median frontal seta (6-C) is long, with 4–5 branches. The outer frontal seta (7-C) has 9 branches. The comb consists of an irregular row of 10–25 scales; each individual scale has a well developed terminal spine (Fig. 10.137). Seta 1-VIII is inserted into the dorsal half of segment VIII, with 2–4 branches. Setae 2-VIII, 3-VIII, 4-VIII and 5-VIII are articulated medioventrally and 2–4 branched. The siphon is very short and conical, and forms a piercing apparatus. The siphonal tuft (1-S) is inserted ventrolaterally near the middle of the siphon, and pecten teeth are absent. In addition to 1-S, two pairs of simple setae and two pairs of curved spine-like setae with hooked ends are present which support the penetration of the siphon into the plant tissue. Abdominal segment X is elongated, much longer than it is wide, and is completely encircled by the saddle. The saddle is covered with short and stout, usually simple spicules, rarely two or three on a common base (Fig. 10.137a). The saddle seta (1-X) is 2–3 branched, inserted quite apart from the posterior margin. The upper (2-X) and lower (3-X) anal setae are multiple-branched, 2-X is half as long as 3-X. The cratal setae (4-X) have 10–14 tufts and 2 precratal setae (4-X) are widely separated. The anal papillae are lanceolate, subequal in length, and shorter than the saddle.

Biology: The species has one generation per year in the north (Service, 1969) and 2–3 generations in the south (Gutsevich et al., 1974). The females deposit the eggs in rounded rafts. The larvae hatch in intervals of up to two weeks after oviposition (Guille, 1975) and usually hibernate in the 3rd or 4th-instar. Larvae and pupae live submerged and obtain oxygen from the aerenchyma of aquatic plants and move very little. Breeding sites may be various permanent water bodies rich in *Acorus* sp.,

Typha sp., *Phragmites* sp., *Glyceria* sp., *Sparganium* sp., *Ranunculus* sp. and *Carex* sp. (Shute, 1933; Natvig, 1948; Guille, 1976). Pupation takes place at the end of May or beginning of June. In Yugoslavia, blood searching females have been registered from June to September, but usually have a seasonal peak during July (Petric, 1989). Females can be very numerous and severe molesters of humans and domestic animals, in the surroundings of fresh waters or slightly saline marches, lakes, old river beds and estuaries. Also frequent indoor feeding on man has been registered in England (Shute, 1933) and occasionally in Portugal (Ribeiro *et al.*, 1988). Nuisance is usually restricted to surroundings of breeding sites but the females can use ascendent air currents to invade, in considerable number, areas up to altitudes of 800–900 m (Gilot *et al.*, 1976). Females prefer to feed on mammals (Service, 1968c; Ribeiro *et al.*, 1988; Petric, 1989) but could also take their blood-meal from birds (Service, 1969) and amphibians (Shute, 1933). Jaenson *et al.* (1986b) found the species in horse stables and human bait collection. In England, the peaks of biting activity occurred after sunset and just after sunrise (Shute, 1933; Service, 1969), while a nocturnal biting activity was typical for a population from Yugoslavia (Petric, 1989). Biting activity could be registered at a temperature between 9 and 26°C and a relative humidity between 30 and 92%. Swarming of males could be observed 1 hour after sunset and at dawn (Marshall, 1938). In the laboratory, copulation was observed in small cages of 40 × 40 × 120 cm. The species has been reported being autogenous, but some females may be unable to develop the first egg batch without taking a blood-meal (Guille, 1975).

Distribution: *Cq. richiardii* is a common species throughout Europe and widely distributed in the western Palaearctic region.

Note on systematics: Guille (1975) stated that some authors consider the nearctic species *Cq. perturbans* as a geographic race of *Cq. richiardii*. Similarity in both morphological and biological characters of the two species should stimulate further investigation.

Medical importance: Females infected with West Nile virus (WN), and Omsk haemorrhagic fever virus (OHF) were detected in wild populations (Detinova and Smelova, 1973). Acording to the same authors, *Cq. richiardii* can transmit OHF and tularemia under laboratory conditions.

10.6. GENUS *ORTHOPODOMYIA* THEOBALD

The palps of the females are 1/3 as long as the proboscis. The vertex is covered with erect forked scales. The postpronotum usually has two setae, and postspiracular setae are absent. Tarsomeres I on the fore and mid legs are distinctly longer than the other four tarsomeres together. The combined length of tarsomeres IV and V is shorter than tarsomere III of the fore and mid legs. The abdomen is parallel sided with a truncated end, the cerci are blunt and moderately projecting. The palps of the males are slender, and about as long as the proboscis. The terminal palpomere (sometimes also the subterminal) is greatly reduced without dense, long setae. The hypopygium is similar to those of the genus *Culiseta* and subfamily Anophelinae. The general colouration of the larvae is red or pink, and violet-blue before pupation. The antennal seta (1-A) is confined to the basal half of the shaft, with four or more branches. The inner and median frontal setae (5-C and 6-C) are long, multiple-branched. The setae of the thorax and abdomen are very long, particularly the lateral setae 6-III to 6-VI. Conspicuous sclerotised plates are present on the dorsal surface of all or at least on one of the segments VI to VIII. The siphon is without a pecten but with a single pair of siphonal tufts (1-S). The siphonal index is at least 2.5, often much

more. The saddle completely encircles the anal segment. The ventral brush is made up of 12 or more cratal setae (4-X). The dorsal pair of the anal papillae is longer than the ventral pair. The larvae develop in tree holes, bamboo stumps, axils of bromeliads (arboreal or dendrolimnocolous species). All species are rare, none of them being known as molestants.

The small genus comprises 24 species spread throughout the Neotropical and Oriental regions, a few of them ranging northwards into the Palaearctic and Nearctic regions, some are isolated on Madagascar and Mauritius, but none are found in continental Africa. *Or. pulcripalpis* is the only member of the genus found in the Palaearctic region (Gutsevich *et al.*, 1974; Dahl and White, 1978).

Orthopodomyia pulcripalpis (Rondani) 1872

Female: It differs from most of the other Palaearctic mosquitoes by its conspicious pattern of white scales on a black background producing several easily recognised characters. The proboscis has black scales and a moderately broad ring of white scales within the apical half. The palps are nearly half as long as the proboscis, black, with a white ring at the base and in the middle, and the apex is white. The antennae are black, the pedicel is covered with white scales, and three to five basal annuli of the flagellum may have a median line of white scales. The head is black, and covered with a mixture of black and white scales. The posterior margin of the eyes dorsally is covered with white scales. The scutum is covered with narrow black scales, and distinctly ornamented with three pairs of narrow, white longitudinal stripes. A dorsocentral pair extends from the anterior margin to the posterior third of the scutum, where it is interrupted. It continues on over the prescutellar dorsocentral area as a wider stripe converging and ending on the scutellum. The second pair of stripes is confined to the posterior submedian area. The third pair of stripes almost completely borders the scutum.

All the legs are covered with black scales with a metallic shine, with white spots on the femoro-tibial and tibio-tarsal articulations. The dorsal surface of the femora is speckled with white scales. Tarsomere IV of the fore legs is not as long as it is broad (Fig. 6.4a). The fore and mid tarsi are entirely black scaled, sometimes with weakly developed, narrow basal white rings, most conspicuous at the base of tarsomere I of the mid legs. Tarsomeres I to IV of the hind legs have white rings at both ends of each tarsomere, and tarsomere V is completely white. The wing veins have dark brown scales, and the bases of Sc and R have patches of silvery white scales. The abdomen is blackish brown, covered with scales of the same colour as the scutum, with moderately broad basal bands of white scales.

Male: The palps are distinctly longer than the proboscis, black scaled, with white rings at the articulations of the palpomeres, and the apical palpomere is entirely white. The basal lobe of the gonocoxite is conical, with 4–5 large spine-like setae (Fig. 10.138). The gonostylus is relatively narrow, and slightly expanded in the basal half. The apical spine of the gonostylus is inserted laterally close to the apex, and subdivided into five finger-like serrations. It is longer than the width of the gonostylus at the point of insertion. Claspettes are absent.

Larva (Fig. 10.139): In all instars the larvae are easily recognised by the absence of pecten teeth and the pinkish colouration. The head is dark and rounded. The antennae are straight, less than 1/4 the length of the head, without spicules. The antennal seta (1-A) is articulated at the end of the basal third of the antennae, with 3–7 short fan-like branches. The mouth parts are transitional between browser and filter feeder. The postclypeal seta (4-C) is long, with 5–7 branches. The frontal setae are long and multiple-branched. The inner (5-C) is 5–8 branched, the median (6-C) is 9–10 branched extending well beyond the anterior margin of the head, and the outer (7-C) is of similar length and branching as 4-C. The thorax has very long lateral setae. The abdomen distinctly narrows towards the end, with long lateral setae. The

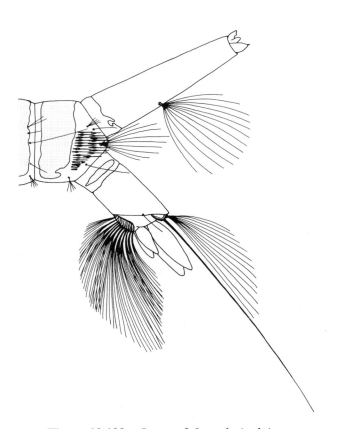

Figure 10.138. Hypopygium of *Or. pulcripalpis*.

Figure 10.139. Larva of *Or. pulcripalpis*.

section of the main tracheal trunks within the thorax and abdominal segments V to VII is enlarged. Sclerotised plates are usually present on abdominal segment VI, and always present on segments VII and VIII (Fig. 8.4a). The plate is broadest on segment VII and extends down to its lateral sides. On segment VIII it is narrower but nearly encircles the segment. The plate on segment VI is small and narrow, and restricted to its dorsal part. The comb consists of 23–30 scales arranged in two rows. Most of the scales have a pointed terminal spine. Seta 3-VIII is strongly developed, resembling the siphonal tuft. The siphonal index is 3.5–4.0. The siphonal tuft (1-S) is composed of 8–13 long branches forming a fan, inserted at the basal 2/5 of the siphon. The pecten is absent, and small sclerites are laterally positioned in front of the saddle. The saddle completely encircles the anal segment, and is narrowed ventrally. The saddle seta (1-X) is short and inserted at the posterior margin of the saddle. The upper anal seta (2-X) forms a short, asymmetrical fan with 9–14 branches, the length of the branches increases progressively towards the lower anal seta (3-X). 3-X is simple and very long. The ventral brush has 12–14 cratal setae (4-X). The anal papillae are lanceolate, and the dorsal pair is twice as long as the ventral pair.

Biology: It is a polycyclic species, the larvae occur throughout the season from May to October in southern Bulgaria (Bozkov *et al.*, 1969) and from June to October in Greece. Hibernation takes place in the fourth larval stage, the larvae can survive the freezing of the water surface (Shannon and Hadjinicolaou, 1937). The larvae breed in water with a high pH value in tree holes (dendrolimnocolous) or holes in the roots of elm, oak, beech, horse-chestnut, olive and plane (Munstermann *et al.*, 1985; Shannon and Hadjinicolaou, 1937). Large tree holes that can support a permanent presence of water are preferred breeding sites. Larvae are often found in association with those

of *An. plumbeus, Oc. pulcritarsis, Oc. geniculatus, Oc. echinus* and *Oc. berlandi* (Rioux, 1958; Coluzzi, 1968; Bozkov *et al.*, 1969). An ornithophilic host preference has been reported (Ribeiro *et al.*, 1988). Any attempt to feed the females on man in the laboratory failed (Shannon and Hadjinicolaou, 1937). According to Gutsevich *et al.* (1974) the females rarely bite man and are most active during the day in shaded places.

Distribution: Palaearctic species, mainly distributed in the Mediterranean region, Black Sea coast and Transcaucasia. In Europe it extends northwards to Belgium and southern England.

Note on systematics: The original spelling *pulcripalpis*, first published by Rondani in 1872, was replaced by the spelling *pulchripalpis* by Verrall in 1901 and adopted by Knight and Stone (1977). The correct original spelling was revived after Snow (1985) and adopted by Ward (1992).

Medical importance: The vectorial capacity of the genus *Orthopodomyia* is not well studied, although many of its members prefer birds as hosts and probably play a role in amplification of avian arboviruses (Zavortink, 1968).

10.7. GENUS *URANOTAENIA* LYNCH ARRIBALZAGA

The species of the genus *Uranotaenia* are generally small dark mosquitoes, which are characterised by having short palps in both sexes, the proboscis is usually swollen at the tip and the antennae of the males are very plumose. The scutum has lines or spots of flat metallic shining scales, the antepronotal lobes are well separated. The pleural setae are reduced in number, and the scales usually form only one or two patches or stripes. The wings exhibit a characteristic venation pattern with the anal vein (A) sharply bent apically, ending before or at the same level as the furcation of the cubitus (Cu). The abdomen of the females is blunt ended, and the cerci are short and rounded. The male genitalia have a short, somewhat conical shaped gonocoxite with a moderately developed basal lobe. The gonostylus is short with a small apical spine, and claspettes are absent. The larvae are small with a dark head and short antennae. The inner (5-C) and median (6-C) frontal setae are stout and spine-like in many species. The abdominal segment VIII is covered with characteristic sclerotised plates on its lateral parts. The pecten teeth and siphonal seta (1-S) are present and the saddle completely encircles the abdominal segment X. The larvae rest with their bodies almost parallel to the water surface, unlike most culicines and may be, at the first glance, confused with anophelines. Eggs may be deposited either in boat shaped rafts or laid singly on the water. Very little is known about the adult feeding behaviour.

Uranotaenia is a relatively large genus, which occurs mainly in regions with tropical climates and only a few species are found in the more temperate zones of the holarctic region. About 207 valid species have been recorded in the genus, which is divided into two subgenera, namely *Uranotaenia* Lynch Arribalzaga and *Pseudoficalbia* Theobald (Peyton, 1972; Knight and Stone, 1977; Ward, 1988, 1992). The only species occuring in the Palaearctic region, *Ur. unguiculata*, is placed in the subgenus *Pseudoficalbia*, which is characterised by having no suture between the prealar area and the mesepisternum.

10.7.1. Subgenus *Pseudoficalbia* Theobald

Uranotaenia (Pseudoficalbia) unguiculata Edwards 1913

Female: A dark mosquito notable for its conspicuous stripes of flat silvery scales along the lateral margin of the scutum. It can be distinguished from all other European species by the shape of its anal vein, which is sharply bent apically and ends slightly before or at the same level as the furcation of the cubital vein (Fig. 6.2a). The proboscis is blackish brown with patches or an indistinct line of pale scales on the ventral surface, and markedly thickened at the apex. The antenna is brown, and the pedicel has light scales. The head is mainly dark scaled with silvery scales along the eye margins and on the occiput. The scutum is covered with blackish or dark brown scales. Distinct lateral stripes of flat silvery scales stretch from the anterior margin to the wing base. A similar stripe runs across the pleurites stretching from the antepronotum to the mesepimeron. The scutellum has blackish brown scales. The legs are largely dark, with patches of white scales at the tips of the femora and tibiae, and pale longitudinal stripes on the anterior surface of the fore and mid femora. The tibiae often have a diffuse pale ring in the middle, with dark brown tarsomeres. The wing veins are dark scaled, and the bases of the subcosta (Sc) and radius (R) have pale scales. The terga are covered with dark brown iridescent scales, often with triangular spots of white scales which can be found mainly on the last segments, and the sterna have light scales.

Male: The palps are considerably shorter than the proboscis. The gonocoxite is broad and short, and irregularly conical shaped (Fig. 10.140). The small flattened basal lobe bears several long, stout setae. The gonostylus is broad and flattened dorsoventrally with the inner margin sigmoid shaped, and the apical spine of the gonostylus is pointed.

Larva (Fig. 10.141): It is easily recognised by its sclerotised plates on the lateral parts of abdominal segment VIII, from which 5–8, usually 6, dark spine-like comb scales arise in a row along the posterior margin. The antennae are short, almost without spicules and with a tiny simple antennal seta (1-A), which arises close to the midpoint of the antennal shaft. The head is

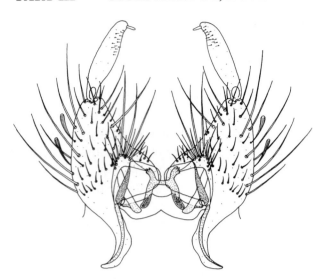

Figure 10.140.   Hypopygium of *Ur. unguiculata*.

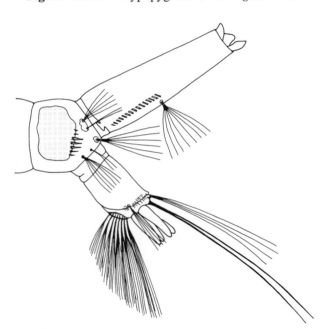

Figure 10.141.   Larva of *Ur. unguiculata*.

dark and slightly broader than it is long with the base of the labrum located more anteriorly than in other species. The mouth brushes are distinctly bent apically. The labral seta (1-C) is strongly developed. This is an adaptation to the feeding behaviour of the larva, which feeds on the surface film from below, while bending its

head backwards with the body in a horizontal position. The inner (5-C) and median (6-C) frontal setae are long, stout and simple, 5-C rarely 2–3 branched. The inner setae are located close together behind the median setae. The outer frontal setae (7-C) have 4–7 branches. Abdominal segment VIII has the above mentioned sclerotised plates. The siphon is nearly conical and slightly tapered towards the apex, with a siphonal index of 3.2–4.0. The pecten consists of 13–20 weakly pigmented teeth, with the distal tooth reaching to the point of insertion of the siphonal tuft (1-S). The tuft has 7–12 branches and arises near the middle of the siphon. The saddle completely encircles abdominal segment X, and its posterior margin is covered with distinct tiny spicules. The saddle seta (1-X) is 3–5 branched, and is about the same length as the saddle. The ventral brush has 8–11 multiple-branched tufts of cratal setae (4-X), and precratal setae are absent. The anal papillae are lanceolate, pointed or rounded at the end and generally shorter than the anal segment .

Biology: The favoured larval breeding sites of *Ur. unguiculata* are pools, ditches or canals with stagnant or little flow of water and a rich growth of aquatic vegetation. They are also common in shallow shores of lakes, which are overgrown with *Lemna* sp., *Scirpus* sp. and *Phragmites* sp. Often they can also be found in shaded localities. The larvae prefer fresh water and are only occasionally found in water with a slight salinity. They often occur together with larvae of *An. hyrcanus, An. sacharovi, Cx. p. pipiens, Cx. modestus* and *Cx. theileri*. Larvae of *Ur. unguiculata* can be found from May until the beginning of October with a peak in August. The adults are most abundant in late summer, this may lead to the presumption that the species hibernates in the adult stage. It seems likely that the females of *Ur. unguiculata* do feed on blood, but rarely bite man or mammals even though they may be capable of doing so.

In a population from Turkmenistan autogeny has been observed (Rioux *et al.*, 1975). Other species of the genus are well known to feed on amphibians (Remington, 1945).

Distribution: *Ur. unguiculata* is a frequent species throughout the Mediterranean region. In Europe, its distribution range stretches as far north as Germany (Becker and Kaiser, 1995). In eastern Europe the species can be found in the southern Ukraine and the Volga delta with further occurrence in middle and southwest Asia to Iran and Pakistan.

Note on systematics: A subspecies of *Ur. unguiculata*, ssp. *pefflyi* is recorded from the Arabian peninsula (Knight and Stone, 1977).

IV

Control of Mosquitoes

11

Biological Control

11.1. INTRODUCTION

Biological control, in the broadest sense, is defined as the reduction of the target population by the use of predators, parasites, pathogens, competitors or toxins from microorganisms (Woodring and Davidson, 1996).

Biological control aims to reduce the target population to an acceptable level and at the same time to avoid side-effects to the ecosystem. As far as mosquito control is concerned, biological control measures should integrate the protection of humans from mosquitoes with the conservation of biodiversity, whilst avoiding toxicological and ecotoxicological effects. As a result, the regulatory power of the ecosystem is maintained by protecting the existing community of mosquito predators.

The use of beneficial organisms for the control of mosquitoes was first recognised in the 19th century, when attempts were made by introducing predators such as dragonflies (Lamborn, 1890). However, the mass breeding and successful introduction of predators such as hydra, flatworms, predacious insects or crustaceans, often brings a range of problems. Such problems did not occur, or to only a limited extent, with the use of fish such as the mosquito fish, *Gambusia affinis*, which was successfully introduced into many countries to control mosquito larvae in the early 1900s (Legner, 1995).

With the discovery and large-scale use of synthetic insecticides in the 1940s and 1950s, the biological control of mosquitoes was no longer considered to be an important matter.

However, the initial euphoria that greeted the success of synthetic insecticides rapidly dissipated as resistance subsequently developed within the target populations. Moreover, despite the beneficial effects of chemical insecticides, they also often have unwanted characteristics, such as their non-selectivity which frequently causes ecological damage. As public awareness of environmental issues increased, regulations controlling the application of chemicals were tightened. As a result, a renaissance of the biological control of mosquitoes took place in the 1960s and 1970s. By 1964, Jenkins had already listed more than 1500 parasites, pathogens and predators which were candidates for biological control. Today, the literature on mosquito antagonists is immense.

The major advantage of biological control measures is that existing predators are conserved, which will then assist the control effort by preying upon newly-hatched mosquito larvae after the control operation. In this way, the efficacy of the current control measures is considerably enhanced.

Apart from the conservation of existing populations of predators, parasites or pathogens, there are two major strategies for the augmentation of populations of mosquito antagonists.

Inoculation means the release of small numbers of predators, parasites or pathogens into the habitat of the target organisms. The antagonists become established, they reproduce and multiply under favourable living conditions in the new habitat. In this way a sustained suppression of the target population can be achieved by successive generations of mosquito enemies. For instance, the inoculation of fish into newly-flooded rice fields is a very common practice in mosquito control.

Inundation means the release of an overwhelming number of predators, parasites, pathogens or their toxins into the mosquito habitat. Such mass release of organisms or applied pathogens (toxins) can have an immediate effect through a significant reduction of the target population. For instance, inundative control is successfully practised with microbial pathogens which are produced in artificial cultures (*e.g. Bacillus thuringiensis* and *B. sphaericus*). Only rarely do the antagonists become established in the habitat, for example *B. sphaericus* is able to recycle under certain conditions.

A prerequisite for the successful use of predators, parasites or pathogens, is a precise knowledge of the biology of the antagonist in question and its interaction with the ecosystem. For example, the introduction of foreign faunal elements as predators risks damaging or displacing existing populations of predators. For instance, introduced fish may reduce numbers of aquatic insects, crustaceans or amphibians which would otherwise be effective predators of mosquito larvae. Rare indigenous species which do not feed on mosquito larvae may also be endangered. A thorough understanding of the predator/prey or parasite/host relationships is therefore of fundamental importance for the successful and ecologically sound use of antagonists. Mosquitoes inhabit very different habitats, and have developed various life strategies by adapting to habitats with very different abiotic and biotic conditions. Antagonists can only successfully reduce a target population if their own life strategy is adapted to the target population.

11.2. PREDATORS

In general, predators of the immature mosquito stages are more effective than predators of the adults (Figs 11.1 and 11.2). As a rule, mosquito larvae and pupae are concentrated at their breeding sites and are more easily available to predators than the widely dispersed adults. Moreover, adult mosquitoes evade many predators by their mostly nocturnal way of life. Mosquitoes have the characteristics of typical r-strategists which especially means a high rate of reproduction, and relatively short life cycle. Predators are particularly effective if they have a similarly high rate of reproduction and/or a high rate of feeding, like fish.

11.2.1. Vertebrate Predators

11.2.1.1. Fish (Osteichthyes)

The best known fish predator of mosquitoes is the so-called mosquito fish, *G. affinis*, which is native to the south-east United States, eastern Mexico and the Caribbean, and the common guppy, *Poecilia reticulata*, which is native to tropical South America. Both fish are effective predators because their upward-facing mouth enables them to consume mosquito larvae living on the water surface. Other biological attributes of these viviparous fish are their high reproduction rate, small size (3–6 cm), and high tolerance to variations in temperature, organic pollution and salinity. *G. affinis* can survive water temperatures below 13°C and even overwinters in areas with short periods of frost, but *P. reticulata* is restricted to subtropical and tropical climates. The latter is used particularly in urban areas against *Cx. p. quinquefasciatus*, which usually occurs in masses in highly polluted water bodies (Sjogren, 1972).

The mosquito fish is the most widely disseminated organism for mosquito control. As early as 1937 Hackett reported on the value of *G. affinis* for the control of malaria in Europe. It is thought that *G. affinis* may have contributed significantly to the reduction of malaria in Turkey and Iran (Tabibzadeh *et al.*, 1970; Inci *et al.*, 1992). In the U.S.A., mosquito fish

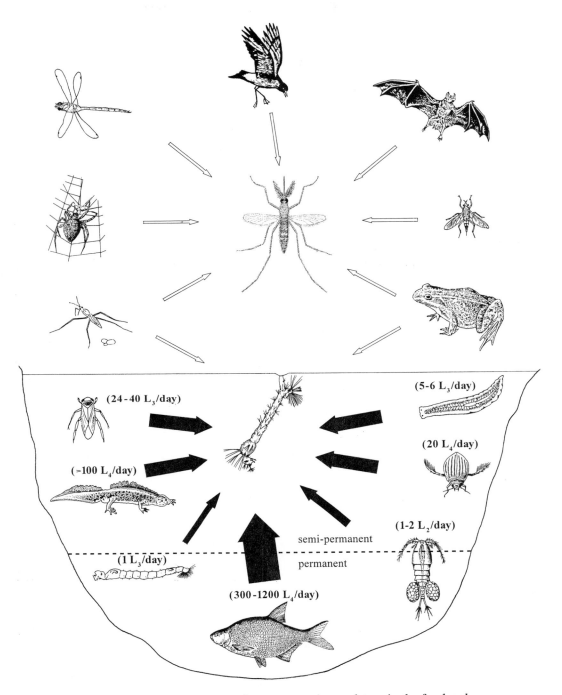

Figure 11.1. Importance of some mosquito predators in the food web.

are bred by abatement districts and released for control. In California they are used successfully in rice fields against the immature stages of *An. freeborni* and *Cx. tarsalis*. Whereas a stocking rate of more than 500 female mosquito fish per hectare in rice fields gave excellent control of *Cx. tarsalis* (Hoy and Reed, 1971; Steward *et al.*, 1983), significant reduction rates against *An. freeborni* were only achieved when more than 4000 fish per hectare were used (Kramer *et al.*, 1987a,b, 1988a,b). The inundative release of 4800 mosquito fish per hectare was effective

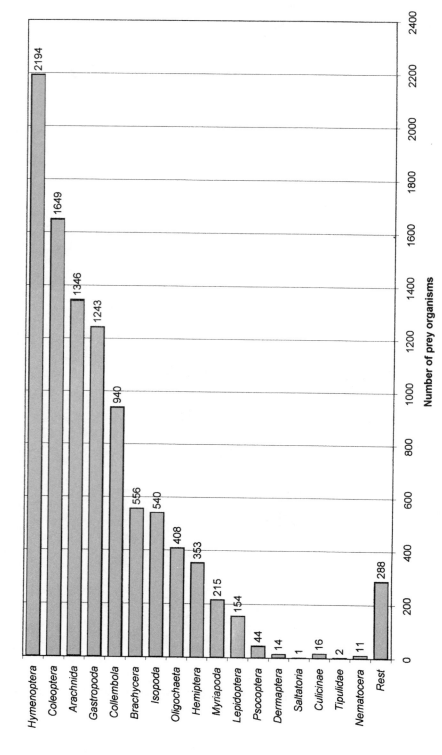

Figure 11.2. Prey of Anura in the Upper Rhine Valley (Blum *et al.*, 1997). The number of Anura was 2163 and the total number of prey organisms sampled was 10 068.

against *Psorophora columbiae* in Arkansas rice fields (Davey and Meisch, 1977a,b,c). In California a reduction rate of more than 70% was observed in a population of *Cx. p. quinquefasciatus* by the inundative release of mosquito fish in urban underground storm drains. Generally, mosquito fish are most effective in water where the vegetation is less dense and the mosquito larvae are more accessible.

The South American *P. reticulata* is a popular aquarium fish which has been introduced into many countries through the ornamental fish trade. This guppy frequently inhabits sewage ditches in subtropical and tropical countries, where it definitively contributes to the reduction of larval populations of *Cx. p.quinquefasciatus*, the main vector of bancroftian filariasis. Sasa and Kurihara (1981) discussed the use of guppies in filariasis control programmes. In Sri Lanka, free-living fish were caught and introduced into mosquito breeding sites (Sabatinelli *et al.*, 1990). In Malaysia, the fish are even used in water containers for the control of *Ae. aegypti*, the vector of dengue and dengue haemorrhagic fever. In South California, Sjogren (1971) and Mian *et al.* (1985; 1986) successfully tested guppies against mosquitoes in sewage plants.

Before the discovery of *B. thuringiensis israelensis*, the use of fish was lauded as being one of the most successful biological weapons for the control of mosquitoes. Many critics, however, disapprove of the use of mosquito fish because of their concern for the native fish fauna. *G. affinis* is omnivorous, and feeds not only on invertebrates which can themselves be useful predators, but also on the eggs and offspring of indigenous fish. The introduction of *G. affinis* can lead to the destruction of extensive aquatic predator populations consisting of water beetles, water bugs, copepods, dragonflies or urodelans. It can greatly repress or even eliminate economically important fish species such as carp. Schoenherr (1981) and Lloyd (1987) reported that more than 30 species of

native fish have been adversely effected by the introduction of *G. affinis*.

For this reason, it is essential to study the biology of predacious organisms with special reference to prey selectivity, reproductive potential and their benefit as predators in relation to costs and the environmental damage they might cause, before any releases are actually made. The ecosystem must also be carefully evaluated to determine the potential effect of the released organisms on the existing biota. This process is even more important if non-indigenous species are to be released.

The use of *G. affinis* for mosquito control in general is no longer recommended by the World Health Organisation because of its aggressiveness towards numerous aquatic organisms, and also because of its doubtful contribution to the control of mosquito-borne diseases (Service, 1983). It remains doubtful whether the fish gives an appropriate level of mosquito control for any reduction in the incidence of malaria.

Fish are also occasionally introduced because of their ability to consume aquatic vegetation. As predators, they can reduce the number of mosquitoes, but as consumers of aquatic plants they can also reduce the niches for their larval development. In California the subtropical cichlids *Tilapia zilli, Orechromis mossambia* and *O. hornorum* were found to be useful.

Because of the risk to an existing community which can result from the introduction of non-indigenous organisms, native fish should be used for control of mosquitoes. In China the grass carp, *Ctenopharygodon idella*, and the common carp, *Cyprinus carpio*, are used in rice fields as young fish. During flooding of the rice fields, the fish not only contribute to the control of mosquito breeding but also feed on rice pests such as grasshoppers which fall on to the water surface. In addition, the fish also manure the rice cultures with their excreta. Furthermore, the grass carp, as a plant feeder, prevents the growth of weeds. The beneficial effects of the grass carp led to an increase in the total rice yield by more than 20%. If the rice fields are drained before the harvest, the fish are kept for another year in permanent main irrigation ditches belonging to the rice farmers. Because of their relatively fast growth, carp can subsequently be used for personal requirements, either as a source of protein or for sale at the market (Xu, pers. comm.).

Representatives of the Cyprinodontidae produce hard-shelled eggs that are resistant to drought. In China, *Oryzias latipes* is used in rice fields to control mosquitoes (Xu, pers. comm.). These widely distributed fish only grow to about 4 cm in length and feed predominantly on insects falling on to the water surface or on mosquito larvae. Despite their small size, one fish eats on average 51 *Anopheles* or 118 *Culex* fourth-instar larvae per day. In Asia, species belonging to the labyrinth fish, such as *Macropodus opercularis*, *M. chinensis*, and *Tanichthys albonubes*, are avid feeders of mosquitoes.

In the Americas the cyprinodontid *Cynolebias bellottii* breeds in temporary waters. The eggs can survive dry periods and so the fish can be successfully used as predators in rice fields (Coykendall, 1980; Walters and Legner, 1980; Gerberich and Laird, 1985).

Shallow floodlands which form mass breeding sites for mosquitoes can be deepened into permanent water bodies, or connected by ditches to fish waters if it is environmentally sound to do so. The fish and their offspring are then able to invade the breeding sites of floodwater mosquitoes such as *Ae. vexans*. In field investigations in Germany, native fish species demonstrated an impressive feeding performance. Fish of approximately 5 cm in length were kept in cages in mosquito breeding sites and fed with fourth-instar larvae of *Ae. vexans*. At a water temperature averaging 22°C, the following feeding rates of fourth-instar larvae/day were recorded: *Cyprinus carpio*: 302; *Carassius carassius*: 238; *Tinca tinca*: 185; *Gasterosteus*

aculeatus: 178; *Abramis brama*: 148; *Rutilus rutilus*: 147; *Alburnus alburnus*: 113; *Leucaspius delineatus*: 99; *Scardinius erythrophthalamus*: 80 and *Gobio gobio*: 63. Feeding rates of more than 1000 *Aedes/Ochlerotatus* larvae within 12 hours were shown by the bigger *C. carassius* and *S. erythrophthalamus* (Gebhard, 1990).

11.2.1.2. Amphibians (Amphibia)

Urodela (newts) and their larvae are important predators of mosquito immature stages (Sack, 1911; Martini, 1920b; Twinn, 1931). In feeding experiments in Europe, *Triturus cristatus* and *T. vulgaris* proved to be voracious consumers of mosquitoes (Kögel, 1984). Whilst two week old larvae of *T. cristatus* devoured about 15 third-instar larvae of *Cx. p. pipiens*, 5–10 week old larvae captured up to 100 fourth-instar larvae per day. The feeding rate of *T. vulgaris* and *T. cristatus* is approximately the same.

In contrast to urodelans, anurans have little effect as predators on mosquitoes. In a three year study in the floodplains of the Rhine river, 2163 anuran specimens of the taxa *Rana arvalis*, *R. temporaria*, *R. dalmatina*, *R. esculenta* s.l., *Hyla arborea*, *Bufo bufo* and *Pelobates fuscus* were subjected to stomach flushing (Blum *et al.*, 1997; Fig. 11.2).

Most of the prey items were beetles (Coleoptera), springtails (Collembola), snails (Gastropoda), spiders (Araneae), ants (Formicidae) and woodlice (Isopoda), which all belong to the epigeal fauna. Only 0.1% of the prey were Culicidae, mostly adults of *Ae. vexans*. The only anuran known to consume Culicidae at a higher rate is *Bombina bombina*. Lac (1958) reported a percentage of 5.7 and 16.7 of *Cx. pipiens* and *An. maculipennis* s.l. larvae, respectively.

11.2.1.3. Birds (Aves)

In general, birds are not considered to be important predators of mosquitoes, although mosquitoes are a source of food for some bird species (Blotzheim, 1985). For example, the duck *Anas platyrhynchos* was repeatedly observed filtering out larvae of *Ae. vexans* from mass breeding sites.

There are two reasons for the relatively insignificant role of birds as predators of adult mosquitoes. Firstly, the activity phases of mosquitoes and birds do not overlap to any great extent. Most mosquitoes are active at dusk, whereas the majority of birds search for food during the day when mosquitoes are resting in vegetation. Secondly, floodwater mosquitoes occur irregularly, only after floods, and are therefore not available as a stable food resource in seasons without floods. For this reason insects such as midges (Chironomidae) which breed mainly in permanent water and appear in large numbers more or less every year, are a more reliable source of food for birds than floodwater mosquitoes.

An analysis of the food of nestling house martins (*Delichon urbica*) in the Upper Rhine Valley by the neck ring method showed that aphids were the main food source in June, whereas staphylinid beetles (Coleoptera) were the most frequent prey in August (Fig. 11.3). Out of a total of 6761 insects which were identified from the accumulation of nestlings' food, only 6 adults belonged to the family Culicidae.

11.2.1.4. Bats (Mammalia)

The majority of insectivorous birds are daylight hunters and look for prey while mosquitoes are usually resting hidden amongst foliage. Bats, in contrast, usually hunt during the night. Therefore the activity patterns of both bats and mosquitoes do overlap. During or after dusk the bats emerge from their roost. At the same time the flight activity of mosquitoes starts to increase. Female mosquitoes search for blood meals whereas male mosquitoes form copulation swarms. Bats therefore could be considered as potential mosquito predators.

Three bat families (Vespertilionidae, Rhinolophidae, and Molossidae) with about 31 species occur in Europe. All European bat species are exclusively insectivorous, but exhibit a large variety of hunting strategies. For example, larger species like the noctule (*Nyctalus noctula*), Leisler's bat (*N. leisleri*), and the serotine (*Eptesicus serotinus*) are hunters in open spaces high above ground. Smaller and slower flying species such as Bechstein's bat (*Myotis bechsteinii*), Natterer's bat (*M. nattereri*), and the long-eared bat species (*Plecotus* spp.) are gleaners, who snatch their prey from foliage of trees and shrubs. Daubenton's bats (*M. daubentonii*) fly closely over the surface of wide and open water bodies, such as rivers or lakes (Arnold *et al.*, 1998). Finally, the smallest bat species like the pipistrelle (*Pipistrellus pipistrellus*) or Nathusius' pipistrelle (*P. nathusii*) hunt along forest margins and the banks of larger water bodies.

Several studies on the diet of bats have been carried out throughout Europe. Information about the composition of insect food eaten by bats was obtained in each case by analysing insect remains in the bats' faeces. All these studies showed that a great variety of nocturnal insects provide food resources for bats, such as moths (Lepidoptera), beetles (Coleoptera), lacewings (Planipennia), caddis-flies (Trichoptera), mayflies (Ephemeroptera), and midges (Chironomidae) (Swift *et al.*, 1985; Rydell, 1986; McAney *et al.*, 1991; Wolz, 1993). Bats mostly catch and ingest flying insects, but gleaners such as the long-eared bats mentioned above or the mouse-eared bat species (*M. myotis* and *M. blythii*) are able to hunt for diurnal insects on vegetation overnight (*e.g.* butterflies) or catch ground-living organisms (*e.g. Isopoda, Chilopoda*).

There are very few records of bats capturing mosquitoes. Yuval (pers. comm.) was able to observe bats hunting for swarming mosquitoes in Israel. The formation of maternity roosts by

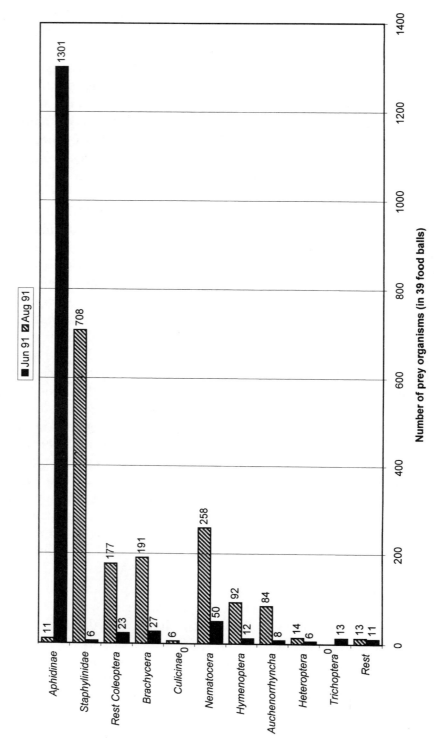

Figure 11.3.   Prey of house martin *Delichon urbica* in the Upper Rhine area.

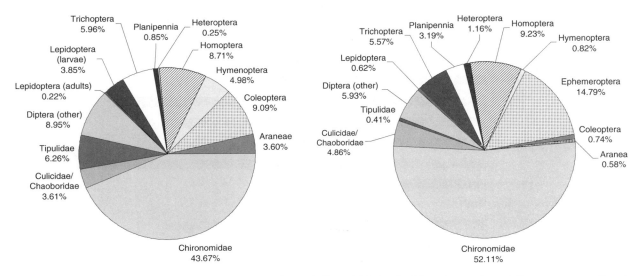

Figure 11.4. Food composition of Daubenton's bats (left) and Nathusius' bats in southwest Germany according to Arnold *et al.*, 2000.

Nathusius' bats is considered to be correlated with the occurrence of mosquitoes in some parts of Germany (Labes, pers. comm.) and in Russia (Sologor and Petrusenko, 1973).

A study of the trophic ecology of Daubenton's and Nathusius' bats in southwest Germany (Arnold *et al.*, 2000) showed that mosquitoes aren't an important food source for these bat species (Fig. 11.4).

Only a few European bat species (such as *M. myotis* or *M. blythii*) rely on a single species or groups of insects as a predominant source of food. The majority of bats can be considered to be opportunistic hunters which focus on insects that occur in large numbers on their hunting grounds (Gloor, 1991).

11.2.2. Invertebrate Predators

Countless invertebrates are known as predators of mosquitoes especially of the larvae. The biology and importance of the predators have been investigated in numerous studies (Lamborn, 1890; Hinman, 1934; Kühlhorn,

1961; Jenkins, 1964; James, 1967; Service, 1977; Kögel, 1984; Collins and Washino, 1985). Although invertebrates have been shown to be effective predators of mosquitoes, they are seldom used in control programmes due to the great difficulties and the high costs involved in mass rearing of these organisms. Nevertheless, their role as consumers of mosquitoes is beyond dispute. Mosquitoes can rarely develop in large numbers in breeding sites where predacious invertebrates are abundant. In this section the different groups of invertebrates and their importance as predators will be discussed (Fig. 11.5).

11.2.2.1. Hydras (Coelenterata)

In feeding experiments, polyps of the genus *Hydra* killed on average between 6 and 21 mosquito larvae per day. Polyps frequently contribute considerably to the reduction of mosquitoes in permanent water bodies (Qureshi and Bay, 1969). Because of their beneficial effects, *Chlorohydra viridissima* (Pallas) were used in

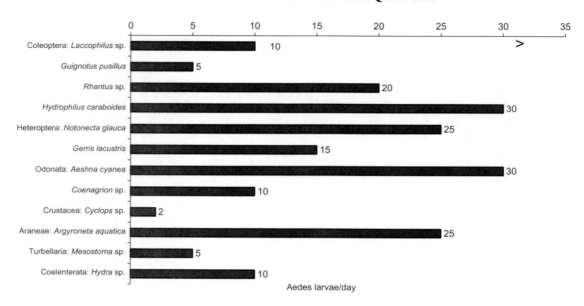

Figure 11.5.   Average feeding rate of some invertebrate predators of mosquito larvae.

a number of field experiments (Lenhoff, 1978; Cress, 1980).

11.2.2.2. Flatworms (Turbellaria)

Flatworms are one of the most thoroughly investigated groups of predators (Legner, 1991, 1995). Some *Mesostoma* species, such as *M. ehrenbergii*, can make a significant contribution to the reduction of mosquitoes. Case and Washino (1979) reported a considerable reduction of *Cx. tarsalis* and *An. freeborni* populations by *Mesostoma* sp. in Californian rice fields, which led to an extensive investigation into the mass rearing of these flatworms.

In central Europe, *Bothromesostoma* sp. sometimes occurs in abundance in ephemeral waters where it can completely suppress *Ae. vexans* populations (Becker, 1984). This flatworm is an efficient predator which feeds mainly on mosquito larvae and has consequently adapted its way of life to that of the prey. Like their prey, the turbellarians can survive dry periods in drought-resistant eggs and hatch during summer floods. At high summer temperatures, they develop within a week to 7 mm long organisms at the reproductive stage

and produce up to 20 eggs. In laboratory experiments a turbellarian consumed up to 67 mosquito larvae during its development. Turbellarians like *Mesostoma* sp. secrete slime to catch mosquito larvae. When the larva sticks to the mucous secretion and tries to escape, the flatworm becomes aware of its prey, smothers it and sucks it up with its pharynx.

11.2.2.3. Freshwater snails (Gastropoda)

Gastropods are considered to be consumers of plants and detritus, and not predators of mosquitoes. However, *Lymnaea (Galba) palustris* was shown to be predacious on mosquito larvae both in the laboratory and in the field (Bishop and Hart, 1931; Baldwin *et al.*, 1955). Their importance as consumers of mosquitoes must, however, be considered as rather slight.

11.2.2.4. Leeches (Hirudinea)

Occasionally, Hirudinea such as *Herpobdella octoculata* are mentioned in the literature as predators of mosquitoes (Kühlhorn, 1961). However, they are of no practical importance in Europe (Baldwin *et al.*, 1955; James, 1965).

11.2.2.5. Spiders and mites (Arachnida)

Spiders are commonly known to be predators of adult mosquitoes particularly at day resting places and shelters for hibernation (Service, 1973). Some species are specialist hunters of aquatic organisms or animals living on the water surface. Representatives of Pisauridae and Lycosidae hunt on the water surface and occasionally also capture mosquito larvae, pupae or emerging adults (Bishop and Hart, 1931). Their importance as predators is, however, considered to be slight (Kühlhorn, 1961).

One of the most effective predators of mosquito larvae is the aquatic spider *Argyroneta aquatica*. This fascinating species builds webs under the water to catch its prey. In aquarium experiments, the maximum feeding rate was 29 fourth-instar mosquito larvae per day (Kögel, 1984). In permanent pools these spiders can lead to a significant reduction of mosquitoes.

Water mites (Hydrachnellae) can also contribute to the suppression of larval mosquito populations. For example, in laboratory experiments the small water mite *Piona nodata* consumed up to 18 mosquito larvae per day (Kögel, 1984). These mites can take up to eight times their own body weight of food within 24 hours (Böttger, 1970). When the adult mosquitoes emerge, some mites attach themselves to the emerging adult and use it as a carrier to disperse to new habitats.

11.2.2.6. Crustaceans (Crustacea)

The tadpole shrimp *Triops cancriformis* is considered to be a predator of mosquito larvae. The shrimp eggs can survive dry periods for several years and the larvae hatch after flooding, often in association with *Aedes/Ochlerotatus* larvae. In California attempts were made to control *Aedes/Ochlerotatus* and *Psorophora* species with *T. longicaudatus* (Tietze and Mulla, 1987, 1991).

Among the crustaceans, copepods are considered to be important predators of mosquito larvae in rice fields (Miura and Takahashi, 1985). It is more their occurrence in enormous numbers than their rate of feeding which makes them important predators of mosquitoes. In laboratory experiments *Megacyclops viridus* and *Acanthocyclops vernalis* consumed on average 1–2 first- and second-instar larvae of *Ae. vexans* per day (Kögel, 1984). Hintz (1951) reported that *Cyclops* sp. caught about 5 first- and second-instar larvae of *Ae. aegypti* per day. Because of their small size, copepods mainly capture early instars of mosquitoes, but as predators they are only effective when present as adults before the mosquito larvae hatch. The use of *Mesocyclops aspericornis*, which has been introduced into artificial containers, wells and terrestrial crab burrows, has led to a reduction of more than 90% of *Ae. aegypti* and *Ae. polynesiensis* in Asia (Riviere *et al.*, 1987a,b; Kay *et al.*, 1992).

11.2.2.7. Insects (Insecta)

Ephemeroptera

The nymphs of Ephemeroptera are not significant as consumers of mosquitoes, although there are references that they can be predacious (Bragina, 1931; Jenkins, 1964; Kühlhorn, 1961). However, *Cloeon dipterum* which commonly occurs in mosquito breeding water feeds mainly on detritus and algae but not on mosquito larvae (Ludwig, 1993).

Odonata

The importance of dragonflies (Odonata) as consumers of mosquitoes has been known for a long time. Both the nymphs and the adults catch mosquitoes (Kögel, 1984; Sebastian *et al.*, 1990). Dragonfly nymphs usually have a long development and therefore occur mainly in permanent waters. In feeding experiments, anisopteran nymphs proved to be extremely voracious. According to Kögel (1984), nymphs of *Aeshna cyanea* consumed up to 100 mosquito larvae each day (average 30 larvae per day). Zygopteran nymphs such as *Coenagrion puella* are usually less efficient as predators (on average, 10 third-instar larvae/day).

In rice fields in Japan up to 208 dragonfly nymphs per m² were recorded. Such mass occurrences classified them among the most important predators of mosquitoes in rice field ecosystems (Mogi, 1978).

Heteroptera

The majority of water bugs (Hydrocorisa and Amphicorisa) are voracious consumers of mosquitoes. However, there is a significant variation in feeding behaviour of different families.

Corixidae: In feeding experiments, one of the most abundant corixids in central Europe, *Sigara striata*, captured 2–3 early larval instars of *Ae. vexans* each day. The larger but less common species, *Corixa punctata* (1 cm), consumed up to 45 larvae per day and even *Cymatia coleoptrata* (6 mm) consumed up to 47 early larval instars of *Ae. vexans* per day (Kögel, 1984). Although the Corixidae are usually common in mosquito breeding sites, their importance as mosquito predators is rather slight because of their mostly omnivorous feeding habits (Washino, 1969).

Naucoridae: One of the most common species in Europe, *Ilyocoris cimicoides*, proved to be extremely voracious. The feeding rate of adult bugs was on average more than 20 mosquito larvae per day, and their nymphs consumed as many as 35 larvae of *Ae. vexans* per day (Kögel, 1984).

Nepidae: The widely distributed water scorpion *Nepa cinerea* consumed 10–18 fourth-instar larvae of *Ae. vexans* daily (Kögel, 1984). This heteropteran inhabits the shallow littoral of stagnant waters where it is certainly one of the most important predators. *Ranatra linearis* can also

Figure 11.6. *Ranatra linearis* feeding on mosquito larvae.

frequently be observed feeding on mosquito larvae (Fig. 11.6).

Notonectidae: Amongst the Heteroptera, notonectids are known for their voracious feeding on mosquito larvae in permanent and semi-permanent waters (Hinman, 1934; Hazelrigg, 1975, 1976; Murdoch *et al.*, 1984; Legner, 1995). For this reason, *Notonecta undulata* and *N. unifasciata* were mass-cultured for use in rice fields in California (Sjogren and Legner, 1989). In experiments with *N. glauca* the daily average feeding rate was 25 third-instar larvae of *Ae. vexans* (Kögel, 1984).

Pleidae: Although *Plea leachi* (size 2–3 mm) is one of the smallest predators, it is able to catch up to 20 mosquito larvae per day (Kögel, 1984).

Gerridae and Hydrometridae: The Gerridae and Hydrometridae live on the water surface, usually in large numbers. They feed mainly on insects which fall onto or emerge from the water surface. *Gerris lacustris* can sometimes be observed seizing mosquito larvae out of the water. While this hunting behaviour is rather an exception in the Gerridae, *Hydrometra stagnorum* is more specialised for pulling mosquito larvae and even pupae out of the water. In laboratory experiments *Hydrometra* sp. captured up to 15 mosquito larvae per day (Pruthi, 1928).

Coleoptera

Because of their abundance and great voracity, many water beetles and their larvae are effective predacious aquatic insects (Baldwin *et al.*, 1955; Trpis, 1960; Kühlhorn, 1961; James, 1966, 1967; Bay, 1972; Service, 1973a; Nelson, 1977; Kögel, 1984). Their ability to inhabit and reproduce in enormous numbers in various mosquito breeding habitats further enhances their significance. The following section briefly outlines the differences between individual representatives of the different water beetle families as mosquito predators.

Dytiscidae: Dytiscids are the most important predators among the water beetles (Nelson, 1977). Not only the large species of the subfamily Dytiscinae, but also the medium sized species of the subfamily Colymbetinae and the small sized species of the subfamilies Hydroporinae, Noterinae and Laccophilinae can show high rates of feeding on mosquito larvae. The large larvae of *Dytiscus marginalis* and *D. circumflexus* feed predominantly on larger creatures such as tadpoles and small fish. However, their young larvae can consume more than 100 *Ae. vexans* fourth-instar larvae per day. Among the medium sized dytiscids, *Rhantus* species are the most efficient predators. In laboratory experiments, the adults of *R. consputus* and *R. pulverosus* captured up to 40 third- and fourth-instar larvae per day (Kögel, 1984). While swimming, they are able to seize the prey with their front legs in a matter of seconds. Not only the adults of *Rhantus* spp. but also their larvae are frequently found in temporary *Aedes/Ochlerotatus* breeding sites, where they feed mainly on mosquito larvae. Depending on the larval stage, they consume between 4 and 6 *Ae. vexans* larvae, up to a maximum of even 20 fourth-instar larvae, per day. The adults of *Agabus* species do not consume as many larvae of *Aedes/Ochlerotatus* as do *Rhantus* spp. adults. *Agabus* spp. are especially abundant in swampy alder forests populated by the immature stages of snow-melt

mosquitoes. Among the dytiscids, the representatives of the Hydroporinae and Laccophilinae achieve very high feeding rates. In laboratory experiments the larvae of *Coelambus impressopunctatus* captured up to 8 third- and fourth-instar larvae of *Ae. vexans*; the larvae of *Hygrotus inaequalis* and *Hyphydrus ovatus* up to 3 *Aedes/Ochlerotatus* larvae, and *Hydroporus palustris* up to a maximum of 10 *Aedes/Ochlerotatus* larvae per day (Kögel, 1984). While the adults of these beetles capture fewer mosquito larvae than do their larvae, the adults of *Laccophilus* spp. are highly voracious. Even the very small species *Guignotus pusillus* (about 2 mm long) can contribute to the reduction of newly-hatched first-instar mosquito larvae. Some beetles of the subfamily Colymbetinae, such as *Colymbetes fuscus* are quoted as potential predators of mosquito larvae (Kühlhorn, 1961; Jenkins, 1964).

Gyrinidae: Because of their mode of living on the water surface, gyrinids are voracious predators mainly of *Anopheles* larvae (Laird, 1947; James, 1966).

Spercheidae: Whilst the adults of this family are not particularly carnivorous, the larvae are considered to be voracious predators with a feeding rate of up to 13 first- and second-instar larvae of *Ae. vexans* per day (Kögel, 1984).

Hydrophilidae: Adult hydrophilids are known to be herbivorous. However, the larvae of some species feed on mosquito larvae and are therefore relatively important predators of mosquitoes (Nielsen and Nielsen, 1953; Hintz, 1951). Remarkably, the larvae of *Helochares obscurus* can capture up to 14 larvae per day (Kögel, 1984). Of all the beetle larvae tested in the laboratory, those of *Hydrophilus caraboides* proved to be the most voracious. In laboratory experiments they consumed a maximum of 67 larvae, with an average of 30 fourth-instar *Ae. vexans* larvae per day. The larvae are frequently observed on the water surface seizing prey with their well-developed mandibles.

Haliplidae: Haliplids are of no importance as predators of mosquito larvae, although they frequently occur in mosquito breeding sites.

Trichoptera

The importance of caddisfly larvae as predators of mosquitoes has been pointed out in many publications (Martini, 1920b; Baldwin *et al.*, 1955; James, 1961, 1966; Service, 1973a). They are the most important predators of snow-melt mosquitoes in semi-permanent water bodies in swampy woodlands. The 2–3 cm long larvae of *Phryganea* sp. and *Limnephilus* sp. have often been observed capturing larvae of snow-melt mosquitoes.

Diptera

Among the Diptera, species of Culicidae and Chaoboridae with carnivorous larvae are particular predators of mosquito larvae. In North America, species of the genus *Toxorhynchites* have been closely studied for decades as antagonists of pest mosquitoes (Gerberg and Visser, 1978; Trpis, 1981; Lane, 1992). The females of these carnivorous mosquitoes, which occur mainly in warm climates, do not suck blood but feed on nectar. They lay their eggs by preference in natural and artificial water containers where the voracious and cannibalistic larvae feed upon other mosquito larvae. They are therefore suitable for the control of mosquitoes breeding in containers. *Toxorhynchites* sp. was reared and released to control *Ae. aegypti* and *Ae. albopictus* which breed predominantly in artificial containers (Riviere *et al.*, 1987b; Miyagi *et al.*, 1992; Tikasingh, 1992). The inundative release or inoculation of *Toxorhynchites* spp. females combined with the application of adulticides can lead to a significant reduction of vector mosquitoes (Focks *et al.*, 1986). The advantage of this process is that the *Toxorhynchites* spp. females, when searching for breeding sites, can naturally disperse into habitats that are difficult

for humans to find and treat. Unfortunately, the level of control is usually not satisfactory because of the often insufficient number of breeding sites used by *Toxorhynchites* spp. for oviposition, the production of a low number of eggs, and the lack of synchrony between predator and prey life cycles. Other mosquitoes which feed on larvae of Culicidae occur mainly in the tropics and belong to the genera *Anopheles, Armigeres, Culex, Eretmapodites* and *Psorophora*.

In Europe, where *Toxorhynchites* species do not occur, other Diptera which are closely related to Culicidae, such as the Chaoboridae, are important predators of mosquitoes. The larvae of Chaoboridae can remain motionless horizontally in the water body, where they seize their prey with the antennae modified for capturing. In the breeding sites of snow-melt mosquitoes, the larvae of *Mochlonyx culiciformis* can contribute substantially to the reduction of mosquito larave. Due to the fact that *M. culiciformis* has adapted its development to that of its prey, its larvae can frequently suppress mosquito populations significantly, particularly if the development of the predator larvae are ahead of that of the mosquito. In laboratory experiments, a fourth-instar larva of *M. culiciformis* captured on average 8 early instars of mosquitoes or one fourth-instar larva per day (Becker and Ludwig, 1983). Taking into consideration the long period of development of about two months and the frequent abundance of *M. culiciformis* larvae in snow-melt waters, the importance as predator of snow-melt mosquitoes is evident (Chodourowski, 1968). In semi-permanent and permanent waters, the larvae of *Chaoborus* spp. are efficient predators and appear mostly in late summer with dense populations. With an average feeding rate of about 4 small mosquito larvae per day, they are not quite so voracious as *M. culiciformis* (Skierska, 1969).

Diptera which can be predacious on adult mosquitoes are the Dolichopodidae, Empididae, Ceratopogonidae and Muscidae (genus *Lispe*) (Lamborn, 1920; Peterson, 1960;

Laing and Welch, 1963; Service, 1965; Clark and Fukuda, 1967).

11.3. PARASITES

Parasites, in this context, are understood as multicellular invertebrates which complete at least one phase of their development within a single host. The most important parasites of mosquitoes are mermithid nematodes.

11.3.1. Nematodes

There are two families of nematodes which are of importance as insect parasites, the Steinernematidae and the Mermithidae (Weiser, 1991).

The Steinernematidae are effective parasites of terrestrial insects, particularly of larvae that develop in the soil. The use of nematodes such as *Steinernema* spp. or *Heterorhabditis* spp. for the control of Diptera including *Musca domestica* is still rather controversial. As for mosquito larvae it was found that they can only be successfully infected in the laboratory.

Mermithid parasites, which occur mainly in water, are more important for the biological control of mosquitoes. Several species of mermithid nematodes have been tested as biological control

agents in various parts of the world. Females lay their eggs in the substrate of mosquito breeding-sites, where they can survive periods of drought. When flooded and when environmental conditions are suitable, the young nematode larvae (10–20 mm) hatch from their eggs and penetrate as pre-parasites through the cuticle of the target larva, by piercing the integument of the host and entering through the narrow opening. The larvae grow into post-parasites inside the host larva in a little over a week, or up to several weeks in cold climates. The post-parasite (1–3 cm long) leaves the host by boring through the larval cuticle, which leads to the death of the host larva (Fig. 11.7). The post-parasites grow to sexual maturity in the substrate of the water body, forming females and males. After copulation, the females lay eggs in the soil of streams or ponds where they remain until the mermithid larvae hatch under certain conditions, migrate to the surface and search for a suitable host. In a few cases infected host larvae are able to pupate and emerge (Blackmore, 1994). In this way infected adults spread the parasites from pool to pool or even reinfect upstream habitats when they die on the edge of aquatic habitats and release the parasite.

Several species of *Romanomermis* (*R. culicivorax, R. iyengari, R. nielseni*) are of great interest for the control of mosquitoes because

Figure 11.7. Post-parasite of *Romanomermis culicivorax* inside an Aedes larva (left) and boring through the cuticle of the larva (right).

their life cycle is completed within only a few weeks and they can be mass reared. Appropriate methods for the mass rearing of *R. culicivorax* have been developed (Petersen, 1980). A moist substrate of sand containing eggs for the inoculation of mosquito breeding sites became available commercially during the 1970s. Unfortunately, these parasites did not become widely used because of difficulties with transportation and with maintenance of the eggs, as well as with sensitivity of the nematodes towards particular environmental conditions such as low water temperature or high salinity.

11.4. PATHOGENS

Macroorganisms such as fish have been used for decades as biological control tools in many mosquito control programmes. However, fish and other predators such as hydra, flatworms, predacious insects or crustaceans and mermithid nematodes have specific ecological requirements and can only be used where their preferred living conditions are met. The life cycle of the predator is frequently not adapted to that of the target organism so that it is unable on its own to bring about an effective reduction of the target population. Mass rearing and release of the predators or parasites is often expensive or even impossible. This limits their large scale use in a number of specific habitats. Special attention has therefore been given to the search for microbial control agents.

Over the past decades efforts on an international scale have led to the discovery of a great variety of pathogens, including entomopathogenic fungi, protozoa, bacteria and viruses (Weiser, 1991; Davidson and Becker, 1996).

11.4.1. Fungi

The fungi that most commonly attack Diptera belong to the following three groups: (1) Mastigomycotina, (2) Entomophthorales and (3) Deuteromycetes. The Mastigomycotina are mainly aquatic organisms. They develop motile zoospores which are able to propel themselves through the water by means of their flagella. Once they have located a suitable host, they penetrate and develop a mycelium. Pathogens of Diptera are known in the groups Chytridiales, Blastocladiales and Lagenidiales.

About 30 species of *Coelomomyces* belong to the Blastocladiales, and infections with *Coelomomyces* spp. are known from more than 50 mosquito species. At first it proved difficult to infect mosquito larvae in the laboratory. Successful infections were possible on a regular basis, when Whisler *et al.* (1974) showed that there is an alternation of hosts in which copepods as well as mosquito larvae are involved (Fig. 11.8). Upon infection, a mycelium forms inside the mosquito larva. Yellowish to brownish coloured sporangia form at the tips of the hyphae, and the haploid zoospores are developed in these by meiosis (Figs 11.9–11.11).

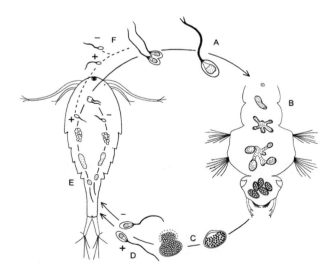

Figure 11.8. Life cycle of *Coelomomyces psorophorae*. (A) Zygote infective for mosquito larvae; (B) development of hyphal bodies (mycelium) and sporangia at the tip of hyphae; (C) release of zoospores; (D) +/− zoospores infect copepode; (E) each zoospore develops to a thallus and gametangia which release isogametes; +/− isogametes fuse to the mosquito infecting zygote (Whisler *et al.*, 1974).

Figure 11.9. Larva of *Ae. vexans* infected with *Coelomomyces psorophorae*.

Figure 11.10. Formation of a sporangium of *Coelomomyces psorophorae* at the tip of a hypha in an anal papilla of an *Ae. vexans* larva.

Figure 11.11. Sporangium of *Coelomomyces psorophorae* (SEM, magnification 850×).

Once the zoospores are released, they infect copepods such as *Acanthocyclops vernalis*, in which a heterothallus is formed which produces isogametes. A biflagellated zygote is formed through fusion of the isogametes. This is the only infectious stage for mosquito larvae.

It penetrates through the cuticle into the larval haemocoel. The initial hopes that a satisfactory parasite for mosquito control had been discovered were dashed by this complex life cycle and associated difficulties in the mass production of the fungi.

Lagenidium giganteum (Lagenidiales), which goes through an asexual and a sexual developmental stage, has shown the most promising results (Lacey and Undeen, 1986; Kerwin and Washino, 1988; Kerwin, 1992). When introduced into a mosquito breeding site, the motile biflagellated zoospores attach themselves to the larval cuticle. Upon penetration, a fungal mycelium forms inside the larval haemocoel, that results in the death of the larva within two to three days. In the cadaver, sporangia are developed in which asexually produced zoospores are formed. The zoospores are released through vesicles on the surface of the larval cadaver. Several further asexual cycles take place, or the fungus undergoes a sexual cycle which terminates with the formation of oospores which are also infectious for mosquito larvae and are characterised by their great persistance at the breeding site (Fig. 11.12).

After it had been found possible to rear *L. giganteum* in an artificial medium, the fungus was released in large-scale field trials (Kerwin and Washino, 1988). High levels of infection were achieved when masses of zoospores were applied. Unfortunately, zoospores are short-lived and disappear from the habitat in the absence of mosquito hosts. Storage of the zoospores is also difficult. On the other hand, the oospores can persist in the environment, but unfortunately they only germinate asynchronously, and consequently high density mosquito populations are only rarely eliminated by a knock-down effect. A combination of different agents which deliver a knock-down and a long-term effect, such as *B. thuringiensis israelensis* preparations together with oospores of *L. giganteum*, could improve the efficacy of the latter in the future.

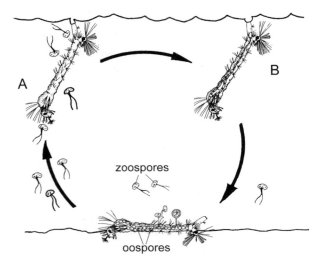

Figure 11.12. Life cycle of *Lagenidium giganteum* (after Woodring and Davidson). (A) biflagellated zoospores (magnified compared to larva) encyst on the cuticle of the mosquito larva; (B) after penetration mycelium develops in the larva and kills it within two to three days. In the cadaver sporangia with exit tubes for the release of asexual zoospores or sexual oospores by fusion of hyphal segments are developed.

The Entomophthorales form various conidia which mainly infect terrestrial insects or their immature stages. *Entomophthora culicis* can infect mosquitoes as they emerge from their pupae (Weiser, 1991).

Deuteromycete fungi such as *Beauveria* sp., *Metarhizium* sp., *Paecilomyces* sp. or *Verticillium* sp. are not host-specific and are not primary pathogens of Diptera. However, their infective conidia can infect insects in stressful situations, for example adults in their shelters during hibernation.

Tolypocladium sp. and *Culicinomyces* sp. fungi are infectious for mosquito larvae and can be produced in artificial media. When the spores of *Culicinomyces clavisporus* are ingested by mosquito larvae, they penetrate the gut wall and proliferate in the larval body. After the death of the larva, the fungus sporulates on the surface of the cadaver. However, effective use of this fungus in mosquito control programmes is unlikely because of its lack of persistence and adequate

recycling ability, difficulties in storage, and the high dosages required.

11.4.2. Protozoa

The protozoa contain the largest number of mosquito parasites. The best studied are the microsporidian intracellular parasites. Larvae infected with microsporidia can easily be recognised even in the field by their milky-white colour. However, despite the great scientific interest of this group of parasites, no one has yet succeeded in using them as microbial control agents. This is mainly because of their complex life cycle, which makes mass production difficult, and their frequently low pathogenicity and persistence.

A common feature of all microsporidia is the development of spores which contain a polar filament internally (Fig. 11.14). This is ejected in the host's gut, and in this way the infectious nucleus is able to penetrate into the host. Two groups of microsporidia are known as mosquito pathogens. The first group, which includes the genera *Nosema* and *Vavraia*, develops asexually and forms only one type of spore. Under laboratory conditions infection of mosquito larvae can be achieved relatively simply by feeding them with spores. The second group, which includes the genera *Amblyospora* and *Parathelohania*, has a complex developmental cycle (Fig. 11.13).

An infected mosquito female transmits the spores in a vertical direction, transovarially to the next generation. In the mosquito larva, the spores can divide meiotically and in this way countless haploid spores arise in the larval fat body and are released when the larva dies (Fig. 11.14).

If they are now ingested by a copepod, they reproduce asexually once again (Andreadis, 1985; Sweeney *et al.*, 1985). If, on the death of the copepod, the haploid spores are released and ingested by a mosquito larva, the sexual phase begins with fusion of the cells, which again form diploid spores. Female mosquito larvae which are only lightly infected can again

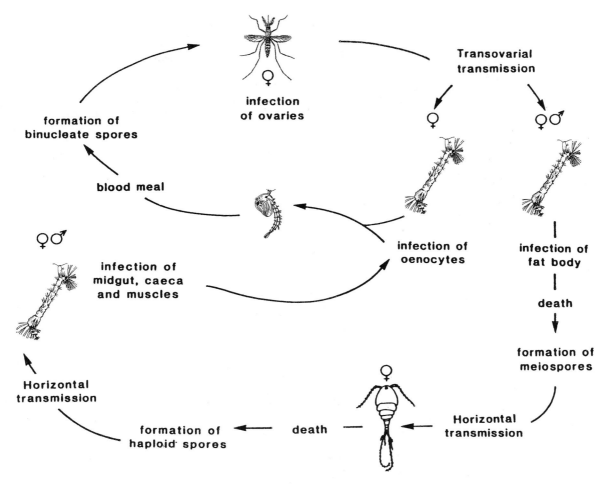

Figure 11.13. Life cycle of *Amblyospora* spp. (after Woodring and Davidson).

transmit the diploid spores transovarially to the larvae of the next generation. This complex developmental cycle prevents any mass-rearing of the parasites and makes it unlikely at present that these micosporidia will be used in control programmes (Sweeney and Becnel, 1991).

11.4.3. Bacteria

Mosquitocidal bacteria have been known since the early 1960s when the first strains of *Bacillus sphaericus* with larvicidal activity were discovered (Kellen and Meyers, 1964). However, these strains were not sufficiently toxic to merit commercial development. The discovery of the gram-positive, endospore-forming soil bacterium *Bacillus thuringiensis* ssp. *israelensis* (*B. thuringiensis israelensis*) in the Negev desert of Israel in 1976 (Goldberg and Margalit, 1977) and of the potent strains of *B. sphaericus* in recent years have inaugurated a new chapter in the control of mosquitoes and blackflies (Singer, 1973; Weiser, 1984; Becker and Margalit, 1996). The new discovered subspecies of *B. thuringiensis* is highly toxic to larvae of most mosquito species and to blackfly larvae. New strains of *B. sphaericus*, such as strain 2362 isolated from an adult blackfly in Nigeria (Weiser, 1984), and strain 2297 isolated in Sri Lanka (Wickramasinghe and Mendis, 1980) are much more potent than the first isolates and are particularly active against larvae

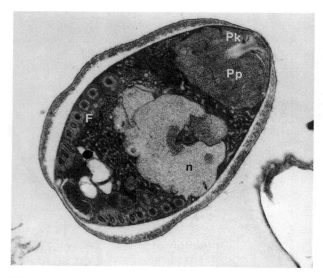

Figure 11.14. Ultrastructure of a spore of *Amblyospora* sp. (F = polar filament, Pk = polar body, Pp = polaroplast, n = nucleus).

Figure 11.15. *B. thuringiensis israelensis* with spore (left) and parasporal body, so-called protein crystal (right). (Micrographs courtesy of J.-F. Charles, Pasteur Institute, Paris).

of *Culex* species and *An. gambiae* (Fillinger *et al.*, 2003).

11.4.3.1. *Bacillus thuringiensis israelensis*

This bacillus produces protein toxins during sporulation that are concentrated in a parasporal body (PSB), called the protein crystal (Fig. 11.15).

These proteins are highly toxic to mosquito and blackfly larvae. The selectivity of the bacillus derives from a variety of factors:

(1) The protein crystal (inactive protoxin) must be ingested by the target insect, and this depends on its feeding habits.
(2) Proteases must then convert the protoxin into biologically active toxins in the alkaline midgut milieu of the target insect.
(3) The toxins must then bind to a cell surface receptor (glycoprotein) of the midgut epithelial cells of the target insect.

This process disturbs the osmoregulatory mechanisms of the cell membrane, thereby swelling and bursting the midgut cells (Fig. 11.16). Non-target organisms do not activate the protoxin into the toxin, or remain undamaged because of the lack of specific receptors on their intestinal cells (Charles et al., 2000; Lüthy and Wolfersberger, 2000).

The insecticidal effect of *B. thuringiensis israelensis* emanates from the parasporal body (PSB), which contains 4 major toxin proteins of different molecular weight, referred to as the Cry4A (125 kDa), Cry4B (135 kDa), Cry10A (58 kDa) and Cry11A (68 kDa) (Delecluse *et al.*, 1996). These toxins bind to specific glycoprotein receptors on the larval midgut brush border (Charles and Nielsen-LeRoux, 1996). A fifth toxin, called the CytA protein (27 kDa), binds to lipids and does not exhibit the specific binding mechanism which the Cry proteins do (Höfte and Whiteley, 1989; Federici *et al.*, 1990; Priest, 1992). Neither the spore nor the living bacilli appear to be involved in the insecticidal process. The more or less spherical PSB is formed at the end of sporulation and consists of three types of protein inclusions separated by thin layers. The largest inclusion is round, of slight electron density, and occupies approximately 50% of the total volume of the PSB. The second type of inclusion (about 20% of the volume of the PSB) is moderately electron-dense and rod-shaped, and the third type of inclusion (about 25% of the PSB) is spherical and highly electron-dense (Ibarra and Federici, 1986; Federici *et al.*, 1990).

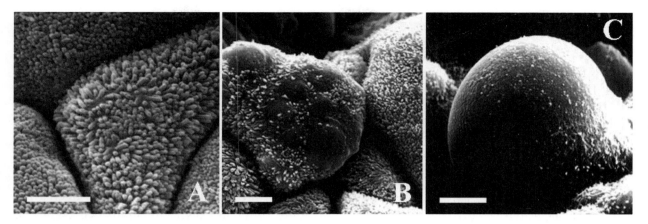

Figure 11.16. (A) midgut epithelium of a healthy *Ae. aegypti* larva; (B) 30 min. after ingestion of the *B. thuringiensis israelensis* protein crystals, swelling of midgut cells and reduction of microvilli; (C) one hour after ingestion cell is about to lyse (micrographs courtesy of J.-F. Charles, Pasteur Institute, Paris).

The mosquitocidal properties of each single solubilised and purified protein have been evaluated in many studies. All tests have shown that each type of protein is mosquitocidal, but none is nearly as toxic as the intact PSB. This high toxicity of the PSB is caused by a synergistic interaction of the 25 kDa protein (split from the 27 kDa protein) with one or more of the higher molecular weight proteins (Ibarra and Federici, 1986; Chilcott and Ellar, 1988; Chang *et al.*, 1993). It is thought that the synergism in the mode of action among the proteins reduces the probability of resistance.

The high toxicity of the PSB to a great variety of mosquito and blackfly species is the most remarkable property of *B. thuringiensis israelensis*. Only at significantly higher dosages are certain other Nematocera species affected, but no other organisms are harmed.

It has been shown in many studies that a plasmid with a molecular weight of approximately 75 MDa plays an essential role in the crystal toxin production. Cloning and characterisation of the different toxin genes have been accomplished (Bourgouin *et al.*, 1986; Thorne *et al.*, 1986; Ward and Ellar, 1988). These results open the possibility of cloning the various toxin genes into host organisms such as

cyanobacteria which can survive and multiply in mosquito breeding sites for a certain time period. In this way a sustained control of mosquito larvae could be achieved and the sequence of retreatments could be reduced. By cloning toxin genes of *B. thuringiensis israelensis* into the genome of *B. sphaericus*, transgenic organisms could be obtained which combine the mosquitocidal properties of both microbials, and the target range of the novel microbial control agent could thereby be increased.

11.4.3.2. *Bacillus sphaericus*

In addition to *B. thuringiensis israelensis*, a second spore-forming bacterium, *B. sphaericus*, has become increasingly important in recent years. The high potential of *B. sphaericus* as a bacterial control agent lies in its spectrum of efficacy and its ability to recycle or to persist in nature under certain conditions, which means that long-term control can be achieved (Hertlein *et al.*, 1979; Mulligan *et al.*, 1980; Lacey, 1990; Ludwig *et al.*, 1994). The timespan between retreatments can thus be extended and personnel costs reduced. This opens up the possibility of a successful and cost-effective control of *Culex* species, particularly of *Cx. p. quinquefasciatus*

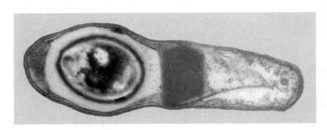

Figure 11.17. *B. sphaericus* with round spore and the parasporal protein inclusion (dark structure on the right site of the spore) which is located in a coated "spore crystal complex" (micrograph courtesy of J.-F. Charles, Pasteur Institute, Paris).

which is the most important vector of lymphatic filariasis and which breeds primarily in highly polluted waterbodies in urban areas.

B. sphaericus can easily be identified by its round spore located terminally in a swollen sporangium (Fig. 11.17). Most recent knowledge indicates that *B. sphaericus* only kills mosquito larvae and, when higher dosages are applied, larvae of Psychodidae, too. Certain mosquito species, such as *Cx. p. quinquefasciatus* and *An. gambiae*, are highly susceptible whereas *Ae. aegypti* larvae are more than 100 times less susceptible. Blackfly larvae as well as other insects, mammals, and other non-target organisms are not susceptible to *B. sphaericus*.

Its efficacy, like that of *B. thuringiensis israelensis*, is based on parasporal protein inclusions which are located in a coated "spore crystal complex". In contrast to *B. thuringiensis israelensis*, *B. sphaericus* has a binary toxin consisting of proteins of two different molecular weights, 51.4 kDa and 41.9 kDa. Both are required for a high level of mosquitocidal activity (Broadwell *et al.*, 1990; Baumann *et al.*, 1991; Berry *et al.*, 1991; Priest, 1992; Davidson and Becker, 1996). Another protein toxin of about 100 kDa is also produced, which is not homologous either to the binary toxin or to *B. thuringiensis israelensis* toxins (Thanabalu *et al.*, 1991). It is thought that receptor binding is similar to that in *B. thuringiensis israelensis* (Davidson, 1988; Davidson and Youston, 1990).

These two microbial control agents were rapidly developed with the support of industry, universities, and national and international organisations such as the World Health Organisation (WHO). Following extensive safety tests and environmental impact studies, the bacilli were quickly put into use. This rapid exploitation was aided by a series of useful properties of the bacterial control agents. In addition to the relative ease with which they can be mass-produced, bacterial control agents are highly efficient, environmentally safe, easy to handle, stable when stored, cost-effective, and suitable for integrated control programmes based on community participation. Furthermore, the costs for development and registration of these agents (about US$500 000) are many times lower than those for a conventional chemical insecticide (about US$20 million). The risk of resistance, especially when *B. thuringiensis israelensis* is used, is much lower than when conventional insecticides are used.

A basic requirement for the successful use of bacterial control agents is the development of effective formulations suited to the biology and habitats of the target organisms. *B. thuringiensis israelensis* preparations can be obtained as water dispersable granules, wettable powders, fluid concentrates, corn cob and ice granules, pellets, tablets or briquettes. The development of suitable formulations based on *B. sphaericus* is also making good progress (Table 11.1).

A few hundred grams or even less of powder, half to two litres of liquid concentrate or a few kilograms of granules per hectare, are usually enough to kill all mosquito larvae. In some situations, a long-term effect can be achieved if larger amounts are used. In recent years, with the production of tablet or briquette formulations, progress has been made towards achieving a long-term effect. Sustained-release floating granules are being developed.

New tablet formulations, based on *B. thuringiensis israelensis* or *B. sphaericus* material sterilised by gradiation to prevent contamination

Table 11.1. Some Commercially Available *Bacillus thuringiensis israelensis* and *Bacillus sphaericus* Products Used in Mosquito and Blackfly Control Programmes

Product	Formulation	Potency (ITU/mg)
B. thuringiensis israelensis (against mosquito and blackfly larvae)		
Aquabac[1]	Primary powder	7 000
Bactimos WP[2]	Wettable powder	5 000
Bactimos PP[2]	Primary powder	10 000
Bactimos G[2]	Granules	200
Ice cubes[3]	Ice granules	45
Teknar HP-D[4]	Fluid concentrate	1200
Teknar TC[4]	Technical powder	10 000
Teknar G[4]	Granules	200
VectoBac 12AS[2]	Fluid concentrate	1 200
VectoBac TP[2]	Technical powder	5 000
VectoBac WDG[2]	Water dispersible granules	3 000
VectoBac DT[2,6]	Tablets	2 250
Bactecide[5]	Water dispersible powder	?
Culinex Tab plus[2,6]	Tablets	2 250
BioTouch[7]	Fluid concentrate	1 000
Bacillus sphaericus (against mosquito larvae)		
Spherimos[2]	Fluid concentrate	120
VectoLex WDG[2]	Water dispersible granules	650 BsITU/mg
Vectolex[2]	Granules	50
Spherico[8]	Fluid concentrate	?

[1]Becker Microbial Products, USA; [2]Valent BioSciences Corp., USA; [3]Icybac GmbH/Phoenix, Germany; [4]Thermo Triology, USA; [5]BioTech International Ltd, India; [6]Culinex GmbH, Germany; [7]Zohar Dalia, Israel; [8]Geratec, Brazil.

of drinking water with spores, can be successfully used for control of container breeding mosquitoes such as *Cx. p. pipiens* or *Ae. aegypti* (Becker *et al.*, 1991; Kroeger *et al.*, 1995).

In addition to commercially available granules based on ground corn cobs, sand granules can also serve as a carrier for wettable powder formulations: 50 kg fire-dried quartz sand (grain size 1–2 mm) with 0.8–1.4 litres of vegetable oil (as a binding material) and 1.8 kg of *B. thuringiensis israelensis* powder (activity 10 000 ITU/mg) should be mixed in a cement mixer. This mixture is sufficient to treat 2–3 hectares. Recently, more cost-effective granules have been developed in the form of ice pellets (Becker, 2003). Ice granules can be easily produced when water suspensions containing the bacterial toxins are frozen into small ice cubes or pearls (3–5 mm) and kept in cold storage rooms until used. The advantages of using ice granules are:

(1) the toxins are bound in the ice pellet, so loss of active material by friction during application is avoided;
(2) as the specific weight of ice is less than that of water; the ice pellets remain in the upper water layer where they release the toxins into the feeding zone of the mosquito larvae as they melt;
(3) the ice pellets penetrate dense vegetation and do not stick to leaves even when it is raining.

The amount of active material per hectare can thus be significantly reduced when compared with granules based on sand.

One limitation to the use of *B. thuringiensis israelensis*, for example against anophelines in rice fields, lies in the relatively brief duration of its activity. Expensive retreatments are frequently necessary. Formulations with a long-term effect, such as sustained-release floating granules, are thus needed.

A particularly attractive feature of *B. sphaericus* is its potential to persist and recycle under certain field conditions. Appropriate formulations have shown a significant residual activity against larvae of *Culex p. pipiens* and *Cx. p. quinquefasciatus* in highly polluted breeding habitats (Hertlein *et al.*, 1979; Davidson and Yousten, 1990; des Rochers and Garcia, 1984; Lacey, 1990; Becker *et al.*, 1995).

When appropriately stored, most preparations based on bacterial toxins can be kept for a long period of time without losing activity. Experience has shown that powder or corn-cob formulations lose little of their activity even after many years in storage. On the other hand, the activity of fluid concentrates may be more labile. Preparations should therefore be retested in bioassays according to WHO guidelines (de Barjac, 1983) when they have been stored for more than a year in temperate climates and for six months in tropical regions.

Standardised methods have been developed to determine the LC$_{50}$ values using standard formulations produced by the Pasteur Institute for comparative purposes (de Barjac, 1983; Dulmage *et al.*, 1990; Navon and Ascher, 2000). The standard powder preparations are IPS 82 for *B. thuringiensis israelensis* tests and SPH 88 for *B. sphaericus* tests. The activity of IPS 82 has been assigned the value of 15 000 ITU/mg and of SPH 88 the value of 1700 ITU/mg. The standards can be obtained from the Pasteur Institute, Paris.

Procedure of the bioassay

The following procedure for bioassays is recommended in order to allow an accurate determination of the potency of the *B. thuringiensis israelensis* and *B. sphaericus* products to be made on a worldwide basis:

50 mg of the standard powder are weighed and poured into a 20 ml penicillin flask, to which 10 ml of deionised water and 15 glass balls (6 mm diameter) are added. This suspension is vigorously homogenised for 10 minutes at 700 strokes/minute, on a crushing vibrator machine. Only one basal dilution is made in a test tube of 22 mm diameter: 0.2 ml of the initial suspension are added to 19.8 ml of deionised water. The test-tube is agitated for a few seconds on an agitator at maximum speed.

From this dilution (50 mg/l), subsequent dilutions are immediately made in plastic cups which have been already filled with 148 ml of deionised water; using precision pipettes or micropipettes. Aliquots of 120 μl, 90 μl, 60 μl, 30 μl and 15 μl are poured into the cups in order to obtain final concentrations of 0.04; 0.03; 0.02; 0.01 mg and 0.005 mg/l respectively of IPS 82 (*B. thuringiensis israelensis*) or SPH 88 (*B. sphaericus*).

Four cups are used for each concentration and for the control.

25 early fourth-instar larvae of *Aedes aegypti* (when *B. thuringiensis israelensis* is to be tested) or 25 early fourth-instar larvae of *Cx. pipiens* (when *B. sphaericus* is to be tested) (each batch of larvae in 2 ml water) are put into each cup by means of a Pasteur pipette. The choice of early fourth-instar is very important, and should be strictly adhered to. A small amount of food (ground mouse diet) is added to each cup in order to avoid excessive mortality caused by nutrition deficiency when the bioassay is run longer than 24 hours.

A comparable initial suspension and series of dilutions are prepared in the same way with unknown preparations to be tested, but with a range of dilutions exceeding that of the standard, to be certain that a reliable regression line can be obtained.

The results of an initial range-finding bioassay using only two widely spaced concentrations of the test material, can be used to select the concentrations used in the full assay more accurately, and as a partial replicate of the full bioassay.

Each series of bioassays will involve at least 500 larvae exposed to the standard treatment, 100 larvae as controls, and 500 to 1000 larvae exposed to the test preparations. All tests should be conducted at 25°C ($\pm$1°C).

The mortality data are recorded after 24 and 48 h by counting both dead and living larvae. The second reading appears suitable in routine work to confirm the previous data, and to check for the possible intervention of factors other than microbial toxins. If some pupae emerge, they have to be taken out and not included in the mortality count.

When control mortalities exceed 5%, the percentages observed in the treated containers should be corrected according to Abbott's formula (Abbott, 1925). Test series with control mortalities greater than 10% should be discarded. Mortality concentration regression lines should be drawn on logarithmic paper. Then the LC_{50} of the series treated with the standard and with the unknown preparations are read and the potency (titre) of the unknown material determined by the following formula:

$$\frac{\text{activity of the standard (ITU)} \times LC_{50} \text{ standard}}{LC_{50} \text{ tested preparation}}$$

The potency or titre of the product is expressed in ITUs (International Toxic Units). For improved precision such bioassays should be repeated on at least three different days, and the standard deviation calculated.

Environmental safety

The exceptional environmental safety of bacterial control agents has been confirmed in numerous laboratory and field tests. The U.S. Environmental Protection Agency (EPA) approved the use of *B. thuringiensis israelensis* as early as 1981. In safety tests on representative aquatic organisms it was shown that in addition to plants and Mammalia none of the taxa tested such as Cnidaria, Turbellaria, Rotatoria, Mollusca, Annelida, Acari, Crustacea, Ephemeroptera, Odonata, Heteroptera, Coleoptera, Trichoptera, Pisces and Amphibia appeared to be affected when exposed in water containing large amounts of bacterial preparations (Table 11.2; Becker and Margalit, 1993).

Even within the Diptera, the toxicity of *B. thuringiensis israelensis* is restricted to mosquitoes and a few nematocerous families (Colbo and Undeen, 1980; Miura *et al.*, 1980; Ali, 1981; Garcia *et al.*, 1981; Molloy and Jamnback, 1981; Margalit and Dean, 1985; Mulla *et al.*, 1982; WHO/IPCS, 1999). In addition to mosquito and blackfly larvae, only those of the closely related Dixidae are sensitive to *B. thuringiensis israelensis*. Larvae of Psychodidae, Chironomidae, Sciaridae, and Tipulidae are generally far less sensitive than those of mosquitoes or blackflies.

In contrast to *B. thuringiensis israelensis*, the toxins of *B. sphaericus* are toxic to a much more restricted range of insects. Blackfly larvae as well as other insects (except for Psychodidae), mammals, and other non-target organisms are not susceptible to *B. sphaericus*.

Toxicological tests were carried out using various mammals. *B. thuringiensis israelensis* when given orally, sub- and percutaneously, intraperitoneally, ocularly, through inhalation and scarification appeared to be innocuous even at high dosages (10^8 bacteria/animal) (WHO, 1982, 2000).

Another important aspect is the widespread occurrence of both bacilli in the soil. They are natural components of the soil micro-ecosystem and not an artificial man-made product where toxic residues may remain after application against pest insects.

Table 11.2. Organisms Not Affected by *B. thuringiensis israelensis*

Taxa	Dosage (ppm)	Species
Cnidaria	100	*Hydra* sp.
Turbellaria	180	*Dugesia tigrina, Bothromesostoma personatum*
Rotatoria	100	*Brachionus calciflorus*
Mollusca	180	*Physa acuta, Aplexa hypnorum, Galba palustris, Anisus leucostomus, Bathyomphalus contortus, Hippeutis complanatus, Pisidium* sp.
Annelida	180	*Tubifex* sp., *Helobdella stagnalis*
Acari	180	*Hydrachnella* sp.
Crustacea	180	*Chirocephalus grubei, Daphnia pulex, Daphnia magna*, Ostracoda, *Cyclops strenuus, Gammarus pulex, Asellus aquaticus, Orconectes limosus*
Ephemeroptera	180	*Cloëon dipterum*
Odonata	180	*Ischnura elegans, Sympetrum striolatum, Orthetrum brunneum*
Heteroptera	180	*Micronecta meridionalis, Sigara striata, Sigara lateralis, Plea leachi, Notonecta glauca, Ilyocoris cimicoides, Anisops varia*
Coleoptera	180	*Hyphydrus ovatus, Guignotus pusillus, Coelambus impressopunctatus, Hygrotus inaequalis, Hydroporus palustris, Ilybius fuliginosus, Rhantus pulverosus, Rhantus consputus, Hydrobius fuscipes, Anacaena globulus, Hydrophilus caraboides, Berosus signaticollis*
Trichoptera	180	*Limnophilus* sp.
Pisces	180	*Esox lucius, Cyprinus carpio, Perca fluviatilis*
Amphibia (larvae)	180	*Triturus alpestris, Triturus vulgaris, Triturus cristatus, Bombina variegata, Bufo bufo, Bufo viridis, Bufo calamita, Rana esculenta, Rana temporaria*

Ease of handling

No special equipment is required for the application of bacterial control agents. Generally, simple knapsack sprayers are adequate for accessible breeding sites. Standard ULV, airblast or mist blower spray equipment may also be used. When dense vegetation or wide spread breeding sites occur, aerial application should be preferred. Rotary seeders or pressurised air sprayers are suitable for the application of granules. Safety precautions as used with toxic chemicals do not have to be considered. Because of the rapid knock-down effect and the high level of efficiency, the success of the treatment can generally be monitored within a few hours of application.

Cost effectiveness

Compared to conventional insecticides, the application of bacterial control agents can be cost-effective. For instance, the German Mosquito Control Association (Kommunale Aktionsgemeinschaft zur Bekämpfung der Stechmückenplage-KABS) mosquito abatement project in Germany effectively suppresses mosquitoes deriving from a catchment area of more than $600 \, km^2$ involving on average $100 \, km^2$ of actual breeding grounds per year. The total annual budget is US$1.5 million. More than 2.5 million residents of the area are protected from an intense nuisance. But environmental considerations, which cannot be expressed in monetary terms, should also

Table 11.3. Susceptibility of Larvae of Nematoceran and Brachyceran Flies to *B. thuringiensis israelensis* (Tested Products: Bactimos WP and VectoBac TP, Potency: 5000 ITU/mg)

Family	Dosage (ppm)	Mortality (%)
Simuliidae		
Simulium damnosum s.l.	0.4	100
Culicidae		
Aedes/Ochlerotatus spp.	0.2	100
Dixidae		
Dixa spp.	2	100
Chaoboridae		
Chaoborus spp.	180	No effect
Psychodidae		
Psychoda alternata	1	100
Sciaridae		
Bradysia sp.	3	90
Chironomidae		
Chironomus sp.	1.8	90
Xenopelopia sp.	4	50
Tipulidae		
Tipula spp.	30	50
Ceratopogonidae	180	No effect
Syrphidae		
Tubifera sp.	180	No effect

be included in these economic calculations (see Chapter 16).

Lack of potential for resistance development

The development of resistance to chemical insecticides represents a serious problem. Bacterial control agents, however, appear less likely to provoke resistance because their mode of action is more complex (Davidson, 1990). However, the resistance of a stored grain pest, *Plodia interpunctella*, to the Lepidoptera-specific *B. thuringiensis kurstaki* has been demonstrated in the laboratory (McGaughey, 1985). Studies have shown that the commercial use of *B. thuringiensis* preparations in agriculture can lead to resistance within a few years. For example, the diamondback moth, *Plutella xylostella*, which was repeatedly treated with

B. thuringiensis kurstaki on farms in Hawaii, was found to be 41 times more resistant than populations that were only minimally exposed to *B. thuringiensis kurstaki* (Tabashnik *et al.*, 1990). Such resistance phenomena have not yet been observed with *B. thuringiensis israelensis*.

Resistance studies have been carried out by the KABS with populations of *Ae. vexans* which were constantly exposed to *B. thuringiensis israelensis* over a period of 10 years and were therefore subjected to constant and intense selection pressure. These mosquitoes were compared with *Ae. vexans* populations taken from a remote location which had never been exposed to *B. thuringiensis israelensis* and had never been under selection pressure. No reduction in the sensitivity of these mosquitoes to *B. thuringiensis israelensis* could be detected (Becker and Ludwig, 1993). Similar results were obtained by Kurtak *et al.* (1989) and by Hougard and Back (1992). They found that after 10 years of intensive application of *B. thuringiensis israelensis* in West Africa the susceptibility of the blackfly, *Simulium damnosum*, had not changed.

The complex mode of action of *B. thuringiensis israelensis* partly explains the relative absence of resistance. The lethal changes in the midgut cells are induced only by the synergistic effects of the different toxin proteins present in the parasporal body of *B. thuringiensis israelensis*. This combination reduces the likelihood of resistance. On the other hand, when the gene encoding a single toxin protein was cloned into a microorganism and then fed to larval mosquitoes, resistance was induced within a few generations (Georghiou and Wirth, 1997).

Resistance is less likely when there is a variable gene pool of target populations. The large size of many mosquito and blackfly populations thus inhibits the development of resistance because only a portion of the population is exposed to the toxin. Floodwater mosquitoes and most blackflies undertake considerable migrations. This behaviour produces a substantial gene flow within their populations which should at least slow the onset of resistance.

However, resistance to *B. sphaericus* has been demonstrated in both the laboratory and the field. In southern France, a population of *Cx. pipiens* developed a level of resistance of more than 16 000 fold after 18 rounds of *B. sphaericus* treatment (Sinegre *et al.*, 1994). Nielsen-LeRoux *et al.* (1995) could demonstrate in a laboratory population of *Cx. pipiens* that resistance at a level of 100 000 fold to *B. sphaericus* binary toxin can be caused due to a change in the receptor on midgut brush-border membranes. In other cases, binding to the receptor took place but there was no toxicity. In all cases resistance was shown to be recessive (Charles and Nielsen-LeRoux, 1996). It seems that the risk of resistance to bacterial toxins is inversely proportional to the complexity of the mode of action, which is definitely less complex with *B. sphaericus* than with *B. thuringiensis israelensis*.

Suitability for integrated control programmes with community participation

Bacterial control agents are particularly well suited for use in integrated programmes because their toxic effect is selective, so they do not affect predatory organisms. These and other predators can therefore be included as additional elements in an integrated control strategy. The effect of predators can continue after the bacterial control agents have been applied. This indirectly produces a sustained suppressive effect (Mulla, 1990).

Bacterial control agents can be mixed with other larvicides, such as preparations that are applied to alter the surface film. This enhances diffusion over the water surface and destroys mosquito pupae which are not affected by microbial toxins (Roberts, 1989).

The WHO "Primary Health Care Concept" increasingly seeks to involve local residents in the search for solutions to health care problems. Bacterial control agents have a considerable safety advantage over synthetic insecticides because neither the operator nor the occupants of treated sites become exposed to potentially dangerous chemicals. For this reason, such preparations are particularly well suited for use by volunteers.

Applications of bacterial toxins do not harm beneficial animals such as honeybees, silkworms, or aquatic animals such as fish, shrimps or oysters. These formulations can therefore be used in ecologically sensitive areas. Because they are biodegradable, no toxic residues remain after their use. Their environmental safety permits bacterial control agents to be accepted both by officials and the general public. Indigenous isolates found in the country in question can be used.

Factors influencing the efficacy of bacterial control agents

In addition to the different susceptibility of various mosquito species to bacterial toxins, a variety of factors can influence their efficacy (Becker *et al.*, 1993; Ludwig *et al.*, 1994). This efficacy depends upon the developmental stage of the target organisms, their feeding behaviour, organic content of the water, the filtration effect of target larvae as well as that of other non-target organisms, photosensitivity and other abiotic factors such as water temperature and depth, the sedimentation rate as well as the shelf-life of the *B. thuringiensis israelensis* and *B. sphaericus* formulations (Mulla *et al.*, 1990; Becker *et al.*, 1992). The long-term effect is also strongly influenced by the recycling capacity of the agent (Aly, 1985; Becker *et al.*, 1995).

Species and instar sensitivity: Larvae lose their sensitivity to bacterial toxins as they develop (Becker *et al.*, 1992). For instance, second-instar *Ae. vexans* are about 11 times more sensitive than fourth-instar larvae at a water temperature of 25°C (second-instars: $LC_{90} = 0.014 \pm 0.007$ mg/l;

fourth-instars: $LC_{90} = 0.149 \pm 0.004$ mg/l). The differences in sensitivity are less at temperatures between 8 and 15°C. Nonetheless, second-instars of *Ae. vexans* are more than twice as sensitive as fourth-instars at a water temperature of 15°C (second-instars: $LC_{90} = 0.062 \pm 0.025$; fourth-instars: $LC_{90} = 0.145 \pm 0.004$ mg/l). In field experiments only half the dosage required to kill third-instar larvae is needed for second-instar larvae. If fourth-instar larvae are dominant, then the dosage must be doubled again. It is therefore recommended that control measures commence while the larvae are at an early developmental stage.

Large differences in sensitivity can also be distinguished between various mosquito species due to differences in feeding habits, ability to activate the protoxin and the binding of the toxin to midgut cell receptors. For instance, larvae of *Cx. p. pipiens* were found to be 2–4 times less susceptible to *B. thuringiensis israelensis* than *Aedes/ Ochlerotatus* larvae of the same instar. By contrast, larvae of *Cx. p. pipiens*

are highly sensitive to the toxins of *B. sphaericus* whereas *Aedes/Ochlerotatus* larvae are much less so.

Temperature: The feeding rates of mosquito larvae are influenced by water temperature (Fig. 11.18). For instance, the feeding rate of *Ae. vexans* decreases as temperature decreases, and this is accompanied by a reduction in consumption of bacterial toxins. In bioassays, second-instar larvae *Ae. vexans* are more than 10 times less sensitive at 5°C than at 25°C (5°C: $LC_{90} = 0.145 \pm 0.065$ mg/l; 25°C: $LC_{90} = 0.0142 \pm 0.007$ mg/l) (Becker *et al.*, 1992). The same effect could be found when fourth-instar larvae were tested. Application of bacterial formulations should be conducted at a temperature threshold of 8°C when *Aedes/ Ochlerotatus* species are to be controlled.

Size of the water body: Because bacterial toxins diffuse throughout the entire body of water, deep water requires higher rates than shallow water of the same surface area. Since larvae of many mosquito species feed near the

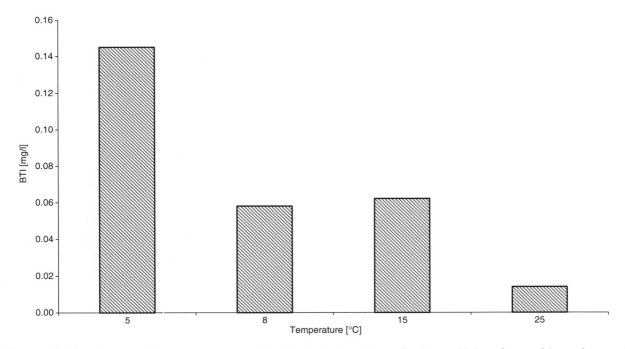

Figure 11.18. Impact of the temperature on the *B. thuringiensis israelensis* sensitivity of second-instar larvae of *Aedes vexans* (values are given as LC_{90}).

surface, effectiveness depends on the concentration and durability of the toxins in the upper 10 cm of the water body.

Larval density: Bioassays with *Ae. vexans* have shown that with increasing numbers of larvae, the amounts of *B. thuringiensis israelensis* applied must be markedly increased (Becker *et al.*, 1992). At a density of 10 fourth-instar larvae per 150 ml water, the LC_{50} value was 0.0162 ± 0.004 mg/l; with 75 larvae, the LC_{50} was about 7 times higher ($LC_{50} = 0.1107 \pm 0.02$ mg/l). The presence of other filter-feeding organisms such as Cladocera causes similar effects. In tests with *B. thuringiensis israelensis*, the LC_{50} and LC_{90} values were 5 and 6 times higher at a density of 90 *Daphnia* sp./150 ml than were those without *Daphnia* sp..

State of nutrition: The state of nutrition and the amount of available food influence the sensitivity of mosquitoes to *B. thuringiensis israelensis*. In laboratory studies two or three times more *B. thuringiensis israelensis* was required for an equal level of mortality in the presence of added food (or polluted water) compared with clean water (Mulla *et al.*, 1990).

Sunlight: Although radiation, such as that from a Cobalt[60] source, is well suited for the sterilisation of *B. thuringiensis israelensis* products without significantly reducing their toxicity (Krieg, 1986; Becker, 2002), strong sunshine appears to reduce their larvicidal effect. For example, *B. sphaericus* preparations were active for more than three times longer in shaded water than in water exposed to the sun (Sinegre, 1990). The LC_{90} values obtained with *B. thuringiensis israelensis* (6000 ITUs/mg) in bioassays with third-instar *Cx. p. pipiens* in sunny sites (6000–12 000 lux for 7 hours) and in shaded sites (<150 lux) at the same time and under identical conditions (t = 25 ± 1°C) were very different (Becker *et al.*, 1992). Whilst the LC_{90} value was 0.054 ± 0.008 ppm under shaded conditions, it was about four times higher at sunny sites ($LC_{90} = 0.235 \pm 0.036$ ppm).

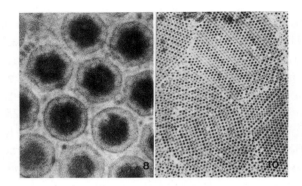

Figure 11.19. Iridescent viruses in the fat body of an *Aedes cantans* larva, 160 000× (left) 7500× (right).

Recycling processes: A particularly attractive feature of *B. sphaericus* is its high efficacy against *Culex* species and its potential to persist and recycle under certain field conditions (Des Rochers and Garcia, 1984). In laboratory tests it was shown that the presence of mosquito larval cadavers in the water contributes to the maintenance of toxic levels of *B. sphaericus*. Larval cadavers seem to contain all the nutrients necessary both for vegetative multiplication of the bacteria and for toxin synthesis associated with the sporulation process. Aly (1985) was also able to demonstrate experimentally the germination of *B. thuringiensis israelensis* in the gut of *Aedes/Ochlerotatus* larvae. When compared with other environmental conditions it seems that larval cadavers are of crucial importance for recycling processes (Becker *et al.*, 1995).

It is important to understand the impact of these factors on routine treatments. Particularly because it allows the correct calculation of the optimal dosage, the selection of the right formulation in various environmental situations, and the optimal timing for application against different mosquito species (Becker and Rettich, 1994).

11.4.4. Viruses

None of the viruses so far known are suitable for the control of Diptera, although a number of viruses have been isolated from different members of the order.

Nuclear polyhedrosis virus and cytoplasmic polyhedrosis virus are known as common pathogens of Lepidoptera, but these viruses are less frequent in Diptera populations and are usually of low pathogenicity and therefore are not used for control.

Iridescent virus infections are most common in mosquito larvae (Fig. 11.19), but less common in blackfly larvae. The infected larvae are blue-green or violet iridescent. None of these viruses nor any of the parvoviruses known to be infectious for *Ae. aegypti, Cx. p. pipiens* or *Simulium vittatum* can yet be used as potential microbial control agents (Weiser, 1991).

Chemical Control

12.1. HISTORY

Man has to compete with insects for many of his fundamental needs, and for this reason battle lines have been drawn out since prehistoric days. There is evidence that humans have been engaged in war with pests for more than 4000 years (Conway, 1976). Astonishingly, although the conflict is old, its dynamics and the behaviour of its main participants are still not well understood. Even the objectives, at least of man, are poorly defined. Only in recent decades have people begun to ask fundamental questions, and to search for sophisticated methods which would respect both nature, and human needs.

In the early days man used inorganic materials to combat insect pests. The Sumerians apparently used sulphur compounds to control insects well before 2500 BC. The Chinese in 1200 BC used plant-derived fumigants as well as mercury and arsenic compounds as insecticides (Retnakaran *et al.*, 1985). Plinius the Elder in his *Historia Naturalis* (70 AD), includes a summary of pest control practices extracted from the Greek literature of the preceding 200–300 years. Most of the methods and materials usually based on superstition and folklore were useless. Frustration with insurmountable insect problems often resulted in the use of illogical methods to overcome the difficulties of plagues and/or damages caused.

The first written accounts of a powder ground from the heads of pyrethrum flowers are by Boccone in 1697 and Buxbaum in 1728 (Ruigt, 1985), who both mentioned that people on the Asian side of the Caucasus were aware of the existence of an insecticidal factor in local pyrethrum species (*Pyrethrum carneum* and *P. roseum*). By the mid-nineteenth century pests were controlled with some degree of success by the use of chemicals. Between 1800 and 1825, pyrethrum, lime and sulphur combinations were recommended as insecticides, arsenic suggested for sheep scab control, and mercuric chloride for bedbug control. From 1825 to 1850, phosphorous paste was the official rodenticide for rats in Prussia, and rotenone was used as an insecticide in Asia (Ware, 1989). Around 1870, a range of secondary plant products—nicotine, derris and quassia—were recommended against winter eggs of the plant louse (*Phylloxera* sp.). With these products, and the use of arsenic Paris green and kerosene emulsion, the era of professional application of pesticides had begun.

The history of mosquito control in modern times goes back to the well known discovery by Ronald Ross in 1897 that mosquitoes are vectors of malaria. Before that, mosquitoes were considered as a plague, a part of the natural order of things, something to be endured.

In his review of the history of malaria control, Boyd (1949) distinguished two eras, essentially before and after the establishment of the role of mosquitoes in pathogen transmission. At this time it would be interesting to explore how art and the practice of mosquito control has fared since Ross's discovery.

Following the report by Gillies (1978), one can recognise several alternating periods of confidence or doubt over the prospects for the attempted elimination of malaria from the world. Each period of optimism has been ushered in by a major advance in knowledge and/or technology, and has been followed by a period of disillusion when the hopes offered by the advances of the previous decades were not fulfilled. And it has remained so until today; periods of optimism have been followed by frustration resulting from the failure to make further progress.

In attempting to review the present status of insecticides used for mosquito control, it may be helpful to briefly examine the recent past.

The first recorded use of a synthetic organic insecticide, dinitro-*o*-cresol, occurred in 1892, and by the 1930s a range of such compounds had been discovered and found limited use (Cremlyn, 1978). From the 1920s onwards, the increasing potency of insecticides as tools for insect control led to their growing predominance as an approach to insect control.

A series of dramatic discoveries during the late 1930s, provided new synthetic insecticides which had enormous potential for wide-spread use, and which further reinforced the emphasis on a chemical approach to insect control.

In 1939 the "wonder" insecticide of the chlorinated hydrocarbon group, dichlorodiphenyltrichloroethane, best known as DDT, was introduced. This was soon followed by the development of organophosphates in Germany, which had been evolving since 1932, when Lange and von Krueger first synthesised organofluorophosphate esters. In the early 1950s carbamates were created in Switzerland. Between 1960–1970s the discovery and development of photostable pyrethroids, mainly in Japan and the UK, led to profound changes in the practice of insecticide utilisation. These exceptionally potent, biodegradable compounds may be used in the field at rates as low as 20 g/ha, which is 10–100 fold lower than more conventional insecticides, and eventually leads to a lower burden of chemicals in the environment. Unfortunately, although they are an impressive achievement, the synthetic pyrethroids are not the perfect answer to the insecticide problem, since they have a high toxicity to many aquatic species, vertebrate and invertebrate, and to many beneficial insects.

The adverse effects of the sometimes indiscriminate and excessive use of DDT were dramatically portrayed by the publication of Rachel Carson's "Silent Spring" (1962). As a result of the problems with chlorinated hydrocarbons, alternative methods were explored. These alternatives include insecticides developed as an outcome of rational leads from basic entomological research on metabolic disrupters, moult inhibitors and behaviour modifiers of insects. Highly potent inhibitors of chitin synthesis in the insect integument were developed in the early 1970s in the Netherlands. Since the target site of action for these chemicals is known and is susceptible to disruption only in certain species at certain times during the life cycle, these materials are thought to have fewer serious deleterious effects on non-target species. A second example is based on the discoveries that insect juvenile hormones regulate many developmental functions in insects and that they are unique to arthropods. This initiated the synthesis of juvenoids in the early 1980s for selective insect control. As with other alternative strategies for insect control, the full potential of juvenoids in selective and environmentally sound insect control, has not been fully realised to date.

The 1960s and 1970s also experienced increased research on genetic control of vectors. Numerous areas of genetic techniques such as translocations, cytoplasmatic incompatibility, meiotic drive and sterile hybrids were investigated, but none of them became usable. Similarly, much research was conducted on chemosterilisants, but none of the compounds became operational in vector control programmes. It is also important to note the extensive research on the evaluation and development of biological agents such as viruses, bacteria, protozoa, fungi, nematodes, and other parasites and pathogens. So far, with the exception of bacterial agents, none of these have emerged as widely useful tools in vector control programmes (Mulla, 1994).

The use of *B. thuringiensis* var. *israelensis* which was introduced into mosquito control programmes at the beginning of the 1980s, is discussed in detail in Chapter 11.

The effects of pesticides on non-target organisms and the environment, have been a source of worldwide contention for more than twenty years, and form the basis for most legislation intended to control or prohibit the use of specific pesticides. The most readily identified adverse effects of pesticides on non-targets were those of the persistent organochlorine insecticides (such as DDT) and their metabolites or conversion products, on certain species of fish and birds. Consequences less readily identifiable include the effects of pesticide residues in food and the environment, on humans and domestic animals.

Due to difficulties in recognising the diversity of the ecological relationships between non-target and the target organisms, notable mistakes have been made in the nonselective use of pesticides. However, these mistakes do not appear to have had permanent or irreversible consequences on non-targets and the environment. Great efforts have been made to safeguard the protection and preservation of non-target organisms.

These environmental problems were not the only ones to become evident as synthetic organic insecticides came into widespread use. Resistance, although not a new phenomenon, greatly and rapidly increased in a number of agriculturally and medically important species. The need for higher dosages and more frequent applications of insecticides to combat these pesticide-induced problems has, on occasions, been disastrous.

As late as 1955, *Plasmodium* spp. infected considerably more than 200 million people causing malaria throughout the world. The annual death rate from this debilitating disease has been reduced from 6 million in 1939 to 2.5 million in 1965 and to less than 2 million today. Through the use of insecticides, similar progress has been made in controlling other important tropical disease. However in spite of the progress made, it appears that there still remains the ever lurking danger to humans from mosquito borne diseases such as malaria, dengue fever, encephalitis, yellow fever, lymphatic filariasis *etc*. The number of deaths resulting from all wars appears insignificant alongside the toll taken by insect-borne diseases.

The contemporary strategy in successfully implementing pest and vector management programmes is to carefully select from among a variety of techniques, the combination of control options which is best suited to achieve the objectives, and at the same time to preserve the ecological balance. This approach, which has to maintain a high flexibility, has gained much support not only among scientists, but also among the general public in the last 20 years, and is referred to as Integrated Pest Management (IPM).

12.2. INSECTICIDES

When millions of humans are killed or disabled annually from insect-borne diseases, and the world toll of insects, vector-borne diseases, weeds, and rodents is estimated at $100 billion annually, it becomes apparent that the control of various harmful organisms is essential for the future development of human health, agriculture and industry. In the process of accommodating these vital human requirements, pesticides have consequently become an indispensable part of the global process. In the 1940s the chemical era was opened with the discovery of synthetic organic insecticides, and a totally new concept of insect control began.

12.2.1. Classification of Insecticides

Insecticides used in mosquito control belong to four major chemical groups, chlorinated hydrocarbons, organophosphates, carbamates, and pyrethroids, and a special class, the so-called insect growth regulators (IGRs).

The first generation of insecticides are the stomach poisons, such as arsenicals. The second generation include the contact insecticides: chlorinated hydrocarbons, organophosphates, carbamates and pyrethroids. After having extensively studied the various physiological effects of the juvenile hormone, Williams in 1967 suggested that their analogues (juvenoids) could be used as insect specific control agents to which pest species may be unable to develop resistance. He referred to this class of compounds as "third-generation pesticides". The fourth generation of insecticides has been derived from the entomopathogenic abilities of some microbials. Commercial preparations based on bacteria have been available since the beginning of the 1980s and form an important part of mosquito control management. The microbial insecticides have a substantial role in mosquito control programmes in Europe and the United States, and are discussed in Chapters 11 and 16.

All the above mentioned compounds include substances which cause mosquito mortality and may be categorised in various ways for different purposes. They could also be classified according to their uses on the stage of insect life cycle acted upon: larvicides or adulticides (Table 12.1), or by their mode of action against the insects: stomach poison, contact poison, fumigant poison, *etc.*

One of the WHO classifications of insecticides is on the basis of the acute toxicity of the pesticide to laboratory animals. The hazard of a compound is assessed by determining the lethal

Table 12.1. Main Uses of Synthetic Organic Insecticides by Chemical Group

Chemical group	Larvicides	Adulticides	
		Space application	Residual application
Chlorinated hydrocarbons			X
Organophosphates	X	X	X
Carbamates		X	X
Pyrethroids	X	X	X

effect on test animals, usually rats, when it is applied to their skin (dermal toxicity) or when it is ingested (oral toxicity). The dose which is statistically determined to be lethal to 50% of the tested animals (LD_{50}) is assumed to be a measure of the hazard to man and other mammals. The WHO latest recommended classification of pesticides by hazard was issued in 1994 (Table 12.2).

Finally, insecticides can be classified as inorganic, natural organic and synthetic organic. Most of the insecticides used today for mosquito control fall in the last category. A list of insecticides which may be used in mosquito control programmes according to the WHO Operational Manual on the application of insecticides for control of the mosquito vectors of malaria and other diseases (1992) is given in Table 12.3.

12.2.2. Insecticide Formulations

The active ingredient of a chemical insecticide is rarely suitable for application in its pure form. Once the insecticide is manufactured in its relatively pure form, the technical grade material is then formulated. It is processed into a usable form for direct application, or for dilution prior to an application. The formulation is the form in which the pesticide is available at the market.

It is usually necessary to add other nonpesticide substances to the active ingredient so that the chemical can be used at the required

Table 12.2. WHO Criteria for Classifying Pesticides by Hazard

| Class of toxicity of product or formulation | LD$_{50}$ for rat (mg/kg body weight) | | | |
| | Oral | | Dermal | |
	Solids	Liquids	Solids	Liquids
IA extremely hazardous	5 or less	20 or less	10 or less	40 or less
IB highly hazardous	>5–50	>20–200	>10–100	>40–400
II moderately hazardous	>50–500	>200–2000	>100–1000	>400–4000
III slightly hazardous	>500–2000	>2000–3000	>1000	>4000
III N negligible hazard	>2000	>3000	—	—

concentration and in an appropriate form, permitting ease of handling, transportation, safety, application, effectiveness, and storage. To meet these needs chemical insecticides may be formulated in a number of ways, to enhance their lethal effectiveness, simplify their preparation for application, and meet different requirements of larviciding and adulticiding. Each type may be supplied in a variety of concentrations. The concentration of a formulation is designated either on a weight to weight basis, when both the active ingredient (a.i.) and the formulation are solid, such as DDT 5% dust, or on a weight to volume basis, when a.i. is solid and the formulation liquid, such as chlorpyrifos 400 g/litre emulsifiable concentrate.

By improvement of the formulations, the effectiveness, persistence and/or safety of insecticides can be increased. An example is the recent development of a slow-release larvicide formulation, designed to last longer when applied into the water. Larviciding, especially with microbial agents or IGRs, could be made more persistent and affordable by appropriately developed formulations which release the agent slowly and at the appropriate depth where it will be most easily encountered by the target larvae.

The safety of space treatments has been increased by development of water based formulations or replacement of emulsifiable concentrates by suspension concentrates.

Technical concentrates (TC/TP) are powders or liquids containing the highest concentration of active ingredient which can be manufactured, packaged, shipped or reformulated into more dilute formulations. Chemicals are rarely used in this form, except for some ultra-low-volume (ULV) formulations used for control of adult mosquitoes.

Solution concentrates (SC) are liquids containing a high concentration of active ingredient that has been diluted with oil or petroleum solvents. These solution concentrates may also be used undiluted for ULV applications, and are commonly used for thermal aerosol application (fogging) when further diluted with highly refined oils or kerosene.

Emulsifiable concentrates (EC) are solution concentrates which are first dissolved in an organic solvent combined with emulsifying agents. This product can then be diluted with water to an appropriate concentration in the spray tank before use. The emulsifying agent or agents cause the technical product and organic

Table 12.3. List of Chemical Insecticides which may be Used in Mosquito Control, WHO (1997c)

Active ingredient	Use[a]	Chemical type or class[b]	Toxicity class
Alphacypermethrin	A	PY	II
Bendiocarb	A	C	II
Bioresmethrin	A	PY	III
Chlorpyrifos	L, A	OP	II
Chlorpyrifos-methyl	L	OP	II
Cypermethrin	A	PY	III N
Cyfluthrin	A	PY	II
DDT	A	OC	II
Deltamethrin	L, A	PY	II
Diazinon	L, A	OP	II
Dichlorvos	L, A*	OP	I B
Diflubenzuron	L	IGR	III N
Etofenprox	L	PY	III N
Fenitrothion	L, A	OP	II
Fenthion	L, A	OP	I B
Jodfenphos	L, A	OP	III N
Lambdacyhalothrin	A	PY	II
Malathion	L, A	OP	III
Methoprene	L	IGR	III N
Methoxychlor	A	OC	III N
Naled	L, A	OP	II
Permethrin**	L, A	PY	II
Pirimiphos-methyl	L, A	OP	III
Propoxur	A	C	II
Pyrethrins	A	n.o.	II
Pyripoxiphen	L	IGR	III
Resmethrin	A	PY	III
Surface film	L	SF	—
Triflumuron	L	IGR	III
Temephos	L	OP	III N

*Fumigant may be dangerous in oil solution.
**May be dangerous in oil solution.
[a]L = larvicide; A = adulticide.
[b]IGR = Insect Growth Regulator; OC = chlorinated hydrocarbons; OP = organophosphates; C = carbamates; PY = pyrethroids; SF = surface film; n.o. = natural organic.

solvent solution to disperse evenly in the water when mixed by stirring. It is sometimes necessary to maintain a level of agitation in order to keep the chemical dispersed in the tank. Emulsifiable concentrates are used in larviciding when dispersion throughout water is required and they may be used undiluted as ULV sprays. They are sometimes also used for residual spraying of impervious surfaces such as houses in urban and semi-urban areas.

Ultra-low-volume formulations (ULV) solution concentrates have, as a rule, a low volatility, high viscosity and high specific gravity. ULV insecticides are applied as a mist or aerosols in space spraying. ULV treatment indicates a technique for applying the minimum amount of liquid per unit area from the air or ground, which by definition is less than 5 litre/ha. This type of operation has resulted in substantial savings in mosquito control programmes through speed and reduction in labour requirements, but demands very accurate application since the droplet size of ULV formulation is of considerable importance. Droplet size affects:

(a) the chances of penetrating through vegetation and other obstacles to reach the target;
(b) the total number of droplets per unit volume of insecticides and the toxic content per average droplet;
(c) the distribution of droplets on treated surfaces;
(d) the time for which a droplet remains airborne.

Both favourable and unfavourable results from using ultra-low-volume insecticide application have been reported. Disparity in results obtained has been due primarily to the weather conditions, insecticide application rate and the degree of insecticide penetration into the target area. As might be expected, applications at higher rates together with good penetration of dwellings have proved the best control of adult vector populations (Gratz, 1991).

Water dispersible powders (WDP/WP) are, as the name suggests, insoluble powders that resemble dust, but which may be dispersed in water. The technical product can be either a solid or liquid. Liquid technical products and concentrates can be absorbed onto an inert solid, and then used as a water dispersible powder. Solid formulations are ground to a fine powder and surface active agents added to promote dispersion of the particles, without excessive foaming, when they are added to water. Wettable powders form a suspension rather than a solution when added to water, and need regular agitation after being added to water. The diluents in some wettable powders are abrasive and can wear down pumps and nozzles, which then have to be regularly checked and replaced (Wade, 1997). Most house spraying is done with water dispersible powders. They are usually applied at high volume rates with low concentrations of the active ingredient.

Soluble powders (SP) are similar to wettable powders, except they form true solutions in water. Because they dissolve completely in water they do not need constant agitation nor do they abrade equipment.

Granules are inert carriers impregnated or coated with the solution concentrate of an insecticide. They are made by applying a liquid formulation of an active ingredient to coarse particles of some type of absorptive material such as attapulgite, clay, or corn cobs, or to a non-sorbent material such as sand or other mineral or non-mineral substance. Even frozen water in the form of ice cubes, may be used as a carrier for some materials such as microbial agents, and has been recently used successfully for aerial treatments of flooded habitats (Becker, 2003). Granular treatments can be made at any time of day since they can be applied aerially in winds up to 20 km/h without significant drift.

Vapourising strips or tablets are special formulations of insecticides, mainly for household use, which are usually made from polymeric materials containing a volatile compound. They release the volatile insecticide when exposed to the air or heat.

Aerosols: The active ingredients must be soluble in the volatile, pressurised petroleum solvent to which a propellant is added. Fluorocarbon propellant was used until recently, but has now been largely replaced by the less detrimental carbon dioxide. When the petroleum solvent is atomised, it evaporates rapidly leaving the micro-droplets of toxicant suspended in the air. Aerosols commonly produce droplets well below 10 μm in diameter. Consequently, aerosols of all varieties should be inhaled as little as possible. The "push-button" variety of aerosol dispenser was first developed during World War II. More recently the total release aerosols have been designed to discharge the entire contents in a single application and are available for home owners as well as for commercial operators. They are effective only against flying and crawling insects and provide little residual effects.

Slow release systems blend an active ingredient with a material from which it will evaporate or be released at a controlled rate. Ingredients that are volatile or subject to degradation, e.g. by sunlight, remain active in these formulations much longer.

Micro-encapsulated products are a form of a slow-release formulation in which the active ingredient is enclosed in a material such as polyamide, neoprene, polyvinyl chloride or polyester. The active ingredient can diffuse from the matrix. Micro-capsule suspensions (CS/MC) are of two types:

(a) porous wall capsules usually containing diazinon, chlorpyrifos, permethrin *etc.*, through which active material slowly leaks over time;

(b) non-porous capsules (*e.g.* fenitrothion), that remain intact until ruptured or burst by tarsal contact with insect pests.

Encapsulated materials generally demonstrate a reduced absolute mortality together with extended residuality (Wade, 1997). To this date, the formulations offering the most lasting residual effect are those in which the active ingredient is present in micro-emulsion together with a surface film forming system "Secondary system EW's" (S-EW).

Pellets and briquettes are larger in size than granules, varying in size from 1.5 cm in diameter for pellets (tablet-like shape) to several cm for briquettes. The size and weight of briquettes vary depending on the carrier and amount of other additives present. Both pellets and briquettes release active material as they dissolve and disintegrate in mosquito breeding water bodies.

Extended residual briquettes (XR) are designed to release effective levels of active ingredient over a longer period into the environment. Release of active material occurs by dissolution of the briquette. XR briquettes are of a special value in permanent, hard to reach breeding sites, or in situations where a pre- or post-flood treatment provides a long-term effect.

12.2.3. Insecticide Application Techniques

The success of an insecticide application programme depends very much on the quality of the equipment used, and the quality of its operation. Manuals on mosquito control refer to a wide range of application equipment for delivering insecticides to the target site. In this section of the chapter, the more commonly used insecticide application equipment in mosquito control programmes is briefly described, according to WHO/CTD/WHOPES (1997).

Hand-operated compression sprayer: These sprayers are designed to apply insecticides onto surfaces with which the insect will come into contact, or into a breeding site. An insecticide formulation and water are either mixed before filling the tank or mixed within it. The tank is then pressurised by a hand-operated plunger. A trigger on the sprayer lance controls the release of spray through a nozzle. Filtering the water while filling the sprayer, regular sprayer maintenance by qualified personnel, and regular nozzle checks and calibration, are essential in maintaining the sprayers' effectiveness (Wade, 1997). Abrasion from particles in the water can cause deterioration of the nozzle, resulting in an excessive increase in the insecticide liquid,

which in turn may result in application of an incorrect or uneven insecticide dose. A drawback of this kind of sprayer is that as the tank is emptied, pressure decreases causing a fall in the discharge rate, unless the sprayer is provided with a pressure control device. Many of the problems with these sprayers at the treatment sites are related to a lack of proper cleaning at the end of each day.

Mist blower (power-operated): This equipment can either be portable or vehicle-mounted. Portable knapsack mist blowers are powered by a two stroke engine producing a high velocity air stream in which the insecticide formulation is atomised as a fine mist. The volume discharge can be adjusted through flow restrictors. At high flow rates large droplets are produced. Large droplets coat surfaces, whereas the smaller droplets act as an aerosol, impacting on insects both in flight or at rest. Although the insecticide formulation is diluted in water, the overall volumes applied with mist blowers are still relatively small. Knapsack mist blowers can cover a large area in a relatively short time, and provide ease of access to areas which vehicle-mounted equipment cannot reach. The disadvantage of the sprayer is the risk of burns from the engine, and discomfort caused by heat, vibration and noise.

Aerosol generators (power-operated): These are cold aerosol ULV sprayers for application of insecticides as technical material, or more usually, diluted in oil or water. Only formulations recommended for ULV use by the manufacturer should be applied by this equipment. The machines can be hand-held or truck mounted, depending on their size. The volume of material used per unit area is much lower than with thermal foggers or mist blowers. By using ULV generators a larger area can be covered more quickly. Portable ULV generators may also be used for indoor treatment, when access is difficult with compression sprayers. With ULV aerosol generators, the calibration and accuracy of the droplet size is particularly important.

Thermal foggers (power-operated): These machines, which are either portable or vehicle mounted, are preferred in some vector control programmes where the very visible fog treatment is perceived as more effective. In most parts of Europe however, the same feature is repulsive for many people, especially those with environmental concerns. The droplet size is far less controlled than with ULV machines, and a wide range of droplet sizes from less than 1–200 μm are produced. In such situations some insecticide will be wasted due to convection currents or early fall out.

Aerial spraying equipment: Large-scale and emergency vector control programmes often employ aircraft to apply insecticides. The aircraft are well suited for the rapid treatment of large areas where there is no access to the target site, or when vegetation is dense. Aerial treatment can be used both for adulticide and larvicide application. Accurate placement of the insecticide formulation is usually more difficult with aircraft than with ground application equipment.

Insecticides which have a rapid knockdown effect on mosquitoes and are relatively harmless to other organisms are best suited for space spraying. Organophosphorous compounds first became widely used for such applications, and later various formulations of carbamates and synthetic pyrethroids have become available. According to the WHO operational manual on the application of insecticides for control of the mosquito vectors of the malaria and other diseases (1996), there are 19 compounds suitable for cold aerosol sprays and thermal fogs (Table 12.4).

Table 12.4. Insecticides Suitable for Application as Cold Aerosol Sprays or Thermal Fogs for Mosquito Control (WHO, 1996)

Insecticide	Chemical[a]	Dosage of a.i.[b] (g/ha)		Toxicity[c] (mg/kg of body weight)
		Cold aerosols	Thermal fogs[d]	
Bendiocarb	C	4–16	—	55
Bioresmethrin	PY	5	10	>7 000
Chlorpyrifos	OP	10–40	150–200	135
Cyfluthrin	PY	1–2	2	250
Cypermethrin	PY	1–3	—	250
Cyphenothrin	PY	2–5	—	318
Deltamethrin	PY	0.5–1.0	—	135
Dichlorvos	OP	150	200–300	56
D-phenothrin	PY	5–10	—	>5 000
Etofenprox	PY	10–20	10–20	>10 000
Fenitrothion	OP	250–300	270–300	503
Lambdacyhalothrin	PY	1.0	1.0	56
Malathion	OP	112–693	500–600	2 100
Naled	OP	56–280	—	430
Permethrin	PY	5	10	500
Pirimiphos-methyl	OP	230–330	180–200	2 018
Propoxur	C	100	—	95
Resmethrin	PY	2–4	—	2 000
Zeta-cypermethrin	PY	1–3	—	106

[a]C = carbamate; OP = organophosphate; PY = pyrethroid.
[b]a.i. = active ingredient.
[c]Toxicity: oral LD_{50} of a.i. for rats. (Toxicity and hazard are not necessarily equivalent.)
[d]The concentration of the finished formulation when applied depends on the performance of the spraying equipment used.

12.3. DIFFERENT CHEMICAL GROUPS OF INSECTICIDES

12.3.1. Chlorinated Hydrocarbons

The chlorinated hydrocarbons are insecticides that contain carbon, chlorine and hydrogen. They are also referred to by other names: organochlorines, chlorinated organics, chlorinated insecticides. The chlorinated hydrocarbons can be divided into subgroups according to structural differences, but they have in common chemical stability, low solubility in water, moderate solubility in organic solvents and a low vapour pressure (Hill and Waller, 1982). The stability and solubility of the chlorinated hydrocarbons means that they are highly persistent, and this may lead to long term contamination of the environment and gradual bioaccumulation in animals at the higher end of food chains. For these reasons they have been banned by most developed countries. The broad spectrum of activity of these compounds, their persistence and their hazard to the environment explain why their use in insect pest management is largely considered inappropriate, although there are still situations where this chemical group is an important control option, *e.g.* in some malaria vector control programmes.

DDT: 1,1,1-trichloro-2,2-bis (p-chlorophenyl) ethane

Dichlorodiphenyltrichloroethane (DDT) is probably the best known and most infamous chemical of the 20th century. It is recognised as the most useful insecticide ever developed. Search for a long lasting insecticide by the Swiss entomologist Paul Müller in 1939 lead to the rediscovery of the insecticidal properties of DDT, which had been first synthesised in 1874 by a German graduate student, Othmar Zeidler. DDT proved to be extremely effective against mosquitoes and flies, ultimately bringing the Nobel Prize in medicine to Müller in 1948, for his life saving discovery. No other compound has revolutionised control of insects so dramatically and in such a short time, and no other insecticide has received such attention and criticism as an environmental hazard.

The great advantages of DDT were its persistence and relative cheapness, so that it could be readily used on a large scale to protect people living around the world. Outstanding success was achieved by residual house-spraying against endophilic malaria vectors. Bruce-Chwatt (1971) pointed out that DDT has saved some 15 million lives by malaria control alone.

As is well known, most insecticides have caused problems during the last decades for two reasons: pest resistance and environmental pollution. The first of these began with DDT-resistant flies in 1947 (Busvine, 1978); a problem which has grown steadily, but so insidiously that it has attracted little attention except from those concerned with pest control. On the other hand, in the 1960s, the problem of environmental residues became front-page news. As could be expected, there was a widespread official reaction in the form of safety regulations. The result was a virtual banning of organochlorines in many countries.

DDT still, however, remains an important component of malaria control in some rural areas or countries where it is available, and where the local vector species are still susceptible (Chavasse and Yap, 1997). Even so, the assessment of the efficacy/resistance to DDT in an area is essential for planning an effective malaria control strategy.

Mode of action

In insects, DDT produces tremors throughout the body, and hyperexcitability, which are subsequently followed by loss of movement (ataxia). Beyond this point an apparent paralysis develops, which may be completed over several hours post-poisoning.

Although the compound produces symptoms consistent with a toxic action on the nervous system, there are differences between DDT and other groups of insecticides, indicating some differences in the mode of action. In some complex manner it disturbs the balance of sodium and potassium ions within the sensory neurons, thereby causing spontaneous firing, and preventing normal passage of impulses. This major symptom of DDT poisoning results from its ability to induce repetitive firing on sense organs, giant axons, presynaptic terminals, and smaller nerve fibres. Whether a given neuron fires repetitively or not depends upon the membrane properties of the nerve cell itself, and on the elevation of the negative after-potential which contributes to the repetitive discharge. The relative effectiveness of this after-potential in generating repetitive firing, depends upon the properties of the sodium gate of the particular fibre. A further aspect of this process is that DDT interferes with the stabilising action of Ca^{2+} at the axonal surface, and therefore leads to membrane destabilisation. The disruption is transmitted to the rest of the nervous system, causing the muscles to twitch, which is usually followed by convulsions and death. DDT is a relatively slow acting insecticide, and has the unusual quality of being more toxic to insects as the ambient temperature is reduced, a negative correlation that is also characteristic of some pyrethroids. Gammon (1978) showed that the chemical had excitatory effects on both peripheral and central neurons and, although peripheral effects were not quantified, the effects on central neurons became more pronounced as the temperature was reduced.

A few striking points concerning DDT should be highlighted, in order to understand some of the well documented problems associated with it. One of the most remarkable points of DDT is its chemical stability. This important property results in having a long half-life in the soil, the aquatic environment, and in plant and animal tissues. It is not readily broken down by microorganisms, enzymes, heat or ultra violet light. Secondly, DDT has been reported in the chemical literature to be probably the most water-insoluble compound ever synthesised. However, it is soluble in fatty tissue and, as a consequence of its resistance to metabolism, it is readily stored in the fatty tissue of any animal ingesting it, either alone or dissolved in food, even when it is part of another animal. It accumulates in every predator, as well as in those that eat plants bearing even traces of DDT. The principal peculiarity of this movement in the food chain is the biological magnification.

A lesson is to be learned from the rise and fall of DDT. There is a need for an informed and cautious approach to pesticide usage. It is essential to conduct basic research before we encourage widespread application of any particular product, to enable us to identify and restrict—if not to avoid—those products that may pose dangers to the environment and to health.

12.3.2. Organophosphates

The chemically less stable organophosphate insecticides (OP) partly replaced the persistent chlorinated hydrocarbons. The term organophosphate is usually used as a generic name to include all insecticides containing phosphorus. The OPs have several other commonly used names: organic phosphorous, phosphorus insecticides, phosphorus esters or phosphoric acid esters. They are all derived from phosphoric acid. The OPs have two distinctive features. As well as having a lower chemical stability, they are generally much more toxic to vertebrates than the organochlorine insecticides. Their lower persistance brought them into use as alternatives to persistent organochlorines, particularly to DDT. But unfortunately, resistance to OPs also developed, and became more widespread as these compounds were more widely used. This situation led to a plea by WHO "to encourage commercial companies to continue the search for compounds suitable for pest control" (WHO, 1976).

Mode of action

OPs exert their toxic action at synapses by inhibiting cholinesterase (ChE), an important enzyme in the nervous system. It is well established that OPs inhibit acetylcholinesterase (AChE), and acetylcholine (ACh) concentration then rises in treated insects (reviewed by Corbett, 1974). Organophosphates interact with AChE in the same way as ACh does, phosphorylating the same serine hydroxyl that ACh acetylates. Organophosphates inhibit the enzyme almost irreversibly, with dephosphorylation occurring over a period measured in weeks or months, contrary to some other chemical groups, such as carbamates, where the action of decarbamilation occurs from minutes to less than 8 hours (Taylor, 1980). This has led some to refer to organophosphates as irreversible inhibitors, and carbamates as reversible ones. This inhibition, resulting in the accumulation of acetylcholine, interferes with the neuromuscular junction, producing rapid twitching of muscles and final paralysis of the insect. However, the precise sequence of events between enzyme inhibition and death is still not fully understood. The excess acetylcholine has widespread harmful effects, which may include the uncontrolled hormone release known to accompany insect poisoning (Maddrell, 1980).

Several insecticides which have been performing an important role in mosquito control worldwide, belong to the group of OPs. Perhaps the most important one is malathion, which is also the oldest and most widely used aliphatic OP.

Malathion: 0,0-dimethyl phosphorodithioate ester of diethyl mercaptosuccinate

$$
\underset{\underset{CO_2CH_2CH_3}{|}}{(CH_3O)_2P\overset{\overset{S}{\|}}{S}CHCH_2CO_2CH_2CH_3}
$$

Malathion was introduced in 1950, and was quickly adopted in agriculture, mosquito control programmes and urban pest management. Because of its fast action and exceptionally low acute toxicity to humans and other warm-blooded animals, it was the insecticide of choice, controlling most species of household insect pests.

As a public health insecticide, malathion has found its most common use as an ultra-low-volume (ULV) formulation, in controlling adult mosquitoes. Its low mammalian toxicity allows the application of the compound in urban areas where dense mosquito populations are a nuisance or threaten human health. Many of the ULV applications of malathion are carried out by aircraft or helicopters. Aerially applied malathion ULV is still effective at the rate of 320 g/ha, in regions where resistance to OPs has not yet occurred. The application equipment has to be calibrated to dispense droplets small enough to remain suspended in the air, but large enough to contain a sufficient amount of insecticide to kill an adult mosquito on which they impinge. Results of ULV aerial treatments worldwide show that the success of ULV application depends very much on weather conditions. Above all wind speed, which has to be less than 3 m/sec, time of application (preferably dusk which coincides with the activity of the species being controlled), and the presence of inversion conditions, should be taken into account.

Malathion is also applied by ULV ground equipment. The maximum vehicle speed of 16 km/h for such treatments, often limits the area that can be treated during the period of mosquito activity. Any increase in the speed of the vehicle should only take place if the insecticide output is also increased correspondingly. Fortunately, most truck-mounted cold aerosol generators used for applying ULV insecticides for adult mosquito control, are also equipped with variable flow control systems. These systems, when properly calibrated, automatically adjust to dispense the correct amount of insecticide in relation to vehicle speed. The effect of operating pressure on the droplet volume median diameter (VMD) at various flow rates for malathion ULV is well documented. Nevertheless, for any particular situation the aerosol droplets have to be collected, counted and measured using an appropriate procedure: collect droplets on Teflon coated plates, count them using a compound microscope at a magnification 200×, take into consideration the insecticide spread factor (for malathion 0.69), and analyse the data according to an available computer programme. Dukes *et al.* (1990) determined that an increase in the flow rate required a corresponding increase in the blower pressure, to maintain the labelled malathion (Cythion) droplet median diameter of 17 μm. For a flow rate of 254.3 ml/min., at the maximum vehicle speed of 32 km/h, a machine pressure of 7–8 psi (48.3–55.2 kPa) is required to consistently achieve the droplet size criteria on the Cythion label.

As the efficacy of ULV pesticide application against mosquitoes is based on the premise of an airborne insecticide droplet impinging on a flying mosquito, this method is often ineffective against indoor resting species, due to the decreased number of droplets reaching individual target mosquitoes. This explanation has been suggested for the failure of various ground and air ULV applications against *Ae. aegypti*, such as a malathion (Cythion) treatment at a rate of 438.8 ml/ha in the Dominican Republic (Perich *et al.*, 1990), which only achieved 91% mortality against a 100% susceptible strain. Similar results have been seen in Puerto Rico (Fox, 1980) and other countries of the region. In Europe the most significant indoor species belong to the genus *Culex*. In this case ULV adulticiding is not recommended due to the behaviour pattern of the species.

Very broad spectrum activity makes this compound undesirable in most natural environments.

Naled: 1,2-dibromo-2,2-dichloroethyl dimethyl phosphate

$$\overset{\displaystyle O}{\underset{\displaystyle Br}{BrCl_2CCHOP(OCH_3)_2}}$$

Naled (Dibrom) has been a widely used brominated aliphatic OP with low mammalian toxicity. Naled was one of the choices for mosquito adult control throughout the United States, but has been withdrawn from further use (Mulla, 1994). In the Americas it was also widely applied in large area mosquito control programmes by ULV or conventional sprays. Mount *et al.* (1968) reported moderate efficacy of naled with caged *Oc. taeniorhynchus* when ULV treatments were applied at a dosage of approximately 9.0 ml a.i./ha. The results presented by Linley *et al.* (1988) showed that ULV treatments with naled, applied by vehicle-mounted equipment at 10.22 ml a.i./ha, and a droplet size of 13.5 μm, did not control *Oc. taeniorhynchus* particularly effectively. In Florida, ULV treatments with naled applied from vehicle-mounted equipment was used by many mosquito control districts predominantly as a measure against mosquitoes, but also with some impact on *Culicoides* spp. (Ceratopogonidae).

A study carried out by Howard and Oliver (1997) based on 11 years' analysis (1984–1994) of data from New York swamp habitats of the primary enzootic vector (*Cs. melanura*) of eastern equine encephalitis (EEE) showed the effect of naled treatments on the vector population. Naled treatments accomplished only short-term reductions in the mosquito abundance. Despite repeated applications, populations of *Cs. melanura* increased 15-fold in some parts of the treated area, which discredits the suggestion that naled applications could reduce the risk of EEE.

Dichlorvos (DDVP): 0,0-dimethyl-0–2,2-dichlorovinyl phosphate

$$\overset{\displaystyle O}{Cl_2C=CHOP(OCH_3)_2}$$

DDVP is also an aliphatic OP with a very high vapour pressure, giving it strong fumigant qualities. Apart from being known as an active ingredient used in vapourising strips, from which it is slowly released, DDVP also serves as an effective mosquito adulticide being applied in mixture with kerosene or diesel (Vapona 7 or Nuvan 7) through thermal fog generators. Thermal fogs using dichlorvos can achieve an immediate and high level of reduction of adult outdoor mosquitoes (mainly of *Aedes/Ochlerotatus* species). The disadvantage of such treatments is generally low persistence, and therefore DDVP treatments must be repeated at fairly short intervals to ensure continued suppression of the adult mosquito population. Bearing in mind that the mammalian toxicity of the product is very high (oral LD_{50} for rats is 56 mg/kg), DDVP is in WHO toxicity class Ib, and its use has been discontinued in most European countries. The other disadvantage, not to be neglected, is its poor, or almost absent selectivity to the entomofauna. Its only advantage is the speed of action which is useful in epidemic situations.

Fenitrothion: 0,0-dimethyl 0-(4-nitro-m-totyl) phosphorothioate

CH₃

O₂N—⟨ring⟩—OP(OCH₃)₂ (S double bond)

Fenitrothion is a phenyl derivate of OPs (with a benzene ring), and this moiety is critical in improving stability compared with the aliphatic OPs.

Fenitrothion under various trade names (Acothion EC 20, Fenitrothion 20 EC, *etc.*) has been extensively used in some European mosquito control programmes, especially as a larvicide in closed water systems (sewage, waste water containers). However in France it has produced resistance in *Aedes/Ochlerotatus* and *Culex* species (Sinegre, 1989), and also in extensively irrigated crops in Spain, where resistant strains of *An. atroparvus* and *Cx. theileri* have developed (Grandes and Sagrado, 1988). Studies conducted by Wesson (1990) have demonstrated that the larvae of 26 strains of *Ae. albopictus* from the United States, Brazil,

southeast Asia and Japan, showed different levels of susceptibility to fenitrothion when it was applied at the baseline susceptibility concentration of 0.02 mg a.i./litre.

Studies in Thailand and Malaysia showed considerable persistence of fenitrothion 40 EC when it was used as an adulticide, especially when sequentially applied indoors. Pant *et al.* (1974) showed that it was possible to control *Ae. aegypti* for as long as six months following two ULV ground applications of fenitrothion. Rivera *et al.* (1993) demonstrated that deposits from airborne sprays of fenitrothion 40 WP applied at a rate of 2 g a.i./m² were effective for control of *An. albimanus* for up to 60 days post treatment.

Fenthion: 0,0-dimethyl-0-[4-(methylthio)-m-tolyl] phosphorothioate

CH₃

CH₃S—⟨ring⟩—OP(OCH₃)₂ (S double bond)

Fenthion is another phenyl derivate which has been used widely for mosquito and fly control (Baytex). Fenthion was commonly used as a perifocal spray in the *Ae. aegypti* eradication campaign in the Americas. It is, however, too toxic to be applied as a larvicide in water that may be used for drinking purposes by man or animals. Fenthion is used as a part of an overall mosquito control programme in some regions of Florida. The product is applied at a rate of 32.5 g/ha as a 1 : 20 mixture with diesel.

Concerns over the potential of fenthion to cause adverse effects on estuarine biota appear to be unclear, although there were records of fenthion acute LC_{50} figures for mosquitoes, which exceeded those for some mysids and pink shrimp. The degree to which the risk might be apparent for non-target estuarine communities, depends on various factors such as mixing and dilution of the product, characteristics of local habitats, weather conditions, potential for cumulative effects of repeated applications during a season *etc.* (Clark *et al.*, 1987).

Wesson (1990) showed in a larval bioassay that fenthion can be successfully used in controlling *Ae. albopictus* at a rate of 0.02 mg a.i./litre. This rate was efficient in suppressing 25 out of 26 tested strains of this species. The LC_{95} values for the remaining strain (originating from the United States) showed a four-fold higher tolerance compared to susceptible strains.

Diazinon: 0-0-diethyl-(2-isopropyl-6-methyl-5-pyrimidinyl) phosphorothioate

Diazinon was probably the first heterocyclic OP derivate from this group, and was made available in 1952. The heterocyclics have generally longer lasting residues than many of the aliphatic or phenyl derivates. Diazinon has been commonly used as a residual fly spray and as an adulticide against indoor mosquitoes. It has a relatively low mammalian toxicity, and has a very good record for urban pest control, largely because it is effective against a wide range of insect pests found around homes.

Chlorpyrifos: [0,0-diethyl-0-(3,5,6-trichloro-2-pyridyl) phosphorothioate]

Chlorpyrifos is a heterocyclic organophosphate and as the active ingredient in the formulation Dursban E 48, has been used as a larvicide to control *Aedes/Ochlerotatus, Anopheles* and *Culex* species in many countries in the tropics, North America and Europe. However, resistance

appeared quickly when it was applied frequently in *Culex* habitats such as in southern France (Sinegre, 1985). In California, Schaefer and Dupras (1970) demonstrated that application of chlorpyrifos resulted in a fairly long-term control of mosquito larvae in polluted water sources. Mosquito abatement districts used chlorpyrifos especially for regular treatments of waste water. However, in 1974 control failures in these habitats were observed (Georghiou *et al.*, 1975). Susceptibility tests with chlorpyrifos conducted by Wesson (1990), showed that some of the 26 *Ae. albopictus* strains (coming from Brazil and southeast Asia) exposed to chlorpyrifos at a concentration of 0.01 mg a.i./ litre, displayed tolerance to this insecticide. However, the majority of strains tested were still susceptible to the chlorpyrifos concentration used.

Chlorpyrifos in various formulations has given very good control and long persistence in a variety of different container types (Glancey *et al.*, 1968; Taylor, 1968). Its mammalian toxicity would exclude it from use in potable water containers, but it would be a useful product in such larval habitats as flower vases at cemeteries, and in small containers maintaining waste water. Chlorpyrifos has also been used for ULV treatment, particularly in the United States

The use of this compound in open water systems, inhabited by fish, is not justifiable because of fish toxicity.

Pirimiphos methyl: 0,[2-(diethylamino)-6-methyl-4-pyrimidinyl] 0,0-dimethyl phosphorothioate

Pirimiphos methyl, also a heterocyclic OP, often known by its trade name Actellic, became the insecticide of choice in some malaria control regions after resistance was recorded in anopheline species to some other OPs, carbamates and to some synthetic pyrethroids. Resistance of *An. sacharovi* in Turkey to phoxim, chlorpyriphos, propoxur, and tolerance to malathion, was reported by Ramsdale (1975), and Davidson (1982). However, surface treatment of pirimiphos-methyl 50% EC at 0.9 g/m² caused a significant decrease in parous rate and 96.9% reduction in resting density. The persistence on different surface materials for seven weeks post-spray resulted in mortality rates of 73 to 98% (Kasap *et al.*, 1992).

For residual mosquito control, pirimiphos methyl may be used at rates of up to 2 g a.i./m². The optimum volume will vary with the surface to be sprayed. Mosquito larvae in shallow water (up to 10 cm deep) may be controlled with 1 part of pirimiphos-methyl 50% EC mixed with 99 parts of water (0.5%) and applied at a volume of 100 l/ha (10 ml/m²). At rates of 0.5 to 0.1% a.i., depending on the water depth, suppression of mosquito larvae, with retreatments after 5–8 days, can be achieved.

Temephos: 0,0′-(thiodi-4-1-phenylene) 0,0,0,0-tetramethyl phosphorothioate

$$(CH_3O)_2PO-\!\!\!\!\bigcirc\!\!\!\!-S-\!\!\!\!\bigcirc\!\!\!\!-OP(OCH_3)_2$$

Temephos is a heterocyclic OP, and is normally available under the trade name Abate. It has been used as a larvicide in mosquito control programmes for a longer period of time than most other products. Due to its very low mammalian toxicity, with an LD_{50} of 2030 mg/kg (WHO class III N), it is used in many different aquatic habitats, including water intended for human consumption. So far, the only insecticide compounds that have been toxicologically approved as safe for use in potable waters are temephos, permethrin, methoprene and products based on *B. thuringiensis israelensis* (WHO, 1991; WHO-IPCS, 1993). The disadvantage of temephos is a disagreeable smell which remains after the treatment of containers.

Any larvicide intended to be used against mosquitoes occurring in ecologically valuable habitats, or against mosquitoes breeding in potable water must have an extremely low level of mammalian toxicity, along with a very low level of toxicity to non-target invertebrates. In this respect temephos has a different profile from some other larvicides (IGRs and microbials) used in the control of mosquito larvae. When applied at a rate of 0.3 litre/ha, temephos EC formulation was very effective against all larval instars of *Aedes/ Ochlerotatus, Culex* and *Anopheles*. However it also affected populations of *Podura aquatica* (Collembola) and immature stages of Odonata, Ephemeroptera, Coleoptera and Diptera: Chironomidae (Zgomba *et al.*, 1983; Zgomba, 1987).

The sand granule formulations containing 1% to 5% of active ingredient, have been successfully used as larvicides in a wide range of flood water situations, controlling *Aedes/ Ochlerotatus* larvae. The application rate of this product depends on the water quality, and ranges from 10 to 20 kg/ha for 1% temephos sand granules. It can be applied by hand, by ground equipment, from a boat, or by helicopter and aircraft.

A number of field trials and programmes have shown the feasibility of larval control of *Ae. aegypti* using a larvicide, or combination of larvicides and appropriate environmental measures. A reduction of adult density by up to 95.4% was achieved when temephos was applied in 1% sand granule formulation to water bodies containing larvae of *Ae. aegypti* in Thailand. When temephos EC was sprayed in India and Burkina Faso, the period of larval control, following a single application, lasted up to eight weeks (Geevarghese *et al.*, 1977; Hervy and Kambou, 1978).

Resistance to temephos has appeared in some areas following intensive use of the compound. It was reported by Sinegre (1984) in some parts of France, in agricultural zones of Spain by Grandes and Sagrado (1988) and in the Dominican Republic by Mekuria *et al.* (1991).

An increased level of tolerance elsewhere, such as in the Caribbean islands (Georghiou *et al.*, 1987) and French Polynesia, has also been reported on several other occasions from mosquito control programmes, but it is assumed that the level of temephos resistance is not high enough to disrupt its operational use. According to Wesson (1990) who examined the OP larval susceptibility of 26 *Ae. albopictus* strains, three U.S. strains and one Japanese strain had LC_{95} values at or above the diagnostic concentration, which for temephos was 0.04 mg a.i./litre. The tolerance level of one strain from the Americas was increased three times. All Brazilian and southeast Asian strains had LC_{95} values below the diagnostic concentration. After reports from several authors, Breaud (1993) reviewed mosquito resistance to temephos in Florida. The most abundant *Culex* and *Aedes/Ochlerotatus* species already showed tolerance to temephos in 1979 (6X and 39X), with a tendency to develop resistance to other larvicides and adulticides from the OP group in the region.

12.3.3. Carbamates

Carbamates are derivates of carbamic acid, and were first introduced in 1951 by the Geigy Chemical Company in Switzerland. Before the first synthetic carbamate was produced, the first natural carbamate was discovered in grapes from African vine in the mid 19th century. Due to its instability, the corresponding dimethylcarbamate was synthesised (Aeschlimann and Reinert, 1931). Initially it was not known that N,N-dimethyl carbamates were generally less toxic to insects than the later synthesised N-methyl carbamates. This latter group was developed subsequently, and became the more important group of these insecticides.

Mode of action

The carbamates' mode of action is basically the same as that of organophosphates. They affect the activity of acetylcholinesterase (AChE), which catalyses the hydrolysis of acetylcholine (ACh), the chemical neurotransmitter which acts at synapses in the nervous system of the insect. Following the inhibition of the enzyme caused by carbamylation, ACh accumulates thus prolonging the action of the neurotransmitter at the cholinergic synapses. The resulting hyperexcitation of the nervous system is accompanied by convulsions, then paralysis and ultimately it leads to death of the treated insect (Eldefrawi, 1985). However in the case of carbamates, the enzyme inhibition is more easily reversed than with organophosphates, and the insects can recover if given too low a dose (Dent, 1991).

Carbamates have a broad spectrum of activity and usually act by contact or stomach action. They have been used effectively against vectors that have developed resistance to the organochlorines and organophosphates.

Propoxur: o-isopropoxyphenyl methyl-carbamate

One of the most commonly used products containing propoxur, as a contact and stomach poison is Baygon. Endophylic mosquitoes can be effectively suppressed by residual insecticide treatments of walls and ceilings of human dwellings with propoxur WP. The residual application may be done using a knapsack sprayer, and will typically provide three months of control. Outdoor area treatment with cold foggers, for short-term protection against exophilic species, is also possible by mixing propoxur WP with the appropriate quantity of diesel oil, petroleum or water, and applied at a rate of 1–1.5 litre solution/ha. To control mosquitoes by ground or aerial ULV application, Baygon UL 120 may also be used. Ground treatments with propoxur ULV may be carried out with undiluted material at a flow rate of 250 ml/min, and a vehicle speed of 8 km/h, or with a correspondingly increased amount at higher speeds. Aerial application can be conducted in two ways:

(a) Conventional: 300 to 1200 ml of the ULV concentrate is diluted in 4–9 litre kerosene/ha. The higher rates and volumes should be used over dense vegetation, but the lower rate and volume may be successfully used in open areas with little vegetation.

(b) ULV: the propoxur ULV concentrate is applied undiluted at 300–1200 ml/ha; the higher rates are recommended for areas with dense vegetation.

Bendiocarb: 2,2-dimethyl-1,3-benzodi-oxol-4-yl methylcarbamate

Bendiocarb has been widely used as a residual household insecticide (Ficam WP 80) by pest control operators against cockroaches and flies, and in mosquito programmes for adult control, both as a residual spray and for ground or aerial ULV application (Ficam ULV).

Evans (1993) showed that deposits of bendiocarb caused very low irritation of mosquitoes which therefore are not repelled. This feature makes bendiocarb suitable for use in programmes in which residual insecticides are selectively applied to the mosquitoes' preferred resting sites within buildings. Such selective treatments are sometimes used in house spraying programmes for vector control, in order to reduce insecticide use. Arredondo-Jimenez *et al.* (1992) reported that selective application of bendiocarb was as effective as complete spraying for the control of *An. albimanus* in Mexico.

Asinas *et al.* (1994) showed that in the Philippines *An. flavirostris* could be controlled by selective application of bendiocarb formulation at a rate of 40 ml/m^2 and a dosage of 400 mg a.i./m^2. Application of 80% bendiocarb water-dispersible powder was carried out with a Hudson X-Pert sprayer fitted with a Tee-jet 8002 nozzle, to give an even spray deposition across the swath. Following the application, the mortality levels of *An. flavirostris*, were in the range of 75 to 100% three months post-treatment.

12.3.4. Pyrethroids

Pyrethroids were a new generation of highly active synthetic insecticides. They emerged from prolonged efforts to improve the biological activity and chemical stability of the natural pyrethrins, long known for their insecticidal effects. Natural pyrethrins consists of a mixture of insecticidal esters, extracted from the flowerheads of *Chrysanthemum* spp.

Commercial production of the so-called "Dalmatian insect powder" ground from flowers of *C. cinerariaefolium* started in Dalmatia, and was widely used by 1840 (Casida, 1980). Flower production then moved to California, and at the beginning of the 20th century cultivation started in Japan, well known for its long existing chrysanthemum horticulture tradition. However between 1935 and 1940 the extract from pyrethrum flowers grown in Kenya replaced the Japanese product, because it contained a higher concentration of pyrethrins. Since the development in 1941 of the pressurised aerosol container for the delivery of pyrethrum, the product could be applied in droplets smaller than 30 μm, which increased the efficacy and cost effectiveness of the pyrethrins. The natural materials are expensive and unstable to UV light and temperature. However because of their low mammalian toxicity they are preferably used for the control of house flies, mosquitoes and other indoor pests. A very efficient way of controlling mosquitoes within homes was developed by incorporating pyrethrum extract into burning coils, the smoke of which is harmless to humans but repels, knocks down or kills mosquitoes (Baillie and Wright, 1985).

Around 1949 the effectiveness of pyrethrins was enhanced by the addition of synergists, such as piperonyl butoxide. Synergists are not insecticidal by themselves, but increase the potency of pyrethrum formulations by inhibiting their biodegradation (Yamamoto, 1973).

Pyrethroids, which are more stable than the pyrethrins, have partially replaced or supplemented the use of the three other major classes of insecticides in several areas of pest control. Insects and other arthropods are particularly susceptible to intoxication by pyrethroids. A promising area of pyrethroid use is in the control of various vectors and nuisance mosquitoes. Besides causing a knockdown effect, pyrethroids may also have a repellent or anti-feeding action.

There is evidence that pyrethroids can also be effectively applied as mosquito larvicides. This allows potential application of permethrin as a larvicide in an integrated mosquito control programme, particularly because permethrin has an approval for use in drinking water at a concentration of 15 g/litre (WHO, 1991). However, the use of pyrethroids in water bodies inhabited by fish is not recommended due to the fish toxicity of these substances.

A large number of pyrethroid compounds have been synthesised during the last few decades, some of which are exceptionally potent. Although it has become clear that the insecticidal activity can be enhanced through various structural modifications, the stereochemical features also appear to be of major importance for the effectiveness of the product. As a result, the research aimed at producing new compounds useful as insecticides has involved not only chemical substitution within molecules of interest, but also resolution and purification of the most active isomers. The relationship of the structure and activity is complex, but for example, the incorporation of a (S)-α-cyano group at the appropriate position is associated with an increase in activity (Elliot and Janes, 1978).

Mode of action

Considerable progress in the synthesis of these materials has been made, since the mode of action of pyrethroids has become better understood (reviewed by Elliott, 1980). Pyrethroids are neurotoxic to insects. Insects treated with pyrethroids show restless behaviour, hyperexcitability, become uncoordinated and are then paralysed; flying insects are in general rapidly knocked down. Exactly which symptoms will occur depends on the type and dosage of the pyrethroid used. They act at the nerve membrane to modify the sodium channels, probably by obstructing protein arrangement changes at the lipid-protein interface. The resulting neurophysiological changes depend on the nerve element, temperature and the structure of the applied compound, but typically include repetitive firing, block of impulse conduction or of neuromuscular transmission. Sensory neurons, neurosecretory cells and nerve endings appear to be particularly sensitive to these effects. The lethal activity of pyrethroids seems to involve action on both peripheral and central neurons, while the knockdown effect is most probably produced by peripheral intoxication.

The close similarity in action between pyrethroids and DDT is an interesting phenomenon which is still far from being understood. Pyrethroids share many characteristics with DDT; one intriguing feature is that they become more toxic to insects as the temperature is lowered (a negative temperature coefficient).

Commercially available insecticides of this group include the well known natural pyrethrins and their "first generation" synthetic analogues such as allethrin and resmethrin (which tend to decompose in sunlight), and a wide range of "second generation" photostable analogues such as permethrin and cypermethrin.

Pyrethroids, as broad-spectrum insecticides, are toxic also to beneficial insects, including honeybees and predators of common pests. In contrast to this, there are also examples when some of the pyrethroids are more toxic to the targetted *Heliothis virescens* (Noctuidae) than to the parasitoid *Campolestis sonorensis* (Ichneumonidae) (Shour and Crowder, 1989). Fish toxicity is of particular concern in view of the potential use of pyrethroids in aquatic habitats for the control of mosquito larvae, against which pyrethroids have demonstrated excellent activity (Mulla *et al.*, 1978). Selective control of mosquito larvae, without affecting fish, can be established by choosing the right concentration of a pyrethroid. Application rates of 2 ppb deltamethrin or 45 ppb permethrin had no effects on reproduction of mosquito fish, while mosquito larvae were effectively eliminated (Mulla *et al.*, 1981).

According to Quelennec (1988) pyrethroids can also be recommended in special circumstances for: (1) disinfection of aircraft—aerosol formulations containing a pyrethroid can be sprayed in flight, (2) personal protection—pyrethroid impregnated nets and clothing, and (3) household formulations—mosquito coils and electric mats containing a pyrethroid.

Resistance

Pyrethroid-resistant strains are known in at least 40 species of arthropods (Georghiou, 1992). This large number is surprising in view of the limited use of pyrethroids in the past. Cross-resistance or multiple resistance generated by DDT and organophosphates is the main mechanism by which this resistance has arisen. Resistance to pyrethroids was exhibited first by those insects which were already resistant to DDT, *e.g.* house flies, (Busvine, 1951; Sawicki, 1978), and some mosquito species and strains (*Cx. tarsalis, Ae. aegypti,* two *Anopheles* species) as reported by Prasittisuk and Busvine (1977) and Plapp and Hoyer (1968).

The majority of pyrethroid resistance data has been derived from studies with the house fly. The most important of several pyrethroid resistance mechanisms is the knock-down resistance, a mechanism which results in broad resistance to pyrethroids, DDT and DDT analogues. Knock-down resistance induced by DDT confers inherent cross-resistance against the knock-down effect of pyrethroids, and vice versa. It is also suspected that knock-down resistance results from a fundamental change in the sodium channel in the nervous system of insects, although the expression of this trait depends to some extent on pyrethroid structure (Miller, 1988). Developing resistance against pyrethroids may curtail the otherwise promising prospects for the use of these compounds as pest-control agents.

Resmethrin

Resmethrin is used in mosquito control programmes, particularly when urban *Culex* species, as vectors of St. Louis Encephalitis (SLE) are controlled by ULV applications in the Americas. Since some of the *Culex* species have developed resistance to compounds from other chemical groups, resmethrin has gained more in significance. Resmethrin can be effectively used as Scourge, containing 12% active ingredient and 36% synergist (piperonyl butoxide). When diluted in mineral oil, it is applied at a rate of 237 ml/min, at a vehicle speed of 16 km/h, and will typically give a satisfactory reduction of the adult *Culex* population. It has been shown that a pair of treatments on successive evenings can give an increase in mortality for a longer period. In addition, Reiter *et al.* (1990) demonstrated an interesting finding, that resmethrin treatment resulted in an impressive reduction of the oviposition rate of *Culex* mosquitoes on the day of application (oviposition rates fell by

74–84% on the night of treatment), and this impact was observable for at least eight days following the application.

Deltamethrin

Deltamethrin (K-Othrine) can be used as an insecticide for ULV application, both by ground or aerial applications. During emergency control of *Ae. aegypti* in the Dominican Republic, 78.5% mortality of caged adult mosquitoes was obtained following one application of deltamethrin formulation (K-Othrine) applied at a dose of 2.1 g a.i./ha. The insecticide was applied using a truck-mounted ULV air generator which generates an air blast of 336 km/h, at a speed of 8 km/h (Tidwell, 1994). Das *et al.* (1986) conducted a study to find out the susceptibility status of multiresistant *An. culicifacies* populations using deltamethrin at a concentration of 0.025%. The strains tested showed very high resistance to DDT and the dieldrin, and well pronounced resistance to malathion. With deltamethrin however, all the tests showed 100% mortality, which implies that DDT and dieldrin resistance did not confer cross-resistance to deltamethrin. By contrast, the organochlorine and malathion resistant *An. gambiae* in Sudan, was cross-resistant to pyrethroids (Davidson and Curtis, 1979).

Permethrin

Permethrin synergised with piperonyl butoxide (Biomist 30:30) applied from aircraft using Micronair AU 4000 rotary atomisers, at rates of 43.9 to 65.8 ml a.i./ha provided efficient control of adult *An. quadrimaculatus* and *Cx. quinquefasciatus* respectively. Ground application

conducted against the same species required an application rate of 1.6–1.8 ml a.i./ sec. The treatments were applied using either a Leco Model 1600 HD aerosol generator, or an 18 Hp Vectec Grizzly Model, driven at 9.3 km/h with a nozzle pressure of 5.5 psi, or at higher speeds ensuring an appropriately increased flow rate (Weathersbee *et al.*, 1986; Efird *et al.*, 1991).

Permethrin impregnated bednets and clothing provide good protection against mosquitoes. In tests conducted in Florida, Schreck *et al.* (1984) reported permethrin-treated clothing alone provided 89.1% protection from *Oc. taeniorhynchus*, compared with 94.4% protection provided by the combination of permethrin treated uniforms and topically applied repellents. A study conducted in Alaska showed that permethrin-treated clothing alone reduced biting of *Cs. impatiens* by 93%, compared with 99% protection obtained with treated clothing and repellent (Lilie *et al.*, 1988).

Synthetic pyrethroids have traditionally been used in conjunction with oil diluents, but due to environmental concerns, other formulations based on water to reduce the environmental impact have been introduced. Meisch *et al.* (1997) have conducted trials with the oil-based Permanone 31–36 (permethrin 31% and piperonyl butoxide 36%) and the water-based Aquareslin (permethrin 20% and piperonyl butoxide 20%) formulations of permethrin at rates of 2.03 and 3.93 g a.i./ha against *An. quadrimaculatus* and *Cx. quinquefasciatus*. Results indicate significantly greater control of both species at the higher application rate for both formulations.

Lambdacyhalothrin

In 1989 the pyrethroid lambdacyhalothrin (Icon) became available. Schaefer *et al.* (1990) conducted laboratory and field evaluation of the compound on *Cx. tarsalis* adults. LC_{95} values under laboratory conditions were as low as 0.0029–0.0063 mg/litre, depending on the strain susceptibility. A 2% lambdacyhalothrin formulation was diluted in mineral oil, and applied as droplets of 15 μm volume median diameter (VMD). In field trials, lambdacyhalothrin was applied as an aerosol at 0.125, 0.25, 0.5 and 1.0% a.i. The highest rate resulted in 90% or greater mortality of all exposed adults. These findings showed that lambdacyhalothrin offers high potential for aerosol application.

12.3.5. Insect growth regulators

All compounds belonging to this group disrupt the normal growth and development of insects, and are termed insect growth regulators (IGRs). They were developed as a result of rational leads from basic entomological research on metabolic disrupters, moult inhibitors and behaviour modifiers of insects. Since the target site of action for these chemicals is known and susceptible to disruption only in certain species at certain times during their life cycle, these materials are thought to have fewer serious detrimental effects on non-target species.

There are two major groups of insect growth regulators, that differ in their modes of action:

(1) The chitin synthesis inhibitors such as diflubenzuron, cyromazine, and triflumuron which interfere with new cuticle formation, resulting in moult disruption, and

(2) the juvenile hormone analogues which interfere with the metamorphic processes affecting development to the adult stage.

Benzoylphenyl ureas (diflubenzuron)

The insecticidal activity of the benzoylphenyl urea analogues was discovered 1970 by the Philips—Duphar Company, the Netherlands. One of the first analogues which was shown to be effective against insects, resulted from the combination of two herbicides, dichlobenil and diuron, in an attempt to create a superherbicide. The resulting compound was totally inactive as a herbicide, but was toxic to insects. Surprisingly, the compound's action was very different from that of other insecticides used at that time. The mortality was invariably connected with the process of moulting. Treated adults were not killed, but their reproductive capacity was strongly reduced. The chemistry and the symptoms of the benzoylphenyl urea analogues were unique, and they therefore became a new class of insecticides.

Mode of action

All studies of the benzoylphenyl urea mode of action, implicate the enzyme chitin synthetase as the actual biochemical moiety which interacts with the toxicant. Post *et al.* (1974) showed that the last step in the chitin synthesis process, the polymerisation of the N-acetylglucosamine units catalysed by the enzyme chitin synthetase, is inhibited.

After the IGR treatment, symptoms are not generally observed until the moulting process is initiated. The degree of the disruption of ecdysis (moulting of the old cuticle) is related to the dosage, and is characteristic for each benzoylphenyl urea analogue, and the insect species. (Ecdysal disruption may occur in the larvalpupal moult, too). The range of effects include:

(1) Ecdysis is prevented completely, so that the insect dies within the old cuticle.

(2) Ecdysis is initiated, but not completed. In some cases the exuvia splits normally but the ecdysis proceeds no further. In others, the moult proceeds until the old cuticle remains attached only to some of the last abdominal segments.

(3) In some cases the old cuticle is almost completely shed from the body but remains attached to the head capsule and mandibular region, and prevents further feeding or development.

Benzoylphenyl ureas exhibit several unique characteristics. In general they are more toxic when they are ingested by the target organism. Application of diflubenzuron, or other IGRs larvicides from the same group, in the field should therefore coincide with active feeding of the target insect. The second characteristic concerns the "activity window" of the benzoylphenyl ureas. Peak activity occurs at distinct times of the insect's development, apparently during peaks in chitin synthesis. In practical terms, susceptibility of larvae is greatest just prior to each moulting. Treatments should therefore be applied either at the time of the greatest susceptibility of the insects, or alternatively the treatment should have a sufficient residual activity to span the next "activity window". The third characteristic concerns the prolonged larval life after a toxic dose is acquired. Even after exposure at the susceptible stage, the insect may live in a moribund state for many days before dying. Tests with various mosquito species by Mulla and Darwazeh (1975) showed little or no adult emergence after diflubenzuron treatments for 27 days. Mortality was recorded during larval and pupal stages as well as during adult eclosion.

A synchronised larval population is not essential for the effective use of these products, since all larval instars are susceptible to diflubenzuron treatment. However it is advisable to conduct treatments during the early larval stages of development. Field tests by Schaefer et al. (1975)

revealed that several *Aedes/Ochlerotatus* spp. and *Cx. tarsalis* resistant to OP insecticides, were effectively controlled in all larval stages by diflubenzuron.

In the aquatic habitats where diflubenzuron has been applied against mosquito larvae, the impact of the active ingredient on a large number of non-target species has been assessed. In general, detrimental effects were detected as temporary, even after repeated treatments. (Ali and Mulla, 1978; Miura and Takahashi, 1975; Zgomba et al., 1983; Zgomba, 1987).

Juvenile Hormone Analogues

The use of the natural juvenile hormone as a selective insecticide is not feasible because of its environmental instability and difficulties of synthesis. The major breakthrough came when Bowers (1969) synthesised substituted aromatic terpenoid ethers that were potent mimics, sometimes several hundred-fold more active than the natural hormone. Since then, more than 500 juvenile hormone analogues (JHA or juvenoids) with different substitutions and varying degrees of insecticidal activity and specificity, have been synthesised. However, major JHAs which have been used so far in controlling disease vectors, are methoprene, hydroprene or phenoxyphenoxy carbamate compounds.

Mode of action

Insect metamorphosis is under hormonal control, and the dramatic series of changes resulting in sequential metamorphosis is orchestrated by the brain through its neurosecretions (Miller, 1980; Richards, 1981). The natural juvenile hormones are secreted into the haemolymph by the corpora allata (two glands at the base of the brain). The titre of juvenile hormone in the haemolymph is high in the early larval stages, drops precipitously during the last larval instar, is virtually absent in the pupae, and then present again in the adult (Akamatsu et al., 1975). The natural juvenile hormone has two distinct biochemical effects: one during the larval stage and the other in the adult. During the larval stage it suppresses metamorphic changes. In the adult, juvenile hormone is required for several reproductive functions such as ovarian development, yolk synthesis, pheromone production and accessory gland development. It is evident that upsetting the titre of juvenile hormone at certain periods during the life history will adversely affect metamorphosis. Moreover, such an induced titre might have a domino effect and disrupt other hormonally controlled functions.

JHAs resemble the natural juvenile hormone in activity, but may or may not be similar in structure. The presence of synthetic hormone mimics in the insect, at a time when the natural juvenile hormone level is low, can result in disruption of normal development.

The morphogenetic effect of the JHA compounds is primarily seen during the larval-pupal transformation, and may result in various degrees of incomplete metamorphosis. They also influence the endocrine physiology of the insect, which may result in abnormal morphogenesis. Methoprene inhibits the release of an executive hormone from the corpora allata early in the last larval instar, but stimulates the glands prior to pupation. JHA can also block embryonic development and so may be ovicidal too. Various types of effects ranging from ovicidal effects to delayed effects during postembryonic life have been reported. Treating young larvae will have very little effect on metamorphosis, since the requirement for larval-larval moult is a high titre of the hormone. However, if the last larval instar is treated, it results in abnormal pupation.

One of the main reasons for juvenile hormone analogues being effective as control agents is their chemical structure (terpenoid), which enables them to penetrate the cuticle with great ease, and exert their effects on the target tissue, the epidermis.

The ecotoxicological effects of JHA have been extensively investigated. The toxicity to vertebrates is extremely low, *e.g.* the oral LD_{50} of methoprene for rats is over 34 500 mg/kg (Siddall, 1976). However, adverse effects on non-target organisms that receive a dose of JHA during their last larval instars, could also be expected. At levels far above the dosage used for mosquito control, methoprene produces a short-term toxicity effect on the water flea, *Daphnia magna*, the side-swimmer *Hyalella azteca*, the tadpole shrimp, *Triops longicaudatus* and some other organisms (Miura and Takahashi, 1975). However high concentrations of methoprene had no adverse effects on *Dugesia dorotocephala*, the planarian which feeds on mosquito larvae and which can be used as a biological control agent. It has therefore been proposed that integrated use of planarians and JHA for the control of mosquitoes is viable (Levy and Miller, 1978).

In operational studies, the most commonly adopted methods for JHA efficacy evaluations are perhaps those which are based on (1) introducing captured third- and fourth-instar larvae into the JHA treated water, collecting pupae after an interval of time (*e.g.* 4 days), and then daily observation of the adult emergence, (2) collection of immature mosquitoes from the treated site, and emergence and mortality evaluation in the laboratory. For both methods, emerged adults are usually categorised as: completely emerged and flying; completely emerged and dead on the water surface; partially emerged with the tarsi still attached to the pupal exuvia; or partially emerged with the abdomen and part of the thorax remaining in the exuvia (Kramer *et al.*, 1993). Additionally the dead pupae also need to be recorded. The percent mortality is calculated as the number of pupae not completely emerging/number of pupae in the sample ×100. There is an alternative but similar calculation which uses the following classification of emergence: dead pupae (DP); dead adults (DA); and live adults (AA). From these data the % Emergence Inhibition (% E. I. = % control) may be calculated using the formula after Klein (1993):

$$\% \text{ E.I.} = (DP + DA)/(DP + DA + AA) \times 100$$

Methoprene

Methoprene, one of the most often used JHA, was synthesised by Henrick and colleagues in 1973 (Baillie and Wright, 1985). More than 20 years of research and field use has shown that methoprene is one of the most environmentally safe mosquito control products. It is available in liquid, pellet and briquette formulations which allow considerable flexibility at the operational level. Liquid methoprene formulations may be applied by standard ground or aerial methods against the second, third and fourth-instar larvae typically found in floodwater breeding sites, within four days after flooding. In areas with dense vegetation or canopy, mixtures of liquid methoprene formulations with sand may be applied with granule application equipment. The persistence of these methoprene formulations is up to ten days. The pellet formulation, releases an effective level of methoprene for up to 30 days in floodwater sites, artificial water containers, tyres, waste treatment ponds, man-made depressions, tree holes *etc.* Depending on the biotic and abiotic factors of the breeding site, the application rate ranges from 3.0–11.5 kg/ha. A study by Kramer *et al.* (1993) demonstrated that methoprene pellets applied at 3.4 kg/ha against *Oc. dorsalis* prior to marsh inundation, provided 99% control up to 42 days, 86.4% up to 131 days, and 66.6% control eight months after the application. Proportions of partially emerged adults increased over the course of the study. Of all the completely emerged mosquitoes from the treated region of the marsh, 66% were found dead on the water surface, compared with only 0.7% from the untreated area.

For long-term control, methoprene briquettes are perhaps the principal choice, especially where access to the breeding site is difficult, and where the area may be flooded repeatedly during the season. Placement of this product may be done at or before the beginning of the mosquito season; *e.g.* prior to the flooding, while the sites are still dry. For control of *Aedes/Ochlerotatus* and *Psorophora* larvae in shallow depressions (up to 50–60 cm depth), one briquette per 18 m² should be applied. Briquettes have to be placed at a higher rate (1 per 9 m²) for control of *Culex, Culiseta, Anopheles, Mansonia* and *Coquillettidia spp.* larvae. The duration of the efficacy depends on water temperature and the number of fluctuations. In continental and Mediterranean climates the effective suppression of mosquitoes can be expected for up to four months. Douglas *et al.* (1994) reported that when solid, sustained-release methoprene formulations were applied, the highest (S)-methoprene residue detected was 6 µg/litre. The majority of samples (85%) contained residues of 1.0 µg/litre. Such low residues do not constitute a significant risk to non-target organisms.

Effectiveness of a sustained-release sand granule formulation of methoprene was also established for *Oc. taeniorhynchus*, a major mosquito pest in coastal areas of the United States (Kline, 1993). In the field, inhibition of emergence in mosquitoes exceeded 90% when these granules were applied post flood at 5.6 kg/ha.

There is also a possibility to use an on-site method of preparing a granular formulation from liquid methoprene formulations. To prepare the granular formulation, it is necessary to mix dry sand with the liquid methoprene formulation in a rotating-type mixer (concrete mixer) for five to ten minutes, until the sand is uniformly coated. Silicon dioxide should then be added and mixed for an additional five minutes. This will provide a dry mixture, which flows freely through standard granule application equipment. The typical application rate of such a formulation is 11–12 kg/ha.

Fenoxycarb: $C_{17}H_{19}NO_4$

Although its mode of action is like that of juvenoids, fenoxycarb actually belongs to the carbamate group of insecticides. Research done by Axtell *et al.* (1980) and Dame *et al.* (1976) has shown that fenoxycarb suppresses the development of *Cx. tarsalis, Cx. pipiens, Cx. quinquefasciatus, Oc. melanimon* and *Ps. columbiae*. None of the fenoxycarb formulations tested by Dorn *et al.* (1981) and Mulla *et al.* (1985) produced harmful effects on mayfly, dragonfly or various beetle larvae. Freshwater zooplankton exposed to the treatment were also unaffected.

Ps. columbiae was exposed to a range of fenoxycarb formulations by Weathersbee III *et al.* (1988). The results showed that 0.5% fenoxycarb sand formulation at a rate of 22 g a.i./ha was effective against the larval population in rice fields. Excellent control (95%) was achieved against larvae introduced at up to 24 hours post treatment, and residual activity remained above 55% for up to 120 hours after treatment. The fenoxycarb formulation based on corn cob also offered excellent initial control (95%) at the same rate (22 g a.i./ha), but the residual activity decreased to 41% by 120 h post treatment.

Pyriproxyfen: 2-[1-methyl-2(4-phenoxyphenoxy) ethoxy] pyridine

Like fenoxycarb, pyriproxyfen is also structurally unrelated to the natural insect juvenile hormone,

but its biological activity is the same as that of the juvenoids (Schaefer *et al.*, 1988). Studies with pyriproxyfen demonstrated very high activity against *Aedes/Ochlerotatus, Culex* and *Psorophora* species, both in laboratory and in field studies, at rates as low as 0.56 g a.i./ha (Mulla *et al.*, 1989; Schaefer *et al.*, 1990). When pyriproxyfen was applied in dairy waste water lagoons at 100 g a.i./ha in single and multiple applications, it resulted in control of *Culex* spp. for a period of 7–68 days. The length of the control period appeared to be related to water quality, with greater residual activity in more polluted sources (Mulligan III and Schaefer, 1990). The assumption was that the active ingredient was adsorbed onto organic debris and remained highly toxic to the larvae, even after the water was replenished with untreated waste water.

In general, IGRs have high levels of activity and efficacy against various species of mosquitoes in a variety of habitats. Additionally, they show a good margin of safety to non-target biota including fish and birds. On the basis of these attributes, IGRs are likely to provide another tool for mosquito control, supplementing organophosphorus, pyrethroid and microbial larvicides.

12.3.6. Novel insecticide classes

Despite the efforts of academia, the pesticide industry, and national and international agencies, the discovery and introduction of new insecticides remains a slow process. Although several novel classes of insecticide have been recently developed for agricultural use, no new classes of insecticides have been developed for mosquito control in the last 2 decades.

The World Health Organisation Pesticide Evaluation Scheme (WHOPES) was set up in 1960 to liase with insecticide manufacturers to help identify new insecticides, both from within existing classes and from totally novel classes. The organisation continues to encourage manufacturers to submit novel compounds for routine evaluation against mosquitoes and other pests.

It is hoped that a better understanding of the needs of the mosquito control community, together with new approaches to novel insecticide discovery, for example based on a better understanding of the mosquito at the genomic level, will in time facilitate the discovery of new compounds with appropriate properties. In the meantime mosquito management organisations need to ensure the correct use of the existing compounds, and their integration with other control techniques, to ensure a sustainable approach to mosquito control.

12.4. MANAGEMENT AND MONITORING OF INSECTICIDE RESISTANCE

When mosquito populations are exposed to selection pressure from insecticides, they may become resistant. Resistance has been defined as "the developed ability in a strain of insects to tolerate doses of toxicants which would prove lethal to the majority of individuals in a normal population of the same species" (Cremlyn, 1978). These strains tend to be rare in a normal population, but widespread use of an insecticide can reduce the normal susceptible population thereby providing the resistant individuals with a competitive advantage. The resistant individuals multiply in the absence of intraspecific competition and, over a number of generations, quickly become the dominant proportion of the population. Hence, the insecticide is no longer effective and the insects are considered to be resistant. A population may even develop "cross resistance" which means that the population is not only resistant to one insecticide of a particular class, but also to other insecticides in the same class, even when it never has been treated with those other insecticides. More severe is the phenomenon of "multiple resistance" where

separate detoxification mechanisms for unrelated insecticides are present, resulting in an insect population that is resistant to different classes of insecticides, which makes its control with insecticides extremely difficult.

Since Melander (1914) first reported insecticide resistance, the number of insect species and mites worldwide that have developed strains resistant to one or more pesticides, has increased to at least 504 and continues to rise. The number of insecticide resistant arthropods of public health importance has risen from two in 1946, to 198 in 1990 (Oppenoorth, 1985; Georghiou, 1990; Georghiou and Lagunes-Tejeda, 1991).

12.4.1. Resistance mechanisms

Resistance mechanisms are generally dependent on single genetic factors. Species that have been under continued selection pressure with one or a range of different insecticides, have often accumulated a number of resistance (R) – genes and corresponding resistance mechanisms, which may lead to cross or multiple resistance.

Several mechanisms seem to be responsible for the development of resistance of insects to insecticides. These involve either the detoxification of the toxic compound by biochemical metabolism, or a tolerance due to a decreased sensitivity to the toxic compound at its site of action. Normally, an insecticide penetrates rapidly through the integument, reaching the site of action. The site may be a vital enzyme, nerve tissue or receptor protein. Insecticide molecules bind to the site, and when they have attained threshold concentrations, they disrupt vital processes and cause the insect's death. Resistance may be selected at each step of this pathway: the integument, where reduced permeability may occur, thus reducing the rate of entry of the insecticide; new or more abundant metabolic enzymes may be selected, causing break down of the insecticide more efficiently;

or altered target sites may be selected to which the insecticide no longer binds. Of these three types of mechanisms, metabolism and insensitivity at the site of action are the most important. A reduction in the rate of cuticular penetration aids both types of mechanisms in a synergistic way (Georghiou, 1994). In addition to those, another form of insecticide resistance is behavioural resistance, where insect behaviour becomes modified so the insect no longer comes into sufficient contact with the insecticide deposit (Miller, 1988).

12.4.2. Resistance surveillance

Resistance monitoring should be an integral part of any mosquito control programme. The susceptibility of mosquitoes should be verified before the start of control operations, to provide baseline data for insecticide selection and choice of technique to be applied. Regular surveillance will allow early detection of resistance, so that resistance management strategies can be implemented or, in the case of late detection, evidence of control failure can justify the replacement of the insecticide. The operational criterion of resistance has usually been taken as the survival of 20% or more of field collected individuals, tested at the currently known diagnostic concentration of the particular insecticide, using *e.g.* WHO test kits. The diagnostic concentration of a particular insecticide is that which has been found to reliably cause complete mortality of strains which have never before encountered insecticide, and are therefore assumed to be susceptible.

Practical tests for detection of specific resistance mechanisms in individual insects have been developed. These include filter paper tests for esterases (Pasteur and Georghiou, 1989) and acetylcholinesterase (AChE) (Dary *et al.*, 1991), microtitre plates for esterases (Dary *et al.*, 1990) monooxygenases, glutathione transferase and others (Georghiou, 1990).

12.4.3. Resistance management

According to Georghiou (1994), factors that influence evolution of resistance can be classified into three categories: genetic, biological and operational. Georghiou and Taylor (1986) have quantified the influence of individual factors and have shown that some are positively correlated with the development of resistance (*i.e.* gene dominance, population isolation, insecticide persistence, *etc.*) while others are negatively correlated (immigration of sensitive populations into areas where resistant population exist, untreated refugia where populations without selection pressure for R-genes can develop). If the relative influence of each factor could be expressed quantitatively, a reliable model might be constructed to predict the risk for resistance in a given situation. That risk can then be reduced through appropriate modification of one or more of the operational factors.

In general, resistance management is intended to prevent or delay as far as possible, the development of resistance to an insecticide, while at the same time maintaining an effective level of mosquito control. Based on three categories of factors that influence the development of resistance, Georghiou (1994) suggested the following approaches to resistance management: management by moderation, management by saturation and management by multiple attack.

Management by moderation: Management by moderation recognises that susceptibility genes are a valuable resource and it attempts to preserve them by limiting the chemical selection pressure that is applied. Measures in this category include the use of low insecticide rates, infrequent applications, non-persistent insecticides, and preservation of refugia.

While management by moderation comes close to meeting environmental standards, it may not be appealing where there is a need to control human disease vectors, or eradicate newly introduced pests. In these cases, the saturation or multiple attack concepts may be more appealing.

Management by saturation: The term "saturation" does not imply saturation of the environment with pesticides. It indicates saturation of the insect's defences by excessive, heavy, or frequent use of insecticides to leave absolutely no survivers. This approach has more merit during the early stages of selection when resistance genes are rare, existing mainly in the heterozygous state.

Another means of suppressing the insect's defences is the use of synergists. Piperonyl butoxide (PB) has been used for many years as a synergist of pyrethrins in household aerosol sprays, and more recently with pyrethroids in the control of mosquitoes and houseflies. By suppressing the insect's mixed function oxidase system, which is involved in the degradation of pyrethroids, PB effectively removes the selective advantage of this mechanism. The approach would not apply where alternative pathways of detoxication are also present (Ranasinghe and Georghiou, 1979).

Management by multiple attack: The multiple attack strategy is based on the premise that control can be achieved through the action of several independently acting stresses, including insecticides, each exerting selection pressure that is below the level which could lead to resistance. This approach includes the application of chemicals in mixtures or in rotation (Georghiou, 1983, 1990; Roush and McKenzie, 1987; Tabashnik, 1989)

The strategy of using mixtures assumes that the mechanisms of resistance to each member insecticide are different and that initially resistance genes exist at such low frequencies that they do not occur together in any single individual within a given population. Any insect that may survive the exposure to one of the insecticides in the mixture, is nonetheless killed by the other.

The strategy of using rotation may be applied in situations where resistant mosquitoes have a lower biotic fitness than susceptible individuals, which results in a gradual decline in the frequency of resistant genes when the selecting insecticide is withdrawn, or is replaced by a neutral insecticide that is not affected by cross-resistance.

Alternatively, different insecticides maybe applied in a mosaic pattern, with the size of the mosaic segments determined on the basis of insect movement and gene flow. The resulting mosaic of different selection pressures may effectively delay the development of resistance throughout the population.

The feasibility of using two insecticides in rotation, mixture, or sequentially for resistance management, has been examined in several laboratories by means of cage experiments (Georghiou, 1983; Cilek and Knapp, 1993; Curtis *et al.*, 1993; McKenzie and Byford, 1993). However the work has led to divergent conclusions, as the outcome of each approach will depend on many factors, including the appropriate choice of insecticides based on their mode of action, the potential mechanisms of resistance to them, the prior exposure of the target population to insecticidal selection pressure, and the presence of a significant fitness differential between resistant and susceptible individuals.

In the field, the relative impact of single use, rotation and mosaic strategies on the rate of onset of insecticide resistance in anopheline mosquitoes has been tested in a large scale project in Mexico. Following extensive baseline studies on resistance mechanisms and gene frequencies, different villages and regions were subjected to different spraying regimes with different classes of insecticides, over a period of several years. At present the data are still being analysed, but interim results indicate that although techniques such as rotation or mosaic treatment do not prevent the development of resistance, they do slow it's onset, as compared with repeated use of a single insecticide (WHO/IRAC, 2003).

13

Physical Control

Physical control of mosquitoes in the broadest sense focuses on a) the reduction of mosquito breeding sources by sanitation, b) the modification of the breeding habitats to discourage mosquito breeding and to encourage the development of mosquito predators (water management); c) the modification of the air-water interface by applying surface layers to prevent the uptake of oxygen by mosquito larvae and pupae; d) the reduction of mosquito–human contact.

13.1. SANITATION

The elimination of unnecessary water sources should be an important aspect of each mosquito control programme, and can be achieved at various levels.

For example, through public information programmes, (e.g. school visits, internet, public libraries, advertisements, posters, leaflets, etc) home-owners and businesses can be encouraged to prevent mosquitoes from breeding on their property by applying appropriate steps. Depending on the region, these measures may include:

a) Removal and disposal of unnecessary water collections (for example, flower pots, cans, discarded tyres). Where appropriate, such containers may simply be inverted, or holes drilled in the bottom.
b) Essential water containers, such as bird baths, outdoor pet dishes, or fire buckets, should be emptied at least once a week and then re-filled with fresh water.
c) Cover rainwater containers with lids.
d) When not in use, swimming pools must be fully drained and kept free of water.
e) Ensure that roof drains and gutters are clear of debris and free-flowing.
f) Ensure that water does not accumulate in cellars.
g) Drain water from tree-holes and fill with sand or cement.
h) Ornamental ponds should be stocked with fish that feed on mosquitoes.
i) Air-conditioning systems should be properly maintained to prevent water accumulating.
j) Small farmers (e.g. those producing salad vegetable close to city areas) should insure that irrigation channels and water storage containers, are designed and maintained to prevent mosquito breeding.
k) On construction sites, water storage containers should be maintained to prevent mosquito breeding.
l) In harbour areas, disused boats should be stored so that rainwater does not accumulate in the boats and encourage breeding.
m) In cemeteries, containers for flowers should not retain water that may support mosquito breeding.

In some countries or municipalities, the public information programme is re-enforced with regulations that place a legal requirement on property owners to prevent mosquito breeding on their premises. The regulations are enforced by public health inspectors that visit premises to ensure compliance.

In addition to education, encouragement and enforcement, the city authorities have a direct role to play in minimising mosquito breeding sites on their own property. This may include:

a) Ensuring that waste water processing plants do not support mosquito breeding.
b) Ensuring that roads, drains, storm-water areas, sewerage systems, street waste containers and other details installed by city authorities, do not become water holding area that can support mosquito breeding.
c) Ensuring that mosquito breeding sites are eliminated in public parks.

13.2. WATER MANAGEMENT

In the early days of mosquito control, large projects were undertaken to drain swamps; practices which nowadays are uncommon due to the importance of wetlands as wildlife habitats.

Improvement and maintaining of ditch systems encourages water flow. This may either leave mosquito larvae stranded on dry land before they can complete their development, or will allow fish to move into mosquito breeding sites and consume the mosquito larvae.

Before the canalization of large river systems, they were often flanked by many permanent and semi-permanent bodies of water, which were refuges for predators. After canalization, and lowering of the water table, many of these semi-permanent water bodies have disappeared and there now remains only temporary shallow water bodies, which are potential mass breeding sites for floodwater mosquitoes. By deepening such areas so that they again become permanent or semi-permanent water bodies, habitats for fish and other predators are formed. The depth of the ponds, or the amount of time they remain filled with water, determines whether a pond for fish, amphibians, water bugs or beetles is created. By excavating an area, and by using the excavated soil to build up the surrounding ground, a topographical contrast is created, so that the flood land is reduced and a defined shoreline provided.

13.3. MODIFICATION OF THE WATER–AIR INTERFACE

13.3.1. Oil

From the time when the Panama Canal was built in the 19th century to the present, petroleum oils have been frequently used to control immature stages of mosquitoes. Yet, the use of oils is extremely limited because of the potential for ecological damage.

The efficacy of oil against larvae and pupae of *Cx. tarsalis*, when applied at 9–18 l/ha, was demonstrated by Mulla and Darwazeh (1981) and Schultz *et al.* (1983). However, the oil film is detrimental to all organisms with similar respiration mechanisms to mosquito larvae. Furthermore, aquatic plants could also be destroyed. The application of such agents for mosquito control should therefore be restricted to breeding sites that are of no value from an environmental point of view. Another disadvantage of oil is that it does not always spread well on heavily polluted water.

13.3.2. Surface films

Surfactants and monomolecular organic surface films act by the physicochemical modification of the air–water interface (Garrett and White, 1977). Larvae and pupae cannot penetrate through the film at the water surface to obtain atmospheric oxygen, and newly emerged adults will drown on the treated water surface. The effectiveness in controlling mosquito larvae and pupae is attributed to a reduction in water surface tension, with a subsequent wetting of tracheal structures resulting in anoxia, rather than chemical toxicity. Large numbers of film-forming surface active agents were evaluated as mosquito larvicides (Mulla, 1967a,b), and some of these were found to have potential for practical mosquito control programmes.

13.3.2.1. Liparol

A self-spreading biodegradable surface film called "Liparol" was developed by Schnetter and Engler (1978). This substance is a mixture of soybean lecithin and paraffin, with carbon chains containing between 12 and 14 carbon atoms.

The lecithin, as a macromolecule with a hydrophilic and hydrophobic end, acts by reducing the surface tension of the water body. The hydrophilic end spreads on the water surface, whereas the hydrophobic end holds the paraffin film. The mode of action is based firstly on the interaction of the hydrophobic end of the lecithin molecule, and the hydrophobic layers within the pupal trumpet or the larval siphon. When the pupa pierces the water surface covered by the film, the lecithin enters the trumpet and covers the hydrophobic layer. When the pupa dives, the hydrophilic end of the lecithin is orientated towards the inner part of the trumpet, allowing water to enter. Due to the reduced surface tension pupae and larvae are unable to pierce the water surface again.

At a rate of 0.6–1.0 ml/m^2, the liparol is effective against larvae in the late fourth-instar and pupal stages, while other organisms that take oxygen from the water are unaffected. In polluted ponds with a clear surface and less oxygen in the water, all pupae and nearly all fourth-instar larvae die within one hour after an application of 0.6 ml/m^2. In polluted ponds as well as in those with high oxygen content, an application rate of 0.8–1.0 ml/m^2 is necessary

for 100% mortality of pupae. First-instar to early fourth-instar larvae are killed only by very high doses, since they can also obtain sufficient dissolved oxygen through the integument. In practice the film is applied as a liquid formulation at an average application rate of 7.5 l/ha. Depending on environmental factors such as water temperature and sunshine, the film is active for only six to ten hours.

Although the organic surface film is relatively selective, it has some adverse effects on other surface breathing arthropods, such as water bugs (*e.g. Notonecta glauca*), which are highly sensitive, and some water beetles (*e.g.* Hydrophilidae and Dytiscidae). Because of the effect on the tension of the water surface, arthropods which live on the water surface, such as the striders (*Gerris* spp.) are also affected (Becker and Ludwig, 1981). Another disadvantage of the method is that application has to be very carefully timed, to coincide with the period when late fourth-instar larvae and pupae occur in the breeding sites.

13.3.2.2. Monomolecular Surface Films (MSF)

The use of the monomolecular film Arosurf MSF (iso-stearyl alcohol 20 E) as a mosquito larvicide was first reported by Garrett (1976). Since then, the efficacy of this larvicide has been demonstrated on many mosquito species in a wide variety of aquatic habitats (Levy *et al.*, 1981, 1982; Mulla *et al.*, 1983; Takahashi *et al.*, 1984). These authors show that the product can be sprayed in polluted water habitats at surface dosages as low as 0.33 ml/m^2, and suppress more than 95% of the immature stages of *Culex* spp. Similar levels of control were achieved by ground and aerial sprayings against immature stages of *Oc. taeniorhynchus* and *Oc. infirmatus* in salt marsh habitats of Florida.

Disruption of the integrity of the film by wind, heavy rain or by high concentration of algae and/or other emergent aquatic vegetation, is usually the cause of poor kill of larvae and pupae.

Quantitative studies by Hester *et al.* (1989) showed that a single treatment of Arosurf MSF applied at a rate 0.94 ml a.i./m^2 in a 1 : 9 Arosurf-water mixture, did not significantly affect the exposed aquatic vegetation. The effects of Arosurf MSF on a variety of aquatic non-target organisms have been studied extensively. There were no effects on mayfly larvae, adult diving beetles and ostracods after treatment at a rate of 9.35 l/ha (Mulla *et al.*, 1983). Field evaluation of the same rate of Arosurf MSF by Takahashi *et al.* (1984) indicated that corixids, notonectids, clam shrimp and an adult beetle were actually affected but all, except for the clam shrimp, had recovered to pretreatment population levels by three days after the treatment.

13.3.3. Polystyrene beads

Layers of expanded polystyrene beads can be used for covering the water surface of underground breeding sites, *e.g.* pit latrines or flooded cellars (Reiter, 1978; Sharma *et al.*, 1985). The persistence of the layer can have an indefinite life, unless the pit is flooded and the beads are swept away. Apart from making it difficult for larvae to respire, thick layers of beads prevent the emergence of adults and obstruct oviposition. The effectiveness of bead layers in preventing mosquito emergence has been repeatedly demonstrated by placing exit traps over latrine pits before and after application of bead layers (Curtis, 1990).

Heating the polystyrene powder to about 100°C softens the plastic and expands the beads by up to 30 times in volume. Beads are heated with steam or boiled in water, sieved and poured into the pit. The beads spread satisfactorily when still wet from the "cooking" process. There is no significant change in mass during the expansion process, so the weight of the polystyrene

required per pit is similar whether it is assessed before or after expansion. The relative merits of factory or on-site expansion of the beads depend on the ease and cost of transport. There are records that 1.5 tonnes of polystyrene were sufficient for the treatment of 3000 pit latrines (500 g/pit latrine) each with a cross-section area of approximately 1 m^2.

13.4. REDUCTION OF HUMAN-MOSQUITO CONTACT

Physical control methods also focus on the reduction of human-mosquito contact and on excluding mosquitoes from the interior of buildings. These methods include the following:

a) wear long trousers and sleeves when outdoors when mosquitoes are most active;
b) install screens of 16 to 18 mesh on windows, doors and porches;
c) do not sit outside after dark in areas where mosquitoes are active;
d) use mosquito netting when sleeping in an unscreened locations.

More examples for personal protection such as impregnated bed nets or repellents are given in chapter 14. There are also a number of commercially available devices that claim to repel or deter mosquitoes, but on the whole these have not been found to be effective. For instance, electronic mosquito repellers that emit high frequency sound have been shown to offer no useful protection. Light traps catch mainly insects which positively react to light, such as chironomids or moths, but not mosquitoes which respond to odours. Furthermore, candles containing the repellent citronella for the use outdoors are also not very effective.

14

Personal Protection

14.1. IMPREGNATED BED NETS

Insecticide impregnated bed nets for the control of malaria and other vector-borne diseases have been used successfully in several regions: China, Thailand, Latin America and in some African countries. Within one decade, work on insecticide impregnated bed nets and curtains has grown from small scale tests, to the operational use of more than 10 million treated nets (WHO, 1996). The concept is to place a small quantity of a fast acting insecticide of low mammalian toxicity directly in the path of the host seeking mosquito. Regarding the safety and acceptability of this method of mosquito protection, it was stated by WHO that properly treated bed nets should pose no hazard to those who use them. As a result of its low mammalian acute oral toxicity, low dermal toxicity, low volatility, and its high efficacy, permethrin has been widely used for bed net impregnation.

The treatment of nets or curtains may be considered as the most rational form of selective insecticidal treatment in the sense that the treated surfaces are those which blood-seeking mosquitoes are bound to encounter in their efforts to approach humans (Mouchet, 1987). According to the costs and availability of fabrics (cotton, polyester, nylon) for bed nets, and the very low dosages of the pyrethroids on nets required for efficient protection, net impregnation seems to be affordable for vector control and, if compared to DDT treatments to protect the same human population, sometimes cheaper.

Since most *Anopheles* species bite at night, it has been assumed that nets must be useful against malaria (WHO/IDRC, 1996; Najera and Zaim, 2002). Moreover, since the majority of the malaria vectors to feed on man indoors, screening is one of the essential measurements (Curtis, 1990). In parts of China and Kenya, evidence has been obtained of insecticide treated nets being effective against malaria. However, it has also been recognised that in many parts of the world, nets will not give sufficient protection to control malaria, because of the poor community participation, inappropriate use of the nets and the exophily of some *Anopheles* vectors. In most of the research studies, impregnation of nets has been done by dipping them into an emulsion made by mixing an emulsifiable concentrate of the pyrethroid with water.

A number of compounds can be employed as effective impregnation insecticides as shown in Table 14.1 (Curtis *et al.*, 1996).

In routine use, pyrethroid threated bed nets need to be re-treated at intervals, normally annually, to maintain an effective dose of insecticide on the fabric. In operational use, however, most nets are not retreated regularly, and as few as 5% of nets in use may carry the correct dose of insecticide. To overcome this problem, and so help ensure the effectiveness of malaria control programmes, WHO has coordinated the development of Long Lasting Nets (LLN). Specialised industrial fabric treatment techniques are used during the manufacture of LLNs, to ensure that the insecticide binds very

Table 14.1.   Insecticides used for the Treatment of Mosquito Nets or Curtains

Insecticide	Formulation[a]	Dose[b] a.i. mg/m^2	Toxicity[c] (Rat oral LD$_{50}$ of a.i.[d]; mg/kg of body weight)
Alphacypermethrin	SC	20–40	79
Bifenthrin	SC	25	55
Cyfluthrin	EW	30–50	250
Deltamethrin	SC	15–25	135
Etofenprox	EC	200	>10 000
Lambdacyhalothrin	CS	10–20	56
Permethrin	EC	200–500	500

[a]Other formulations may also be appropriate. SC = suspension concentrate; EW = Emulsion oil-in-water emulsion; CS = microencapsulated; EC = emulsifiable concentrate.
[b]Doses refer to synthetic netting. Higher doses may be required for cotton netting.
[c]Toxicity is not necessarily equivalent to hazard.
[d]a.i. = active ingredient.

tightly to the textile fibres. Despite routine use and regular washing, LLNs remain active for the life of the net. It is planned that LLNs will be the basis of malaria prevention programmes, particularly in Africa.

14.2. REPELLENTS AGAINST MOSQUITOES

14.2.1. Repellents on Skin or Clothing

Repellents are the most common means of personal protection against blood seeking arthropods and for the prevention of arthropod-borne disease transmission. Previous work has concentrated mainly on simple solutions of topical repellents and the chemical treatment of clothing (Rutledge *et al.*, 1978). Current protection is based on controlled release arthropod repellent formulations, and the impregnation of fabrics with permethrin. Repellents may be applied to the skin, clothing or in some cases to screens, to prevent or deter arthropods from attacking humans.

The use of topical repellents plays an important role in the protection of people in areas where no large-scale mosquito control methods are carried out. Several repellents have been developed for self protection against mosquitoes (benzyl benzoate, butyl ethyl propanediol, dibuthyl phthalate, dimethyl phthalate, ethyl hexanediol, butopyronoxyl, 2-chlorodiethylbenzamide, acylated 1,3 aminopropanole derivates) among which the best known is probably deet (N,N-diethylmethyl-3-methylbenzamide).

Repellents are available in a variety of formulations (liquids, lotions, solid waxes, creams, foams, impregnated wipe-on towelettes, soaps), and can be dispensed from tubes, pressurised cans, roll-ons *etc*. When applying a repellent, the biting habits of different mosquitoe species should be considered. For general protection against mosquitoes all exposed parts of the body, such as arms, legs, face (except the eye area) should be treated.

Impregnating clothing with a repellent provides extra protection, and temporary protection can be achieved with a spray-on application. For longer lasting effects, clothes should be dipped in a solution of a repellent and a pyrethroid.

Different authors have reported various durations of repellent activity of several deet formulations. Gupta *et al.* (1987) demonstrated that a controlled release cream formulation of 33% deet was not more effective than the standard liquid formulation against several species of Australian mosquitoes. Annis (1990) presented a new extended duration repellent formulation (EDRF) of deet which was significantly more effective than the liquid formulation. Although the EDRF provided significantly greater protection than the liquid formulation between 6 and 12 h post application, the degree of protection began to decline after 8 h. To provide protection against disease transmission, depending on the formulation, reapplication would be necessary before significant degradation in repellency occurs, which has been reported to be 6 to 8 h post treatment. Hoch *et al.* (1995) reported a new extended duration topical insect/arthropod repellent formulation of deet which is a multipolymer sustained-release formulation, and has numerous advantages over available repellents including lower deet concentration and extended protection time. Tests of the new formulation showed 95% or better biting protection from *Ae. aegypti* and *An. stephensi* up to 6 and 8 h, respectively. However, questions have been raised about its safety and effectiveness against some *Anopheles* species and its aggressiveness to paint, varnish and some hard plastics.

Owing to the public concerns over safety of synthetic chemicals, interest in botanical repellents has increased lately, and has stimulated a re-examination of the repellent data going back as far as 1939, when MacNay delivered a paper on 38 botanical repellents against northern American mosquitoes. Reanalysis of the MacNay data has been done by Rutledge and Gupta (1996), who used statistical analysis according to contemporary standards. According to these authors, the data showed that pyrethrum extract (1:1 and 1:3 in olive oil) was one

of the most active repellents, providing protection from *Oc. stimulans*, *Ae. vexans* and *Oc. trichurus* for about 4.5 h, among several other materials which acted as repellents for shorter periods of time. Prior to the advent of synthetic repellents, pyrethrum and citronella oil were widely used in repellent lotions, sprays, smokes and candles. Although pine tar oil, thymol and geraniol (among others from the MacNay list) have had comparatively little use as repellents, their protection periods (above three hours) did not differ significantly from those of pyrethrum and citronella oil in the study of Rutledge and Gupta (1996).

Several investigators have reported that repellents can act as attractants when present at low concentrations, deposits or residues. Dubitskyi (1966) found that vapours of dimethyl phthalate, deet and some other compounds were attractant to *An. hyrcanus*, *Ae. cinereus*, *Ae. vexans*, *Oc. caspius* and *Cx. modestus*. On the other hand, many materials that are normally thought to be attractants have been shown to be repellent at high concentrations. Smith *et al.* (1970) reported that lactic acid was attractant to *Ae. aegypti* at concentrations normally present on the skin and in the breath, but repellent at a higher concentration (4 mg/cm^2). Thus, it is vital that the labels of commercial repellents should be amended to include instructions to wash off or reapply the repellent when it is no longer effective.

The long term usage of deet is not recommended, particularly for children, and as a result major chemical companies are actively involved in the development of synthetic alternatives to deet and/or improved formulations to prolong efficacy or increase consumer acceptability. When repeated applications of a repellent are required, one of the natural products, such as qwenling should be used. Qwenling has been derived from the lemon eucalyptus plant in China and contains an active ingredient p-menthane-3,8-diol (PMD). Neem oil has also

Table 14.2. Active Ingredients Suitable for Personal Protection in Household Insecticide Products

Product	Active ingredient	Concentration range (%)	Toxicity[a] (oral LD_{50})[b]
Mosquito coil	d-allethrin	0.1–0.3	500
	d-trans allethrin	0.05–0.3	500
	transfluthrin	0.02–0.05	>5000
	prallethrin	0.03–0.08	460
Electric vapouriser mat	d-allethrin	25–60 mg/mat	500
	d-trans allethrin	15–30 mg/mat	500
	transfluthrin	6–15 mg/mat	>5000
	prallethrin	6–15 mg/mat	460
	S-bioallethrin	15–25 mg/mat	700
Liquid vapouriser	d-allethrin	3–6	500
	transfluthrin	0.8–1.5	>5000
	d-transfluthrin	1.5–6.0	500
	prallethrin	0.6–1.5	460
	S-bioallethrin	1.2–2.4	700

[a]Toxicity and hazard are not necessarily equivalent.
[b]Oral LD_{50} of active ingredient for rats (mg/kg of body weight).

shown a good repellent effect against *Anopheles* mosquitoes, and according to WHO (1997c) it is safe for use.

14.2.2. Mosquito Coils

Mosquito coils are made from a mixture of active and inert ingredients, and are most commonly used in Asia, Africa and the western Pacific region. The active ingredients usually used in mosquito coils belong to the pyrethroids (Table 14.2) because they provide quick knockdown action, and present no significant hazard in use. The active ingredients in the smoke, released when the coil is burnt, deter mosquitoes from the immediate vicinity of the burning coil, or even from entering houses. Depending on the size of the room in which the coil is used, and the type of active ingredient in it, the mosquito biting rate can be reduced by up to 80% (Chavasse and Yap, 1997).

14.2.3. Vapourising Mats

These devices consist of a mat impregnated with an insecticide, and an electrically powered heater to vapourise the insecticide. Mats are typically made of porous fibres impregnated usually with a pyrethroid, stabilisers, slow-releasing agents, perfumes and some colouring agents. Depending on the type of the heater, the optimum temperature is between 110°C and 160°C. When the mat is heated, insecticide vapour is released to provide a low aerial concentration of insecticide. This induces behaviour changes of flying insects through a sequence of sub-lethal effects including deterring them from entering the room, bite inhibition and knock-down. Continued exposure results in the death of the insect. The advantage of using mats over coils is that they do not produce unpleasant smoke. The disadvantage is that electricity is required.

14.2.4. Liquid Vapouriser

The principle of this product is similar to that of the vapourising mat. A liquid vapouriser has to be plugged into an electric socket. It consists of a heater and a bottle of liquid insecticide. The liquid, which is a mixture of a pyrethroid or a natural repellant (plant extracts) and a solvent, is drawn up through a wick, typically made from materials such as carbon, ceramic or fibre. The end of the wick is positioned within the heating portion of the device, and when heated, the insecticide vapourises from the wick. A bottle of liquid insecticide with its wick is usually designed to last up to 360 hours, which requires replacement of the bottle every one or two months.

Mosquito coils, vapourising mats and liquid vapourisers may be considered as insecticide formulations, with the pyrethroids being the most widely used active ingredients for this use. The suggested active ingredients suitable for these devices are given in Table 14.2. There is a developing market and many different brands need to be monitored and approved by national authorities before introduction into a country. Worldwide surveys have shown that consumers spend a considerable amount of money each year on such personal protection devices. Due to the risk of allergic reactions and peripheral nervous system disturbances these products should not be used in rooms where children sleep. Therefore, the emphasis should be made on promoting the use of a safe and effective wide scale mosquito control programme whenever possible, instead of individual protection, which should be considered only as an additional tool or as an integral part of a larger mosquito control programme.

15

Integrated Pest Management

What are the costs in terms of the effects on the environment, human health, wildlife, beneficial insects and the safety of food and water, if people continue to use pesticides at the current rate? Mankind has recognised the costs in the last decades, and will continue to suffer the consequences unless people reduce their dependence on non-selective pesticides as being the only solution to mosquito problems. The society must advocate specific, carefully planned pesticide use, integrated with other control measures. Integrated pest management (IPM) devises a workable combination of the best components of all control methods applicable to, for example, a mosquito problem. For this reason IPM in mosquito control, which could be also called Integrated Mosquito Management (IMM), is the application of sound ecological principles to maintain mosquito populations below the acceptable nuisance and health injury level. It is more precisely defined as the rational combination of all available control methods aimed to maintain mosquito population at acceptable levels in the most effective, economical and safe manner.

IPM emerged as a strategy for pest control as a result of a significant change in attitude, prompted by the excessive use of chemical pesticides during the 1940–1960 period in Europe and the United States. During this time the success of chemical control with the use of persistent insecticides was so spectacular, that other control techniques were not taken into consideration.

15.1. MOSQUITO MANAGEMENT RESEARCH

Although there may have been times when mosquito management was carried out on the basis of intuition, local anecdotal experience, or of dogma, the discipline is now becoming increasingly evidence-based. Hypotheses concerning the relative performance of different management techniques for example, are now tested rigorously before being considered for use. Such research on mosquito management has typically been carried out by perhaps three separate types of organisations.

Public sector mosquito control organisations in specific countries or regions, will often have their own internal research group. Such groups will address particular local problems or questions that have arisen with respect to their own operational mosquito management programmes. This research is primarily intended for internal use and consumption, but will filter out into a wider domain, through publications, conferences and the internet, where it can be accessed by other organisations.

University researchers may attract funding from national or international bodies, typically to address fundamental questions relating to the biology, significance, or management of mosquitoes. This basic research may not immediately benefit practical programmes, but will help provide a solid base on which operational and practical decisions can be made. This research will almost always move into the public domain, thus benefiting a wider audience.

Private organisations, such as those involved in the discovery and development of novel insect control products will also be involved in research into new mosquito management technologies, improved delivery systems, and into the potential environmental impact of these new technologies. In addition to in-house research, private organisations will also typically fund some experiential programmes in universities for example. Although much of this private funded research remains within the organisation, nonetheless enough will move to a wider audience to benefit others involved in mosquito management.

Although these three basic types of research organisation tend to have different short-term objectives, between them they provide coverage of most of the key areas of mosquito management. However there is no formal body that oversees or manages this work, and as a result it does not grow at an even pace. The emergence of new technologies in one area will catalyse a flurry of new research projects, until the next breakthrough triggers work in a different area. To some extent it will be the responsibility of the individual mosquito control organisation to ensure that they have the information they require to deliver an effective management programme, and to try to remedy any shortfalls or imbalances in information.

Of course there have been imbalances in the knowledge and practice of mosquito management. At times research has focussed more on vector control, to the detriment of nuisance mosquito management. At other times, programmes have perhaps been overly dependent on pesticides and paid scant attention to the integration of a range of control techniques. Increasingly however, as the different players in mosquito management attend the same conferences, read the same journals, and access the same web sites, there is emerging a gradual consensus on the main challenges facing the discipline.

15.2. SOCIO-ECONOMIC ASPECTS OF MOSQUITO CONTROL

The optimal management strategy can only be determined with reference to the ecology of the mosquitoes, and their interaction with other components such as habitat management. However, a management strategy developed solely on the basis of mosquito ecology and their breeding sites would have little relevance to the real world, where these factors cannot be considered in isolation from the wider and integral social and economic context.

Socio-economic aspects in relation to mosquito management consider the social context within which a management strategy is to fit, the constraints imposed by the cultural and economic situation of the region on the type of management practices that are appropriate, and the subsequent likelihood of their adoption. Therefore, it is essential to identify the type of region that the mosquito management project aims to assist, and in doing so to gain an understanding of the community's regional needs and resources, their own perception of the problem and objectives, as well as encouraging and enlisting their support and collaboration in the project's implementation. The adoption of control strategies will not only depend on the region, community or abatement district, but also on their perception of the costs, not necessarily in monetary forms but in time and effort. The real availability of a technique must be considered at the outset as a major factor in the development of a strategy. Without consideration of this and other socio-economic factors, strategies for control may be developed that have little relevance to the country or region.

There is also a tendency within insect pest management, to develop and employ control techniques that can provide only a short-term solution to the problem. Such techniques can work effectively only over a limited time scale, because as soon as their use becomes widespread, pests will adapt to them and render them useless. This may provide work for researchers and short-term economic gains for commercial companies, but it does not ultimately solve pest problems, or contribute a great deal to the development of sound insect pest management strategies. Despite this, many techniques gain acceptance within the framework of mosquito management, because each new technique is claimed to be another weapon to be added to the mosquito control armoury.

15.3. ACTION THRESHOLDS AS A COMPONENT OF INTEGRATED MOSQUITO MANAGEMENT

Classical IPM has a number of key components, such as identifying pests and understanding their biology and behaviour, monitoring pest levels, establishing a level of infestation above which control measures are justified, and using a mix of appropriate control measures to reduce pest damage.

The IPM concept has been subsequently adjusted to suit other types of pest control, including nuisance mosquito control. In nuisance pest control, most of the core components of IPM have been readily adopted, with the exception of defining a threshold infestation level. Although the threshold concept fits agriculture well, there have been difficulties in establishing and using threshold infestation levels for nuisance pests. Nonetheless in mosquito control, the fundamental need for such a threshold remains. Should nuisance mosquito control aim to reduce biting levels by 75%, or by 95%, to less than 5 bites per night, to less than 1 bite per night, or even to zero? Without defining such a level, mosquito control may, on the one hand, be over-using insecticides (resulting in unnecessary expense, and possible environmental damage), or may on the other hand be failing to provide the standard of control expected by residents and visitors. Defining such tolerance or threshold levels should therefore be a central part of the planning of mosquito control programmes.

In the United States there are several examples of the use of action thresholds for mosquito control activities. For example, Headlee (1932) concluded from his research in New Jersey that the tolerance for mosquito bites was no more than 4 bites per night. More recently Robinson and Atkins (1983) studied the knowledge and attitudes of residents in Virginia to mosquitoes. Their data showed that 50% of the residents considered that 3 or more mosquito bites per night constituted a problem. By contrast, surveys of residents in Texas revealed that a mean of 5.7 bites/night was "no problem"; 7.7 bites/night was a "mild" problem; while 11.5 bites/night was considered a "severe" problem (John, 1987). It should also be recognised that mosquito bites at night (*e.g.* biting *Cx. p. pipiens*) when people are trying to sleep, have a different "weight" as compared to bites outdoors during daytime or dusk (*e.g.* bites by *Aedes/Ochlerotatus* spp.).

Most mosquito control programmes in Europe are funded by local or national taxes, and is carried out typically by public sector organisations on behalf of the taxpayer. It is therefore quite appropriate that the taxpaying public should play a part in defining the thresholds and standards of mosquito control programmes. To date however, relatively few mosquito control organisations have actively involved the public in setting standards for mosquito control. In Italy though, there have been very considerable efforts to survey the public in order to establish their tolerance of mosquito biting, and then relate this to light trap catches and treatment timing. In Germany it has been shown that the floodwater mosquito *Ae. vexans* is a nuisance problem in human settlements when more than 50 specimens are caught in dry ice baited traps about 2 km away from the settlements. This process has enabled local mosquito control organisations to ensure that their programmes provide the level of control expected by the community.

Surveys of the local community have to be carried out in order to gather information on mosquito biting tolerance. Such surveys should be very carefully designed and standardised, so that information can be compared between different regions across the country, and compared between years. Surveys should be carried out in both rural and urban areas, and in areas with and without current mosquito problems. Such surveys should be repeated on a regular basis, as community tolerance of mosquito nuisance is likely to change over time.

Experience has shown that a community's tolerance of mosquito bites tends to drift downwards over time. In the Rhine valley for example, the level of mosquito biting that was considered as acceptable by the community 10 years ago, is considered unacceptable nowadays. Mosquito control efforts therefore have to be steadily intensified over time. These increased efforts are also likely to cost more; initially reducing mosquito bites from 20 bites per night to 5/night, is less expensive than the subsequent reduction from 5 bites/night to 1/night. Forward-looking mosquito control organisations need to recognise and anticipate such changes.

The level of reduction in mosquito biting required by the community, may affect not just the intensity of the mosquito control efforts, but also the type of mosquito control techniques used. For example a 75% reduction in biting by urban *Culex* spp. may be achieved by the careful use of larvicides alone. However if the community need a 95% reduction, then this may require much more radical measures, such as engineering works to improve urban drainage, and statutory measures to remove breeding sites on private property.

However setting such unambiguous targets for control is not just a help for the mosquito control organisation itself. With closer scrutiny of public expenditure in Europe, setting and adhering to thresholds and targets enables the municipality and the community as a whole, to assess both the technical and financial effectiveness of its local mosquito control organisation.

15.4. INFORMATION DISSEMINATION

The techniques which have been used for teaching and education, by slides, films, and different personal contacts, are available to a restricted number of people and are affecting only a limited circle of a community, and do not last long enough. We need to search for more effective methods of communication through mass media or the Internet, in order to reach the general public. A great variety of public education media are currently available, and are being successfully employed to disseminate the necessary information to the target audience. The value of new approaches has to be brought to all of those concerned.

Improvements in understanding the actual communication networks used by community participants, may reduce the risk of communication gaps occurring. Carefully chosen continuing, compelling TV programmes with the desired "message" should be brought to people at peak viewing hours. One such example is the Italian TV service, RAI, which has introduced a popular weekly science programme eagerly watched by a large audience, regardless of the education level (Curtis, 1990).

An inadequate base of knowledge can also delay the adoption of a new technology. The application of computer technology in integrated pest/mosquito management is a point in question. However, despite the tremendous potential of computer modelling and system science for use in the design, implementation and day-to-day operation of integrated mosquito management, relatively little of that potential has yet been recognised. Mosquito problems in Europe are almost always nuisance, economic and to a lesser extent health problems, albeit with sociological and, at times, psychological components. Their effective management must be based on a sound ecological and economic understanding of the problem. Public willingness to fund mosquito control is influenced to a large degree by public perception of the benefits of doing so, and the credibility of the applicable science.

15.5. IPM TECHNIQUES

IPM in a mosquito control strategy might include several techniques:

Water management and breeding site reduction: Its goal is to eliminate permanent breeding sites by a range of means, which are often regarded as source reduction measures. These measures consist of simple actions such as cleaning street gutters, or major remodelling of landscapes.

Monitoring: A monitoring programme involves seasonal information on the trends and abundance of both the larval and adult mosquito populations. Dipping provides an indication of larval and pupal populations, whereas landing rate counts or human bait catches (HBC), along with trapping techniques such as CO_2 baited traps, measure relative adult abundance. Monitoring is necessary for the affirmation of any programme but is usually labour- and cost-intensive. Monitoring of post-treatment mosquito populations can also be used to compare the benefits of different control strategies, such as larviciding versus adulticiding.

Biorational larvicides: These are the alternatives to conventional larvicides. Biorational control uses the insect's own biology, physiology and ecology to develop an effective control methodology. Biorational control agents belong to three general categories:

(i) Biologicals—the mosquitoes' natural enemies (e.g. mosquito fish) are introduced or their number increased in larval habitats, in order to suppress the mosquitoes.

(ii) Microbials—bacteria (*Bacillus thuringiensis* ssp. *israelensis*), fungi (*Lagenidium* sp.) or other pathogens.

(iii) Biochemical agents—IGRs, such as chitin synthesis inhibitors (diflubenzuron) or juvenoids (methopren). These materials are highly selective, have low vertebrate toxicity and are applied in low volumes.

Selective use of adulticides: Current adulticides include organophosphates, carbamates, botanicals (*e.g.* pyrethrum) and pyrethroids. Their principal advantage is that they provide rapid control. However the high costs of repeated applications, *e.g.* as a result of re-infestation from untreated areas, make adulticides too expensive. More important, both adulticides and conventional larvicides are broad spectrum insecticides that affect non-target organisms, and as a result are coming under more mistrust and observation for both health and environmental reasons.

Although in general the role of larviciding in mosquito control is significant, it should nonetheless be said that the reduction of vector populations cannot rely solely upon this method, regardless of the efficacy obtained through the larvicide application. The reduction of vector populations like *Ae. aegypti*, for example, as a result of community participation programmes, is not practical for use in emergency control situations, because such programmes do not target adult mosquitoes that may already be infected with a virus (Gratz, 1991).

Table 15.1 lists insecticides which could be successfully applied as larvicides, whereas Table 15.2 lists adulticides suitable for interior treatment against mosquitoes, derived from WHO (1997c).

15.6. INSECTICIDE REGISTRATION AND USAGE

In most countries, insecticide approval and use is regulated by law. National regulatory

Table 15.1. Insecticides Suitable as Larvicides for Mosquito Control

Insecticide	Chemical type[a,b]	Dosage of a.i.[c] (g/ha)	Formulation[d]	Duration of effective action (weeks)	Toxicity[e] oral LD$_{50}$ of a.i.[c] for rats (mg/kg of body weight)
B. thuringiensis ssp. *israelensis*	MI	[f]	AQ,GR,WP,WT	1–2	>30 000
B. sphaericus	MI	[f]	AQ,GR,WP	1–2	>5 000
Chlorpyrifos	OP	11–25	EC,GR,WP	3–17	135
Chlorpyrifos-methyl	OP	30–100	EC,WP	2–12	>3 000
Deltamethrin	PY	2.5–10[g]	EC	1–3	135
Diflubenzuron	IGR	25–100	GR	2–6	>4 640
Etofenprox	PY	20–50	EC	5–10	>10 000
Fenitrothion	OP	100–1000	EC,GR	1–3	503
Fenthion	OP	22–112	EC,GR	2–4	586
Fuel oil	—	[h]	soln	1–2	—
Malathion	OP	224–1000	EC,GR	1–2	2 100
Methoprene	IGR	100–1000	Slow release suspension	2–6	34 600
Permethrin	PY	5–10	EC	5–10	500
Phoxim	OP	100	EC	1–6	1 975
Pirimiphos-methyl	OP	50–500	EC	1–11	2 018
Pyriproxyfen	IGR	5–10	EC,GR	4–12	>5 000
Temephos	OP	56–112	EC,GR	2–4	8 600
Triflumuron	IGR	40–120	EC,WP	2–12	>5 000

[a]Pyrethroids are not normally recommended for use as larvicides because they have a broad spectrum impact on non-target arthropods and their potency may readily potentiate larval selection for pyrethroid resistance.
[b]IGR = insect growth regulator; MI = microbial insecticide; OP = organophosphate; PY = synthetic pyrethroid.
[c]a.i. = active ingredient.
[d]AQ = aqueous; EC = emulsifiable concentrate; GR = granules; soln = solution; WP = wettable powder; WT = water dispersible tablet.
[e]Toxicity and hazard are not necessarily equivalent.
[f]Dosage according to the formulation used.
[g]The lowest levels are recommended for fish-bearing waters.
[h]Apply at 142–190 l/ha, or 19–471/ha if a spreading agent is added.

agencies will require substantial evidence that a particular pesticide is suitable for the intended use, before allowing it to be sold and used. This evidence typically covers several categories, including the physico-chemical properties of the pesticide, the impact of the material on the public and the environment, and its effectiveness against the target pest. These regulations apply to manufacturers, distributors, and to the final user. They also apply to almost all pesticides, including biological agents such as *B. thuringiensis* ssp. *israelensis*. Pesticide users in public sector organisations, as well as in the private sector, will all be typically obliged to comply with these regulations. Within the European Community for example, active ingredients are regulated under the European Biocides Directive (98/8/EC, BPD), while individual formulations are registered at national level.

The cost of generating the data to satisfy regulatory organisations, as well as the cost of registration fees in each country, can be considerable. This investment tends to limit the availability of any particular insecticide to those countries in which the supplier expects a

Table 15.2. Insecticides Suitable as Adulticides for Interior Treatment against Mosquito Vectors

Insecticide	Chemical type[a]	Dosage of a.i.[b] (g/m^2)	Duration of effective action (months)	Insecticide action	Toxicity[c] (mg/kg of body weight)
Alpha-Cypermethrin	PY	0.02–0.03	4–6	Contact	79
Bendiocarb	C	0.1–0.4	2–6	Contact and airborne	55
Bifenthrin	PY	0.025–0.05	2–4	Contact	55
Carbosulfan	C	1–2	2–3	Contact and airborne	250
Chlorpyrifos-methyl	OP	0.33–1	2–3	Contact	>3 000
Cyfluthrin	PY	0.02–0.05	3–6	Contact	250
Cypermethrin	PY	0.5	4 or more	Contact	250
DDT	OC	1–2	6 or more	Contact	113
Deltamethrin	PY	0.01–0.025	2–3	Contact	135
Etofenprox	PY	0.1–0.3	3–6 or more	Contact	>10 000
Fenitrothion	OP	2	3–6	Contact and airborne	503
Lambda-Cyhalothrin	PY	0.02–0.03	3–6	Contact	56
Malathion	OP	2	2–3	Contact	2 100
Permethrin	PY	0.5	2–3	Contact	500
Pirimiphos-methyl	OP	1–2	2–3 or more	Contact and airborne	2 018
Propoxur	C	1–2	3–6	Contact and airborne	95

[a]C = carbamate; OC = chlorinated hydrocarbon; OP = organophosphate; PY = synthetic pyrethroid.
[b]a.i. = active ingredient.
[c]Toxicity: oral LD_{50} of a.i. for rats. (Toxicity and hazard are not necessarily equivalent.)

worthwhile return on this initial investment. Any particular regional mosquito control organisation may therefore not have the full potential range of mosquito insecticides available to them, and will be restricted to those products that have actually been registered for mosquito control in their own country. Where potentially useful insecticides have not been registered in a particular country, then it may be possible for the local mosquito control organisation to encourage the supplier to go through the registration process. Sometimes the mosquito control organisation has also been able to assist with any efficacy or eco-toxicology tests required for national registration, to the mutual benefit of the local community, the mosquito control organisation, and the supplier.

For the end user of insecticides, the legislation typically puts a wide range of requirements in place, and these must be followed. The requirements will typically be found on the insecticide product label, and mosquito control staff at all levels should be familiar with the details of the label requirements. Most mosquito control organisations will train their staff carefully to ensure an understanding of all aspects of the label requirements, and many will operate surveillance to ensure that staff actually comply with the regulations whilst working in the field. Such label requirements will typically specify a wide range of conditions including the following:

Use of personal protective equipment

Staff involved in mixing and application of insecticide are likely to be required to wear certain protective equipment, such as coveralls, gloves and face-shield, to prevent personal contamination. Such items should be provided, replaced and laundered by the employer, where appropriate. Even in the absence of label recommendations, use of minimum protection, such as a coverall, is good practice.

Personal hygiene

Immediate removal by washing of any direct contamination of the eyes and skin is essential. In addition, face and hands should be washed before meals and other breaks, and before leaving work at the end of the day.

Application equipment and dosages

Insecticide labels will normally specify the application equipment to be used, *e.g.* granule applicator, ULV sprayer. Such equipment should be maintained in good working order, and calibrated at regular intervals. Dilution of the pesticide, where required, will be explained fully and should be followed carefully. Only through careful calibration and monitoring of equipment performance is it possible to ensure that the correct quantity of insecticide is applied, and so achieve high efficacy whilst at the same time avoiding significant environmental impact.

Timing and frequency of treatments

Insecticide labels will normally provide guidance on when treatment should be applied for maximum efficacy, *e.g.* against certain developmental stages of the insect, or under certain meteorological conditions. In the interest of resistance management, there may also be restrictions on the frequency of pesticide use. All such conditions should be followed carefully.

Pesticide storage

There will normally be requirements to ensure that pesticides are stored safely and securely, and so avoiding possible safety incidents resulting from unauthorised access by the public, or from fire or extreme weather conditions.

Disposal of insecticide washing and empty containers

To avoid possible pollution or contamination, empty containers and washings from the application equipment should be disposed of in accordance with local regulations. In some cases it may be possible to use washings as the diluent for the next batch of insecticide to be prepared.

Clarification of details of pesticide label requirements can normally be obtained from the supplier, or the national regulatory agency. Useful information may also be obtained from a variety of other sources, for example the International Programme on Chemical Safety. Human and environmental safety of mosquito control activities has been very good in recent years, and it is the responsibility of all those involved in this important work, to ensure that this situation is maintained.

16

Implementation and Integration of Mosquito Control Measures into Routine Treatments

The most successful approach for attacking a nuisance or vector organism is when an integrated control strategy is implemented, in which all appropriate technological and management techniques are utilised, to bring about an effective decline of target populations in a cost-effective manner (WHO, 1983).

An integrated control strategy includes biological, chemical, physical and environmental management components.

With the development and successful use of powerful chemical insecticides (*e.g.* DDT) most control programmes after the second World War were based on the sole use of these insecticides. However, the onset of resistance against chemical insecticides and the increasing environmental and public health concerns, renewed the interest in integrated control strategies involving environmentally safe measures.

A single method may provide adequate control in a certain situation, however, the utilisation of a range of control and management methods within an integrated approach may be more cost-effective, cause less environmental damage, be effective for a longer time and be better suited at the community level (Rafatjah, 1982; WHO, 1982, 1983). In addition, the use of alternative strategies can reduce

the risk of developing resistance and strengthen interdisciplinary collaboration.

The most common mosquito control methods are:

(1) *Chemical control*: Residual spraying of adulticides such as DDT, malathion, bendiocarb, permethrin or related compounds is the most frequently practised method against endophilic and endophagic vector mosquitoes (Zaim, 2002). Mosquitoes are repelled from entering houses, or killed when they come into contact with the insecticides which are mainly sprayed on the interior walls of the houses or used to impregnate bednets (Brown, 1976; Lacey and Lacey, 1990). Space spraying against resting or flying mosquitoes can be implemented when exophilic and exophagic mosquitoes such as floodwater mosquitoes occur in masses, thus avoiding a nuisance in a defined area. However, space spraying usually protects against nuisance mosquitoes for a short time only, especially when species of a high migratory capacity are involved, which can easily reinvade the treated areas as the effect of the insecticide decreases. Chemicals can also be applied to mosquito breeding sites when they are accessible and defined. Developments towards more

425

environmentally friendly insecticides such as insect growth regulators (IGRs) can reduce the ecological damage to wildlife in ecologically sensitive wetlands.

(2) *Physical control measures* include: (a) environmental management *e.g.* reduction of breeding sources, water management, draining or filling of areas, or vegetation management, to create conditions unfavourable for mosquito breeding. Initial costs of environmental management may be high, but cost-effective on a long-term basis; (b) the use of surface layers such as monomolecular films or polystyrene beads which prohibit breathing of the immature stages of mosquitoes at the water surface. Additionally they can also reduce the deposition of eggs on the water surface (Curtis *et al.*, 1990; Reiter, 1978); (c) the reduction of human-mosquito contact, *e.g.* the use of bednets impregnated with pyrethroids. Treated bed nets may act as baited insecticidal traps attracting and killing anthropophilic mosquitoes (Schreck and Self, 1985; MacCormack and Snow, 1986).

(3) *Genetic control* of pest species has been used successfully in agricultural and veterinary pest control programmes. However, so far this technique has achieved very limited success in mosquito control. Several techniques can be used: (a) the sterile-male release technique which is based on the induction of sexual sterility in males through the use of radiation or chemical sterilants, and the consequent inundating of natural populations with modified males (Knipling, 1955, 1959); (b) cytoplasmic incompatibility between certain mosquito populations which has been ascribed to the presence of a rickettsial endosymbiont, *Wolbachia* spp., in the gonads (Portaro and Barr, 1975; Clements, 1992); (c) chromosomal translocations which may cause an increase in sterility of the progeny (Curtis, 1968a,b; Rai and McDonald, 1971; Laven *et al.*, 1972; Hallinan *et al.*, 1977). It has been demonstrated that meiotic drive genes in mosquitoes can distort the normally equal sex ratio in favour of males

(Wood and Newton, 1991). Recent developments in molecular genetics, such as the ability to introduce recombinant DNA constructions into the genome of organisms (transformation), offer possibilities for genetic control. It is hoped to genetically engineer mosquitoes such that they are refractory to infection by parasites (*e.g. Plasmodium* spp.), or unable to transmit them.

(4) *Biological control* includes natural and applied biological control. Naturally biological control is the reduction of mosquitoes through naturally occurring biotic agents (Woodring and Davidson, 1996). Applied biological control is planned human intervention to add biological control agents to a breeding site (augmentation). Potential biological control agents are pathogens and predators. Also aquatic plants such as ferns (*e.g. Azolla* sp.) or blue green algae (*e.g. Anabaena* sp.) can have a negative impact upon mosquito oviposition and the survival of the developmental stages of mosquitoes.

With the discovery of microbial control agents (*e.g. Bacillus thuringiensis israelensis* or *B. sphaericus*) a new era in biological control began. Under certain conditions the control of mosquitoes by naturally occurring biotic agents can now be supported efficiently by planned human intervention. The application of protein toxins from bacilli does not harm the non-target organisms in the mosquito breeding sites (*e.g.* fish and predacious insects), which can then further contribute to the natural reduction of the target population.

The environmental compatibility, and the high acceptance of microbial control agents by the community as a basis for an intensive participation, are important arguments for the application of these biological methods. Furthermore, by using larvicides, the mosquito problems are tackled at source. The nuisances and/or the vectors are fought, where they are most concentrated. They are killed in the breeding sites, before they disseminate out into a much larger area as adult mosquitoes. Thus in contrast to the application of adulticides, usually a smaller, more defined area has to be treated, resulting in a crucial economic advantage for the use of larvicides. However, microbial control agents or other larvicides are not practical for use against mosquitoes whose breeding areas are difficult to access.

Any control programme basically needs certain core elements, such as an effective management and administration including appropriate facilities, a personnel structure with clear responsibilities, and target-specific control tools which are used within an appropriate control strategy which the local community supports, and in which it is involved. The strategy has to be adapted to the specific needs of each particular situation during the operation, *e.g.* selection of a suitable formulation which fits the ecological conditions and the bionomics of the target organisms. The operation has to be guided by target surveillance and an environmental monitoring programme, as well as by a goal-oriented research and educational programme. Intensive public relations work increases the involvement and support of the general public.

16.1. PREREQUISITES FOR THE SUCCESSFUL IMPLEMENTATION OF THE PROGRAMME

The following steps for the implementation have to be conducted:

(1) a baseline collection of entomological and ecological data which should include:
—the occurrence of mosquito species with special regard to their vector capacity and/or nuisance potential;
—the monitoring of the population dynamics of the most important mosquito species in relation to abiotic (*e.g.* climatic factors, such as rainy and dry seasons, or temporary flooding periods) and biotic conditions (the assessment of predators of the mosquito immature stages, which can assist in forecasting the development of mosquito populations);
—information about mosquito migration, biting and resting habits are essential to define the intervention area;
(2) mapping and individual numbering of each significant breeding site are essential when larvicides are used;
(3) selection of appropriate tools and application systems to fight the target organisms;
(4) the effective dosage has to be assessed to assure a cost-effective operation;
(5) design of the control strategy should be based on the results obtained in pilot studies;
(6) training of the field staff;
(7) governmental application formalities;
(8) intensive public relations work.

16.1.1. Entomological Studies

Precise knowledge of the bionomy of the relevant mosquito species, is essential to fulfill the requirements of an economical and ecological sound control programme. Thus a monitoring programme for adult mosquitoes should be conducted to determine the species composition, abundance and phenology (population dynamics) in relation to the climatic conditions, as well as the spatial and temporal distribution related to migration. In different habitats human bait catches should be carried out to determine the most significant anthropophilic mosquito species. The migration behaviour of the most relevant species should be taken into account for the design of the strategy. For monitoring the adult mosquito populations, CO_2 baited suction traps are typically used to catch adult female mosquitoes (see section 4.3). Samples are usually taken at regular intervals from late afternoon until the next morning, during the main period of flight activity. After collecting the traps, the mosquitoes are counted and identified. When traps are located in an isolated mass breeding site and in concentric circles of 2, 5, and 10 or more kilometres in diameter, the horizontal dispersal behaviour of the emerging adults can be determined in relation to wind speed, temperature or vegetation (Petric *et al.*, 1999).

In a variety of typical breeding sites within the control area, larval collections should be taken at regular intervals to assess the species composition and abundance. Usually 10 dips or more with a standard dipper are taken per breeding site to obtain average and comparable data. The larval instars are recorded and a sufficient number of larvae is taken to the laboratory to determine species composition.

16.1.2. Mapping of the Breeding Sites

Precise mapping and individual numbering of each significant breeding site enables rapid and effective communication between field staff, and so provides a solid basis for a successful operation. Several other studies are also carried out during mapping: (1) characterisation of the breeding sites according to their productivity (mosquito densities) and mosquito population dynamics; and (2) assessment of the ecological conditions of the major breeding sites, *e.g.* plant associations, or occurrence of predators, or rare and sensitive organisms.

Geographic Information Systems

Geographic Information Systems (GIS) have become very widely used by professionals in research, government and industry, in dealing with spatially related data.

A GIS consists of an organised collection of computer hardware, software, and geographic data, designed to efficiently collect, store, update, manipulate, analyse and display all forms of geographically referenced information.

Modern information technology allows the integration of GIS systems with database technology, and with digital mobile field data collection systems supported by a Global Positioning System (GPS).

The ability to link information about where things are, with the information about what things are like, provides the user with a better understanding of spatial phenomena and their relationships, that may not be apparent without such advanced techniques of query, selection, analysis and display.

Special care should be taken in designing and establishing a high quality GIS, to ensure that it contains all the relevant data for a specific purpose, including both geographic data (where is the feature—location and geometry) and properties (what is it like—attribute data).

Typical applications of GIS are:

—Mapping locations of certain features
—Mapping quantities and densities
—Finding out properties of distinct areas using database queries
—Mapping change by comparing how features change over a period of time, in order to forecast future conditions.

An extensive overview on GIS can be found in Burrough *et al.* (1998).

Application of GIS and information technology to mosquito control

GIS and information technology can greatly improve the survey, logistics and documentation of mosquito control operations. The possible applications range from direct digital site mapping using GPS assisted mobile devices, to timely aggregation of operational reports.

A spatially referenced database containing all features of interest, is the basis for all further data collection and analysis. This spatial element enables thematically related features (*e.g.* population densities of certain species, flooding areas, plant associations and vegetation type, zones of nuisance or disease) to be organised in separate layers of information, which can then be analysed and displayed in a user defined context.

The following is an outline of possible applications:

- Analysis and query of available digital maps, aerial photos or satellite imagery, and thematic maps (*e.g.* hydrology, flooding zones, wetland inventories), in order to determine potential larval habitats.
- Spatial analysis to determine relationships between nuisance or disease, and breeding sites (calculation of buffer-zones, map- and database-query).
- GPS-assisted field data collection and breeding site inspection (detailed habitat mapping, larval and pupal survey). The use of handheld systems that synchronise data with the main database, allows accurate and timely processing of results and database updates.
- Forecasting of time and location of appropriate control activities, based on correlations between the spatial occurrence of triggering events for larval development (*e.g.* water levels and flooding areas, local weather data, the potential of larval development sites, and the results of current survey data).
- Preparation of operational maps to improve logistics, to calculate the quantities of control materials and manpower required, and to calculate the duration and costs of treatment.
- Storage of historical site profiles and related attribute data on the basis of operational maps, enables future potential larval development, resulting from dynamic triggering events, to be predicted.
- GPS-assisted operations allow the tracking and direct digital documentation of field activities (*e.g.* aerial application).
- Reports and documentation of survey and control activities can be assisted by user-defined database and map queries, which give immediate access to information stored in the database. The results of the queries can then be visualised and printed in the form of standardised thematic maps, graphics or tables.

For more detailed and up to date information on this fast developing technology, it is recommended to look at the webpages of commercial GIS-developers or mosquito control agencies.

16.1.3. Selection of Appropriate Tools

Data are collected in the implementation phase, to ensure the appropriate choice of control tools, of application techniques, and of the correct formulations and effective dosages, depending on the characteristics of the insecticide as well as on the biotic and abiotic conditions at the application site.

The optimal use of the product requires homogenous dispersal of the material in a recommended dosage over the target area in a certain period of time. For instance, the volume of water used for a dilution depends on the spray system (size of the nozzles, pressure of the system and speed of application) to obtain the desired dosage per area. The rate of emission in relation to the speed of application has to be calibrated before routine treatments, to ensure application of the correct dosage.

The selected equipment must be correctly adjusted and operated. Insufficient knowledge and lack of training on how to use and maintain the application equipment, and on how to supervise spraying teams, can be important factors that lead to a poor application of the mosquito control products.

16.1.4. Effective Dosage Assessment

The efficacy of a control agent such as *B. thuringiensis israelensis* is influenced by a great variety of biotic and abiotic factors: the susceptibility of the target mosquito species, the stage of development, the feeding behaviour of mosquito larvae, the density of larval mosquito populations, the temperature and quality of water, the intensity of sunlight, and the presence of filter-feeding non-target organisms (Lacey and Oldacre, 1983; Mulla *et al.*, 1990; Becker *et al.*, 1992; Becker and Rettich, 1994). The characteristics of the formulations in use, such as potency, settling rate and shelf-life, can also influence the effectiveness of microbial control agents.

It is therefore important to understand the impact of these factors on routine treatment, especially as it affects the calculation of the dosage, the selection of the right formulation for each particular environmental situation, and the appropriate timing for the treatment.

Assessment of the Potency of the Products and Effective Dosages

Before new microbial formulations are used in the field, their efficacy is evaluated against *Ae. aegypti* and indigenous mosquito species in the laboratory (minimum effective dosage) and field. All bioassays are run according to World Health Organisation (WHO) guidelines (WHO, 1981). The potency of the product (International Toxic Units = ITU) is determined by the formula given in the chapter on biological control.

Assessment of the Minimum Effective Dosage (LC_{99})

Different instars of the indigenous mosquito species are collected in the field and their sensitivity is determined by bioassays. The procedure for bioassays has to be adapted to the specific needs of the study. Water from the larval habitat should be used instead of distilled water, to avoid any dramatic change in the living conditions of the larvae which might affect the evaluation of the susceptibility of the tested species. Various concentrations are used to achieve mortality rates between 10% and less

than 100%. Each formulation is run at least 3 times at 5–7 different concentrations. The LC_{99} values determined for field-collected larvae is defined as the minimum effective dosage, and serves as a guideline for the assessment of the field trials.

Assessment of the Optimum Effective Dosage in the Field

Based on the results obtained in the laboratory, the optimum effective dosage for field applications is determined. A series of small field tests in various larval habitats are conducted to determine the optimum effective dosage against naturally occurring larvae of the indigenous mosquito species. For the field tests, concentrations of 1, 2, 4, and even 8 times the value of the minimum effective dosage (LC_{99} value obtained in the bioassays) are used. Each concentration should be tested in triplicate in the characteristic breeding sites and untreated breeding sites should serve as a control.

In most countries, the dosage for each mosquito control product is set by the national regulatory agency, not by the local mosquito control organisation. The national regulatory agency will in turn typically select dosages that have been established by the manufacturer.

16.1.5. Design of the Control Strategy

The control strategy for large-scale operations should be determined according to several considerations:

(1) The migration behaviour of the target mosquitoes. The objective of the strategy is to keep mosquitoes away from human settlements, and so the migratory behaviour of the species in question needs to be considered. Species such as *Ae. vexans* that readily migrate, should be controlled even in breeding sites that are far away from settlements (*Ae. vexans* can migrate more than 15 km when

population pressure is high). Domestic mosquitoes (*Cx. p. pipiens*) that migrate not more than a few hundred metres are controlled only within the settlements and within a radius of 500 m. Snow-melt mosquitoes such as *Oc. cantans* usually do not migrate further than 2 km, and prefer to stay in densely vegetated areas. These mosquitoes may be controlled in buffer zones of about 2 km in diameter around villages.

(2) The potential mosquito productivity of a breeding site is a criterion for the relevance of that breeding site.

(3) The climatic conditions (changes of the water level, length of rainy and dry season), which influence the population dynamics of the mosquitoes.

(4) The population dynamics of the target organisms determine the timing of the treatment which will have the greatest impact on the target organisms.

(5) The residual effect of the control agent is relevant for the sequence of re-treatments.

(6) Adaptation of the control techniques to the ecological conditions. According to ecological conditions, such as water level and vegetation growth, the most suitable formulation and application equipment should be selected.

(7) Development of an integrated control strategy, including predators, environmental management, and community participation. The cooperation between political decision makers, public authorities, scientists and the general public, are of great importance for sustaining success in mosquito control.

16.1.6. Training of Field Staff

A crucial component for a successful control programme are adequately trained and efficient field teams. On the one hand, the control programme must be tightly organised, and on the other hand, the control team must be flexible enough to respond to each individual situation with the most appropriate techniques and at the correct time. The field staff have to be trained in the biology and ecology of the target mosquitoes and their identification, the principles of the methodology, the mode of action of the control agents, the application techniques and how to carry out the treatment which is most appropriate for each situation. The teams should meet regularly to exchange experiences from the field and for further training.

16.1.7. Governmental Application Requirements

Governmental approval for the use of a mosquito control agent is a legal requirement for the operation. The application for approval will contain data on the materials and formulations in use, strategy of control, dosages, application techniques, assessment of the threshold for the control operation, assessment of the mosquito density and ecological considerations.

16.1.8. Public Information Systems

Appropriate information will increase the public acceptance of the treatments (see section 15.4). At regular intervals the media should be informed about the actual operations and the results of the control.

16.1.9. Community Participation

Mosquito control is particularly successful when local people are involved in the search for solutions to mosquito problems especially when they are related to mosquitoes breeding inside the settlements (Becker, 1992). The responsibility for the definition of the strategy, the planning, and the regional organisation of the control programme rests with the programme authorities, but even at the planning stage the expected cooperation of the community must be taken into account (Halstead *et al.*, 1985; Yoon, 1987).

The programme authorities have to evaluate how the control strategy may be best achieved through community participation. The successful interdigitation of the vertical (regional authorities) and the horizontal organisational (community) structures through community participation means that the people become "actors" instead of "spectators" (Diesfeld, 1989). The programme has to enable people to contribute to the solution of their mosquito problem related to their own settlement. This can be achieved by a comprehensive campaign of instruction dealing with the biology of the mosquitoes and with straightforward methods of control explained by means of web sites, leaflets, television, daily newspapers, circulars, posters, videos, slides and demonstrations. Thus "help through self-help" can be achieved. It is important to keep the level of motivation high over a long period. People are frequently aware of the problem, but they take no action or become involved just for a short time.

Community participation can be best and most meaningfully achieved in the control of mosquitoes that breed in or close to human settlements, such as *Cx. pipiens*, *Cs. annulata*, *Ae. albopictus* or *An. plumbeus*. The following methods can be used:

(1) Source reduction, by elimination of unnecessary water bodies;
(2) The weekly replacement of water in bowls and jugs as well as cleaning of containers to remove and destroy mosquito eggs;
(3) The careful covering of water containers, *e.g.* with lids or netting to prevent egg laying and larval breeding;
(4) The release of larvivorous fish into water reservoirs;
(5) The filling of potholes with concrete or by draining the water to avoid the collection of water;

(6) The application of larvicides if none of these methods is appropriate (*e.g.* fizzy tablets like Culinex tablets based on *B. thuringiensis israelensis* or *B. sphaericus*, methoprene briquettes or Abate granules). There are many arguments in favour of introducing the use of microbial control agents with community participation.

16.2. ROUTINE TREATMENTS

In the following examples, the principles of the implementation of a mosquito-control programme in different European countries will be documented.

16.2.1. Mosquito Control in Croatia

In Croatia 50 species of mosquitoes have been recorded so far. Geographically Croatia can be divided into three different regions (Pannonian plains, the mountain region and the Adriatic coast) which are inhabited by different mosquito species.

In the Pannonian plains 25 mosquito species can be found. The major nuisance species is *Ae. vexans* mainly associated with *Oc. sticticus*, *Ae. cinereus*, and *An. messeae* which breeds in the inundation areas of the rivers Sava, Drava and Danube.

In the mountain areas (Gorski Kotar, Lika, Velebit, North Dalmatia) species such as *Oc. nigrinus*, *Oc. punctor*, *Oc. geniculatus*, and *An. maculipennis s.s.* occur in moderate numbers.

In the Mediterranean area 20 species have been identified so far, with the most important ones being *Cx. p. pipiens* (the main nuisance species in human settlements), *Oc. caspius*, *Oc. mariae*, *Oc. leucomelas*, and *Cs. annulata*.

Until the 1950s malaria was transmitted in Dalmatia by *An. sacharovi* and *An. labranchiae*.

Mosquito control operations are usually organised and financed at local levels (cities, towns, villages) such as in Osijek, Vukovar,

Slavonski Brod, Vinkovci, Županja, Bilje, Beli Manastir, Pula, Poreč, Umag, Split, and Zagreb. Regional initiatives are aiming at merging the ongoing programmes into larger operational units.

Aerial spraying against adult mosquitoes is restricted by the Croatian government. For ground application Malathion, Baygon, Icon, AquaReslin are used. More and more larval control is based on the use of *B. thuringiensis israelensis*.

16.2.2. Mosquito Control in Czech and Slovak Republics

In the Czech Republic, the Elbe Lowland (along an approximate 150 km stretch of the Elbe River from the town of Melnik to Kradec Kralove) and the Morava Lowland (the banks of the Morava river and its tributaries from the town of Olomouc to Breclav, ca. 200 km^2), have a total population of about 500 000 people in need of effective mosquito control.

In the Slovak Republic, Rye Island—north bank (ca. 70 km) of the Danube river from Bratislava to Komarno (the "Mosquito City"), the inundation areas of the Vah river, and the lowlands of East Slovakia (inundated by the Bodrog and Latorica rivers), all experience mosquito problems.

The Elbe and Morava lowlands are regularly flooded after the thaw of snow in early spring, while in summer some areas are irregularly flooded once or twice a year (there are some years with no flood at all), and in addition there are artificially flooded meadows or forests (for agriculture or forestry purposes).

The remnants of the Danube flood-plain forests are usually flooded two to four times each summer. The extent of the flood water depends on the snow-melt in the Alps and on rainfall, and it is necessary to constantly monitor the water flow in the Danube. Floods may be partly regulated by the Gabcikovo dam systems, if necessary.

In the Elbe and Morava lowlands, the snow-melt mosquito *Oc. cantans* and the flood-water mosquitoes, *Ae. vexans, Oc. sticticus,* and *Ae. rossicus*, the latter in Southern Moravia and the Rye Island only, are the dominant nuisance species causing great discomfort to inhabitants, and serious economic damage by impacting on tourism and property values. In some regions severe mosquito nuisance may make agricultural or forestry work temporarily impossible, and in some localities (natural deer parks) it may cause serious loss of income.

Females of the autogenic *Cx. p. pipiens* biotype *molestus* are an important mosquito pest in major cities of both countries. They breed especially inside buildings (*e.g.* in flooded warm cellars).

Though mosquitoes in most parts represent only a nuisance, the fact that in the southern part of Moravia mosquitoes are also vectors of the Tahyna virus (which can cause a flu-like illness or very rarely meningoencephalitis) should not be overlooked. More than 90% of forest workers in the Breclav region exhibited high titres of Tahyna virus antibodies, according to studies made in the early 1990s. The Calovo and West Nile viruses have also been isolated in some regions of former Czechoslovakia, although no epidemics have been recorded yet.

Abundant breeding sites of *Anopheles* mosquitoes are widespread in southern Moravia, southern and eastern Slovakia, but no authochtonous malaria cases have occurred since its eradication in 1952.

Until 1989, aerial adulticiding (with organophosphates) or granular application of various larvicides (organophosphates or *B. thuringiensis israelensis*) were carried out in selected areas.

Since 1989 only small scale adulticiding using pyrethroids has been carried out. Field larviciding is practically non-existent. In 1997 (after catastrophic floods in Moravia) tens of thousands of hectares were aerially treated by cold aerosol containing pyrethoids (permethrin). Autogenic *Cx. p. pipiens* biotype *molestus* breeding sites (inside houses) are eliminated by drying-out (if technically possible) or they are treated by OP larvicides.

16.2.3. Mosquito Control in France

Organised mosquito control has a long history in France, which reflects the severe problems caused by these insects. Mosquito control in the areas of France in which mosquitoes cause the greatest medical and nuisance problems, is the responsibility of five independent organisations. EID Méditerranée (Entente Interdepartementale pour la Demoustication) based in Montpellier covers the French part of the Mediterranean basin, EID Ain Isère-Rhône-Savoie based in Chindrieux covers the Rhone-Alps region, EID Atlantique based in St-Crépin covers the Atlantic coast, two abatement districts in Alsace protects the human population on the French side of the Upper Rhine Valley and one other is active in Paris.

EID Méditerranée was founded in 1958, and has grown into a respected institution in southern France, since the development of tourism and economic growth in that region would have not be possible without the activities of this organisation. The EID Atlantique was established in 1963, the EID Rhone-Alps in 1966, the Alsace region started the mosquito control activities in 1984, and Paris district in 1992. Initially sustained by the central French government until the 1980s, the control operations are now paid by the local (Départments and communities) and regional administrations.

Most effort is directed at larval control, either through physical methods, management of wetlands and taking into account the whole fluvial hydrosystem, or by use of larvicides.

Correlation of larval abundance and water fluctuation, followed by ecological maps and dominant vegetation of the sites, provided the basis for implementation of larval control.

From the mid 1980s, chemical insecticides were partly replaced with biological larvicides. By the year 1997 some of the organisations, such as EID Rhone-Alps and in the Alsace region, based their larval control exclusively on biological larvicides (Fig. 16.1).

Each French mosquito district has its own particular environmental conditions for mosquito development, which require special approaches. The Mediterranean organisation has developed control programmes for 350 000 ha from Marseille to the Spanish border; the Rhone-Alps carries out mosquito control across 250 000 ha; the Atlantic organisation covers 100 000 ha; Alsace approximately 8000, and the Paris district covers 2000 ha. In addition to flood areas and/or wetlands, numerous *Culex* generations may also develop in a wide range of breeding sites (*e.g.* water containers). These urban breeding sites require different treatments. Some are physically eliminated, some are made unsuitable for mosquito development through different management, while others are sprayed with conventional larvicides or biological control agents (*B. sphaericus* and *B. thuringiensis israelensis*).

16.2.4. Mosquito Control in Germany

The control of mosquitoes in Germany has a long history. In the 1920s and 1930s breeding sites were treated with petroleum oils (Britz, 1986; Becker and Ludwig, 1983). During the 1950s and 1960s adulticides were used. In the early 1970s, the mosquito population was extremely high because of frequent fluctuations of the water level of the Rhine. The people in the villages couldn't spend any length of time outside their houses. There was an attack rate of more than 500 female mosquitoes per minute. As a reaction to this natural disaster 44 towns and communities on both sides of the River Rhine merged their interest in a united mosquito control programme, the KABS (Kommunale Aktionsgemeinschaft zur Bekämpfung der Stechmückenplage e.V.)

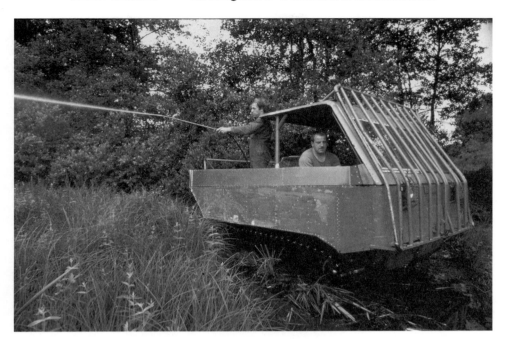

Figure 16.1. Application of larvicides by an amphibious vehicle. Photo courtesy of E.I.D. Alpes.

which was founded in 1976. Nowadays more than 100 cities and municipalities along a 310 kilometre stretch of the Upper Rhine River, with a total population of 2.7 million people, have joined forces to control the mosquitoes, mainly *Ae. vexans* over a breeding area of some 600 km^2 of the Rhine's flood-plain. The budget of the programme is approximately 1.9 million Euro a year which results in overall costs per person per year of approximately 1 Euro. The structure of the KABS is shown in Fig. 16.2.

The overall concept is integrated biological control (IBC); this means integration of the protection of humans against mosquitoes and the conservation of biodiversity. When the ecosystem is compared with a web and each group of organisms represents one mesh, the strategy of the KABS aims at the removal or reduction of one single mesh which respresents the mosquitoes without cutting other meshes in the "food web".

This goal could only be reached optimally when biological control methods are used. The conservation and encouragement of predators is

an important goal of the programme. Therefore, microbial and biological methods are integrated with environmental management (*e.g.* improvement of the ditch system for regulation of the water level and provision of permanent habitats for aquatic predators such as fish).

For the successful implementation and use of microbial control agents the following prerequisites were necessary (see Chapter 16.1): entomological studies, precise mapping and numbering of all major breeding sites, assessment of the effective dosage in bioassays and in small field tests, adaptation of the application technique to the requirements in the field, design of the control strategy as well as training of the field staff and governmental application formalities.

For more than two decades *B. thuringiensis israelensis* and *B. sphaericus* have been successfully used in Germany as biological control agents against floodwater mosquitoes (*e.g. Ae. vexans*) and *Culex* mosquitoes (*e.g. Cx. p. pipiens* biotype *molestus*). Over 2000 km^2 of breeding areas have been treated with *B.*

Organisation of the KABS

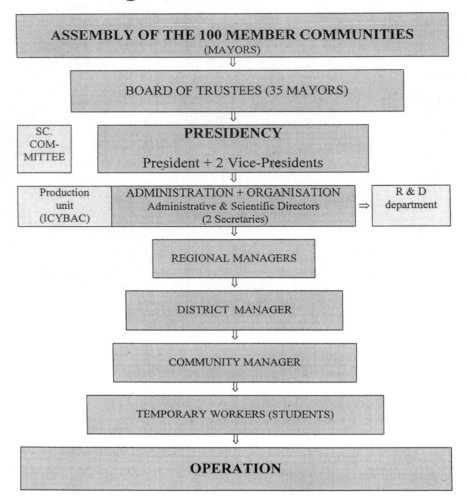

Figure 16.2. Structure of the KABS.

thuringiensis israelensis, resulting in a reduction of the mosquito population year by year of more than 90%. The environmental safety and efficacy of the bacilli toxins are derived from a variety of factors which are described in Chapter 11.

The flood plains of the Rhine are usually inundated two to four times each summer. The extent of the flooding depends on the snow-melt in the Alps and on rainfall, and it is constantly necessary to monitor the water flow in the Rhine and in the flood plain. During flooding, *Ae. vexans* and other floodwater mosquito larvae hatch within minutes or hours at temperatures exceeding 8°C. Before control measures are to be conducted, the larval density

and the larval stages are checked by means of ten sample scoops at representative breeding sites, in order to justify the action being undertaken and to establish the correct dosage and the best formulation to use. One day after application, spot sample scoops are taken at the reference breeding sites to check mosquito density and thereby establishing the efficacy of the treatment.

According to the extent of the flooding, 20–30% of the potential breeding areas of 600 km^2 have to be treated regularly by the 400 collaborators of the KABS. For treating first- and second-instar larvae, 250 g of powder formulation (3 000 ITU/mg) or less than 1 litre of liquid

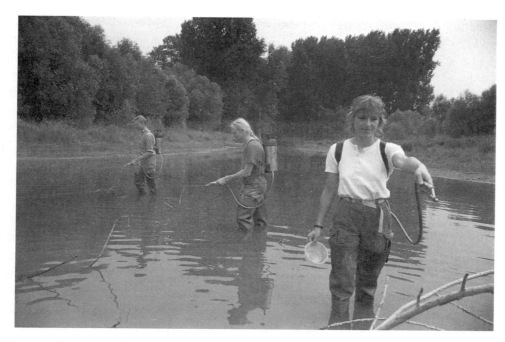

Figure 16.3. Ground application of microbial control agents by means of knapsack sprayers.

concentrate (1200 ITU/mg) are dissolved in 9–10 litres of filtered pond water for each hectare treated and applied by knapsack sprayers. For deeper breeding sites or when late instar larvae are present, the dosage is doubled (Fig. 16.3).

During the most severe floods, usually a third of the area has to be treated with *B. thuringiensis israelensis* granules which are dispensed with the aid of a helicopter (dosage 10–20 kg/hectare). From 1981 to 2002, more than 100 tonnes of *B. thuringiensis israelensis* powder and fluid concentrate and more than 1000 tonnes of *B. thuringiensis israelensis* ice and sand granules mixed with quartz sand, vegetable oil and *B. thuringiensis israelensis* powder, were used, to treat over 2000 km^2 of breeding area. Since 1997 the more cost-effective *B. thuringiensis israelensis* ice granules have been applied by helicopters to thousands of hectares of floodwater mosquito breeding sites (Fig. 16.4).

Control of urban mosquito species is mainly carried out by householders or inhabitants. To assist with this, KABS provides information on the biology of *Cx. p. pipiens* biotype *molestus* and the appropriate control measures.

Figure 16.4. Aerial application of Bti-ice granules.

Culinex® tablets have been particularly successful. They kill *Culex* larvae in water containers for a period of several weeks. In drainage systems and large cesspools with eutrophic water bodies, *B. sphaericus* as a liquid or powder formulation is applied against *Culex* larvae. Each year about 1 million Culinex-*B. thuringiensis israelensis/ B. sphaericus* tablets are successfully applied against *Cx. p. pipiens*, especially in rainwater containers.

Monitoring the Programme

Some 8% of the KABS budget is invested in monitoring mosquito populations, mosquito resistance and environmental impact. All the studies carried out to date have shown that the introduction of *B. thuringiensis israelensis* and *B. sphaericus* has reduced the numbers of nuisance mosquitoes to a tolerable level, but that the diversity and beauty of the ecosystem as a whole has not been damaged.

Monitoring mosquito abundance: To monitor the abundance of adult mosquitoes, a sufficient number of traps are placed at comparable sites throughout the entire inundation area. These are monitored twice a month from April to September. On each occasion for a whole night, the mosquito density is sampled by means of carbon dioxide light-traps. Catches in areas where no control measures have been undertaken serve as points of reference (100% of the mosquito population) for catches from areas being treated, in order to determine the success of the measures (mortality rate in percent). It has been shown that since the widespread application of *B. thuringiensis israelensis* in 1981, mass occurrences of mosquitoes have been successfully averted. Naturally, these control measures have had an extremely positive reception among the local people.

Monitoring the environmental impact: It has been essential to document the environmental impact of *B. thuringiensis israelensis* and *B. sphaericus* applications, in order to provide a scientific basis for rebuting the arguments commonly brought against mosquito control by its opponents. Before large-scale application of microbial control agents was undertaken, the most important members of various aquatic groups (*Cnidaria* to *Amphibia*) were screened in the laboratory and in small-scale field trials for their susceptibility to microbial control agents. This study showed that in addition to mosquitoes (Culicidae) and black flies (Simuliidae), only a few species of midges (Chironomidae) were affected by *B. thuringiensis israelensis*. For the most part, these midges were much less susceptible to *B. thuringiensis israelensis* than the target organisms. *B. sphaericus* is toxic to an even narrower range of insects: certain mosquito species, such as *Culex* species, are highly susceptible, *Aedes/ Ochlerotatus* species are much less susceptible, and black fly larvae as well as other insects (exception: *Psychodidae*) and non-target organisms are not susceptible.

The development of insects in treated and untreated water is regularly monitored using emergence traps. The occurrence and abundance of insects in treated areas is assessed by regular light trap catches. All investigations have shown that while the numbers of *Aedes/ Ochlerotatus* mosquitoes are drastically reduced, all other insects continue to develop in the water and, as winged adults, provide a food resource for birds, amphibians and bats.

Monitoring the resistance: Mosquito populations are checked at regular intervals for the development of resistance (Becker and Ludwig, 1993). No resistance has been detected after 10 years of treatment with *B. thuringiensis israelensis*. To prevent resistance to *B. sphaericus* developing in *Culex*, *B. sphaericus* and *B. thuringiensis israelensis* are used alternately in the control management plan for this species.

16.2.5. Mosquito Control in Greece

One of the largest mosquito control projects in Europe as far as treatment areas are concerned, is the project related to rice field mosquito control in the area of Thessaloniki (a city of 1.2 million inhabitants in northern Greece). The project started in 1997, and consists of larviciding with the organophosphate temephos, using three helicopters (Hiller UH-12E). GIS data on mosquito larvae productivity gathered systematically (by 30 technicians for 18 000 ha of rice cultures) for the individual parcels of the rice fields (15 ha each on an average) exhibit very distinct space-temporal patterns for the years 1998–2001 (depending on different water management practices). Besides the project in Thessaloniki, control activities were recently implemented in the areas of the prefecture of Serres, Chalkidiki (14 municipalities), Larissa (2 municipalities), Kavala (3 municipalities), Pieria (11 municipalities), Imathia (3 municipalities), Pella (7 municipalities), and Lamia. Altogether more than 100 000 ha of breeding sites are treated with temephos (Abate 1SG, Abate 50 EC), *B. thuringiensis israelensis* (Vectobac and Teknar), Diflubenzuron (IGR, DU-DIM 10WP). *B. thuringiensis israelensis* is mainly used in ecologically sensitive areas.

In seven out of the fifty departments across Greece, there is currently a fully running network of 46 adult sampling stations. Twenty out of the 50 mosquito species found in Greece are monitored weekly including the three main anthropophilic species *Oc. caspius*, *An. sacharovi* and *Cx. p. pipiens*. Quantitative data over time provides considerable insight into the effectiveness of the abatement programme. Thus in the Thessaloniki area, the overall average July nuisance at sunset was approximately 60 bites/15 min in 1996 (baseline year, before the project implementation) versus 2 bites/15 min in 2001.

16.2.6. Mosquito Control in Hungary

There are several regions in which abundant mosquito populations occur in Hungary. The regular flooding areas along the Danube, the Tisza river and their tributaries, and the shores and marsh area around Lake Balaton and Lake Velence. The temperate central European climate offers favourable conditions for the development of large mosquito populations. So far there have been 44 species recorded in the country. *Aedes/Ochlerotatus* species are prolific in all breeding sites, and additionally *Coquillettidia* species are abundant in the thickly reeded marsh areas at Lake Balaton. *Culex p. pipiens* is frequently abundant in villages.

The area of inundation and other potential breeding sites along the rivers and lakes is estimated to be about 2500 km². The country's nature reserves are usually connected to the rivers, and during the overflows they become ideal breeding sites for mosquitoes. In such habitats larval samples may contain 1000–2000 larvae per litre. Along the river Tisza the area has been uniquely endowed with more than 170 ox-bow lakes. Since most of these lakes are situated in the country's nature reserves, the lakes are an important destination for eco-tourists.

Over recent decades many breeding sites in Hungary have disappeared while new ones have been created. Mosquito abundance has not decreased but has rather increased in many areas, apparently due to the deterioration of the water ecosystems and to the formation of new environments favourable for mosquito breeding.

Current mosquito control in Hungary is carried out by several private abatement or pest control organisations which perform the activity in conjunction with urban pest control. Recently attempts have been made to introduce novel technologies and strategies that involve regional cooperation between municipalities and public health authorities. Mapping of the breeding sites, and monitoring of all developing mosquito stages is a high priority for the mosquito control organisations. Some organisations are working on various programmes to improve the formulations available for mosquito control. However at present the main products used are either organophoshates or pyrethroids, and to a lesser extent some IGRs and products based on *B. thuringiensis israelensis*.

In 2001 the Hungarian Mosquito Control Association was founded and in 2002 the major companies dealing with mosquito control joined forces to implement control activities based on the most environmentally safe biological control tools in order to reduce the use of chemical insecticides.

16.2.7. Mosquito Control in Italy

As for most of the European countries, organised mosquito control activities in Italy suffered through a lack of general interest during the malaria post-eradication period 1950–1980. In this period programmes were active but scattered on local bases and mainly organised on adult control. With the increasing socio-economic conditions the demand for mosquito control on large scale areas started in northern Italy in the middle of the 1980s. At the same time two main factors determined a new way to implement mosquito control: the general awareness of the ecological impact of traditional mosquito control methods and the discovery and rapid market availability of selective microbial agents active against mosquito larvae.

Therefore several programmes were implemented mainly based on regional and local public financial support.

One of the first consortia to be established in 1987 was in the Bologna province, an irrigated agricultural area of about 900 km^2 in the Po plain where a group of 20 municipalities decided to merge their interests in a united mosquito control programme. Parts of the budget are covered by voluntary donations from private citizens. From the beginning major attention was devoted to larval control with occasional adult control in defined areas. As a result of specific research, surveys, and monitoring programmes, the control program has adopted larviciding as the only control method.

In the campaigns against the two major target species *Oc. caspius* and *Cx. p. pipiens*, 95% of the total larvicide products used are based on *B. thuringiensis israelensis*, while the remaining 5% consists of Temephos which is used in catch basins only.

According to data collected weekly by the adult mosquito monitoring programmes using CO_2 traps at 23 fixed stations, and from surveys based on public opinion, the results are highly satisfactory. A tolerance threshold based on CO_2 trap catches has been developed.

Breeding sites have been mapped and coded, and their characteristics digitalised using Map Info. This enables field technicians to conduct both ecological surveys and control on operational activities, updating the info on a daily basis. This data may be used for daily bulletin reports, and can be highly valuable when cost/benefit analysis of the programme has to be performed.

One particular achievement obtained in this programme has been the development of a special larvicide applicator. This device consists of a remote-controlled hydraulic arm fitted with larvicide nozzles and a camera, that enables a single operator to gain access, to see into, and to treat difficult breeding sites in urban areas and in dense vegetation. The use of this device increased the overall efficacy (Fig. 16.5).

Breeding habitats are regularly treated with *B. thuringiensis israelensis* at intervals of 5–7 days depending on the water temperature. The mosquito season of the region usually requires 20–24 treatments when *Cx. p. pipiens* is controlled.

Another programme started in 1991 in the Emilia-Romagna (Po delta region) to protect the tourist resorts of Comacchio and Ravenna. The area includes 47 km of coast with an expected yearly presence of 500 000 tourists. Before the operation the programme has to be evaluated by a commission including the institution of the Po Delta Natural Park and representatives of the Local Public Health Bureau as well as of the

Figure 16.5. Use of a special larvicide applicator equipped with a remote-controlled hydraulic arm fitted with larvicide nozzles and a camera. (Photo courtesy of Centro Agricoltura Ambiente, Italy).

Regional Agency for Environmental Protection. The strategy is based on an integrated approach combining the use of larvicides in natural, rural and urban areas as well as adulticides in urban areas. Adult mosquito densities estimated by three-weekly collections by using CO_2 traps, indicate that the current measures have achieved a reduction of 95% compared to the mosquito density as detected at the beginning of the project. The main target species of this region are, in order of importance, *Oc. caspius, Cx. p. pipiens* biotype *molestus, Oc. detritus*, and *Cx. modestus*.

Breeding sites have been digitally mapped and stored on Map Info since the beginning of the project, but are currently under transfer to Arc View programme.

In the Piedmont region, following the approval of the Regional Law n.75/1995 "Contribution to Local Institutions financing mosquito control interventions" several programmes were started and Piedmont rapidly became the region with the most organised mosquito control programme in Italy (Table 16.1).

In order to obtain the regional financial support the following is required: a standard mapping of the breeding sites, the storage of data in a dedicated software specifically developed

Table 16.1. Programmes operating in 2002 in the Piedmont Region

LOCALITY	area (ha)	inhabitants
Acqui Terme	3,300	20,100
Avigliana	2,500	11,100
Basaluzzo	30,000	55,000
Province of Biella	42,000	127,000
Casale M.to	74,000	109,000
Casorzo	3,900	2,500
Castello d'Annone	7,400	6,700
Castiglione T.se	14,000	104,000
Leini	17,400	96,000
Montalto Dora	9,400	43,300
Novara	4,000	102,200
Rivarolo C.se	20,400	47,400
Torino	13,000	900,900
Torrazza P.te	26,600	48,800
Province of Vercelli	5,000	48,000

inside Arc View, the adoption of larval control operations only, the weekly monitoring of adult populations by CO_2 baited traps.

The total contral area is 273 000 ha with a resident population of 1 722 766.

Other programmes are initiated in Lombardia, Veneto, Friuli Venezia-Giulia, Emilia-Romagna, Liguria, Toscana, Lazio,

Basilicata, Sardegna, Sicilia regions dealing with indigenous as well as exotic mosquito species.

More programmes are expected to be established in the near future particularly in central and southern Italy to protect tourist economy, and in northern Italy to reduce *Aedes albopictus* which is rapidly becoming the main noxious species in urban areas. Currently numberless localities have been colonised by the species which find particularly favourable habitats in residential areas. To achieve a good control level high financial support is needed in order to conduct source reduction campaigns and larviciding. *Ae. albopictus* is still spreading in Italy and the bordering countries must be considered at risk of introduction and therefore surveillance systems should be adopted to prevent further dispersal.

16.2.8. Mosquito Control in Poland

In the past mosquito control activities in Poland were strongly connected with the need to reduce the numbers of anophelines, because of their role as malaria vectors.

Malaria was endemic in several areas in Poland during the years after World War I. In 1921, 52 965 malaria cases were recorded, mostly in the eastern and south-eastern parts of the country. Anti-malarial action organised since 1925 in the areas where endemic foci existed, involved not only health care and medical treatment, but also control measures against mosquitoes with Paris Green (Szata, 1997). After World War II, malaria was present on the Baltic coast, in the Mazury-region (northeastern Poland) and near Warsaw. The serious outbreak of the disease in the northern part of the country in the 1940s and 1950s necessitated anopheline mosquito control. In the years 1948–1951 the control measures against both larvae and adult mosquitoes in the region of Gdanska Bay and Zulawy were undertaken as a part of the anti-malarial campaign. The measures consisted of residual treatment of resting places outside and inside buildings, space spraying, and treatment of larval breeding places with DDT. Physical modification of the environment (drainage, levelling of ground, and filling in small ponds) was also included.

In 1952 in Podgrodzie resort (near Szczecin) the first large-scale action against mosquitoes was undertaken, because of the presence of mosquitoes in enormous numbers. Screening surveys carried out at the beginning of July, before the control had started, showed the presence of anophelines (mostly *An. messeae*) and *Cx. p. pipiens* mosquitoes. Buildings (outdoors and indoors), forest areas near the resort, marshes, ponds, and part of Szczecinski Bay were all treated with DDT (Brodniewicz, 1953). In 1956–1958 on the Karsiborz Island in Szczecinski Bay, control measures against *Oc. communis* and *Oc. annulipes* were undertaken. DDT was used in the forest environment and indoors in residents' houses (Bosak *et al.*, 1961). After this action was completed with significant reduction of the mosquito population, similar activities were undertaken in 1958 and 1959 in several resorts in Szczecinski Bay. However, it was pointed out that general assessment of this large-scale mosquito control was ambiguous. Although the control actions were evaluated as being very efficient, in several areas the dose of DDT applied greatly exceeded that recommended by WHO, and negative impacts on non-target organisms were observed (Krzeminski and Brodniewicz, 1969).

As a result, large-scale mosquito control action in natural biotopes was not recommended the following year, but abatement operations in small, selected areas in tourist resorts were continued, mostly against nuisance mosquitoes. For chemical treatments the chlorinated hydrocarbons were used: DDT, BHC, Dieldrin, and in several cases—pyrethroids (Krzeminski and Brodniewicz, 1969).

In the 1960s, however, malaria was eradicated in Poland and the measures against mosquitoes were then undertaken to reduce their numbers to a level at which they no longer caused a nuisance by their bites, especially during periods of synchronised mass emergence.

In the 1970s and 1980s, according to fragmentary data, control of mosquitoes in Poland was focused on adult mosquitoes only and was carried out in selected places as necessary, with residual insecticides. The preparations used were mostly pyrethroid insecticides: permethrin, deltamethrin, cypermethrin, and also etofenprox. Being relatively inexpensive—products containing the organophosphate DDVP were also applied.

The flood, which took place in central Europe in the summer of 1997 resulted in the mass breeding of mosquitoes all over the flooded areas, including Poland, where mosquito control was carried out on a large scale in 6 of the 26 provinces affected by floods, situated mainly along the Odra river. In total 10 000 ha were treated, partially by aerial spraying from helicopters, partially by spraying from the ground. Mostly adulticides were used: AquaReslin Super (permethrin); Trebon 10 S.C. (etofenprox); K-Othrine flow (deltamethrin); and Alfasep Super Kill (cypermethrin) (Gliniewicz *et al.*, 1998).

It must be pointed out that the 97 catastrophic flood and the following huge mosquito problems increased the public's interest in mosquito control. As a result, large-scale control activities were re-initiated in several regions.

Since 1998 mosquitoes have been controlled in the town of Wroclaw, where every year mass emergence of nuisance mosquitoes has caused problems. The control measures have been directed against larvae and adult insects; the larvicides used are Simulin (based on *B. thuringiensis israelensis*) and Dimilin 25 GR (diflubenzuron); adulticides—Aqua Reslin Super and Trebon Mega S.C. (etofenprox). According to the surveillance data, the main nuisance species in Wroclaw has been *Culex p. pipiens*. Approximately 1600 ha have been treated. The efficiency of control was 85% in 1998 and 70% in 1999 (Rydzanicz *et al.*, 2000).

The town of Szczecin and the islands of Uznam (town Swinoujscie) and Karsiborz, also experienced problems in the 1990s resulting from the mass occurrence of nuisance mosquitoes, and the mosquito control measures have therefore continued. In Szczecin only adults were controlled (mainly with Trebon Mega 10 S.C.), while in Swinoujscie and Karsiborz, both *B. thuringiensis israelensis* (Simulin) as a larvicide and etofenprox (Trebon Mega 10 S.C.) as an adulticide were used.

In the town of Gorzow Wielkopolski, situated in the western part of Poland, the mosquito control programme was started in 2001. The treated area included 1250 ha of the city area, suburbs and marshy ground on river banks. Adult mosquitoes were controlled with Trebon Mega S.C., while larval breeding places were treated with *B. thuringiensis israelensis* agent (Simulin). The preparations were applied by helicopters, and from the ground with knapsack sprayers. Local authorities recognised the programme as being effective.

In 2000, a mosquito abatement programme based on *B. thuringiensis israelensis*, was started in Krynica Morska, a tourist resort, located on Vistula-Spit, in an area of protected landscape. Mosquito breeding sites were identified and mapped. Qualitative and quantitative studies on the mosquito fauna showed that *Oc. cantans* was the predominant species (84.7%), followed by *Oc. communis* (Kubica-Biernat *et al.*, 2000).

In April 2001 the first control measures were undertaken. Vectobac 12 AS and Vectobac TP—*B. thuringiensis israelensis* preparations were used as larvicides. The efficiency of the control was about 99% (with Vectobac 12 AS) and 97% (with Vectobac TP) (Kubica-Biernat *et al.*, 2000).

16.2.9. Mosquito Control in Spain

In Spain, organised mosquito control started in the early 1900s following the discovery of mosquitoes as being vectors of malaria. Nevertheless, interest in mosquito studies ceased as soon as the disease was eradicated in this country in 1963 (Pletsch, 1965).

Spain's economic development resulted in an increasing quality of life, and it was increasingly recognised by the public and by public administrators that mosquito nuisance should be considered as a limiting factor for the development of the country. This fact was especially important in towns and cities close to mosquito breeding places, and in areas where tourism was the most important industry.

The first current mosquito control service (MCS) was created in 1982, in Roses Bay and lower Ter River. In the following years, three more services appeared under the direction of public administrations. In 1983 the MCS of the Baix Llobregat was created, followed by the MCS of Huelva in 1985 (Anonimous, 1989), and the MCS of the Ebro Delta in 1991. Three of the MCS are situated in the northeast of Spain, in Catalonia, along the 350 km of Catalan coast, and the other is located in the southwest of Spain, in Andalusia.

All the MCS operate on the basis of integrated pest management focused on larval control, have complete independence, and are related to different local public administrations.

The first MCS of Roses Bay and lower Ter River was created in 1982 for two reasons: to limit the nuisance of mosquitoes, and to demonstrate to the tourist industry that a salt marsh area can be compatible with human activities. This area lies between the Pyrenees and the Montgrí massif, forming a bay with swamps, salt marshes and lagoons. The area covers about 34 000 ha of which 7200 ha are natural mosquito breeding sites. The region also contains more than 700 ha of rice fields and about 7000 septic tanks, in one of the most important tourist areas in Catalonia. *Ochlerotatus caspius, Oc. detritus* and *Ae. vexans* are the most important species in natural breeding sites, while *An. atroparvus* is the major species of the rice fields. Natural areas are mostly swamps, salt marshes and pasture meadows with variable water levels due to rain, sea intrusions or man-made causes. This area, including a natural park, is exclusively treated by biological agents (*B. thuringiensis israelensis*) while the most important *Cx. p. pipiens* breeding sites (the septic tanks spread over the residential areas) are treated with Pyryproxifen.

The MCS of the Baix Llobregat region east of Barcelona, was founded in 1983. This abatement district has to avoid mosquito nuisance along the river area where important tourist resorts exist. The service is responsible for fifteen municipalities stretching south of Barcelona, including the airport. The area of the municipalities involved is 25 000 ha, including 9000 ha of the river delta.

The species causing the nuisance are mainly *Cx. p. pipiens* and *Oc. caspius*. Although nineteen species have been identified, the aforementioned are the most abundant, and so are most important. Natural areas of this region have been disappearing progressively due to the pollution and construction activities in the vicinity of Barcelona. Only 340 ha are potential natural breeding sites.

While the natural zones are only a problem where rainfall or sea intrusions occur, there is potentially continuous *Cx. pipiens* development between May and November in the 290 km of polluted ditches in the district, and to a lesser extent even during the winter period because of underground sites. This species is treated by *B. sphaericus*.

The MCS of the Ebro Delta, covering 7 municipalities and more than 32 000 ha, performs an extremely valuable service in Catalonia. Because of the dramatic nuisance caused in villages and tourist areas by the enormous numbers of mosquitoes, the establishment of a mosquito control service was inevitable by 1991.

This delta is almost covered by more than 22 000 ha of rice fields and 10 000 ha of natural areas that form the Natural Park created in 1983. Most of the problems are related to natural or artificial fluctuations of the water level that cause the appearance of *Oc. caspius,*

Oc. detritus, An. atroparvus and Cx. modestus. Cx. p. pipiens problems occur in ditches and septic tanks in all villages throughout the delta.

The abundant mosquito larval population in the natural area is controlled by *B. thuringiensis israelensis*. For the control of rice field mosquitoes, a buffer zone around each village is treated with Temephos.

The region of Huelva established a control service in 1985, and covers an area of 11 municipalities with more than 130 000 ha. This region is located on the southwest coast of Spain and partially extends along the tidal salt marshes between Portugal and the Lower Guadalquivir River marsh. This natural area was drastically modified by the rise of an industrial complex destroying more than 400 ha of salt marshes. The pollution of the swamp water and the creation of physical barriers have further complicated the already difficult mosquito control on the salt marshes. Nowadays, 100% of the tidal marsh area is protected by environmental policy.

Oc. caspius and *Oc. detritus* are the most important species on the Huelva salt marshes. This area covers 16 000 ha, and is treated by both the conventional insecticide Temephos, and by the biological agent *B. thuringiensis israelensis*.

Cx. p. pipiens breeding sites in Huelva are mainly in rain ditches and gutters. In the city of Huelva, 60 000 gutters are controlled each year by *B. sphaericus*. Treatments are usually repeated on a biweekly basis.

16.2.10. Mosquito Control in Sweden

The first Swedish abatement district (Biologisk Myggkontroll-Nedre Dalälven, BMK-ND) was created in September 2000, as a response to many years of complaints about mosquito nuisance by the local population of the lower portion of the Dalälven river (Nedre Dalälven) in central Sweden. In the initial phase, the seven municipalities with part of their population in the Nedre Dalälven area, as well as four regional governments, guaranteed the financial basis of the project. Further financial support was provided by the Swedish government. A first estimate of the size of the potential mosquito breeding areas was 10 km^2. It was not possible, within a reasonable time frame, to use traditional ecological mapping for such a large and often inaccessible area. Thus, from the very beginning, high resolution orthophoto maps in combination with direct DGPS were used to create precise maps of breeding site location and size.

Approximately 10 km^2 of the Dalälven flood plain between the city of Avesta and the Sea of Bothnia provide temporary wetland areas. These areas are usually flooded once or twice during spring and summer. The extent of the flood depends on the snow-melt in the Scandinavia Mountain Range and on rainfall in the very large catchment area of the river Dalälven. It is constantly necessary to monitor the water level in the river and in the flood plain from the beginning of the snow-melt.

Some 24 different species of mosquitoes have been recorded in the Nedre Dalälven area, and the majority of species attack people. Determining the seasonal and spatial distribution of mosquito populations is an integral component of providing solutions to the periodic infestations of mosquitoes in the Nedre Dalälven area. Distribution patterns and seasonal dynamics of adult mosquitoes in this area have shown that the predominant species is *Oc. sticticus* (80%) followed by *Ae. rossicus* (8%) and *Ae. cinereus* (7%). These three species constitute the key nuisance species during the spring and summer floods. In spring, *Oc. communis, Oc. punctor* and *Oc. intrudens* can also be found in high numbers in the wetlands. *Oc. sticticus* causes the problems due to its mass occurrence and dispersal ability. It feeds viciously on humans and can almost prevent outdoor activities. Up to 60 000 mosquitoes per trap per night were recorded, showing that the mosquito populations can be very large. The extensive flight range of *Oc. sticticus* extends the affected area to about 100 km² or about 10 times the size of the wetland breeding areas. The seasonal dynamics of mosquitoes in the area are mainly influenced by the flooding regime of the Dalälven River which is controlled by a large number of waterpower stations.

Malaria disappeared from Sweden fifty years ago, however, mosquito-borne viruses can still be found. Of the 24 mosquito species recorded in the area, five are involved in virus transmission: *Oc. communis* and *Oc. punctor* are potential vectors for Inkoo virus; *Cx. torrentium*, *Cs. morsitans* and *Ae. cinereus* are the proven vectors of Sindbis virus causing Ockelbo disease. Four of these vector species were frequently found during survey, and one isolate of Sindbis virus from *Ae. cinereus* has been found in the Nedre Dalälven area.

The control of *Ochlerotatus* and *Aedes* mosquitoes is based on the use of *B. thuringiensis israelensis* products. As a prerequisite the following activities were conducted: (1) mapping of 1000 ha of potential breeding areas for floodwater mosquitoes, (2) governmental application for the use of *B. thuringiensis israelensis* from helicopters, (3) calibration of the helicopter spraying equipment, (4) biweekly adult mosquito monitoring covering the whole Nedre Dalälven area, (5) training of the local fire-brigade to perform field work with larval sampling, (6) creation of a frequently updated homepage to disseminate accurate information to the public (www.sandviken.se/ mygg), and (7) performing a 100% successful treatment against *Oc. sticticus* larvae in flooded wetlands.

All major non-profit organisations concerned about eventual negative impacts on non-target organisms have been invited to discuss the development of the mosquito control operations in order to minimise ecological side effects. Several research projects have been set up to analyse the patterns of mosquito interaction with other faunal components in the wetland areas, as well as long-term studies of any side-effects of *B. thuringiensis israelensis* on non-target organisms.

16.2.11. Mosquito Control in Switzerland

Switzerland has several wetland conservation areas at the foothills on both sides of the Alps. They are located at the end of valleys where the rivers have formed plains or deltas before flowing into lakes. These wetland areas are flooded periodically during the spring and summer months by snow-melt and increasingly heavy precipitations along the rim of both sides of the Alps.

Although relatively small in size, these wetland areas exhibit a unique fauna and flora, influenced by the special climatic conditions. Many are under federal protection. Especially the natural reserve Bolle di Magadino in the south of Switzerland, which is located along one of the main route for migratory birds. Another area of interest, exhibiting an extraordinary dynamic is situated at the upper end of the Lac de la Gruyère.

The frequently flooded wetland zones have become major breeding sites for floodwater mosquitoes such as *Ae. vexans* and *Oc. sticticus*. The adult mosquitoes with a flying range of 10 km or more have created a serious nuisance for the nearby residential areas and for tourism, which is a vital economic source of income for Switzerland. The past has shown that mosquitoes on the search for blood meals can considerably affect the quality of life.

Conservation areas require the mutual acceptance and support by the local population as well as the authorities. Emissions in both directions have to be kept to acceptable levels. In 1988 the first long-term mosquito control project in the plain of Magadino was initiated (Lüthy, 2001). Since the mosquito control programme using exclusively products based on *B. thuringiensis israelensis*, was so successful, a second project followed in 1995 at the Lac de la Gruyère. It is likely that other areas such as the natural reserve "Les Grangettes" at the upper end of Lake Geneva will follow.

Reports have to be submitted regularly to working groups which have been established within the two mosquito control projects. Members of the working groups represent the federal government, the cantons, the communities and the tourist organisations. The main topics discussed within both working groups are the safety aspects and the economic feasibility of the mosquito control. Annual authorisation for the mosquito control has to be issued by the federal government.

Over the past decade the control practices have been refined and optimised. Although an integrated approach is promoted, the use of *B. thuringiensis israelensis* has remained so far the main viable option. *B. thuringiensis israelensis* is presently the safest mosquito control agent (WHO, 1999), which is commercially produced. Other approaches of mosquito control were either not efficient enough or would have interfered with the ecosystem.

Aerial application of *B. thuringiensis israelensis* by helicopter has proved to be the only efficient method of reaching the larval breeding sites. Granular formulations of *B. thuringiensis israelensis* are ideal for use in the flooded zones, which are predominantly covered by dense vegetation. The granular formulations have an excellent shelf life over a period of 2–3 years. With enough stock on hand a quick operation can be guaranteed.

Mapping of the breeding sites and timing are the most important elements for successful larviciding. Timing includes consideration of the climatic factors, the expected movements of the water levels in the wetland zones, and the availability of manpower and equipment.

The economy of sustained mosquito control

We have to be aware that all mosquito control projects are on a long-term basis since permanent eradication of mosquito populations within a given wetland ecosystem is impossible. Even with a successful control programme the mosquito population will return the next season at full strength. Thus, experience has clearly shown that the treatments must not be discontinued in order to cope with the unique recovery potential of mosquito populations.

An economic evaluation of a mosquito control project is based on three factors, costs, benefit and risks, which are all closely interdependent (Fig. 16.6).

The costs of control programmes

Integrated mosquito control programmes centered around the use of *B. thuringiensis israelensis* are at first glance expensive compared to chemical practices. Therefore the economic aspects become an important issue once the mosquitoes have been reduced to acceptable levels and once the residents have become accustomed to a mosquito-free environment.

The maintenance of a mosquito-free environment, an intact ecosystem as well as continuous monitoring for potential risks represent the main cost factors.

Table 16.2 demonstrates that the direct operational costs are easy to determine. The integration of the risk assessment is already more difficult since it is not directly related to the size of the breeding sites. Estimated costs for risk assessment may amount to one third of the operational costs.

The benefit of mosquito control consists of a financially measurable part and of components, which relate to improved quality of life of the population within the range of mosquitoes.

In Table 16.2 a cost–benefit analysis for the two mosquito control projects in Switzerland is presented. It shows that the size of the population affected and the significance of

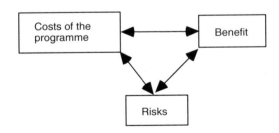

Figure 16.6. Considerations when evaluating a mosquito control project.

Table 16.2. Cost-benefit Analysis of Two Mosquito Control Programmes in Switzerland[1]

	Control project: Plain of Magadino	Control project: Lac de la Gruyère
A: General parameters		
Area with breeding sites to survey	300–500 ha	100 ha
Permanent local residents	35 000	4000
Lodgings by tourists in camping sites	4 50 000 (official figure)	Few overnight lodgings
Reduction of lodgings due to the presence of mosquitoes	>10% equals 45 000 persons[2]	Loss mainly by restaurants Estimate: 1250 diners at 20€
Spending per tourist/day	50€	
B: Losses and additional expenses due to the presence of mosquitoes		
Loss per season	2 250 000€	
Loss by restaurants		25 000€
Costs for mosquito protecting agents	70 000€	8000€
2€ per resident		
Total financial losses	2 950 000€	33 000€
C: Cost for the mosquito control programmes		
Survey and treatments	32 000€	23 000€
Cost per resident	0.9€	5.75€
D: Benefit/cost ratio	92	1.5

[1]The figures presented here are based on our annual reports.
[2]This figure is based on the reduction of lodgings within the mosquito zone 1987 and 1988 before the start of the mosquito control programme.

the area for tourism are main factors. The population within the zone affected by mosquitoes in the Plain of Magadino (Ticino) amounts to about 35 000 people whereas at the lake of Gruyère only 4000 people are affected. In Ticino, tourism is one of the major sources of income whereas at the lake of Gruyère, tourism within the mosquito-affected zones is of minor importance.

Table 16.2 shows clearly that the benefit-cost ratio strongly depends on the number of residents which live within the perimeter that is affected by mosquitoes. The tourist industry is the second key factor on which mosquitoes can have a negative impact, which resulted even in unemployment before mosquito control was initiated.

In any case mosquito control has paid off even under circumstances where the number of residents is small and the tourist activities in the mosquito affected zones are limited.

The increase in quality of life for the residents during the mosquito plagues are not included in the cost benefit analysis. During the relatively short summer period the people refuse to spend their evenings behind mosquito screens. They wish to spend their evenings outside in a mosquito-free environment.

Risk assessment

Deliberate interactions with ecosystems always include potential risks. A zero risk does not exist even if control measures are applied which are fully based on biological means. Continuous risk assessment is required since negative impacts, as we have learnt from the use of chemical pesticides, may surface only after years or decades.

Risk assessments are as a rule expensive if on-site research is involved. Such projects should be financed by the district and/or federal governments.

Within these *B. thuringiensis israelensis* control projects a risk assessment project was initiated. The study involves the fate and dynamics of the spores in the ecosytem with special emphasis on the soil, the aquatic sediment and the water. Furthermore an answer is expected on the potential of the horizontal transfer of delta-endotoxin genes and possible consequences. More information will be also gained on the fate of the delta-endotoxin following deliberate release. The Swiss Federal Government financially supports the project.

No field resistance of *Aedes vexans* larvae against the delta-endotoxin of *B. thuringiensis israelensis* after a long-term application of 14 years with two to three treatments per season could be observed.

Legal aspects

The interesting question about liability of mosquito plagues, which originate from breeding sites inside protected wetland areas, has not been discussed. On the other hand liability is clearly regulated for anthropogenic emissions with damaging impacts on protected ecosystems. The same jurisdiction should be applied to emissions caused by mosquitoes in the surroundings of protected areas. It has been experienced that negative impact on the quality of life and on the economy can be considerable.

16.2.12. Mosquito Control in Yugoslavia

The province of Vojvodina has been subject to a continuous mosquito control programme for the last 30 years. Although other areas, such as the Adriatic coast, also provide favourable mosquito breeding conditions, a consistent approach to mosquito control in these parts of the country has, unfortunately, not been developed so far.

The province of Vojvodina lies in the northernmost part of Yugoslavia, covering a surface area of about 20 000 km². The south-eastern part of the Pannonian Plain bears several plateaus and hills ranging between 150 m to 600 m above sea level, but the remaining area is flat with an average elevation of 80 m above sea level. Three large rivers enrich it, the Danube, Tisza and Sava, as well as several smaller ones with a total length in Vojvodina of 597 km.

Several natural protected areas and swamps are connected to the river lowlands and hilly regions on the river banks. All these protected areas provide very productive breeding sites of a great variety of mosquito species. The climate is temperate central European, characterised by wide fluctuations in air temperature. The yearly average temperature is 11°C. Average precipitation fluctuates between 550–670 mm per year.

The number of mosquito species registered in Vojvodina so far is 32 out of the 51 known in former Yugoslavia. *Aedes/ Ochlerotatus* mosquitoes that comprise more than 90% of trap catches and almost 100% of human bait catches, are dominant in the floodplains. The number of females landing to feed on one forearm of a human can exceed 300 mosquitoes in five minutes. The low parts of the plain are occasionally flooded. Each rise of the Danube water level of between 250 cm and 650 cm at the Novi Sad measuring station, triggers hatching of *Ae. vexans, Oc. sticticus, Ae. cinereus* and *Ae. rossicus* larvae in billions. Many extensive marshes, ox-bow-swamps and a complex system of irrigation channels are present, which provide numerous and extensive mosquito breeding sites. Deep depressions of the old river meanders, which contain water the whole year round, show typical zones of submerged, floating and emergent hydrophyte vegetation. The main species in the last two biotopes are *An. messeae, An. maculipennis s.s., Cx. modestus* and *Cx. p. pipiens*. The banks of the rivers, meanders and marshes are usually lined by willow, poplar and alder trees. Brambles, various shrubs and climbers make a thick coverage of the ground and an excellent day-time resting-place for the hordes of floodwater mosquitoes. Extensive pastures and neglected fields overgrown with dense grassy, weedy and bushy vegetation stretch in close proximity to swampy areas. The stagnant and slowly flowing water on the alkaline and saline soils with a relatively high concentration of dissolved salts, offer good conditions for *An. atroparvus, Oc. caspius* and *Oc. dorsalis*. On the Danube flood-plains the most abundant *Aedes/Ochlerotatus* species are *Ae. vexans* and *Oc. sticticus* that occur in very large numbers two to five times per year, depending on the Danube fluctuation pattern. The network of irrigation channels has been neglected for decades. It is blocked by aquatic vegetation and cannot collect and drain water quickly enough to prevent formation of puddles and marshes suitable for mosquito development. In such conditions the previously mentioned species of *Anopheles* and *Culex* mosquitoes are most dominant. Saline waters close to several spas are breeding sites of *Oc. caspius, Oc. dorsalis* and *An. atroparvus*. Marshes with dense aquatic vegetation provide ideal breeding sites for *Cq. richiardii*. Urban environments up to 10 km from major rivers are often invaded by masses of dispersing *Aedes/Ochlerotatus* females. Inhabitants are also continuously bitten by several generations of *Cx. p. pipiens*.

Malaria was eradicated from the region after the Second World War. The last cases of endemic *Plasmodium vivax* malaria were recorded in 1955 close to Nis in the south of Serbia, in the vicinity of rice fields that have been breeding sites of *An. maculipennis s.s.* and *An. messeae* (Guelmino, 1956). In the last decade, the average number of imported malaria was about 40 cases per year. Owing to global climatic changes influenced by "El Nino" and "La nina" activity, warming of the Pannonian Plain climate is forecast. The average number of *An. maculipennis* complex in the Koviljski rit nature reserve has increased in the last few years,

and could increase the risk of malaria re-emergence in Vojvodina province. Apart from malaria, several surveys of Tahyna virus have been conducted in Serbia. The highest percentage of positive sera was found in the southern part of the river Tisza lowlands (Gligic and Adamovic, 1976; Adamovic and Gligic, 1980).

Mosquito control in Vojvodina has been organised since 1976 under the umbrella of the Province government. The Faculty of Agriculture (University of Novi Sad) was the founder and organiser of the control programme until 1985, when insecticide purchase and pest control organisations (PCO) were transferred to trained personnel in the local mosquito abatement district. The Faculty of Agriculture remained involved in education, monitoring and the evaluation of insecticide properties, dosages and the effectiveness of control measures. Since 1980, when *B. thuringiensis israelensis* was first introduced as a larvicide in parallel with organophosphates and IGR compounds, *B. thuringiensis israelensis* has been a subject of continuous research and application programmes at various scales. Year by year different *B. thuringiensis israelensis* formulations have been tested, but not used on a large scale in Vojvodina mainly for economic reasons. From 1993 onwards, the municipality of Novi Sad has started to support the "Environmentally friendly approach in mosquito control". The project's main goal is to provide a base for rational biological mosquito control, mainly by the use of *B. thuringiensis israelensis* products. Monitoring of larval and adult floodwater and house mosquitoes was established, together with mapping of their breeding sites. Development of software supporting the implementation of biological control of floodwater mosquitoes has also been conducted. GIS software programmes have been designed to serve as a database for the information on physico-chemical characteristics of mosquito breeding sites, plant associations, limnofauna, egg, larvae and adult density at breeding sites, and to follow the speed and direction of female mosquito dispersal. The programme can also simulate different levels of flooding, give precise information about areas to be treated and record effectiveness of control measures.

A network of CO_2 baited traps has been established since 1994, which samples mosquitoes on a weekly basis from April to October. Additional traps are used temporarily in the region of Novi Sad municipality, particularly during periods of mosquito dispersal, or to evaluate control measures. Furthermore, 30 traps are continuously used in several localities of Vojvodina (Sombor, Subotica, and Pancevo) to assess the dispersal potential of the major nuisance species. As a result of the project activities, since 1997 at least one generation of floodwater mosquitoes is controlled solely by *B. thuringiensis israelensis* application. Other seasonal outbreaks of floodwater mosquitoes are suppressed by combined methods (*B. thuringiensis israelensis* for larviciding and ULV pyrethroids/organophosphates for adulticiding) using air or ground equipment.

References

Abbott, W. S., 1925. A method of computing the effectiveness of an insecticide. *J. Econ. Entomol.* 18: 265–267.

Adamovic, Z. R., 1980. Über die Verbreitung und Bevölkerungsdichte von *Anopheles atroparvus* Van Thiel (Diptera: Culicidae) in Serbien und Mazedonien, Jugoslawien. *Anz. Schädlingskde., Planzenschutz, Umweltschutz* 53: 83–86.

Adamovic, Z. R., and Gligic, A., 1980. Habitats and distribution of the mosquito species processed in the Arbovirus isolation attempts in Serbia (Yugoslavia), in: *Arboviruses in the mediterranean countries.* (J. Vesenjak-Hirjan, J. S. Porterfield, and E. Arslanagic eds.), Zbl. Bakt. Suppl. 9, Gustav Fischer Verlag, Stuttgart, pp. 251–256.

Adhami, J., and Murati, N., 1987. Presence of the mosquito *Aedes albopictus* in Albania. *Revista Mjekesöre* 1: 13–16.

Aeschlimann, J. A., and Reinert, M., 1931. Acethylcholinesterase and anticholinesterases, in: *Comprehensive insect physiology, biochemistry and pharmacology.* (G. A. Kerkut, and L. I. Gilbert, eds.), Pergamon Press, pp. 115–130.

Agoulon, A., Desireux, M., Brutus, L., Guilloteau, J., and Marjolet, M., 1997. Genetique des populations de moustiques *Aedes detritus* A et B et *Aedes caspius* sur le littoral atlantique et mise en evidence de l'autogenese. *Bull. Soc. Zool. France* 122(2): 157–158.

Aitken, T. H. G., 1954. The Culicidae of Sardinia and Corsica. *Bull. Ent. Res.* 45: 437–494.

Akamatsu, Y., Dunn, P. E., Kézdy, J., Kramer, K. J., Law, J. H., Rubenstein, D., and Sanburg, L. L., 1975. Biochemical aspects of juvenile hormone actions in insects, in: *Control Mechanisms in Insect Development.* (R. H. Meints, and E. Davies, eds.), Plenum Press, New York, pp. 123–149.

Alekseev, E. V., 1989. Bloodsucking mosquito *Aedes (Rusticoidus) krymmontanus* Alekseev, sp. n., a relic of the entomofauna of the crimea (in Russian). *Parazitologyia* (Leningrad) 23: 173–178.

Ali, A., 1981. *Bacillus thuringiensis* serovar. *israelensis* (ABG-6108) against chironomids and some nontarget aquatic invertebrates. *J. Invert. Pathol.* 38: 264–272.

Ali, A., and Mulla, M. S., 1978. Effects of chironomid larvicides and diflubenzuron on non-target invertebrates in residental-recreational lakes. *Environ. Ent.* 7: 21–27.

Aly, C., 1985. Germination of *Bacillus thuringiensis* var. *israelensis* spores in the gut of *Aedes* larvae (Diptera: Culicidae). *J. Invertebr. Pathol.* 45: 1–8.

Amerasinge, F. P., and Ariyasena, T. G., 1990. Larval survey of surface water-breeding mosquitoes during irrigation development in the Mahaweli project, Sri Lanka. *J. Med. Entomol.* 27: 789–802.

Amichot, M., Castella, C., Cuany, A., Berge, J. B., and Pauron, D., 1992. Target modification as a molecular mechanism of pyrethroid resistance in *Drosophila melanogaster. Pestic. Biochem. Physiol.* 44: 183–190.

Andersson, I. H., and Jaenson, T. G. T., 1987. Nectar feeding by mosquitoes in Sweden, with special reference to *Culex pipiens* and *Culex torrentium. Med. Vet. Entomol.* 1: 59–64.

Andreadis, T. G., 1985. Experimental transmission of a microsporidian pathogen from mosquitoes to an alternate copepode host. *Proc. Nat. Acad. Sci. USA* 82: 5574.

Annis, B., 1990. Comparison of the effectivness of two formulations of DEET against *Anopheles flavirostris. J. Am. Mosq. Control Assoc.* 6(3): 430–432.

Anonymous, 1987. Guidelines for the safe handling of pesticides during their formulation, packing, storage and transport. GIFAD, Avenue Albert Lancaster, 79A, 1180 Brussels, Belgium.

Anonymous, 1989. Control de les poblacions de mosquits al Baix Llobregat. Estudi de l'aplicació conjunta de diferents tècniques. CCBL i MMAMB. Barcelona, pp. 192.

Apfelbeck, V., 1928. Beiträge zur Kenntnis wenig bekannter Stechmücken. *Zeitschr. wiss. Insektenbiol.,* Bd. IV: 28–31.

Apfelbeck, V., 1931. Zur Biologie der in Jugoslawien beobachteten dendrolimnokolen Stechmücken (Culicidae). *Acta Societatis Entomologicae Jugoslavicae.* V–VI (1930–1931), 1/2: 49–61.

Aranda, C., Panyella, O., Eritja, R., and Castella, J., 1998. Canine filariasis. Importance and transmission in the Baix Llobregat area, Barcelona (Spain). *Vet. Parasitol.* 77: 267–275.

Arnaud, J. D., Rioux, J. A., Croset, H., and Guilvard, E., 1976. *Aedes (Ochlerotatus) surcoufi* (Theobald, 1912). Retablissement du binome; analyse morphologique position au sein du complexe holarctique *"excrucians".* *Ann. Parasitol. hum. et comp.* 51(4): 477–494, illus.

Arnold, A., Braun, M., Becker, N., and Storch, V., 1998. Beitrag zur Ökologie der Wasserfledermaus (Myotis daubentonii) in Nordbaden.—*Carolinea*, 56: 103–110; Karlsruhe.

Arnold, A., Braun, M., Becker, N., and Storch, V., 2000. Zur Nahrungsoekologie von Wasser- und Rauhhautfledermaus in den nordbadischen Rheinauen. *Carolinea* 58: 257–263; Karlsruhe.

Arredondo-Jimenez, J. I., Brown, D. N., Rodriguez, M. H., and Loyola, E. G., 1992. The control of *Anopheles albimanus* in southern Mexico by selective spraying of bendiocarb, in: *Proc. XIII Int. Congr. Trop. Med. Malaria*, Vol. 2, Mahidol University, Bangkok, p. 165.

Artemiev, M. M., 1980. Anopheles mosquitoes—main malaria vectors in the USSR, in: *International scientific project on ecologically safe methods for control of malaria and its vectors.* The USSR state commitee for science and technology (GKNT)/United Nations Environment Programme (UNEP). Collected Lectures 2: 45–71.

Asinas, C. Y., Hugo, C. T., Boase, C. J., and Evans, R. G., 1994. Evaluation of selective spraying of bendiocarb (Ficam VC) for the control of *Anopheles flavirostris* in the Philippines. *J. Am. Mosq. Control Assoc.* 10(4): 496–501.

Aspöck, H., 1965. Studies of Culicidae (Diptera) and consideration of their role as potential vectors of arboviruses in Austria. XII Int. Congr. Ent., London, 767–769.

Aspöck, H., 1979. Biogeographie der Arboviren Europas. Beitr. z. Geoökologie d. Menschen, 3. Geomed. Symp. Geograph. Z., Beiheft 51: 11–28.

Aspöck, H., 1996. Stechmücken als Virusüberträger in Mitteleuropa. *Nova Acta Leopoldina* NF 71, Nr. 292: 37–55.

Axtell, R. C., Rutz, D. A., and Edwards, T. D., 1980. Field tests of insecticides and insect growth regulators for the control of *Culex quinquefaciatus* in an aerobic animal waste lagoons. *Mosq. News* 40(1): 36–42.

Baillie, A. C., and Wright, W., 1985. Biochemical Pharmacology, in: *Comprehensive Insect Physiology, Biochemistry and Pharmacology*, Volume 11. (G. A. Kerkut, and L. I. Gilbert, eds.), Pergamon Press, Oxford, pp. 324–356.

Baldwin, W. F., James, H. G., and Welch, H. E., 1955. A study of predators of mosquito larvae and pupae with a radioactive tracer. *Canad. Ent.* 87: 350–356.

Baqar, S., Hayes, C. G., Murphy, J. R., and Watts, D. M., 1993. Vertical transmission of West Nile virus by *Culex* and *Aedes* species mosquitoes. *Am. J. Trop. Med. Hyg.* 48(6): 757–762.

Bar-Zeev, M., Maibach, H. I., and Khan, A. A., 1977. Studies on the attraction of *Aedes aegypti* (Diptera: Culicidae) to man. *J. Med. Entomol.* 14: 113–120.

Barber, M. A., and Rice, J. B., 1935. Malaria studies in Greece: the malaria infection rate in nature and in the laboratory of certain species of *Anopheles* of east Macedonia. *Ann. Trop. Med.* 29: 329–348.

Bardos, V., and Cupkova, B., 1962. The Calovo virus—the second virus isolated from mosquitoes in Czechoslovakia. *J. Hyg. Epid. Microbiol. Immunol.* 6: 186–192.

Bardos, V., and Danielova, V., 1959. The Tahyna virus—a virus isolated from mosquitoes in Czechoslovakia. *J. Hyg. Epid. Microbiol. Immunol.* 3: 264–276.

Barlow, F., 1985. Chemistry and formulation, in: *Pesticide application: principles and practice.* (P. T. Haskell, ed.), Clarendon Press, Oxford, pp. 1–34.

Barr, A. R., 1958. The mosquitoes of Minnesota (Diptera: Culicidae). *Univ. Minn. Agric. Exp. Stn. Tech. Bull.* 228, 154 pp.

Barr, A. R., and Azawi, A., 1958. Notes on the oviposition and the hatching of eggs of *Aedes* and *Psorophora* mosquitoes (Diptera: Culicidae). *Univ. Kans. Sci. Bull.*

Barraud, P. J., 1934. *The fauna of British India, including Ceylon and Burma.* Family Culicidae. Tribes Megarhinini and Culicini. Diptera. Vol. V. Taylor and Francis, London, 463 pp., illus.

Bassi, D. G., Finch, M. F., Weathersbee, A. A., Stark, P. M., and Meisch, M. V., 1987. Efficacy of Fenoxycarb (Pictyl) against *Psorophora columbiae* in Arkansas ricefields. *J. Am. Mosq. Control Assoc.* 3(4): 616–618.

Bates, M., 1940. The nomenclatur and the taxonomic status of the mosquitoes of the *Anopheles maculipennis* complex. *Ann. Ent. Soc. Am.* 33: 343–356.

Bates, M., 1949. Ecology of anopheline mosquitoes, in: *Malariology, a comprehensive survey of all aspects of this group of diseases from a global standpoint.* (M. F. Boyd, ed.), Saunders Company, London, pp. 302–330.

Baumann, P. M., Clark, A., Baumann, L., and Broadwell, A. H., 1991. *Bacillus sphaericus* as a mosquito pathogen: Properties of the organism and its toxins. *Microbiol. Revs.* 55: 425–436.

Bay, E. C., 1972. Biological control and its applicability to biting flies, in: *Proc. Symp. on Biting Fly Control and Environmental Quality.* (A. Hudson, ed.), Univ. Alberta, Edmonton, Ottawa. DR 217, May 16–18, 1972, Inf. Canada, pp. 65–70.

Beach, R., 1978. The required day number and timely induction of diapause in geographic strains of the mosquito *Aedes atropalpus. J. Insect. Physiol.* 24: 448–455.

Beaty, B. J., and Marquardt, W. C., 1996. The biology of disease vectors. University Press of Colorado, Colorado, USA, 632 pp.

Becker, N., 1984. Ökologie und Biologie der Culicinae in Südwest-Deutschland. Ph.D. thesis, University of Heidelberg, 404 pp.

Becker, N., 1989. Life strategies of mosquitoes as an adaptation to their habitats. *Bull. Soc. Vector Ecol.* 14(1): 6–25.

Becker, N., 1992. Community participation in the operational use of microbial control agents in mosquito control programs. *Bull. Soc. Vector Ecol.* 17(2): 114–118.

Becker, N., 2002. Sterilisation of *Bacillus thuringiensis israelensis* products by gamma radiation. *J. Am. Mosq. Control Assoc.* 18(1): 57–62.

Becker, N., 2003. Ice granules containing endotoxins of microbial agents for the control of mosquito larvae – a new application technique. *J. Am. Mosq. Control Assoc.* 19(1): 63–66.

Becker, N., and Kaiser, A., 1995. Die Culicidenvorkommen in den Rheinauen des Oberrheingebiets mit besonderer Berücksichtigung von *Uranotaenia* (Culicidae, Diptera)—einer neuen Stechmückengattung für Deutschland. *Mitt. dtsch. Ges. allg. angew. Ent.* 10: 407–413.

Becker, N., and Ludwig, H. W., 1981. Untersuchungen zur Faunistik und Ökologie der Stechmücken (Culicinae) und ihrer Pathogene im Oberrheingebiet. *Mitt. dtsch. Ges. allg. angew. Ent.* 2: 186–194.

Becker, N., and Ludwig, H. W., 1983. Mosquito Control in West Germany. *Bull. Soc. Vector Ecol.* 8(2): 85–93.

Becker, N., and Ludwig, M., 1993. Investigations on possible resistance in *Aedes vexans* field populations after a 10-year application of *Bacillus thuringiensis israelensis. J. Am. Mosq. Control Assoc.* 9(2): 221–224.

Becker, N., and Margalit, J., 1993. Control of Dipteran pests by *Bacillus thuringiensis,* in: *Bacillus thuringiensis: Its uses and future as a biological insecticide.* (P. Entwistle, M. J. Bailey, J. Cory, and S. Higgs, eds.), John Wiley and Sons, Ltd., Sussex, England, pp. 147–170.

Becker, N., and Rettich, F., 1994. Protocol for the introduction of new *Bacillus thuringiensis israelensis* products into the routine mosquito control program in Germany. *J. Am. Mosq. Control Assoc.* 10(4): 527–533.

Becker, N., Djakaria, S., Kaiser, A., Zulhasril, O., and Ludwig, H. W., 1991. Efficacy of a new tablet formulation of an asporogenous strain of *Bacillus thuringiensis israelensis* against larvae of *Aedes aegypti. Bull. Soc. Vector Ecol.* 16(1): 176–182.

Becker, N., Zgomba, M., Ludwig, M., Petric, D., and Rettich, F., 1992. Factors influencing the activity of *Bacillus thuringiensis* var. *israelensis* treatments. *J. Am. Mosq. Control Assoc.* 8(3): 285–289.

Becker, N., Ludwig, M., Beck, M., and Zgomba, M., 1993. The impact of environmental factors on the efficacy of *Bacillus sphaericus* against *Culex pipiens. Bull. Soc. Vector Ecol.* 18(1): 61–66.

Becker, N., Zgomba, M., Petric, D., Beck, M., and Ludwig, M., 1995. Role of larval cadavers in recycling processes of *Bacillus sphaericus. J. Am. Mosq. Control Assoc.* 11(3): 329–334.

Becker, N., Jöst, A., Storch, V., and Weitzel, T., 1999. Exploiting the biology of urban mosquitoes for their control, in: *Proc. 3rd Int. Conf. Insect Pests in Urban Environment,* Prague, Czech Rep., (W. H. Robinson, F. Rettich, and G. W. Rambow, eds.), pp. 425–430.

Belkin, J., 1962. *The mosquitoes of the South Pacific* (Diptera: Culicidae), Vol. 1, Univ. Calif. Press, 608 pp.

Bellini, R., Carrieri, M., Burgio, G., and Bacchi, M., 1996. Efficacy of different ovitraps and binomial sampling in *Aedes albopictus* surveillance activity. *J. Am. Mosq. Control Assoc.* 12: 632–636.

Berlin, G. W., 1969. Mosquito studies (Diptera: Culicidae). XII. A revision of the Neotropical subgenus *Howardina* of *Aedes. Contr. Am. Ent. Inst.* 4(2):1–190, illus.

Berry, C., Hindley, J., and Oei, C., 1991. The *Bacillus sphaericus* toxins and their potential for biotechnological development, in: *Biotechnology for biological control of pests and vectors.* (K. Maramorosch, ed.), CRC Press, Boca Raton, FL, pp. 35–51.

Bickley, W. E., 1980. Notes on the status of *Aedes cinereus hemiteleus* Dyar. *Mosq. Syst.* 12: 357–370.

Bidlingmayer, W. L., 1964. The effect of moonlight on the flight activity of mosquitoes. *Ecology* 45(1): 87–94.

Bidlingmayer, W. L., 1975. Mosquito flight paths in relation to the environment. Effect of vertical and horizontal visual barriers. *Ann. Ent. Soc. Am.* 68: 51–57.

Bidlingmayer, W. L., 1985. The measurement of adult mosquito population changes—some considerations. *J. Am. Mosq. Control Assoc.* 1: 328–347.

Bidlingmayer, W. L., and Evans, D. G., 1987. The distribution of female mosquitoes about a flight barrier. *J. Am. Mosq. Control Assoc.* 3(3): 369–377.

Bishop, S. S., and Hart, R. C., 1931. Notes on some natural enemies of the mosquitoes in Colorado. *J. New York Ent. Soc.* 39: 151–157.

Blackmore, M. S., 1994. Mermithid parasitism of adult mosquitoes in Sweden. *Am. Midl. Nat.* 132: 192–198.

Blotzheim, U. N. G. von, 1985. *Handbuch der Vögel Mitteleuropas.* Aula-Verlag, Wiesbaden, 14 Vols.

Blum, S., Basedow, Th., and Becker, N., 1997. Culicidae (Diptera) in the Diet of Predatory Stages of Anurans (Amphibia) in Humid Biotopes of the Rhine Valley in Germany. *J. Vector Ecol.* 22(1): 23–29.

Bohart, R. M., 1948. The subgenus *Neoculex* in America north of Mexico. *Ann. Ent. Soc. Am.* 41: 330–345.

Bohart, R. M., and Washino, R. K., 1978. *Mosquitoes of California.* 3rd edition, Univ. Calif. Div. Agr. Sci., Berkeley, Publ. No. 4084, 153 pp.

Boorman, J. P. T., 1961. Observations on the habits of mosquitoes of Plateau Province, Northern Nigeria, with particular reference to *Aedes (Stegomyia) vittatus* (Bigot). *Bull. Ent. Res.* 52: 709–725.

Boreham, P. F. L., and Atwell, R. B. (eds.), 1988. *Dirofilariasis.* CRC Press, Boca Raton, Florida.

Borg, A., and Horsfall, W. R., 1953. Eggs of floodwater mosquitoes. II. Hatching stimulus. *Ann. Ent. Soc. Am.* 46: 472–478.

Bosak, T., and Dworak, Z., 1961. Zwalczanie plagi komarów na wyspie Karsiborz w osiedlach ludzkich i przyleglych terenach otwartych. *Prz. Epid* 15(1): 59–66.

Böttger, K., 1970. Die Ernährungsweise der Wassermilben (Hydrachnellae, Acari). *Int. Rev. Ges. Hydrobiol.* 55(6): 895–912.

Bourgouin, C., Klier, A., and Rapoport, G., 1986. Characterization of the genes encoding the haemolytic toxin and the mosquitocidal delta-endotoxin of *Bacillus thuringiensis israelensis. Mol. Gen. Genet.* 205: 390–397.

Bourguet, D., Fonseca, D., Vourch, G., Dubois, M.-P., Chandre, F., Severini, C., and Raymond, M., 1998. The acetylcholinesterase gene *Ace*: a diagnostic marker for *Cx. pipiens* and *Cx. quinquefasciatus* forms of the *Culex pipiens* complex. *J. Am. Mosq. Contr. Assoc.* 14(4): 390–396.

Bowers, W., 1969. Juvenile hormone: Activity of aromatic terpenoid ethers. *Science* 164: 323–325.

Boyd, M. F., 1949. *Malariology. A comprehensive survey of all aspects of this group of diseases from a global standpoint.* By sixty-five contributors, edited by M. F. Boyd. Vol. 1, 787 pp., illus.

Bozicic-Lothrop, B., 1988. Comparative ecology of *Aedes dorsalis* complex in the Holarctic. *Proc. Calif. Mosq. Contr. Assoc.* 56: 139–145.

Bozkov, D., 1961. Komarni larvi ot Burgasko i Strandza planina. Kompleksni izsledvaniya na prirodnite ognisha na zaraza v rayona na Strandza planina. Blgarska Akademia na Naukite, Sofia, pp. 147–153.

Bozkov, D., 1966. Krovososushiye komary (Diptera: Culicidae) Bolgarii. Entomologicheskoe Obozrenie, Akademia Nauk SSSR XLV, 3: 570–574.

Bozkov, D., Hristova, T., and Canev, I., 1969. Stechmücken an der bulgarischen Schwarzmeerküste. *Bull. Inst. Zool. et Mus.* 29: 151–166.

Bragina, A., 1931. Beiträge zur Kenntnis der mazedonischen Culiciden nebst einigen Bemerkungen allgemeiner Art. *Arch. Schiffs- u. Tropenhyg.* 35: 345–353.

Breaud, T. H., 1993. Insecticide resistance in Florida mosquitoes: a review of published literature. *J. Florida Mosq. Control Assoc.* 64(1): 14–21.

Briegel, H., 1973. Zur Verbreitung der Culicidae (Diptera: Nematocera) in der Schweiz. *Rev. Suisse Zool.* 80(2): 461–472.

Briegel, H., and Kaiser, C., 1973. Life-span of mosquitoes (Diptera: Culicidae) under laboratory conditions. *Gerontologia* 19: 240–249.

Britz, L., 1986. Zur Kenntnis der Stechmückenfauna (Diptera: Culicidae) des DDR-Bezirkes Leipzig. *Angew. Parasitol.* 27(2): 91–103.

Broadwell, A. H., Baumann, L., and Baumann, P., 1990. Larvicidal properties of the 42 and 51 kilodalton *Bacillus sphaericus* proteins expressed in different bacterial hosts: Evidence for a binary toxin. *Curr. Microbiol.* 21: 361–366.

Brodniewicz, A., 1953. Dezynsekcja Miasteczka Dzieciecego w Podgrodziu kolo Szczecina. Warszawa.

Brogdon, W., 1988. Microassay of acetyl-cholinesterase activity in small portions of single mosquito homogenates. *Comp. Biochem. Physiol.* 90C: 145–150.

Brown, A. W. A., 1976: How have entomologists dealt with resistance? *Proc. Amer. Phytopath. Soc.*, 3: 67.

Bruce-Chwatt, L. J., 1971. Insecticides and the control of vector-borne diseases. *Bulletin of the World Health Organization* 44: 419–424.

Bruce-Chwatt, L. J., and de Zulueta, J., 1980. *The rise and fall of malaria in Europe.* A historico—epidemiological study. University Press, Oxford, 240 pp.

Bruce-Chwatt, L. J., Draper, C. C., Avradamis, D., and Kazandzoglou, O., 1975. Sero-epidemiologica: surveillance of disappearing malaria in Greece. *J. Trop. Med. Hyg.* 78: 194–200.

Brust, R. A., and Costello, R. A., 1969. Mosquitoes of Manitoba. II. The effect of storage temperature and relative humidity on hatching of eggs of *Aedes vexans* and *Aedes absserratus* (Diptera: Culicidae). *Can. Entomol.* 101: 1285–1291.

Brust, R. A., and Munstermann, L. E., 1992. Morphological and genetical characterization of the *Aedes (Ochlerotatus) communis* complex (Diptera: Culicidae) in North America. *Ann. Ent. Soc. Am.* 85(1): 1–10.

Bullini, L., and Coluzzi, M., 1978. Applied and theoretical significance of electrophoretic studies in mosquitoes (Diptera: Culicidae). *Parasit.* 20: 7–21.

Bullini, L., and Coluzzi, M., 1982. Evolutionary and taxonomic inferences of electrophoretic studies in mosquitoes, in: *Developments in the genetics of diseases vectors*. Stipes Publishing Company, Champaign, Il., pp. 465–482.

Burgess, L., 1959. Techniques to give better hatches of the eggs of *Aedes aegypti*. *Mosq. News* 19(4): 256–259.

Burrough, P., McDonnell, A., and Rachael, A., 1998. *Principles of Geographical Information Systems*, Oxford University Press.

Busvine, J., 1951. Mechanism of resistance to insecticides in housflies. *Nature* 168: 193–195.

Busvine, J. R., 1978. The future of insecticidal control for medically important insects. Medical Entomology Centenary, Symposium proceedings, The Royal Society of Tropical Medicine and Hygiene, pp. 106–111.

Butterworth, D. E., 1979. Separation of aedine eggs from soil sample debris using hydrogene peroxide. *Mosq. News* 39(1): 139–141.

Callot, J., 1943. Sur *Culex hortensis* et *Culex apicalis* a Richelieu (Indre-et-Loire). *Ann. Parasit. hum. comp.* 19: 129–141, illus.

Cambournac, F. J. C., and Hill, R. B., 1938. The biology of *Anopheles maculipennis*, var. *atroparvus* in Portugal. *Proc. Int. Congr. Trop. Med. Mal.* 2: 178–184.

Cambournac, F. J. C., and Hill, R. B., 1940. Observation on the swarming of *Anopheles maculipennis*, var. *atroparvus*. *Am. J. Trop. Med.* 20(1): 133–140.

Cantile, C., di Guardo, G., Eleni, C., and Aruspici, M., 2000. Clinical and neuropathological features of West Nile virus equine encephalomyelitis in Italy. *Equine vet. J.* 32(1): 31–35.

Carlson, J. O., 1995. Molecular genetic manipulation of vectors, in: *The biology of disease vectors*. (B. J. Beaty, and W. C. Marquardt, eds.), University Press of Colorado, Colorado, USA, pp. 215–228.

Carpenter, S. J., and LaCasse, W. J., 1955. *Mosquitoes of North America (north of Mexico)*. Univ. Calif. Press. vi + 360 pp., illus., 127pls.

Carpenter, S. J., and Nielsen, L. T., 1965. Ovarian cycles and longevity in some univoltine *Aedes* species in the Rocky Mountains of western United States. *Mosq. News* 25: 127–134.

Carson, R., 1962. *Silent spring*. Houghton Mifflin Co., Boston, 368 pp.

Case, T. J., and Washino, R. K., 1979. Flatworm control of mosquito larvae in rice fields. *Science* 206(4425): 1412–1414.

Casida, J. E., 1980. Pyrethrum flowers and pyrethroid insecticides. *Environ. Health Persp.* 34: 189–202.

CDC (Centers for Disease Control), 1987. Imported and indigenous dengue fever—United States. *MMWR* 36: 551–554.

Chadee, D. D., and Corbet, P. S., 1987. Seasonal incidence and diel patterns of oviposition in the field of the mosquito, *Aedes aegypti* (L.) (Diptera: Culicidae) in Trinidad, West Indies: a preliminary study. *Ann. Trop. Med. Parasit.* 81: 151–161.

Chadee, D. D., and Corbet, P. S., 1990. A night-time role of the oviposition site of the mosquito *Aedes aegypti* (L.) (Diptera: Culicidae). *Ann. Trop. Med. Parasit.* 84: 429–433.

Chang, C., Yu, Y.-M., Dai, S.-M., Law, S. K., and Gill, S. S., 1993. High-level cryIVD and cytA gene expression in *Bacillus thuringiensis* does not require the 20 kilodalton protein, and the coexpressed gene products are synergistic in their toxicity of mosquitoes. *Appl. Environ. Microbiol.* 59: 815–821.

Chapman, H. C., 1960. Observation on *Aedes melanimon* and *Aedes dorsalis* in Nevada. *Ann. Entomol. Soc. Am.* 53(6): 706–708.

Chapman, H. C., 1962. A survey for autogeny in some Nevada mosquitoes. *Mosq. News* 22: 134–136.

Chapman, H. C., and Barr, A. R., 1964. *Aedes communis nevadensis*, a new subspecies of mosquito from western North America (Diptera: Culicidae). *Mosq. News* 24: 439–447.

Chapman, R. F., 1982. *The insects: structure and function*. Hodder and Stoughton, London, 919 pp.

Charles, J.-F., Delécluse, A., and Nielson-LeRoux, C., 2000. *Entomopathogenic bacteria: from laboratory to field application*. Kluwer Academic Publishers, Dordrecht/Boston/London, 524 pp.

Charles, J.-F., and Nielsen-LeRoux, C., 1996. Les bacteries entomopathogenes: mode d'action sur les larves de moustiques et phenomenes de resistance. *Ann. Inst. Pasteur*, Actualites, 7(4): 233–245.

Chavasse, D. C., and Yap, H. H., 1997. Chemical methods for the control of vectors and pests of public health importance. World Health Organization, Division of Control of Tropical Diseases, WHO Pesticide Evaluation Scheme.

Chevilon, C., Pasteur, N., Marquine, M., Heyse, D., and Raymond, M., 1995. Population structure and dynamics of selected genes in the mosquito *Culex pipiens. Evolution* 49(5): 997–1007.

Chilcott, C. N., and Ellar, D. J., 1988. Comparative toxicity of *Bacillus thuringiensis* var. *israelensis* crystal proteins *in vivo* and *in vitro. J. Gen. Microbiol.* 134: 2551–2558.

Chinaev, P. P., 1964. On the autogenous development of exophilous mosquitoes in Uzbekistan. *Zool. Zh.* 43: 939–940.

Chippaux, A., Rageau, J., and Mouchet, J., 1970. Hibernation de l'arbovirus Tahyna chez *Culex modestus* Fic. en France. *C. R. Acad. Sci. (D.)* (Paris) 270: 1648–1650.

Chodorowski, A., 1968. Predator-prey relation between *Mochlonyx culiciformis* and *Aedes communis. Pol. Arch. Hydrobiol.* 15: 279–288.

Christophers, S. R., 1929. Note on a collection of anopheline and culicine mosquitoes from Madeira and the Canary Islands. *Ind. J. Med. Res.* 16: 518–530.

Christophers, S. R., 1933. *The fauna of British India, including Ceylon and Burma*. Diptera. Family Culicidae. Tribe Anophelini. Vol. IV, London, 371 pp., illus.

Christophers, S. R., 1960. *Aedes aegypti* (L.), *the yellow fever mosquito. Its life history, bionomics, and structure*. Cambridge Univ. Press, 739 pp.

Cianchi, R., Urbanelli, S., Sabatini, A., Coluzzi, M., Tordi, M. P., and Bullini, L., 1980. Due entita riproduttivamente isolate sotto il nombre di *Aedes caspius* (Diptera: Culicidae). Atti XII Congresso Nazionale Italiane Entomologica, Roma 2: 269–272.

Cianchi, R., Sabatini, A., Bocollini, D., Bullini, L., and Coluzzi, M., 1987. Electrophoretic evidence of reproductive isolation between sympatric populations of *Anopheles melanoon* and *Anopheles subalpinus*. Proc. Third Int. Congr. Mal. Bab. 156. Annency, 1987.

Cilek, J. E., and Knapp, F. W., 1993. Enhanced diazinon susceptibility in pyrethroid resistant horn flies (Diptera: Muscidae): Potential for insecticide resistance management. *J. Econ. Entomol.* 86: 1303–1307.

Clarke, J. L., 1943a. Studies of the flight range of mosquitoes. *J. Econ. Entomol.* 36: 121–122.

Clarke, J. L., 1943b. Preliminary progress report. Do male mosquitoes fly as far as females? Is the flight range of all mosquitoes the same? *Mosq. News* 3: 16–21.

Clark, J. R., Borthwick, P. W., Goodman, L. R., Patrick, J. M., Jr., Lores, E. M., and Moore, C., 1987. Effects of aerial thermal fog application of fenthion on caged pink shrimp, mysids and sheepshead minnows. *J. Am. Mosq. Control Assoc.* 3(3): 466–472.

Clark, T. B., and Fukada, T., 1967. Predation of *Culicoides cavaticus* (Wirth and Jones) larvae on *Aedes sierrensis* (Ludlow). *Mosq. News* 27: 424–425.

Clements, A. N., 1963. *The physiology of mosquitoes*. Pergamon Press, Oxford, 395 pp.

Clements, A. N., 1992. *The biology of mosquitoes*. Vol. 1, Development, Nutrition and reproduction. Chapman and Hall, London, 509 pp.

Colbo, A. H., and Undeen, A. H., 1980. Effect of *Bacillus thuringiensis* var. *israelensis* on non-target insects in stream trials for control of Simuliidae. *Mosq. News* 40: 368–371.

Collins, F. H. and Washino, R. K., 1985. Insect predators, in: *Biological Control of Mosquitoes*. (H. C. Chapman, ed.), *Am. Mosq. Control Assoc. Bull.* 6: 25–42.

Coluzzi, M., 1960. Alcuni dati morfologici i biologici sulle forme italiane di *Anopheles claviger* Meigen. *Riv. Malariologia* 39: 221–235.

Coluzzi, M., 1961. Sulla presenza di *Culex (Culex) theileri* Theobald in Italia centrale, meridionale et in Sicilia. *Bull. Soc. Ent. Ital.* 91: 55–57.

Coluzzi, M., 1962. Le forme di *Anopheles claviger* Maigen indicate con i nomi *missirolii* e *petragnanii* sono due specie riproduttivamente isolate. R. C. accad. *Lincei* 32: 1025–1030.

Coluzzi, M., 1968. Nuove segnalazioni di culicidi in Sicilia. Estratto dal Bollettino della Societa Entomologica Italiana. Vol. XCVIII, N. 9–10: 126–128.

Coluzzi, M., and Bullini, L., 1971. Enzyme variants as markers in the study of precopulatory isolating mechanisms. *Nature* 231: 455–456.

Coluzzi, M., and Contini, C., 1962. The larvae and pupae of *Mansonia buxtoni* (Edwards), 1923 (Diptera: Culicidae). *Bull. Ent. Res.* 53: 215–218, illus.

Coluzzi, M., and Sabatini, A., 1968. Divergenze morfologiche e barriere di sterilita nel complesso *Aedes mariae* (Diptera: Culicidae). *Riv. Parasit.* 29: 49–70, 5pls.

Coluzzi, M., Sacca, G., and Feliciangeli, E. D., 1965. Il complesso *An. claviger* nella sottoregione mediterranea. Cah. ORSTOM Ser. Entomol. med. 3(3/4): 97–102.

Coluzzi, M., Sabatini, A., Bullini, L., and Ramsdale, C., 1974. Nuovi dati sulla distribuzione delle specie del complesso *mariae* del genere *Aedes. Riv. Parassitol.* 35: 321–330.

Coluzzi, M., di Deco, M., and Gironi, A., 1975. Influenza del fotoperiodo sulla scelta del luogo di ovideposizione in *Aedes mariae* (Diptera: Culicidae). *Parasitologia* 17: 121–130.

Coluzzi, M., Bianchi-Bullini, A. P., and Bullini, L., 1976. Speciazione nel complesso *mariae* del genre *Aedes* (Diptera: Culicidae). Atti Associazione Genetica Italiana. Vol. XXI: 218–223.

Conway, G. R., 1976. Man versus pests, in: *Theoretical ecology.* (R. M. May, ed.), Blackwell Scientific Publications, Oxford, pp. 257–281.

Corbet, P. S., 1964. Autogeny and oviposition in arctic mosquitoes. *Nature* 203(4945): 668.

Corbet, P. S., 1965. Reproduction of mosquitoes in the High Arctic. *Proc. XII Int. Congr. Ent. (Lond.)* 12: 817–818.

Corbet, P. S., and Danks, H. V., 1975. Egg-laying habits of mosquitoes in the high arctic. *Mosq. News* 35: 8–14.

Corbett, J. R., 1974. *The biochemical mode of action of pesticides.* Academic Press, London, 330 pp.

Coykendall, R. L. (ed.). 1980. *Fishes in Californian mosquito control.* Calif. Mosq. Control Assoc. Press, Sacramento, USA, 63 pp.

Crampton, J. M., 1992. Potential application of molecular biology in entomology, in: *Insect Molecular Science.* (J. M. Crampton, and P. Eggleston, eds.), San Diego Academic Press, London, pp. 4–20.

Crampton, J. M., and Eggleston, P., 1992. Biotechnology and the control of mosquitoes, in: *Animal parasite control utilizing biotechnology.* (W. K. Young, ed.), CRC Press Inc, Uniscience Volumes, Boca Raton, FL., pp. 333–350.

Crampton, J. M., Morris, A., Lycett, G., Warren, A., and Eggleston, P., 1990. Transgenic mosquitoes: A future vector control strategy? *Parasit. Today* 6: 31–36.

Crans, W. J., 1970. The blood feeding habits of *Culex territans* Walker. *Mosq. News* 30: 445–447.

Cranston, P. S., Ramsdale, C. D., Snow, K. R., and White, G. B., 1987. *Key to the adults, larvae and pupae of british mosquitoes* (Culicidae). Freshw. Biol. Assoc., Scient. Publ. 48, 152 pp.

Cremlyn, R., 1978. *Pesticides. Preparation and mode of action.* John Wiley and Sons, New York, 240 pp.

Cress, F. C., 1980. Other mosquito predators. *Calif. Agric.* 34(3): 20.

Croset, H., Papierok, B., Rioux, J. A., Gabinaud, A., Cousserans, J., and Arnaud, D., 1976. Absolute estimates of larval populations of culicid mosquitoes: comparison of "capture-recapture", "removal" and "dipping" methods. *Ecol. Ent.* 1: 251–256.

Cuany, A., Handani, J., Berge, J., Fournier, D., Raymond, M., Georghiou, G. P., and Pasteur, N., 1993. Action of esterase B 1 on chlorpyrifos in organophosphate resistant *Culex* mosquitoes. *Pestic. Biochem. Physiol.* 45: 1–6.

Curtis, C. F., 1968a. A possible genetic method for the control of insect pests with special reference to tsetse flies, *Glossina* spp. *Bull. Entom. Res.* 57: 509–523.

Curtis, C. F., 1968b. Possible use of translocations to fix desirable genes in insect pest populations. *Nature* 218: 368–369.

Curtis, C. F., 1990. *Appropriate Technology in Vector Control.* CRC Press Inc., Boca Raton, Florida, 233 pp.

Curtis, C. F., Morgan, P. R., Minjas, J. N., and Maxwell, C. A., 1990. Insect proofing of sanitation systems, in: *Appropriate technology in vector control.* (C. F. Curtis, ed.), CRC Press Inc., pp. 173–186.

Curtis, C. F., Hill, N., and Kasim, S. H., 1993. Are there effective resistance management strategies for vectors of human disease? *Biol. J. Linn. Soc.* 48: 3–18.

Curtis, C. F., Myamba, J., and Wilkes, T. J., 1996. Comparison of different insecticides and fabrics for anti-mosquito bednets and curtains. *Med. Vet. Entomol.* 10: 1–11.

Dahl, C., 1974. Circumpolar *Aedes* (*Ochloretatus*) species in north Fennoscandia. *Mosq. Syst.* 6(1): 57–73, illus.

Dahl, C., 1975. Culicidae (Diptera: Nematocera) of the Baltic Island of Oland. *Ent. Tidskr.* 96(3–4): 77–96, illus.

Dahl, C., 1977. Verification of *Anopheles (Ano.) messeae* Falleroni (Diptera: Culicisae) from southern Sweden. *Ent. Tidskr.* 98: 149–152.

Dahl, C., 1980. Postembryonic organization of the genital segments in trichoceridae, tipulidae, and anisopodidae (Diptera: Nematocera). *Zool. Scripta* 9: 165–185.

Dahl, C., 1984. A SIMCA pattern recognition study in taxonomy: Claw shape in mosquitoes (Culicidae, Insecta). *Syst. Zool.* 33(4): 355–369.

Dahl, C., 1988. Taxonomic studies on *Culex pipiens* and *Culex torrentium*, in: *Biosystematics of haematophagous insects.* (M. W. Service, ed.). Systematics Association Special Volume No. 37, Clarendon Press, Oxford, UK, pp. 149–175.

Dahl, C., 1997. Diptera Culicidae, mosquitoes, in: *Aquatic insects of North Europe—A taxonomic handbook.* (A. Nilsson, ed.), Vol. 2, Apollo Books, Stenstrup, Denmark, pp. 163–186.

Dahl, C., 2000. Feeding in nematoceran larvae: ecology, behavior, mechanisms and principles. *Proc. 13th Europ. SOVE Meeting, Ankara.* (S. S. Caglar, B. Alten, and N. Özer, eds.), Society for Vector Ecology, pp. 21–27.

Dahl, C., and White, G. B., 1978. Culicidae, in: *Limnofauna europaea,* 2nd edition. (J. Illies, ed.), Gustav Fischer Verlag, Stuttgart and New York, Swets and Zeitlinger B. V., Amsterdam, pp. 390–395.

Dahl, C., Widahl, L.-E., and Nilsson, C., 1988. Functional analysis of the suspension feeding system in mosquitoes (Diptera: Culicidae). *Ann. Ent. Soc. Am.* 81: 105–127.

Dalla Pozza, G., and Majori, G., 1992. First record *of Aedes albopictus* establishment in Italy. *J. Am. Mosq. Control Assoc.* 8: 318–320.

Dame, D. A., Lowe, R. E., Wichterman, G. J., Cameron, A. L., Baldwin, K. R., and Miller, T. W., 1976. Laboratory and field assessment of insect growth regulators for mosquito control. *Mosq. News* 40: 462–472.

Danielova, V., 1992. Relationships of mosquitoes to Tahyna virus as determinant factors of its circulation in nature. Academia Publishing House of the Czechoslovak Academy of Sciences, Prague.

Danielova, V., and Ryba, J., 1979. Laboratory demonstration of transovarial transmission of Tahyna virus in *Aedes vexans* and the role of this mechanism in overwintering of this arbovirus. *Folia Parasitologica* 26: 361–366.

Danilov, V. N., 1974. On the restoration of the name *Aedes (O.) mercurator* for a mosquito known in the USSR as *Aedes riparius ater* Gutsevich (Diptera: Culicidae) (in Russian). *Parazitologiya* 8: 322–328, illus.

Danilov, V. N., 1979. Nomenclature, synonymy, differential diagnosis, and distribution of *Aedes (Ochlerotatus) albineus* Seguy (Diptera) (in Russian). *Vestn. Zool. Kiev.* (1): 29–35. [Translation 1628 (T1628) Med. Zool. Dep. U. S. Nav. Med. Res. Unit No. 3 Cairo, Egypt.]

Danilov, V. N., 1984. *Culiseta (Culicella) setivalva* Maslov as a synonym of *C. (C.) fumipennis* Stephens (in Russian). *Parazitologiya* 18: 313–317.

Danilov, V. N., 1987. Mosquitoes of the subgenus *Aedes* (Diptera: Culicidae) of the USSR fauna. II. *Aedes dahuricus* sp. n. (in Russian). *Vestn. Zool.* (4): 35–41.

Danks, H. V., and Corbet, P. S., 1973. A key to all stages of *Aedes nigripes* and *Ae. impiger* (Diptera: Culicidae) with a description of first-instar larvae and pupae. *Can. Ent.* 105: 367–376.

Darsie, R. F., and Samanidou-Voyadjoglou, A., 1997. Keys for the identification of the mosquitoes of Greece. *J. Am. Mosq. Contr. Assoc.* 13(3): 247–254.

Darsie, R. F., and Ward, R. A., 1981. Identification and geographical distribution of the mosquitoes of North America, north of Mexico. *Mosq. Syst.* Suppl. 1: 1–313.

Dary, O., Georghiou, G. P., Parson, E., and Pasteur, N., 1990. Microplate adaptation of Gomori's assay for quantitative determination of general esterase activity in single insects. *J. Econ. Entomol.* 83: 2187–2192.

Dary, O., Georghiou, G. P., Parson, E., and Pasteur, N., 1991. Dot blot test for identification of insecticide-resistant acetylcholinesterase in single insects. *J. Econ. Entomol.* 84: 28–33.

Das, M., Srivastava, S. P., Khamre, J. S., and Deshpande, L. B., 1986. Susceptibility of DDT, dieldrin and malathion resistant *Anopheles culicifacies* populations to deltamethrin. *J. Am. Mosq. Control Assoc.* 2(4): 553–555.

Davey, R. B., and Meisch, M. V., 1977a. Control of dark rice-field mosquito larvae, *Psorophora columbiae* by mosquitofish, *Gambusia affinis* and green sunfish, *Lepomis cyanellus*, in Arkansas rice fields. *Mosq. News* 37: 258–262.

Davey, R. B., and Meisch, M. V., 1977b. Low maintenance production studies of mosquitofish, *Gambusia affinis* in Arkansas rice fields. *Mosq. News* 37: 760–763.

Davey, R. B., and Meisch, M. V., 1977c. Dispersal of mosquitofish, *Gambusia affinis* in Arkansas rice fields. *Mosq. News* 37: 777–778.

Davidson, E. W., 1984. Microbiology, pathology and genetics of *Bacillus sphaericus* biological aspects which are important to field use. *Mosq. News* 44: 147–152.

Davidson, E. W., 1988. Binding of the *Bacillus sphaericus* toxin to midgut cells of mosquito larvae: Relationship to host range. *J. Med. Entomol.* 25: 151–157.

Davidson, E. W., 1990. Development of insect resistance to biopesticides. Proc. Second Sympos. on Biocontrol, Brasilia, Oct. 1990, p. 19.

Davidson, E. W., and Becker, N., 1996. Microbial control of vectors, in: *The biology of disease vectors.* (B. J. Beaty, and W. C. Marquardt, eds.), University Press of Colorado, Colorado, USA, pp. 549–563.

Davidson, E. W., and Yousten, A. A., 1990. The mosquito larval toxin of *Bacillus sphaericus*, in: *Bacterial control of mosquitoes and blackflies*: *biochemistry, genetics and applications of* Bacillus thuringiensis israelensis *and* Bacillus sphaericus. (H. de Barjac, and D. Sutherland, eds.), Rutgers Univ. Press, New Brunswick, NJ, pp. 237–255.

Davidson, G., 1982. The agricultural usage of insecticides in Turkey and resurgence of malaria, in: *Proceedings of an International Workshop "Resistance to insecticides used in public health and agriculture"*, 22–26 February, 1982, Sri Lanka. National Science of Sri Lanka, pp. 122–129.

Davidson, G., and Curtis, C. F., 1979. Insecticide resistance and the upsurge of malaria. Annu. Rep. London School Trop. Med. Hyg., 1978–79, pp. 78–82.

Davies, R. G., 1992. *Outlines of entomology*. Chapman and Hall, London, 408 pp.

Davis, E. E., and Sokolove, P. G., 1975. Temperature response of the antennal receptors in the mosquito *Aedes aegypti. J. Comp. Physiol.* 96: 223–236.

De Barjac, H., 1983. Bioassay procedure for samples of *Bacillus thuringiensis israelensis* using IPS-82 standard. WHO Report TDR/VED/SWG (5)(81.3), Geneva, World Health Organization.

Delecluse, A., Barloy, F., and Rosso, M.-L., 1996. Les bacteries pathogenes des larves de dipteres: structure et specificite des toxines. *Ann. Inst. Pasteur,* Actualites, 7(4): 217–231.

Denholm, I., and Rowland, M. W., 1992. Tactics for managing pesticide resistance in arthropods: Theory and practice. *Ann. Rev. Entomol.* 37: 91–112.

Dent, D., 1991. *Insect Pest Management.* C.A.B. International, Redwood Press, 604 pp.

Des Rochers, B., and Garcia, R., 1984. Evidence for persistence and recycling of *Bacillus sphaericus. Mosq. News* 44: 160–165.

Detinova, T. S., and Smelova, V. A., 1973. K voprosu o medicinskom znatcheniy komarov (Diptera: Culicidae) fauni Sovyetskogo Soyuza. *Med. parazitol. (Moskva).* 42(4): 455–471.

Devonshire, A. L., and Field, L. M., 1991. Gene amplification and insecticide resistance. *Ann. Rev. Entomol.* 36: 1–23.

Diesfeld, H. J., 1989. *Gesundheitsproblematik der dritten Welt.* Wiss. Buchges., Darmstadt, 161pp.

Dixon, R. O., and Brust, R. A., 1972. Mosquitoes of Manitoba. III. Ecology of larvae in the Winnipeg area. *Can. Ent.* 104: 961–968.

Dolbeskin, B. V., Gorickaya, B., and Mitrofanova, Y., 1930. Beschreibung einer neuen Art aus der Gattung *Aedes* (n. sp.) aus Osteuropa. *Parasitol. sbornik Zool. Mus. AN SSSR* 1: 253–260, 2pls.

Dorn, S., Frischknecht, M. L., Martinez, V., Zurfluch, R., and Fisher, V., 1981. A novel non neurotoxic insecticide with broad activity spectrum. *Z. Pflanzenkr. Pflanzenschutz.* 88: 269–275.

Douglas, R. H., Judy, D., Jacobson, B., and Howell, R., 1994. Methoprene concentrations in freshwater microcosms treated with sustained-release Altosid formulations. *J. Am. Mosq. Control Assoc.* 10(2): 202–210.

Dubitskyi, A. M., 1966. Positive reaction of mosquitoes and biting midges to repellents. *Izv. Akad. Nauk. Kaz. S.S.R. Ser. Biol.* 1: 53–56.

DuBose, W. P., and Curtin, T. J., 1965. Identification keys to the adult and larval mosquitoes of the Mediterranean area. *J. Med. Ent.* 1(4): 349–355.

Dukes, J. C., Hallmon, C. F., Shaffes, K. R., and Hester, P. G., 1990. Effects of pressure and flow rate on Cythion droplet size produced by three different ground ULV aerosol generators. *J. Am. Control Assoc.* 6(2): 279–282.

Dulmage, H. T., Correa, J. A., and Gallegos-Morales, G., 1990. Potential for improved formulations of *Bacillus thuringiensis israelensis* through standardization and fermentation development, in: *Bacterial control of mosquitoes and blackflies: biochemistry, genetics and applications of* Bacillus thuringiensis israelensis *and* Bacillus sphaericus. (H. de Barjac, and D. Sutherland, eds.), Rutgers Univ. Press, New Brunswick, NJ, pp. 16–44.

Eads, R. B., 1972. Recovery of *Aedes albopictus* from used tires shipped to United States ports. *Mosq. News* 32(1): 113–114.

Eckstein, F., 1918. Zur Systematik der einheimischen Stechmücken. 1. vorläufige Mitteilung: die Weibchen. *Zbl. Bakt.,* Abt. 1, Orig. 82: 57–68.

Eckstein, F., 1919. Zur Systematik der einheimischen Stechmücken. 2. vorläufige Mitteilung: die Larven. *Zbl. Bakt.,* Abt. 1, Orig. 83: 281–294.

Eckstein, F., 1920. Zur Systematik der einheimischen Stechmücken. 3. vorläufige Mitteilung: die Männchen. *Zbl. Bakt.,* Abt. 1, Orig. 84: 223–240.

Edwards, F. W., 1920. Mosquito notes. *Bull. Ent. Res.* 10: 129–137, illus.

Edwards, F. W., 1921. A revision of the mosquitoes of the Palearctic region. *Bull. Ent. Res.* 12: 263–351, illus.

Edwards, F. W., 1932. *Genera Insectorum. Diptera. Fam. Culicidae, Fascicle 194,* Bruxelles, Belgium: Desmet-Verteneuil, Imprimeur-Editeur, 258 pp.

Edwards, F. W., 1941. *Mosquitoes of the Ethiopian Region.* III.—Culicine adults and pupae. British Museum (Natural History), London. 499 pp., illus., 4pls.

Efird, P. K., Inman, A. D., Dame, D. A., and Meisch, M. V., 1991. Efficacy of various ground-applied cold aerosol adulticides against *Anopheles quadrimaculatus. J. Am. Mosq. Control Assoc.* 7: 207–209.

Eldefrawi, A. T., 1985. Acethylcholinesterases and anti-cholinesterases, in: *Comprehensive insect physiology, biochemistry, and pharmacology,* Vol. 12, Pergamon Press, Oxford, pp. 115–131.

Eldrige, B. F., and Edman, J. D., 2000. *Medical Entomology.* Kluwer Academic Publishers, Dordrecht, Boston, London, 659 pp.

Elliott, M., 1980. Established pyrethroid insecticides. *Pestic. Sci.* 11: 119–128.

Elliot, M., and Janes, N. F., 1979. Recent structure-activity correlations in synthetic pyrethroids, in: *Advances in Pesticide Science.* (H. Geissbühler, ed.), Pergamon Press, Oxford, pp. 166–173.

Ellis, R. A., and Brust, R. A., 1973. Sibling species delimination in the *Aedes communis* (DeGeer) aggregate (Diptera: Culicidae). *Can. J. Zool.* 51: 915–959.

Eltari, E., Zeka, S., Gina, A., Sharofi, F., and Stamo, K., 1987. Epidemiological data on some foci of haemorrhagic fever in our country (in Albanian). *Revista Mjekesore* 1: 5–9.

Encinas Grandes, A., 1982. Taxonomia y biologia de los mosquitos del area Salmantina (Diptera: Culicidae). Consejo Superior de Investigaciones Cientificas. Centro de Edafologia y Biologia Aplicada Ediciones Universidad de Salamanca, 437 pp.

Eritja, R., Aranda, C., Padros, J., Goula, M., Lucientes, J., Escosa, R., Marques, E., and Caceres, F., 2000. An annotated checklist and bibliography of the mosquitoes of Spain (Diptera: Culicidae). *Europ. Mosq. Bull.* 8: 10–18.

Espmark, Ä., and Niklasson, B., 1984. Ockelbo disease in Sweden: Epidemiological, clinical, and virological data from the 1982 outbreak. *Am. J. Trop. Med. Hyg.* 33: 203–1211.

Evans, A. M., 1938. *Mosquitoes of the Ethiopian Region.* II.—Anophelini adults and early stages. British Museum (Natural History), London, 404 pp., illus.

Evans, B. R., and Brevier, G. A., 1969. Measurements of field populations of *Aedes aegypti* with the ovitrap in 1968. *Mosq. News* 29: 347–353.

Evans, R. G., 1993. Laboratory evaluation of the irritancy of bendiocarb, lambda-cyhalotrin and DDT to *Anopheles gambiae. J. Am. Mosq. Control Assoc.* 9: 285–293.

Falleroni, D., 1926. Fauna anophelica italiana et suo "habitat" (paludi, risae, canali). Metodi di lotta contro la malaria. *Riv. Malariol.* 5: 553–593.

Farnham, A. W., 1977. Genetics of resistance of housflies (*Musca domestica* L.) to pyrethroids. I. Knock-down resistance. *Pestic. Sci.* 7: 278–282.

Fay, R. W., and Eliason, D. A., 1966. A preferred oviposition site as a surveillance method for *Aedes aegypti. Mosq. News* 26: 531–535.

Federici, B. A., Lüthy, P., and Ibarra, J. E., 1990. Parasporal body of *Bacillus thuringiensis israelensis*: Structure, protein composition, and toxicity, in: *Bacterial control of mosquitoes and blackflies: biochemistry, genetics and applications of* Bacillus thuringiensis israelensis *and* Bacillus sphaericus. (H. de Barjac, and D. Sutherland, eds.), Rutgers Univ. Press, New Brunswick, NJ, pp. 45–65.

Feng, L. C., 1938. The tree hole species of mosquitoes of Peiping, China. *Suppl. Chin. Med. J.* 2: 503–525, 5pls.

Ferrari, J. A., and Georghiou, G. P., 1990. Esterase B 1 activity variation within and among insecticide resistant, susceptible and heterozygous strains of *Culex quinquefasciatus* (Diptera: Culicidae). *J. Econ. Entomol.* 83: 1704–1710.

Ferrari, J. A., and Georghiou, G. P., 1991. Quantitative genetic variation of esterase activity associated with a gene amplification in *Culex quinquefasciatus. Heredity* 66: 265–272.

Ferre, J., Real, M. D., Van Rie, J., Jensens, S., and Peferoen, M., 1991. Resistance to the *Bacillus thuringiensis* bioinsecticide in a field population of *Plutella xylostella* is due to a change in a midgut membrane receptor. *Proc. Nat. Acad. Sci. USA* 88: 5119–5123.

Feyereisen, R., Loener, J. E., Farnsworth, D. E., and Nebert, D. W., 1989. Isolation and sequence of cDNA encoding a cytochrome P450 from and insecticide resistant strain of the house fly, *Musca domestica. Proc. Nat. Acad. Sci. USA* 86: 1465–1469.

Filipe, A. R., 1972. Isolation in Portugal of West Nile virus from *Anopheles maculipennis* mosquitoes. *Acta Virol.* (Praha) 16: 361.

Filipe, A. R., 1980. Arboviruses in Portugal, in: *Arboviruses in the mediterranean countries.* (J. Vesenjak-Hirjan, J. S. Porterfield, and E. Arslanagic, eds.), Zbl. Bakt. Suppl. 9, Gustav Fischer Verlag, Stuttgart, pp. 137–141.

Filipe, A. R., 1990. Arboviruses in the Iberian peninsula. *Acta Virol.* (Praha) 34(6): 582–591.

Fillinger, U., Knols, B.G., and Becker, N., 2003. Efficacy and efficiency of new *Bacillus thuringiesis* var. *israelensis* and *Bacillus sphaericus* formulations against Afrotropical anophelines in Western Kenya. *Tropical Medicine and International Health*, 8(1): 37–47.

Finney, D. J., 1971. *Probit analysis*, 3rd ed. Cambridge Univ. Press.

Focks, D. A., Sacket, S. R., Klotter, K. O., Dame, D. A., and Carmichael, G. T., 1986. The integrated use of *Toxorhynchites amboinensis* and ground level ULV insecticide application to suppress *Aedes aegypti* (Diptera: Culicidae). *J. Med. Ent.* 23: 513–519.

Forattini, O. P., 1996. *Culicidologia medica.* Volume 1. Principios gerais morfologia glossario taxonomico. Editora da Universidade de Sao Paulo—Edusp, 548 pp.

Ford, M. G., Hollman, D. W., Khambay, B. P. S., and Sawicki, R. M. (eds.), 1987. *Combating resistance to xentobiotics.* Ellis Horwood, Chichester, England, 320 pp.

Forrester, N. W., Cahill, M., Bird, L. J., and Layland, J. K., 1993. Management of pyrethroid and endosulfan resistance in *Helicoverpa armigera* (Lepidoptera: Noctuidae) in Australia. *Bull. Ent. Res.*, Suppl. 1: 132 pp.

Fox, I., 1980. Evaluation of ultra-low volume aerial and ground application of malathion against natural populations of *Aedes aegypti* in Puerto Rico. *Mosq. News* 40: 280–283.

Francki, R. I. B., Fauquet, C. M., Knudson, D. L., and Brown, F., 1991. Classification and nomenclature of viruses. Fifth report of the International Committee on Taxonomy of Viruses. *Archives of Virology*, Suppl. 2, Springer Verlag, Wien, 452 pp.

Francy, D. B., Jaenson, T. G. T., Lundström, J. O., Schildt, E.-B., Espmark, A., Henriksson, B., and Niklasson, B., 1989. Ecologic studies of mosquitoes and birds as hosts of Ockelbo virus in Sweden, and isolation of Inkoo and Batai viruses from mosquitoes. *Am. J. Trop. Med. Hyg.* 41: 355–363.

Freier, J. E., and Francy, D. B., 1991. A duplex cone trap for the collection of adult *Aedes albopictus*. *J. Am. Mosq. Control. Assoc.* 7: 73–79.

French, W. L., Baker, R. H., and Kitzmiller, J. B., 1962. Preparation of mosquito chromosomes. *Mosq. News* 22: 377–383.

French-Constant, R. H., Rocheleau, T. A., Steichen, J. C., and Chambers, A. E., 1993. A point mutation in a *Drosophila* GABA receptor confers insecticide resistance. *Nature* (London) 363: 449–451.

Frohne, W. C., 1953. Mosquito breeding in Alaskan salt marshes, with special reference to *Aedes punctodes* Dyar. *Mosq. News* 13: 96–103.

Gabinaud, A., Vigo, G., Cousserans, J., Rioux, M., Pasteur, N., and Croset, H., 1985. La mammophilie des populations de *Culex pipiens pipiens* L., 1758 dans le Sud de la France; variations de ce caractere en fonction de la nature des biotopes des developpement larvaire, des caracteristiques physio-chimiques de leur eaux et de saisons. Consequences practiques et theoriques. Cah. ORSTOM, ser Ent. med. et Parasitol. 23(2): 123–132.

Gaffigan, T. V., and Ward, R. A., 1985. Index to the second supplement to "A catalog of the mosquitoes of the world" (Diptera: Culicidae), with corrections and additions. *Mosq. Syst.* 17: 52–63.

Galliard, H., 1931. *Culex brumpti*, n. sp. moustique nouveau trouve en Corse. *Ann. Parasit. hum. comp.* 9: 134–139.

Gammon, D. W., 1978. Effects of DDT on the cockroach nervous system at three temperatures. *Pestic. Sci.* 9: 95–104.

Garcia, R., Des Rochers, B., and Tozer, W., 1981. Studies on *Bacillus thuringiensis* var. *israelensis* against mosquito larvae and other organisms. *Proc. Calif. Mosq. Vector Control Assoc.* 49: 25–29.

Garnham, P. C. C., 1966. *Malaria Parasites and other Haemosporidia*. Blackwell Scientific Publications, Oxford, 1114 pp.

Garnham, P. C. C., 1980. Malaria in its various vertebrate hosts, in: *Malaria*, Volume 1. (J. P. Kreier, ed.), Academic Press, New York, pp. 95–144.

Garnham, P. C. C., 1988. Malaria parasites of man: life-cycles and morphology (excluding ultrastructure), in: *Malaria Principles and Practice of Malariology*, Volume 1. (W. H. Wernsdorfer and I. McGregor, eds.), Churchill Livingstone, Edinburgh, pp. 61–96.

Garrett, W. D., 1976. Mosquito control in the aquatic environment with monomolecular organic surface films. Naval Research Labor, (Washington, DC), Report 8020, 13 pp.

Garret, W. D., and White, S. A., 1977. Mosquito control with monomolecular organic surface films: I-selection of optimum film-forming agent. *Mosq. News* 37: 344–348.

Gebhard, H., 1990. Stechmückenbekämpfung mit Fischen. Ph.D. thesis, University of Heidelberg, 238 pp.

Geeverghese, G., Dhanda, V., Rango Rao, P. N., and Deobhankar, R. B., 1977. Field trials for the control of *Aedes aegypti* with Abate in Poona city and suburbs. *Indian J. Med. Res.* 65(4): 466–473.

Georghiou, G. P., 1977. The insects and mites of Cyprus. With emphasis on species of economic importance to agriculture, forestry, man and domestic animals. Kiphissia, Athens, Greece, pp. 93–94.

Georghiou, G. P., 1983. Management of resistance in arthropods, in*: Pest resistance to pesticides.* (G. P. Georghiou, and T. Saito, eds.), Plenum Press, New York, pp. 769–792.

Georghiou, G. P., 1986. The magnitude of the resistance problem, in: *Pesticide resistance: strategies and tactics for management.* (E. H. Glass, ed.), Nat. Acad. Sci., Washington, DC, pp. 14–43.

Georghiou, G. P., 1987. Insecticides and pest resistance: the consequences of abuse. Faculty Research Lecture, Academic Senate, University of California, Riverside, 27 pp.

Georghiou, G. P., 1990. Overview of insecticide resistance, in: *Managing resistance to agrochemicals— from fundamental research to practical strategies.* (M. B. Green., H. M. LeBaron, and W. K. Moberg, eds.), Am. Chem. Soc. Symp. Ser. 421, Washington, DC, pp. 18–41.

Georghiou, G. P., 1994. Principles of insecticides resistance management. *Phytoprotection* 75(Suppl.): 51–59.

Georghiou, G. P., and Lagunes-Tejeda, A., 1991. The occurrence of resistance to pesticides in arthropods. Food Agric. Organ. UN. Rome AGPP/MISC/91-1, 318 pp.

Georghiou, G. P., and Taylor, C. E., 1976. Pesticide resistance as an evolutionary phenomenon. Proc. XVth Int. Congr. Entomol. Washington, DC, pp. 759–785.

Georghiou, G. P., and Taylor, C. E., 1977. Genetic and biological influences in the evolution of insecticide resistance. *J. Econ. Entomol.* 70: 653–658.

Georghiou, G. P., and Taylor, C. E., 1986. Factors influencing the evolution of resistance. In: *Pesticide resistance: strategies and tactics for management.* (E. H. Glass, ed.), Nat. Acad. Sci., Washington, DC, pp. 157–169.

Georghiou, G. P., and Wirth, M., 1997. The influence of single vs multiple toxins of *Bacillus thuringiensis* subsp. *israelensis* on the development of resistance in *Culex quinquefasciatus* (Diptera: Culicidae). *Appl. Environ. Microbiol.* 63(3–4): 1095–1101.

Georghiou, G. P., Ariaratnam, V., Pasternak, M. E., and Chi, S. L., 1975. Evidence of organophosphorus multiresistance in *Culex pipiens pipiens* in California. *Proc. Pap. Calif. Mosq. Control Assoc.* 43: 41–44.

Georghiou, G. P., Lagunes-Tejeda, A., and Baker, J. D., 1983. Effect of insecticide rotations on evolution of resistance, in: *5th Int. Congr. Pestic. Chem. (IUPAC), Kyoto.* (J. Miayamoto, and P. C. Kearney, eds.), Pergamon Press, Oxford, UK, pp. 183–189.

Gerberg, E. J., 1970. Manual for mosquito rearing and experimental techniques. *J. Am. Mosq. Control Assoc.* 5: 1–109.

Gerberg, E. J., and Visser, W. M., 1978. Preliminary field trial for the biological control of *Aedes aegypti* by means of *Toxorhynchites brevipalpis*, a predatory mosquito larva. *Mosq. News* 38: 197–200.

Gerberich, J. B., and Laird, M., 1985. Larvivorous fish in the biocontrol of mosquitoes, with a selected bibliography of recent literature, in: *Integrated mosquito control methodologies*, Volume 2. (M. Laird, and J. Miles, eds.), Academic Press, London, pp. 47–58.

Gillett, J. D., 1955. Variation in the hatching-response of *Aedes* eggs. *Bull. Ent. Res.* 46: 241–253.

Gillett, J. D., 1972. *Common African mosquitos and their medical importance.* 106 pp., illus., London.

Gillett, J. D., 1983. Abdominal pulses in newly emerged mosquitoes *Aedes aegypti. Mosq. News* 43: 359–361.

Gillies, M., 1972. Some aspects of mosquito behavior in relation to the transmission of parasites. *Zool. J. Linn. Soc.*, Suppl. No. 1(51): 69–81.

Gillies, M. T., 1978. Confidence and doubt in the history of anopheline control. Medical Entomology Centenary, Symposium proceedings, The Royal Society of Tropical Medicine and Hygiene, pp. 26–33.

Gillies, M. T., and de Meillon, B., 1968. *The Anophelinae of Africa south of the Sahara* (Ethopian zoogeographical

region). 2nd ed., South Afr. Inst. Med. Res. Publ. 54: 343 pp.

Gilot, B., Ain, G., Pautou, G., and Gruffaz, R., 1976. Les Culicides de la Region Rhone-Alpes: bilan de dix annees d'observation. *Bull. Soc. ent. France.* 81: 235–245.

Gjullin, C. M., and Stage, H. H., 1950. Studies on *Aedes vexans* (Meig.) and *Aedes sticticus* (Meig.), flood water mosquitoes, in the Lower Columbia River Valley. *Ann. Ent. Soc. Am.* 43: 262–275.

Gjullin, C. M., Hegarty, C. P., and Bollen, W. B., 1941. The necessity of a low oxygen concentration for the hatching of *Aedes* eggs (Diptera: Culicidae). *J. Cell. Comp. Physiol.* 17: 193–202.

Gjullin, C. M., Sailer, R. I., Stone, A., and Travis, B. V., 1961. *The mosquitoes of Alaska.* U.S. Dep. Agric. Handb. 182: 1–98.

Glancoy, B. M., White, A. C., Husman, C. N., and Salmeda, J., 1968. Low volume application of insecticides for control of adult mosquitoes. *Mosq. News* 216: 356–359.

Glick, J. I., 1992. Illustrated key to the females of *Anopheles* of southwestern Asia and Egypt (Diptera: Culicidae). *Mosq. Syst.* 24(2): 125–153.

Gligic, A., and Adamovic, Z. R., 1976. Isolation of Tahyna virus from *Aedes vexans* mosquitoes in Serbia. *Mikrobiologija* 13(2): 119–129.

Gliniewicz, A., Krzeminska, A., and Sawicka, B., 1998. Mosquito control in Poland in the areas affected by flood in July/August 1997. *Acta Parasitol Port.* 5(1): 39.

Gloor, S., Stutz, H. P. B., and Ziswiler, V., 1995. Nutritional habits of the Noctule bat *Nyctalus noctula* (Schreber, 1774) in Switzerland. *Myotis* 32–33: 231–242.

Goldberg, L. H., and Margalit, J., 1977. A bacterial spore demonstrating rapid larvicidal activity against *Anopheles sergentii, Uranotaenia unguiculata, Culex univittatus, Aedes aegypti* and *Culex pipiens. Mosq. News* 37: 355–358.

Gould, F., Martinez-Ramirez, A., Anderson, A., Ferre, J., Silva, F. J., and Moar, W. J., 1992. Broad-spectrum resistance to *Bacillus thuringiensis* toxins in *Heliothis virescens. Proc. Nat. Acad. Sci. USA* 89: 7986–7990.

Grandes, A. E., and Sagrado, E. A., 1988. The susceptibility of mosquitoes to insecticides in Salamanca Province, Spain. *J. Am. Mosq. Control Assoc.* 4(2): 168–172.

Gratz, N. G., 1991. Emergency control of *Aedes aegypti* as a disease vector in urban areas. *J. Am. Mosq. Control Assoc.* 7: 69–72.

Gratz, N. G., 1999. Emerging and resurging vector-borne diseases. *Ann. Rev. Entomol.* 44: 51–75.

Graziosi, C., Sakai, R. K., and Romans, P., 1990. Method for *in situ* hybridization to polytene chromosomes from ovarian nurse cells of *Anopheles gambiae* (Diptera: Culicidae). *J. Med. Entomol.* 27: 905–912.

Green, C. A., 1972. Cytological maps for the practical identification of females of the three freshwater species of the *Anopheles gambiae* complex. *Ann. Trop. Med. Parasitol.* 66: 143–147.

Green, C. A., and Hunt, R. H., 1980. Interpretation of variation in ovarian polytene chromosomes of *Anopheles funestus* Giles, and *An. parensis* Gillies. *Genetica* 51: 187–195.

Gresikova, M., Sekeyova, M., Batikova, M., and Bielikova, V., 1973. Isolation of Sindbis virus from the organs of a hamster in east Slovakia, in: *1. Internationales Arbeitskolloquium über Naturherde von Infektionskrankheiten in Zentraleuropa, 17.–19. April 1973, Illmitz and Graz*, pp. 59–63.

Guelmino, D., 1956. The sequence of physical and chemical changes in the biotops of anofelines larves. *Recueil des travaux de l'Acad. Serbe des Sc. LII—Institut de recherches medicales.* No. 2: 79–87.

Guille, G., 1975. Recherces eco-ethologiques sur *Coquillettidia (Coquillettidia) richiardii*, (Ficalbi), 1889 (Diptera: Culicidae) du littoral Mediterraneen Francais. I.—Techniques d'etude et morphologie. *Ann. Sci. Nat. Zool.* 12th serie, 17: 229–272.

Guille, G., 1976. Recherces eco-ethologiques sur *Coq. richiardii* du littoral Mediterraneen Francais. *Ann. Sci. Nat.* 18: 5–112.

Gupta, R. K., Sweeney, A. W., Rutledge, L. C., Cooper, R. D., Frances, S. P., and Westrom, D. R., 1987. Effectivness of controlled-release personal-use arthropod repellents and permethrin-impregnated clothing in the field. *J. Am. Mosq. Control Assoc.* 3: 556–560.

Gutsevich, A. V., 1976. On polytypical species of mosquitoes (Diptera: Culicidae). I. *Anopheles hyrcanus* (Pallas, 1771). *Parazitologiya* 10: 148–153.

Gutsevich, A. V., 1977. On polytypical species of mosquitoes (Culicidae). II. *Aedes caspius* (Pallas, 1771). *Parazitologiya* 11(1): 48–51.

Gutsevich, A. V., and Dubitzkiy, A. M., 1987. New species of mosquitoes in the fauna of the USSR. *Mosq. Syst* 19: 1–92.

Gutsevich, A. V., and Goritskaya, V. V., 1970. Occurrence of the mosquito *Aedes thibaulti* (Diptera, Culicidae) in the Soviet Union. *Parazitologiya* 4: 72–73.

Gutsevich, A. V., Monchadskii, A. S., and Shtakel'berg, A. A., 1971. Fauna SSSR Vol. 3 (4). Family Culicidae. 384 pp., Leningrad. Akad. Nauk. SSSR. Zool. Inst. N. S. No. 100. (English translation: Israel Program for Scientific Translations, Jerusalem 1974.)

Hackett, L. W., 1937. *Malaria in Europe.* An ecological study. Oxford Univ. Press, 336 pp.

Hackett, L. W., and Bates, M., 1938. The laboratory for mosquito research in Albania. *Trans. Third Congr. Trop. Med. Hyg.* 2: 113–123.

Hackett, L. W., and Missiroli, A., 1935. The varieties of *Anopheles maculipennis* and their relation to the distribution of malaria in Europe. *Riv. Malariol.* 14: 45–109.

Hallinan, E., Lorimer, N., and Rai, K. S., 1977. Genetic manipulation of *Aedes aegypti*, in: *A cytogenetic study of radiation induced translocations in DELHI strain.* Proc. XV Internat. Cong. Entomol. Washington, DC, Entomology Society of America, pp. 117–128.

Halouzka, J., Pejcoch, M., Hubalek, Z., and Knoz, J., 1991. Isolation of Tahyna virus from biting midges (Diptera: Ceratopogonidae) in Czecho-Slovakia. *Acta Virol.* 35: 247–251.

Halstead, S. B., 1980. Dengue haemorrhagic fever—a public health problem and a field for research. *Bull. WHO* 58(1): 1–21.

Halstead, S. B., 1982. WHO fights dengue haemorrhagic fever. *WHO Chron.* 38(2): 65–67.

Halstead, S. B., 1992. The XXth century dengue pandemic: need for surveillance and research. *World Health Stat. Q.* 45: 292–298.

Halstead, S. B., Walsh, J. A., and Warren, K. D., 1985. Good health at low cost. *Proceedings of a conference sponsored by the Rockefeller foundation*, Bellagio, Italy.

Hammon, W. McD., and Reeves, W. C., 1945. Recent advances in the epidemiology of the arthropod-borne encephalitis. *Am. J. Public Health* 35: 994–1004.

Hammon, W. McD., Reeves, W. C., and Sather, G., 1952. California encephalitis virus, a newly described agent. II. Isolations and attempts to identify and characterize the agent. *J. Immunol.* 69: 493–510.

Hannoun, C., Panthier, R., Mouchet, J., and Eouzan, J.-P., 1964. Isolement en France du virus West-Nile a'partir de malades et du vecteur *Culex modestus* Ficalbi. *C. R. Acad. Sci. (D.) (Paris)* 259: 4170–4172.

Hara, J., 1958. On the newly recorded mosquito, *Aedes (Aedes) rossicus*, D., G. and M., 1930 with the keys to the species belonging subgenus *Aedes* known from Japan (Diptera: Culicidae). Taxonomical and ecological studies on mosquitoes of Japan (Part 10). *Jap. J. Sanit. Zool.* 9: 23–27, illus.

Harbach, R. E., 1985. Pictorial key to the genera of mosquitoes, subgenera of *Culex* and the species of *Culex (Culex)* occurring in southwestern Asia and Egypt, with a note on the subgeneric placement of *Culex deserticola* (Diptera: Culicidae). *Mosq. Syst.* 17(2): 83–107.

Harbach, R. E., 1988. The mosquitoes of the subgenus *Culex* in southwestern Asia and Egypt (Diptera: Culicidae). *Contrib. Am. Entomol. Inst. (Ann Harbour)* 24(1): 1–240.

Harbach, R. E., and Kitching, I. J., 1998. Phylogeny and classification of the Culicidae. *Syst. Entomol.* 23: 327–370.

Harbach, R. E., and Knight, K. L., 1980. *Taxonomists' glossary of mosquito anatomy.* Plexus Publishing, Inc., New Jersey, 413 pp.

Harbach, R. E., and Knight, K. L., 1981. Corrections and additions to *Taxonomists' glossary of mosquito anatomy. Mosq. Syst.* 13: 201–217.

Harbach, R. E., Harrison, B. A., and Gad, A. M., 1984. *Culex (Culex) molestus* Forskal (Diptera: Culicidae): neotype designation, description, variation and taxonomic status. *Proc. Entomol. Soc. Wash.* 86: 521–542.

Harbach, R. E., Dahl, C., and White, G. B., 1985. *Culex (Culex) pipiens* Linnaeus (Diptera: Culicidae). Concepts, type designations and description. *Proc. Entomol. Soc. Wash.* 87: 1–24.

Harris, H., and Hopkinson, D. A., 1976. *Handbook of enzyme electrophoreses in human genetics.* North Holland Publishing Comp., Amsterdam, Oxford, 512 pp.

Harrison, B. A., 1972. A new Interpretation of affinities within the *Anopheles hyrcanus* complex in Southeast Asia. *Mosq. Syst.* 4: 73–83.

Hayes, R. O., Holden, P., and Mitchell, C. J., 1971. Effects on ultra-low volume applications of malathion in Hale County, Texas IV. Arbovirus studies. *J. Med. Ent.* 8(2): 183–188.

Hazelrigg, J. E., 1975. Laboratory colonization and sexing of *Notonecta unifasciata* (Guerin) reared on *Culex peus* Speiser. *Proc. Calif. Mosq. Control Assoc.* 43: 142–144.

Hazelrigg, J. E., 1976. Laboratory rate of predation of separate and mixed sexes of adult *Notonecta unifasciata* (Guerin) on fourth-instar larvae of *Culex peus* (Speiser). *Proc. Calif. Mosq. Control Assoc.* 44: 57–59.

Headlee, N. J., 1932. The development of mechanical equipment for sampling the mosquito fauna and some results of it's use. *Proc. Annu. Meet. N. J. Mosq. Exterm. Assoc.* 19: 106–128.

Hearle, E., 1926. The mosquitoes of the Lower Fraser Valley, British Columbia, and their control. *Nat. Res. Counc. Can. Rep.* 17: 1–94.

Hearle, E., 1929. The life history of *Aedes flavescens* Müller. *Trans. R. Soc. Can., Third Ser.* 23: 85–101.

Hemingway, J., Smith, C., Jayawardena, K. G. I., and Herath, P. R. J., 1986. Field and laboratory detection of the altered acetylcholinesterase resistance genes which confer organophosphate and carbamate resistance in mosquitoes (Diptera: Culicidae). *Bull. Entomol. Res.* 76: 559–565.

Hennig, W., 1966. *Phylogenetic systematics* (translated by D. D. Davis, and R. Zangerl), University of Illinois Press, Urbana, 263 pp.

Hennig, W., 1973. Diptera (Zweiflügler), in: *Handbuch der Zoologie* (J.-G. Helmcke, D. Starck, and H. Wermuth, eds.) IV (2), 31: 1–19.

Hertlein, B. C., Levy, R., and Miller, T. W., Jr., 1979. Recycling potential and selective retrieval of *Bacillus sphaericus* from soil in a mosquito habitat. *J. Invertebr. Pathol.* 33: 217–221.

Hervy, J. P., and Kambou, F., 1978. Village-scale evaluation of Abate for larval control of *Aedes aegypti* in Upper Volta, WHO/VBC 78: 694.

Hester, P. G., Dukes, J. C., Levy, R., Ruff, J. P., Hallmon, C. F., Olson, M. A., and Shaffer, K. R., 1989. Field evaluation of the phytotoxic effects of Arosurf MSF on selected species or aquatic vegetation. *J. Am. Mosq. Control Assoc.* 5(2): 272–274.

Hill, D. S., and Waller, J. M., 1982. *Pests and diseases of tropical crops.* Vol. 1: *Principles and methods of control.* Longman, London and New York.

Hillis, D. E., 1996. *Molecular systematics.* 2nd ed., Sunderland, Mass., 655 pp.

Hinman, E. H., 1934. Predators of the Culicidae (Mosquitoes). I. The predators of larvae and pupae, exclusive of fish. *J. Trop. Med. Hyg.* 37(9): 129–134.

Hintz, H. W., 1951. The role of certain arthropods in reducing mosquito populations of permanent ponds in Ohio. *Ohio J. Science* 51(5): 277–279.

Hoch, A. L., Gupta, R. K., and Weyandt, T. B., 1995. Laboratory evaluation of a new repellent camouflage face paint. *J. Am. Mosq. Control Assoc.* 11(2): 172–175.

Höfte, H., and Whiteley, H. R., 1989. Insecticidal crystal proteins of *Bacillus thuringiensis. Microbiol. Rev.* 53: 242–255.

Horsfall, R. W., 1955. *Mosquitoes, their bionomics and relation to desease.* Ronald Press Co., New York. 723 pp.

Horsfall, W. R., 1956a. A method for making a survey of floodwater mosquitoes. *Mosq. News* 16(2): 66–71.

Horsfall, W. R., 1956b. Eggs of floodwater mosquitoes. III. Conditioning and hatching of *Aedes vexans. Ann. Ent. Soc. Am.* 49: 66–71.

Horsfall, W. R., and Fowler, H. W., 1961. Eggs of floodwater mosquitoes. VIII. Effect of serial temperatures on conditioning of eggs of *Aedes stimulans* Walker (Diptera: Culicidae). *Ann. Ent. Soc. Am.* 54: 664–666.

Horsfall, W. R., and Ronquillo, C. M., 1970. Genesis of the reproductive system of mosquitoes. II. Male of *Aedes stimulans* (Walker). *J. Morph.* 131: 329–358.

Horsfall, W. R., Lum, P., and Henderson, L., 1958. Eggs of floodwater mosquitoes, V. Effect of oxygen on hatching of intact eggs. *Ann. Ent. Soc. Am.* 51.

Horsfall, W. R., Fowler, H. W., Moretti, L. J., and Larsen, J. R., 1973. *Bionomics and embryology of the inland floodwater mosquito* Aedes vexans. University of Illinois Press, Urbana, 211 pp.

Hougard, J.-M., and Back, C., 1992. Perspectives on the bacterial control of vectors in the tropics. *Parasitol. Today* 8: 364–366.

Howard, J. J., and Oliver, J., 1997. Impact of Naled (Dibrom 14) on the mosquito vectors of eastern equine encephalitis virus. *J. Am. Mosq. Control Assoc.* 13(4): 315–325.

Hoy, J. B., and Reed, D. E., 1971. The efficacy of mosquitofish for control of *Culex tarsalis* in California rice fields. *Mosq. News* 31: 567–572.

Huang, Y. M., 1972. Contributions to the mosquito fauna of Southeast Asia. XIV. The subgenus *Stegomyia* of *Aedes* in Southeast Asia. I—The *scutellaris* group of species. *Contr. Am. ent. Inst.* 9(1): 1–109, illus.

Huang, Y. M., 1977. Medical entomology studies—VIII. Notes on the taxonomic status of *Aedes vittatus* (Diptera: Culicidae). *Contr. Am. Ent. Inst* 14(1): 113–132, illus.

Huang, Y. M., 1979. Medical entomology studies—XI. The subgenus *Stegomyia* of *Aedes* in the Oriental region with keys to the species (Diptera: Culicidae). *Contr. Am. Ent. Inst* 15(6): 1–79.

Hubalek, Z., and Halouzka, J., 1999. West Nile Fever—a reemerging mosquito-borne viral disease in Europe. *Emerging Infectious Diseases* 5: 643–650.

Hunt, R. H., 1973. A cytological technique for the study of *Anopheles gambiae* complex. *Parasitologia* 15: 137–139.

Ibarra, J. E., and Federici, B. A., 1986. Isolation of a relatively nontoxic 65-kilodalton protein inclusion from the parasporal body of *Bacillus thuringiensis* subsp. *israelensis*. *J. Bacteriol.* 165(2): 527–533.

Ikeshoji, T., and Mulla, M. S., 1970. Oviposition attractants for four species of mosquitoes in natural breeding waters. *Ann. Ent. Soc. Am.* 63(5): 1322–1327.

Inci, R., Yildirim, M., Bagei, N., and Inci, S., 1992. Biological control of mosquito larvae by mosquitofish (*Gambusia affinis*) in the Batman-Siirt Arva, Turkiye. *Parazitoloji Dergisi* 16: 60–66.

Ishii, T., 1991. The *Culex pipiens* complex. An old but new insect pest. SP World, Osaka 18: 12–15.

Jaenson, T. G. T., 1985. Attraction of mammals to male mosquitoes with special reference to *Aedes diantaeus* in Sweden. *J. Am. Mosq. Control Assoc.* 1(2): 195–198.

Jaenson, T. G. T., 1987. Overwintering of *Culex* mosquitoes in Sweden and their potential as reservoirs of human pathogens. *Med. Vet. Entomol.* 1(2), 151–156.

Jaenson, T. G. T., Lokki, J., and Saura, A., 1986a. *Anopheles* (Diptera: Culicidae) and malaria in northern Europe, with special reference to Sweden. *J. Med. Entomol.* 23(1): 68–75.

Jaenson, T. G. T., Niklasson, B., and Henriksson, B., 1986b. Seasonal activity of mosquitoes in an Ockelbo disease endemic area in central Sweden. *J. Am. Mosq. Contr. Assoc.* 2(1): 18–28.

Jakob, W. L., and Brevier, G. A., 1969a. Application of ovitraps in the US. *Aedes aegypti* eradication program. *Mosq. News* 29: 55–62.

Jakob, W. L., and Brevier, G. A., 1969b. Evaluation of ovitraps in the US. *Aedes aegypti* eradication program. *Mosq. News* 29: 650–653.

James, H. G., 1961. Some predators of *Aedes stimulans* (Walk.) and *Aedes trichurus* (Dyar) (Diptera: Culicidae) in woodland pools. *Can. J. Zool.* 39: 533–540.

James, H. G., 1965. Predators of *Aedes atropalpus* (Coq.) (Diptera: Culicidae) and of other mosquitoes breeding in rock pools in Ontario. *Can. J. Zool.* 43: 155–159.

James, H. G., 1966. Insect predators of univoltine mosquitoes in woodland pools of the pre-cambrian shield in Ontario. *Canad. Ent.* 98: 550–555.

James, H. G., 1967. Seasonal activity of mosquito predators in woodland pools in Ontario. *Mosq. News* 27(4): 453–457.

Jenkins, D. W., 1964. Pathogens, parasites and predators of medically important arthropods: annotated list and bibliography. *Bull. WHO* 30(Suppl.): 1–150.

Jetten, T. H., and Takken, W., 1994. *Anophelism without malaria in Europe*. A review of the ecology and distribution of the genus *Anopheles* in Europe. Wageningen Agric. Univ. Papers 94-5. 69 pp.

Jetten, T. H., Martens, W. J. M., and Takken, W., 1996. Model simulation to estimate malaria risk under climate change. *J. Med. Entomol.* 33(3): 361–371.

John, K. N., Stoll, J. R., and Olson, J. K., 1987. The public's view of mosquito problems in an organised control district. *J. Am. Mosq. Cont. Assoc.* 3: 1–7.

Joslyn, D. J., and Fish, D., 1986. Adult dispersal of *Ae. communis* using Giemsa self-marking. *J. Am. Mosq. Control Assoc.* 2: 89–90.

Joubert, M. L., 1975. L'arbovirose West Nile, zoonose du midi mediterraneen de la France. *Bull. Acad. nat. Med.* 159(6): 499–503.

Jourbert, L., Oudar, J., Hannoun, C., Beytout, D., Corniou, B., Guillon, J. C., and Panthier, R., 1970. Epidemiologie du virus West Nile: Etude d'un foyer en Camargue. IV. La meningoencephalomyelite du cheval. *Ann. Inst. Pasteur* 118: 239–247.

Judson, C. L., 1960. The physiology of hatching of aedine mosquito eggs: Hatching stimulus. *Ann. Ent. Soc. Am.* 53.

Jupp, P. G., 1996. *Mosquitoes of Southern Africa. Culicinae and Toxorhynchitinae.* Ekogilde publishers, Hartebeespoort, South Africa, 156 pp.

Kaiser, A., Jerrentrup, H., Samanidou-Voyadjoglou, A., and Becker, N., 2001. Contribution to the distribution of European mosquitoes (Diptera: Culicidae): four new country records from northern Greece. *Europ. Mosq. Bull.* 10: 9–12.

Karabatsos, N., 1985. International catalogue of arboviruses: including certain other viruses of vertebrates, 3rd ed., Am. Soc. Trop. Med. Hyg., 147 pp.

Kasap, H., Kasap, M., Akbaba, M., Alpetekin, D., Demirhan, O., Lüleyap, Ü., Pazarbasi, A., Akdur, R., and Wade, J., 1992. Residual efficacy of Pyrimiphos Methyl (Actellic™) on *Anopheles sacharovi* in Cukurova, Turkey. *J. Am. Mosq. Control Assoc.* 8(1): 47–51.

Katterman, A. J., Jayawardena, K. G. I., and Hemingway, J., 1992. Purification and characterisation of a carbosylesterase involved in insecticide resistance from the mosquito *Culex quinquefasciatus*. *J. Biochem.* 287: 355–360.

Katterman, A. J., Karunaratne, S. H. P., Jayawardena, K. G. I., and Hemingway, J., 1993. Quantitative differences between populations of *Culex quinquefasciatus* in both the esterases A2 and B2 which are involved in insecticide resistance. *Pestic. Biochem. Physiol.* 47: 142–148.

Kay, B. H., Cabral, C. P., Sleigh, A. C., Brown, M. D., Ribeiro, Z. M., and Vasconcelos, A. W., 1992. Laboratory evaluation of Brazilian Mesocyclops (Copepoda: Cyclopidae) for mosquito control. *J. Med. Entomol.* 29: 599–602.

Kazantsev, B. N., 1931. Cvetoviye variacii bukharskiyh *Aedes caspius.* Parazitologicheskiy Sbornik Zoologicheskogo Instituta Akademii Nauk SSSR. Vol. 2: 85–90.

Kellen, W. R., and Meyers, C. M., 1964. *Bacillus sphaericus* Neide as a pathogen of mosquitoes. *J. Invert. Pathol.* 7: 442–448.

Kellogg, F. E., 1970. Water vapour and carbon dioxide receptors in *Aedes aegypti. J. Insect Physiol.* 16: 99–108.

Kerwin, J. L., 1992. EPA registers *Lagenidium giganteum* for mosquito control. *Soc. Inv. Path. Newsl.* 24(2): 8–9.

Kerwin, J. L., and Washino, R. K., 1988. Field evaluation of *Lagenidium giganteum* and description of a natural epizootic involving a new isolate of the fungus. *J. Med. Entomol.* 25: 452–460.

Keshishian, M., 1938. *Anopheles sogdianus* sp. nov. A new species of the *Anopheles* mosquito *An. sogdianus* sp. nov. in Tadjikistan. *Med. Parasit.*, Moscow 7: 888–896, illus.

Kettle, D. S., 1995. *Medical and Veterinary Entomology.* CAB International, 2nd ed., Oxon, UK, 725 pp.

Kidwell, M. G., and Ribeiro, J. M. C., 1992. Can transposable elements be used to drive disease refractoriness genes into vector populations? *Parasit. Today* 8: 325–329.

Kirchberg, E., and Petri, K., 1955. Über die Zusammenhänge zwischen Verbreitung und Überwinterungsmodus bei der Stechmücke *Aedes (Ochlerotatus) rusticus* Rossi. Beiträge zur Kenntnis der Culicidae. III. *Z. angew. Zool.* 42: 81–94.

Kirkpatrick, T. W., 1925. *The mosquitoes of Egypt.* Egyptian Govt. Anti-Malaria Commision, Cairo, 224 pp.

Kline, D. L., 1993. Small plot Evaluation of a sustained release sand granule formulation of methoprene (San 810 I 1.3 GR) for control of *Aedes taeniorhynchus. J. Am. Mosq. Control Assoc.* 9(2): 155–158.

Knight, K. L., 1951. The *Aedes (Ochlerotatus) punctor* subgroup in North America (Diptera, Culicidae). *Ann. Ent. Soc. Am.* 41: 87–99, illus.

Knight, K. L., 1978. Supplement to a catalog of the mosquitoes of the world (Diptera: Culicidae). Thomas Say Found., *J. Ent. Soc. Am.*, Vol. 6 (Suppl.), 107pp.

Knight, K. L., and Laffoon, J. L., 1970. A mosquito taxonomic glossary. III. Adult thorax. *Mosq. Syst. Newsletter.* 2(3): 132–146.

Knight, K. L., and Laffoon, J. L., 1971. A mosquito taxonomic glossary V. Abdomen (exept female genitalia). *Mosq. Syst. Newsletter* 3(1): 8–24.

Knight, K. L., and Stone, A., 1977. *A catalog of the mosquitoes of the world* (Diptera: Culicidae). 2nd ed., Thomas Say Found., *J. Ent. Soc. Am.*, Vol. 6, xi + 611p.

Knipling, E. F., 1955. Possibilities of insect control or eradication through the use of sexually sterile males. *J. Econ. Entomol.* 48: 459–462.

Knipling, E. F., 1959. Sterile-male method of population control. *Science* 130: 902–904.

Knoz, J., and Vanhara, J., 1982. The action of water management regulations in the region of South Moravia on the population of haematophagous arthropods in lowland forests. *Scripta Fac. Sci. Nat. Univ. Purk. Brun.* 12 (7): 321–334.

Kögel, F., 1984. Die Prädatoren der Stechmückenlarven im Ökosystem der Rheinauen. Ph.D. thesis, University of Heidelberg, 347 pp.

Korvenkontio, P., Lokki, J., Saura, A., and Ulmanen, I., 1979. *Anopheles maculipennis* complex (Diptera: Culicidae) in northern Europe: species diagnosis by egg structure and enzyme polimorphismus. *J. Med. Entomol.* 16(2): 169–170.

Kozuch, O., Labuda, M., and Nosek, J., 1978. Isolation of Sindbis virus from the frog *Rana ridibunda. Acta Virol.* (Praha) 22: 78.

Kramer, V. L., Garcia, R., and Colwell, A. E., 1987a. A preliminary evaluation of the mosquito fish and the inland silverside as mosquito control agents in wild rice fields. *Proc. Calif. Mosq. Vect. Control Assoc.* 55: 44.

Kramer, V. L., Garcia, R., and Colwell, A. E., 1987b. An evaluation of the mosquitofish, *Gambusia affinis*, and the inland silverside, *Menidia beryllina*, as mosquito control agents in California wild rice fields. *J. Am. Mosq. Control Assoc.* 3: 626–632.

Kramer, V. L., Garcia, R., and Colwell, A. E., 1988a. An evaluation of *Gambusia affinis* and *Bacillus thuringiensis* var. *israelensis* as mosquito control agents in California wild rice fields. *J. Am. Mosq. Control Assoc.* 4: 470–478.

Kramer, V. L., Garcia, R., and Colwell, A. E., 1988b. *Gambusia affinis* and *Bacillus thuringiensis* var. *israelensis* used jointly for mosquito control in wild rice. Mosq. Contr. Res. Univ. Calif., 1987 Annu. Rept.: 18–20.

Kramer, V. L., Carper, E. R., and Beesley, C., 1993. Control of *Aedes dorsalis* with sustained-release methoprene pellets in saltwater marsh. *J. Am. Mosq. Control Assoc.* 9(2): 127–130.

Krieg, A., 1986. *Bacillus thuringiensis*, ein mikrobielles Insektizid. *Acta Phytomedica* 10, 191 pp.

Kroeger, A., Dehlinger, U., Burkhardt, G., Anaya, H., and Becker, N., 1995. Community based dengue control in Columbia: people's knowledge and practice and the potential contribution of the biological larvicide *B. thuringiensis israelensis. Trop. Med. Parasitol.* 46: 241–246.

Kruppa, T. V., 1988. Vergleichende Untersuchungen zur Morphologie und Biologie von drei Arten des *Culex pipiens*—Komplexes. Ph.D. thesis, University of Hamburg, 140 pp.

Krzeminski, J., and Brodniewicz, A., 1969. Plaga komarów w uzdrowiskach i kapieliskach na polskim wybrzezu Baltyku w swietle badan w ostatnim dwudziestoleciu, *Balneologia Polska* 14(1–2): 93–104.

Kühlhorn, F., 1961. Untersuchungen über die Bedeutung verschiedener Vertreter der Hydrofauna und -flora als natürliche Begrenzungsfaktoren für Anopheles-Larven (Diptera: Culicidae). *Z. Angew. Zool.* 48: 129–161.

Kubica-Biernat, B., Gliniewicz, A., Kowalska, B., and Stanczak, J., 2000. Development of the mosquito integrated biological control (IBC) program in the Vistula-Spit (Mierzeja Wislana) region, northern Poland. *Proc. 13th Eur. SOVE Meeting*, 24–29 Sept. 2000, Belek, Turkey: 217.

Kubica-Biernat, B., Gliniewicz, A., Kowalska, B., and Stanczak, J., 2001. Mosquito abatement program in the Vistula-Spit (Mierzeja Wislana) region, northern Poland. *Mat. 3rd Int. Con. Vec. Ecol.*, 16–21 Sept. 2001, Barcelona, Spain: 58.

Kunz, C., 1969. Arbovirus-B-Infektionen, in: *Die Infektionskrankheiten des Menschen und ihre Erreger*, Bd. II. (A. Grumbach, and W. Kikuth, eds.), G. Thieme, Stuttgart, pp. 1595–1628.

Kurtak, D., Back, C., and Chalifour, A., 1989. Impact of Bti on black-fly control in the onchocerciasis control programm in West Africa. *Israel J. Entomol.* 23: 21–38.

Labuda, M., 1969. *Aedes (Ochlerotatus) zamitii*, member of *Aedes mariae* complex (Diptera: Culicidae) in Yugoslavia. Bioloski vestnik, Lubljana 2: 23–27.

Labuda, M., Kozuch, O., and Gresikova, M., 1974. Isolation of West Nile virus from *Aedes cantans* mosquitoes in west Slovakia. *Acta Virol.* (Praha) 18: 429–433.

Lac, J., 1958. Beitrag zur Nahrung der *Bombina bombina* L. *Biologia* (Bratislava) 13: 844–853 (Czech., with German summary).

LaCasse, W. J., and Yamaguti, S., 1950. *Mosquito fauna of Japan and Korea*. Off. Surgeon, 8th U.S. Army, Kyoto, Honshu. 34th ed., 268 pp., illus.

Lacey, L. A., 1990. Persistence and formulation of *Bacillus sphaericus*, in: *Bacterial control of mosquitoes and blackflies: biochemistry, genetics and applications of* Bacillus thuringiensis israelensis *and* Bacillus sphaericus. (H. de Barjac, and D. Sutherland, eds.), Rutgers Univ. Press, New Brunswick, NJ, pp. 284–294.

Lacey, L. A., and Lacey, C. M., 1990. The medical importance of riceland mosquitoes and their control using alternatives to chemical insecticide. *J. Am. Mosq. Contr. Assoc.*, Suppl. 2, 1–93.

Lacey, L. A., and Oldacre, S., 1983. The effect of temperature, larval ages and species of mosquito on the activity of an isolate of *Bacillus thuringiensis* var. *darmstadtiensis* toxic for mosquito larvae. *Mosq. News* 43: 176–180.

Lacey, L. A., and Undeen, A. H., 1986. Microbial control of black flies and mosquitoes. *Ann. Rev. Entomol.* 31: 265–296.

Laing, J. E., and Welch, H. E., 1963. A dolichopodid predacious on larvae of *Culex restuans* Theob. *Proc. Entomol. Soc. Ontario* 93: 89–90.

Laird, M., 1947. Some natural enemies of mosquitoes in the vicinity of Palmalmal, New Britain. *Trans. Roy. Soc. N. Z.* 76(3): 453–476.

Lambert, M., Pasteur, N., Rioux, J., Delabre-Belmonte, A., and Balard, Y., 1990. *Aedes caspius* (Pallas, 1771) et *Aedes dorsalis* (Meigen, 1830) (Diptera: Culicidae). Analyses morphologique et genetique de deux populations sympatriques. Preuves de l'isolement reproductif. *Ann. Soc. Ent. France* 26(3): 381–398.

Lamborn, R. H., 1890. *Dragon flies vs. mosquitoes*. Can the mosquito pest be mitigated? Studies in the life history of irritating insects, their natural enemies, and artificial checks by working entomologists. D. Appleton Co., New York, 202 pp.

Lamborn, W. A., 1920. Some further notes on the tsetse flies of Nyasaland. *Bull. Entomol. Res.* 11(2): 101–104.

Lane, C. J., 1992. *Toxorhynchites auranticauda* sp. n., a new Indonesian mosquito and the potential biocontrol agent. *Med. Vet. Entomol.* 6: 301–305.

Lane, J., 1982. *Aedes (Stegomyia) cretinus* Edwards 1921 (Diptera: Culicidae). *Mosq. Syst.* 14: 81–85

Laven, H., Cousserans, J., and Guille, G., 1972. Eradicating mosquitoes using translocations: A first field experiment. *Nature* 236: 456–457.

Legner, E. F., 1991. Formidable position on turbellarians as biological mosquito control agents. *Proc. Calif. Mosq. Vect. Contr. Assoc.* 59: 82–85.

Legner, E. F., 1995. Biological control of diptera of medical and veterinary importance. *J. Vector Ecol.* 20(1): 59–120.

Lehane, M. J., 1991. *Biology of blood-sucking insects*. Harper Collins Academic, London, UK, 288 pp.

Leiser, L. B., and Beier, J. C., 1982. A comparison of oviposition traps and New Jersey light traps for *Culex* population surveillance. *Mosq. News* 42: 391–395.

Lemenager, D. C., Bauer, S. D., and Kauffman, E. E., 1986. Abundance and distribution of immature *Culex tarsalis* and *Anopheles freeborni* in rice fields of the Sulter-Yuba M. A. D.: 1. Initial sampling to detect major mosquito producing rice fields, augmented by aldult light trapping. *Proc. Calif. Mosq. Vect. Control. Assoc.* 53: 101–104.

Lenhoff, H. M., 1978. The hydra as a biological mosquito control agent. Mosq. Contr. Res. Univ. Calif. Annu. Rept., pp. 58–61.

Levy, R., and Miller, T. W., Jr., 1978. Tolerance of the planarian *Dugesia dorotocephala* to high concentrations of pesticides and growth regulators. *Entomophaga* 23: 31–34.

Levy, R., Chizzonite, J. J., Garret, W. D., and Miller, T. W. Jr., 1981. Ground and aerial application of a monomolecular organic surface film to control salt-marsh mosquitoes in natural habitats of southwestern Florida. *Mosq. News* 41: 291–301.

Levy, R., Chizzonite, J. J., Garret, W. D., and Miller, T. W. Jr., 1982. Efficacy of the organic surface film isosteryl alcochol containing two oxyethylene groups for control of *Culex* and *Psorophera* mosquitoes: Laboratory and field studies. *Mosq. News* 42: 1–11.

Lillie, T. H., Schreck, C. A., and Rahe, A. J., 1988. Effectivness of personal protection against mosquitoes in Alaska. *J. Med. Entomol.* 25: 475–478.

Linley, J. R., Parson, R. E., and Winner, R. A., 1988. Evaluation of ULV naled applied simultaneously against caged adult *Aedes taeniorhynchus* and *Culicoides furens*. *J. Am. Mosq. Control Assoc.* 4(3): 326–333.

Lloyd, L., 1987. An alternative to insect control by "mosquitofish" *Gambusia affinis*, in: *Proc. 4th Symposium Arbovirus Research in Australia* (1986) (T. D. St. Georg, B. H. Kay, and J. Blok, eds.) Q.I.M.R., Brisbane, 4: 156–163.

Logan, J. A. (ed.), 1953. The Sardinian project. An experiment in the eradication of an indigenous malarious vector. Amer. J. Hyg. Monogr. Ser. No. 20. Johns Hopkins Pr., Baltimore, 415 pp.

Lok, J. B., 1988. Dirofilaria spp.: taxonomy and distribution, in: *Dirofilariasis*. (P. F. L. Boreham, and R. B. Atwell, eds.), CRC Press, Boca Raton, Florida.

Lozano, A., 1980. Arboviruses in Spain, in: *Arboviruses in the Mediterranean Countries*. (J. Vesenjak-Hirjan, J. S. Porterfield, and E. Arslanagic, eds.), Zbl. Bakt. Suppl. 9, Gustav Fischer Verlag, Stuttgart, pp. 143–144.

Ludwig, H. W., 1993. *Tiere in Bach, Fluß, Tümpel, See*. BLV-Bestimmungsbuch, BLV Verlagsgesellschaft mbH, München; Wien; Zürich, 255 pp.

Ludwig, M., Beck, M., Zgomba, M., and Becker, N., 1994. The impact of water quality on the persistence of *Bacillus sphaericus*. *Bull. Soc. Vector Ecol.*, 19(1): 43–48.

Lüthy, P., Annual reports from 1988 to 2000 on the mosquito control in the plain of Magadino, Ticino, Switzerland.

Lüthy, P., Annual reports from 1995 to 2000 on the mosquito control at the lake of Gruyère, Canton Fribourg, Switzerland.

Lüthy, P., 2001. La lotta biologica control le zanzare alle Bolle di Magadino, in: *Contributo alla conoscenza delle Bolle di Magadino*. (N. Patocchi, ed.), pp. 139–145.

Lüthy, P., and Wolfersberger, M. G., 2000. Pathogenesis of *Bacillus thuringiensis* toxins, in: *Entopathogenic bacteria form laboratory to field application.* (J.-F. Charles, A. Delecluse, and C. Nielson-LeRoux, eds.), Kluwer Academic Publishers, Dordrecht, Boston, London, pp. 167–180.

Luh, P., and Shih, J., 1958. New *Aedes* and *Armigeres* subspecies from Guangxi (in Chinese). *Military Med.* 1: 222–226.

Lundström, J. O., 1994. Vector competence of western European mosquitoes for arboviruses: A review of field and experimental studies. *Bull. Soc. Vect. Ecol.* 19: 23–36.

Lundström, J. O., 1999. Mosquito-borne viruses in Western Europe: A review. *J. Vect. Ecol.* 24(1): 1–39.

Lundström, J. O., Vene, S., Espmark, A., Engvall, M., and Niklasson, B. 1991. Geographical and temporal distribution of Ockelbo disease in Sweden. *Epidem. Infect.* 106: 567–574.

Lundström, J. O., Turell, M. J., and Niklasson, B., 1993. Viremia in three orders of birds (Anseriformes, Galliformes and Passeriformes) inoculated with Ockelbo virus. *J. Wildl. Dis.* 29: 189–195.

Lvov, D. K., 1956. Ueber die Artselbststaendigkeit von *Aedes esoensis* Yam. (Diptera: Culicidae). *Revta. Ent. U.R.S.S.* 35: 929–934, illus.

Lvov, D. K., Skvortsova, T. M., Brerezina, L. K., Gromashevsky, V. L., Yakolev, B. I., Gushchin, B. V., Aristova, V. A., Sidorova, G. A., Gushchina, E. L., Klimenko, S. M., Lvov, S. D., Khutoretskaya, N. I., Myasinkova, A., and Khizhnyakova, T. M., 1984. Isolation of Karelian fever agent from *Aedes communis* mosquitoes. *Lancet* II: 399–400.

MacCormack, C., and Snow, R. W., 1986. Gambian cultural preferences in the use of insecticide treated bed nets. *J. Trop. Med. Hyg.* 89: 295.

MacDonald, G., 1957. *The epidemiology and control of malaria.* University Press, London, Oxford.

Maddrell, S. H. P., 1980. The insect neuroendocrine system as a target for insecticides, in: *Insect neurobiology and pesticide action.* Society of Chemical Industry, London, pp. 329–334.

Madon, M. B., Mulla, M. S., Shaw, M. W., Kluh, S., and Hazelrigg, J. E., 2002. Introduction of *Aedes albopictus* (Skuse) in southern California and potential for its establishment. *J. Vector Ecol.* 27(1): 149–154.

Magnarelli, L. A., 1979. Diurnal nectar-feeding of *Aedes cantator* and *Ae. sollicitans* (Diptera: Culicidae). *Environ. Entomol.* 8: 949–955.

Malkova, D., Danielova, V., Holubova, J., and Marhoul, Z., 1986. *Less known arboviruses of Central Europe. A new arbovirus Lednice.* Academia Publishing House of the Czechoslovak Academy of Science, Prague, 75 pp.

Mandoul, M., Dubos, M., Moulinier, C., and De Cournuaud, M., 1968. Une enquete sur les culicides de Charente-Maritime; presence du genre *Mansonia* Blanchard 1901: *M. (Coquillettidia) richardii* (Ficalbi) 1896 et *M. (C.) buxtoni* (Edwards) 1923. *Bull. Soc. Path. exot.* 61: 282–288, illus.

Manson-Bahr, P. E. C., and Bell, D. R., 1987. *Manson's Tropical Diseases.* Bailliere-Tindall, London.

Marchant, P., Eling, W., van Gemert, G.-J., Leake, C. J., and Curtis, C. F., 1998. Could British mosquitoes transmit falciparum malaria? *Parsitology Today* 14(9): 344–345.

Marchi, A., and Munstermann, L. E., 1987. The mosquitoes of Sardinia: species records 35 years after the malaria eradication campaign. *Med. Vet. Ent.* 1: 89–96.

Margalit, J., and Dean, D., 1985. The story of *Bacillus thuringiensis israelensis* (B.t.i.). *J. Am. Mosq. Control Assoc.* 1: 1–7.

Marshall, J. F., 1938. *The British mosquitoes.* Brit. Mus. (Nat. Hist.), London, xi + 341 pp.

Marshall, J. F., and Staley, J., 1933. *Theobaldia (Culicella) litorea* (Shute) n. sp. (Diptera: Culicidae). *Parasitology* 25: 119–126.

Martini, E., 1920a. Macedonische Culicinae. *Zeitschr. wiss. Insectenbiologie* (1919) 15: 119–120.

Martini, E., 1920b. Über Stechmücken, besonders deren europäische Arten und ihre Bekämpfung. *Beih. Arch. Schiffs- u. Tropenhyg.* 24(1): 1–267.

Martini, E., 1926. Ueber die Steckmücken der Umgebung von Saratow. *Arb. Biol. Wolga-Sta.* 8: 189–227.

Martini, E., 1927. Über zwei neue Stechmücken aus Anatolien. *Arch. Schiffs- und Tropenhyg.* 31: 386–390, illus.

Martini, E., 1931. Culicidae, in: *Die Fliegen der palaearktischen Region.* (E. Linder, ed.), Volumes 11 and 12, Stuttgart, 398 pp.

Martini, E., 1937. Praktische Fragen in der Stechmückenbekämpfung. *Verh. dtsch. Ges. f. angew. Ent.*: 40–59.

Martini, E., Missiroli, A., and Hackett, L. W., 1931. Versuche zum Rassenproblem des *Anopheles maculipennis. Arch. f. Schiffs- u. Tropenhyg.* 35: 622–643.

Maslov, A. V., 1967. Bloodsucking mosquitoes of the subtribe *Culiseta* (Diptera: Culicidae) of the world fauna. Akad. Nauk S.S.S.R., Opred. 93: 1–182, illus.

Mattingly, P. F., 1953. A change of name among the British mosquitoes. *Proc. R. Ent. Soc. London.* B. 22: 106–108.

Mattingly, P. F., 1954. Notes on the subgenus *Stegomyia* (Diptera: Culicidae), with a description of a new species. *Ann. Trop. Med. Parasit.* 48(3): 259–270.

Mattingly, P. F., 1955. Le sous-genre Neoculex (Diptera: Culicidae) dans la sous-region mediterraneenne. I.—Espece, sous-espece et synonymie nouvelles. *Ann. Parasit. hum. comp.* 30: 374–388, illus.

Mattingly, P. F., 1965. The culicine mosquitoes of the Indomalayan Area. Part VI: Genus *Aedes* Meigen, subgenus *Stegomyia* Theobald (Groups A, B and D). Brit. Mus. Nat. Hist., London, 67 pp.

Mattingly, P. F., 1969. *The biology of mosquito-borne disease.* Am. Elsevier Publ. Co., Inc., New York., 184 pp.

Mattingly, P. F., 1970. Mosquito eggs X. Oviposition in Neoculex. *Mosq. Syst. Newsl.* 2(4): 158–159.

McAney, C., Shiel, C., Fairley, J., 1991. The analysis of bat droppings. Occasional publication of the Mammal Society 14: 1–48, London.

McGaughey, W. H., 1985. Insect resistance to the biological insecticide *Bacillus thuringiensis. Science* 229: 193–195.

McGaughey, W. H., and Whalon, M. E., 1992. Managing insect resistance to *Bacillus thuringiensis* toxins. *Science* (Washington, DC) 258: 1451–1455.

McGovern, T. P., Schreck, C. E., and Jackson, J., 1984. Mosquito repellents: N,N-dimethylbenzamides, N,N-dimethylbenzeneacetamides, and other selected N,N-dimethylcarboxamides as repellents for *Aedes aegypti, Anopheles quadrimaculatus* and *Anopheles albimanus. Mosq. News* 44: 11–16.

McIntosh, B. M., 1975. Mosquitoes as vectors of viruses in Southern Africa. Dept. Agr. Techn. Services, Entomology Memoire No. 3. Pretoria, South Africa, 37 pp.

McIver, S. B., 1982. Sensilla of mosquitoes (Diptera: Culicidae). *J. Med. Entomol.* 19: 489–535.

McKenzie, C. L., and Byford, R. L., 1993. Continous, alternating, and mixed insecticides affect development of resistance in the horn fly (Diptera: Muscidae). *J. Econ. Entomol.* 86: 1040–1048.

McLaughlin, G. A., 1973. History of pyrethrum, in: *Pyrethrum, the Natural Insecticide.* (J. E. Casida, ed.), Academic Press, New York, pp. 3–16.

McLintock, J., Burton, A. N., McKiel, J. A., Hall, R. R., and Rempel, J. G., 1970. Known mosquito hosts of western equine virus in Saskatchewan. *J. Med. Ent.* 7(4): 446–454.

Medschid, E., 1928. Über *Aedes lepidonotus* Edw. und *Aedes refiki* n. sp. *Arch. Schiffs- u. Tropenhyg.* 32: 306–315.

Meisch, M. V., Mekk, C. L., Brown, J. R., and Nunez, R. D., 1997. Field efficacy of two formulations of Permanone against *Culex quinquefasciatus* and *Anopheles quadrimaculatus. J. Am. Mosq. Control Assoc.* 13(4): 311–314.

Mekuria, Y., Gwinn, T. A., Williams, D. C., and Tidwell, M. A., 1991. Insecticides Susceptibility of *Aedes aegypti* from Santo Domingo, Dominican Republic. *J. Am. Mosq. Control Assoc.* 7(1): 69–73.

Mian, L. S., Mulla, M. S., and Chaney, J. D., 1985. Biological strategies for control of mosquitoes associated with aquaphyte treatment of waste water. University of California, Mosquito Control Research, Ann. Report, pp. 91–92.

Mian, L. S., Mulla, M. S., and Wilson, B. S., 1986. Studies on potential biological control agents of immature mosquitoes in sewage wastewater in southern California. *J. Am. Mosq. Control. Assoc.* 2: 329–335.

Michelsen, V., 1996. Neodiptera: New insights into the adult morphology and higher level phylogeny of Diptera (Insecta). *Zool. Journ. Linnean Soc.* 112: 71–102.

Mihalyi, F., 1955. *Aedes hungaricus* n. sp. (Culicidae, Diptera). *Ann. Hist.-Nat. Mus. Nat. Hung.*, ser. nov. 6: 343–345, illus.

Mihalyi, F., 1959. Die tiergeographische Verteilung der Stechmückenfauna Ungarns. *Acta zool. ent. Acad. Scient. Hung.* 4: 393–403.

Mihalyi, F., 1961. Description of the larva of *Aedes (Ochlerotatus) hungaricus* Mihalyi (Diptera: Culicidae). *Acta Zool. Acad. Sci. Hung.* 7:231–233.

Milankov, V., Petric, D., Vujic, A., Vapa, L., and Kerenji, A., 1998. Variation of the siphonal characters in sympatric populations of *Aedes caspius* (Pallas, 1771) and *Aedes dorsalis* (Meigen, 1830). *Europ. Mosq. Bull.* 2: 20–23.

Miller, B. R., Crabtree, M. B., and Savage, H. M., 1996. Phylogeny of fourteen *Culex* mosquito species, including the *Culex pipiens* complex, inferred from the internal transcribed spacers of ribisomal DNA. *Insect Molecular Biol.* 5(2): 93–107.

Miller, T. A., 1980. *Neurohormonal Techniques in Insects.* Springer Verlag, New York, Heidelberg and Berlin, 282 pp.

Miller, T. A., 1988. Mechanisms of resistance to pyrethroid insecticides. *Parasitology Today* 4(7): S8–S12.

Minar, J., 1978. Aktivnost nekotorih vidov komarov (Culicidae) v klimatichestih usloviah Chehii (Czechoslovakia). *Parazitologia*. Akademia Nauk SSSR. XII (3): 226–232.

Minar, J., 1981. Results of Czechoslovak-Iranian entomological expeditions to Iran. Diptera: Culicidae, Oestridae. *Acta Ent. et Mus. Nat.*, Prague 40: 83–84.

Mitchell, C. J., 1995. Geographic spread of *Aedes albopictus* and potential for involvement in arbovirus cycles in the Mediterranean basin. *J. Vect. Ecol.* 20: 44–58.

Mitchell, C. J., and Briegel, H., 1989. Inability of diapausing *Culex pipiens* (Diptera: Culicidae) to use blood for producing lipid reserves for overwinter survival. *J. Med. Entomol.* 26(4): 318–326.

Mitchell, C. J., Lvov, S. D., Savage, H. M., Calisher, C. H., Smith, G. C., Lvov, D. K., and Gubler, D. J., 1993. Vector and host relationships of California serogroup viruses in western Siberia. *Am. J. Trop. Med. Hyg.* 49: 53–62.

Miura, T., and Takahashi, R. M., 1975. Effects of the IGR, TH-6040, on non-target organisms when utilized as a mosquito control agent. *Mosq. News* 35: 154–159.

Miura, T., and Takahashi, R. M., 1985. A laboratory study of crustacean predation on mosquito larvae. *Proc. Calif. Mosq. Contr. Assoc.* 52: 94–97.

Miura, T., Takahashi, R. M., and Mulligan, F. S., 1980. Effects of the bacterial mosquito larvicide, *Bacillus thuringiensis* serotype H-14 on selected aquatic organisms. *Mosq. News* 40: 619–622.

Miyagi, I., Toma, T., and Mogi, M., 1992. Biological control of container-breeding mosquitoes, *Aedes albopictus* and *Culex quinquefasciatus*, in a Japanese island by release of *Toxorhynchites splendens* adults. *Med. Vet. Entomol.* 6: 290–300.

Mogi, M., 1978. Population studies on mosquitoes in the rice field area of Nagasaki, Japan, especially on *Culex tritaeniorhynchus. Trop. Med.* 20: 173–263.

Mogi, M., Choochote, W., Khambooruang, C., and Suwanpanit, P., 1990. Applicability of presence-absence and sequential sampling for ovitrap surveillance of *Aedes* (Diptera: Culicidae) in Chiang Mai, northern Thailand. *J. med. Entomol.* 27: 509–514.

Mohrig, W., 1969. Die Culiciden Deutschlands. *Parasitol. Schriftenreihe*, Heft 18, 260 pp.

Molloy, D., and Jamnback, H., 1981. Field evaluation on *Bacillus thuringeinsis* var. *israelensis* as a blackfly biocontrol agent and its effect on non-target stream insects. *J. Econ. Entomol.* 74: 314–318.

Monath, T. P., 1988. *The arboviruses: Epidemiology and ecology*. Vols 1–5, CRC Press, Boca Raton, FL.

Monchadskii, A. S., 1951. The larvae of bloodsucking mosquitoes of the USSR and adjoining countries (Subfam. Culicinae). Tabl. anal. Faune U.R.S.S. 37: 1–383, illus.

Moore, C. G., 1999. *Aedes albopictus* in the United States: Current status and prospects for further spread. *J. Am. Mosq. Control Assoc.* 15: 221–227.

Moore, C. G., and Mitchell, C. J., 1997. *Aedes albopictus* in the United States: Ten year presence and public health implications. *Emerg. Infect. Diseases* 3:329–334.

Mouchet, J., 1987. Delthametrin impregnated bednets, an alternative for mosquito and malaria control, in: VIII Congreso Latinoamericano de Parasitologia, 1. Congreso Guatemalteco de Parasitologia y Medicina Tropical, Guatemala, 201 pp.

Mouchet, J., Rageau, J., Laumond, C., Hannoun, C., Beytout, D., Oudar, J., Corniou B., and Chippaux, A., 1970. Epidemiologie du virus West Nile: etude d'un foyer en Camargue. V. Le vecteur: *Culex modestus* Ficalbi (Diptera: Culicidae). *Ann. Inst. Pasteur* 118: 839–855.

Mount, G. A., Lofgren, C. S., Pierce, N. W., and Husman, C. N., 1968. Ultra-low volume nonthermal aerosols of malathion and naled for adult mosquito control. *Mosq. News* 28: 99–103.

Mulla, M. S., 1967a. Biological activity of surfactants and some chemical intermediates agains pre-imaginal mosquitoes. *Proc. Calif. Mosq. Contr. Assoc.* 35: 111–117.

Mulla, M. S., 1967b. Biocidal and biostatic activity of aliphatic amines against southern house mosquito larvae and pupae. *J. Econ. Entomol.* 60: 515–522.

Mulla, M. S., 1990. Activity, field efficacy, and the use of *Bacillus thuringiensis israelensis* against mosquitoes, in: *Bacterial control of mosquitoes and blackflies: biochemistry, genetics and applications of Bacillus thuringiensis israelensis and Bacillus sphaericus.* (H. de Barjac, and D. Sutherland, eds.), Rutgers Univ. Press, New Brunswick, NJ.

Mulla, M. S., 1994. Mosquito control then, now, and in the future. *J. Am. Mosq. Control Assoc.* 10(4): 574–584.

Mulla, M. S., 1995. The future of insect growth regulators in vector control. *J. Am. Mosq. Control Assoc.* 11(2): 269–274.

Mulla, M. S., and Darwazeh, H. A., 1975. Activity and longevity of insect growth regulators against mosquitoes. *J. Econ. Ent.* 68: 791–794.

Mulla, M. S., and Darwazeh, H. A., 1981. Efficacy of petroleum larvicidal oils and their impact on some aquatic non-target organisms. *Proc. Calif. Mosq. Vector Control Assoc.* 49: 84–87.

Mulla, M. S., Navvab-Gojrati, H. A., and Darwazeh, H. A., 1978. Toxicity of mosquito larvicidal pyrethroids to four species of fresh-water fishes. *Environ. Ent.* 7: 428–430.

Mulla, M. S., Darwazeh, H. A., and Dhillon, M. S., 1981. Impact and joint action of decamethrin and permethrin and freshwater fishes on mosquitoes. *Bull. Environm. Contam. Toxicol.* 26: 689–695.

Mulla, M. S., Federici, B. A., and Darwazeh, H. A., 1982. Larvicidal efficacy of *Bacillus thuringiensis* serotype H-14 against stagnant water mosquitoes and its effects on non-target organisms. *Env. Entomol.* 11: 788–795.

Mulla, M. S., Darwazeh, H. A., and Luna, L. L., 1983. Monolayer films as mosquito control agents and their effects on non-target organisms. *Mosq. News* 43: 489–495.

Mulla, M. S., Darwazeh, H. A., Les, E., and Kennedy, B., 1985. Laboratory and field evaluation of the IGR Fenoxycarb against mosquitoes. *J. Am. Mosq. Control Assoc.* 1: 442–448.

Mulla, M. S., Darwazeh, H. A., and Schreiber, E. T., 1989. Impact of new insect growth regulators and their formulations on mosquito larval development in impoundment and floodwater habitats. *J. Am. Mosq. Control Assoc.* 5: 15–20.

Mulla, M. S., Darwazeh, H. A., and Zgomba, M., 1990. Effect of some environmental factors on the efficacy of *Bacillus sphaericus* 2362 and *Bacillus thuringiensis* (H-14) against mosquitoes. *Bull. Soc. Vector Ecol.* 15: 166–175.

Mullen, G. R., 1975. Predation by water mites (Acarina: Hydrachnellae) on the immature stages of mosquitoes. *Mosq. News* 35(2): 168–171.

Mulligan III., F. S., and Schaefer, C. H., 1990. Efficacy of a juvenile hormone mimic, Pyriproxyfen (S-31183), for mosquito control in dairy wastewater lagoons. *J. Am. Mosq. Control Assoc.* 6(1): 89–92.

Mulligan III., F. S., Schaefer, C. H., and Wilder, W. H., 1980. Efficacy and persistence of *Bacillus sphaericus* and *B. thuringiensis* H-14 against mosquitoes under laboratory and field conditions. *J. Econ. Entomol.* 73: 684–688.

Munsterman, L. E., 1995. Mosquito systematics, current status, new trends, associated complications. *J. Vector Ecol.* 20(2): 129–138.

Munstermann, L. E., Marchi, A., Sabatini, A., and Coluzzi, M., 1985. Polytene chromosomes of *Orthopodomyia pulcripalpis* (Diptera: Culicidae). *Parassitologia* 27: 267–277.

Murdoch, W. W., Bence, J. R., and Chesson, J. A., 1984. Effects of the general predator, Notonecta (Hemiptera) upon a freshwater community. *J. Anim. Ecol.* 53: 791–808.

Murlis, J., 1986. The structure of odour plumes, in: *Mechanisms in insect olfaction.* (T. L. Payne, M. C. Birch, and C. E. J. Kennedy, eds.), Clarendon Press, Oxford.

Murphy, F. A., Fauquet, C. M., Bishop, D. H. L., Ghabrial, S. A., Jarvis, A. W., Martelli, G. P., Mayo, M. A., and Summers, M. D., 1995. *Virus Taxonomy—Classification and Nomenclature of Viruses.* Sixth Report of the International Committee on Taxonomy of Viruses. Springer-Verlag, Wien, New York, 586 pp.

Najera, J. A., and Zaim, M., 2002. Malavia vector control. Decision making criteria and procedures for judicious use of insecticides. WHO/CDS/WHOPES/2002.5, 106 pp.

Natvig, L. R., 1948. Contributions to the knowledge of the Danish and Fennoscandian mosquitoes—Culicini. *Suppl. Norsk ent. Tidsskr.* I, 567 pp.

Navon, A., and Ascher, K. R. S., 2000. *Bioassays of entomopathogenic microbes and nematodes.* CABI Publishing, UK., 324 pp.

Nayar, J. K., and Sauerman, D. M., 1975. The effects of nutrition on survival and fecundity in Florida mosquitoes. Part 2: Utilization of a blood meal for survival. *J. Med. Entomol.* 12: 99–103.

Nelson, F. R. S., 1977. Predation on mosquito larvae by beetle larvae, *Hydrophilus triangularis* and *Dytiscus marginalis. Mosq. News* 37: 628–630.

Nicolescu, G., 1998. A general characterisation of the mosquito fauna (Diptera: Culicidae) in the endemic area for West Nile virus in the south of Romania. *Europ. Mosq. Bull.* 2: 13–19.

Nielsen, L. T., 1957. Notes on the flight ranges of Rocky Mountain mosquitoes of the genus *Aedes. Proc. Utah Acad. Arts. Sci. Letters* 34: 27–29.

Nielsen-LeRoux, C., Charles, J. F., Thiery, I., and Georghiou, G. P. 1995. Resistance in a laboratory population of *Culex quinquefasciatus* (Diptera: Culicidae) to *Bacillus sphaericus* binary toxin is due to a change in the receptor on midgut brush-border membranes. *Eur. J. Biochem.* 228: 206–210.

Nielson, E. T., and Nielson, A. T., 1953. Field observations on the habits of *Aedes taeniorhynchus. Ecology*, 34(1): 141–156.

Niklasson, B., Espmark, Ä., LeDuck, J. W., Gargan, T. P., Ennis, W. A., Tesh, R. B., and Main, Jr., A. J., 1984. Association of a Sindbis-like virus with Ockelbo disease in Sweden. *Am. J. Trop. Med. Hyg.* 33: 1212–1217.

Norder, H., Lundström, J. O., Kozuch, O., and Magnius, L. O., 1996. Genetic relatedness of Sindbis virus strains from Europe, Middle East and Africa. *Virology* 222: 440–445.

O'Brien, R. D., Hilton, B. D., and Gilmour, L., 1966. The reaction of carbamates with cholinesterase. *Mol. Pharmacol.* 2: 593–605.

O'Meara, G. F., 1985. Gonotrophic interactions in mosquitoes: Kicking the blood-feeding habit. *Florida Entomologist* 68(1): 122–133.

O'Meara, G. F., Vose, F. E., and Carlson, D. B., 1989. Environmental factors influencing oviposition by *Culex* (*Culex*) (Diptera: Culicidae) in two types of traps. *J. Med. Entomol.* 26: 528–534.

Oljenicek, J., and Zoulova, A., 1994. Variation in the morphology of male antenna among strains of *Culex molestus* and *C. pipiens* mosquitoes. *Akaieka Newsletter* 16(1): 1–4, (Japan).

Oosterbrook, P., and Cortney, G., 1995. Phylogeny of the nematocerous families of Diptera (Insecta). *Zool. Journ. Linnean Soc.* 115: 267–311.

Oppenoorth, F. J., 1985. Biochemistry of genetics of insecticides, in: *Comprehensive insect physiology, biochemistry and pharmacology.* (G. A. Kerkut, and L. I. Gilbert, eds.), Pergamon Press, Oxford, pp. 731–775.

Pandazis, G., 1935. La faune des Culicides de Grece. *Acta lnst. Mus. Zool. Univ. Athens* 1: 1–27.

Pant, C. P., Mathis, H. L., Nelson, W., and Phantumachinda, B., 1974. A large-scale field trial of ultra-low volume fenitrothion for sustained control of *Aedes aegypti, Bulletin World Health Organisation* 51: 409–419.

Panthier, R., Hannoun, C., Beytout, D., and Mouchet, J., 1968. Epidemiologie du virus West Nile. Etude d'un foyer en Camargue. III. Les maladies humaines. *Ann. Inst. Pasteur* 115: 435–445.

Papadopoulos, O., 1980. Arbovirus problems in Greece, in: *Arboviruses in the Mediterranean Countries.* (J. Vesenjak-Hirjan, J. S. Porterfield, and E. Arslanagic, eds.), Zbl. Bakt. Suppl. 9, Gustav Fischer Verlag, Stuttgart, pp. 117–121.

Papaevangelou, G., and Halstead, S. B., 1977. Infections with two dengue viruses in Greece in the 20th century. Did dengue hemorhagic fever occur in the 1928 epidemic? *J. Trop. Med. Hyg*: 80: 46–51.

Papierok, B., Croset, H., and Rioux, J. A., 1975. Estimation de l'effectif des populations larvaires d' *Aedes (O.) cataphylla* Dyar, 1916 (Diptera: Culicidae), II, Methode utilisant le'coup de louche' ou "dipping". Cah. ORSTOM, ser. *Entomol. Med. Parasit.*, 13, 47–51.

Parker, K. R., and Mant, M. J., 1979. Effects of tsetse salivary gland homogenate on coagulation and fibrinolysis. *Thrombosis and Haemostasis* 42: 743–751.

Parrish, D. W., 1959. The mosquitoes of Turkey. *Mosq. News* 19: 264–266.

Pasteur, N., and Georghiou, G. P., 1989. Improved filter paper test for detecting and quantifying increased esterase activity in organophosphate resistant mosquitoes (Diptera: Culicidae). *J. Econ. Entomol.* 82: 347–353.

Pasteur, N., Rioux, J.-A., Guilvard, E., Pech-Perieres, M.-J., and Verdier, J.-M., 1977. Existence chez *Aedes (Ochlerotatus) detritus* (Haliday 1833) (Diptera: Culicidae) de Carmargue de deux formes sympathriques et sexuellement isolees (especes jumelles). *Ann. Parasitol.* 52: 325–337.

Pasteur, N., and Sinegre, G., 1978. Chlorpyrifos (Dursban®) resistance in *Culex pipiens pipiens* L. from Southern France: inheritance and linkage. *Experientia* 34: 709–710.

Perich, M. J., Tidwell, M. A., Williams, D. C., Sardelis, M. R., Pena, C. J., Mandeville, D., and Boober, L. R., 1990. Comparison of ground and aerial ultra-low volume applications of Malathion against *Aedes aegypti* in Santo Domingo, Dominican Republic. *J. Am. Mosq. Control Assoc.* 6: 1–7.

Petersen, J. J., 1980. Mass production of the mosquito parasite *Romanomermis culicivorax*: effect of density. *J. Nematology* 12: 45–48.

Peterson, B. V., 1960. Notes on some natural enemies of Utah black flies (Diptera: Simuliidae). *Canad. Entomol.* 92: 266–274.

Petrić, D. 1989. Seasonal and daily mosquito (Diptera: Culicidae) activity in Vojvodina. Ph.D. thesis. University of Novi Sad. Yugoslavia, 134 pp.

Petrić, D., Pajovic, I., Ignjatovic Cupina, A., and Zgomba, M., 2001. *Aedes albopictus* (Skuse, 1984), a new mosquito species (Diptera: Culicidae) in the entomofauna of Yugoslavia (in Serbian). Abstract Volume. Symposia of Serbian Entomologist's 2001. Entomological Society of Serbia, Goc, pp 29.

Petrić, D., Zgomba, M., Bellini, R., Veronesi, R., Kaiser, A., and Becker, N., 1999. Validation of CO_2 trap data in three European regions. *Proc. 3rd Int. Conf. Insect Pests in Urban Environment*, Prague, Czech Rep., (W. H. Robinson, F. Rettich, and G. W. Rambow, eds.) pp. 437–445.

Peus, F., 1929. Beiträge zur Faunistik und Ökologie der einheimischen Culiciden. I. Teil. *Z. Desinfektor* 21: 76–98.

Peus, F., 1930. Bemerkungen über *Theobaldia subochrea* Edw. (Diptera: Culicidae). *Mitt. dtsch. ent. Ges.* 1: 52–60, illus.

Peus, F., 1933. Zur Kenntnis der Aedes-Arten des deutschen Faunengebietes. (Diptera: Culicidae). Die Weibchen der *Aedes communis*-Gruppe. *Konowia* 12: 145–159, illus.

Peus, F., 1935. *Theobaldia* (Subg. *Culicella*) *ochroptera* sp. n., eine bisher unbekannte Stechmücke. *Mar. Tierw.* 1: 113–121, illus.

Peus, F., 1937. *Aedes cyprius* Ludlow (=*A. freyi* Edwards) und seine Larve (Diptera: Culicidae). *Arch. Hydrobiol.* 31: 242–251.

Peus, F., 1940. Die Stechmückenplage und ihre Bekämpfung, II. Teil. Die *Aedes*-Mücken. *Z. f. Hygienische Zoologie* 32: 49–79.

Peus, F., 1951. Stechmücken. Die neue Brehm-Bücherei 22, 80 pp.

Peus, F., 1954. Über Stechmücken in Griechenland. (Diptera: Culicidae). *Bonn. zool. Beitr.* 1: 73–86.

Peus, F., 1967. Culicidae, in: *Limnofauna Europaea*, 1st edition. (J. Illies, ed.), Gustav Fischer Verlag, Stuttgart.

Peus, F., 1970. Bemerkenswerte Muecken am Tegeler Fliess. *Berliner Naturschutzblaetter.* 18–26, illus.

Peus, F., 1972. Über das Subgenus *Aedes* sensu stricto in Deutschland. (Diptera: Culicidae). *Z. angew. Ent.* 72(2): 177–194.

Peyton, E. L., 1972. A subgeneric classification of the genus *Uranotaenia* Lynch Arribalzaga, with a historical review and notes on other categories. *Mosq. Syst.* 4(2): 16–40.

Peyton, E. L., Campbell, S. R., Candeletti, T. M., Romanowski, M., and Crans, W. J., 1999. *Aedes (Finlaya) japonicus japonicus* (Theobald), a new introduction into the United States. *J. Am. Mosq. Contr. Assoc.* 15(2): 238–241.

Pilaski, J., 1987. Contributions to the ecology of Tahyna virus in Central Europe. *Bull. Soc. Vector Ecol.* 12(2): 544–553.

Pilaski, J., and Mackenstein, H., 1985. Nachweis des Tahyna-Virus bei Stechmücken in zwei verschiedenen europäischen Naturherden. *Zbl. Bakt. Hyg.*, I Abt. Orig. B 180: 394–420.

Pires, C. A., Ribeiro, H., Capela, R. A., and Ramos, H. Da Cunha., 1982. Research on the mosquitoes of Portugal (Diptera: Culicidae) VI—The mosquitoes of Alentejo. *Ann. Inst. Hig. Med. Trop.* 8: 79–102.

Plapp, F. W., and Hoyer, R. F., 1968. Possible pleiohopism of gene conferring resistance to DDT, DDT analogs and pyrethroids in the housfly and *Culex tarsalis. J. Econ. Entomol.* 61: 761–765.

Pletsch, D., 1965. Informe sobre una misión efectuada en España en Septiembreen–Noviembre de 1963 destinada a la certificación de la erradicación del paludismo. *Revista de Sanidad e Higiene Pública.* 7,8,9: 309–355.

Portaro, J. K., and Barr, A. R., 1975. "Curing" Wolbachia infections in *Culex pipiens. J. Med. Entomol.* 12: 265.

Post, L. C., DeJong, B. J., and Vincent, W. R., 1974. 1-(2,6-disubstituted benzoyl)-3 phenylurea insecticides: Inhibiters of chitin synthesis. *Pestic. Biochem. Physiol.* 4: 473–483.

Postiglione, M., Tabanli, B., and Ramsdale, C. D., 1973. The *Anopheles* of Turkey. *Riv. Parassit.* 34: 127–159.

Prasittisuk, C., and Busvine, J. R., 1977. DDT-resistant mosquito strains with cross-resistance to pyrethroids. *Pest. Sci.* 8: 527–533.

Pratt, H. D., and Jakob, W. L., 1967. Oviposition trap reference handbook. *Aedes aegypti* handbook series No. 6, National Communicable Disease Centre, 33 pp.

Pratt, J. J., Heterick, R. H., Harrison, J. B., and Haber, L., 1946. Tires as a factor in the transportation of mosquitoes by ships. *Mil. Surgeon* 99: 785–788.

Price, G. D., Smith, N., and Carlson, D. A., 1979. The attraction of female mosquitoes (*Anopheles quadrimaculatus* Say) to stored human emanations in conjunction with adjusted levels of relative humidity, temperature and carbon dioxide. *J. Chem. Ecol.* 5: 383–395.

Priest, F. G., 1992. Biological control of mosquitoes and other biting flies by *Bacillus sphaericus* and *Bacillus thuringiensis. J. Appl. Bacteriol.* 72: 357–369.

Provost, M. W., 1953. Motives behind mosquito flight. *Mosq. News* 13: 106–109.

Pruthi, H. S., 1928. Some insects and other enemies of mosquito larvae. *Indian J. Med. Res.* 16: 153–157.

Pucat, A. M., 1965. The functional morphology of the mouthparts of some mosquito larvae. *Quaest. Entomol.* 1: 41–86.

Puri, B., Henchal, E. A., and Burans, J., 1994. A rapid method for detection and identification of flaviviruses by polymerase chain reaction and nucleic acid hybridization. *Arch. Virol.* 134: 29–37.

Quelennec, G., 1988. Pyrethroids in the WHO Pesticide Evaluation Scheme (WHOPES). *Parasit. Today.* 4(7): 15–17.

Qureshi, A. H., and Bay, E. C., 1969. Some observations on *Hydra americana* Hyman as a predator of *Culex peus* Speiser mosquito larvae. *Mosq. News* 29(3): 465–471.

Rabb, R. L., 1972. Principles and concepts of pest management, in: *Implementing practical pest management strategies*, proceedings of a USDA Cooperative Extension Service workshop. Purdue University, West Lafayette, Indiana, pp. 6–29.

Rafatjah, H. A., 1982. Prospects and progress on IPM in worldwide malaria control. *Mosq. News* 42: 41–97.

Rai, K. S., 1995. Genetic control of vectors, in: *The biology of disease vectors.* (B. J. Beaty, and W. C. Marquardt, eds.), University Press of Colorado, Colorado, USA, pp. 564–574.

Rai, K. S., and McDonald, P. T., 1971. Chromosomal translocations and genetic control of *Aedes aegypti*, in: *The sterility principle for insect control or eradication.* Vienna: Internat. Atomic Energy Agency Press, pp. 437–452.

Ramos, H. Da Cunha., 1983. Contribuicao fara o estudo dos mosquitos limnodendrofilos de Portugal *Garcia de Orta, Serie de Zoologia* 11: 133–154.

Ramos, H. Da Cunha., Ribeiro, H., Pires, C. A., and Capela, R. A., 1977/78. Research on the mosquitoes of Portugal, (Diptera: Culicidae) II—The mosquitoes of Algarve. *Ann. Inst. Hig. Med. Trop.* 5(1–4): 238–256, 1977/78.

Ramos, H. Da Cunha, Ribeiro, H., Pires, C. A., and Capela, R. A., 1982. Research on the mosquitoes of Portugal (Diptera: Culicidae) VII—Two new Anopheline records. *Ann. Inst. Hig. Med. Trop.* 8: 103–109.

Ramos, H. Da Cunha., Lucientes, J., Blasco-Zumeta, J., Osacar, J., and Ribeiro, H., 1998. A new mosquito record for Spain (Diptera: Culicidae). *Abstracts of the XIth European SOVE meeting, Acta Parasitologica Portuguesa* 5: 21.

Ramsdale, C. D., 1975. Insecticide resistance in the *Anopheles* in Turkey. *Trans. R. Soc. Trop. Med. Hyg.* 69: 226–235.

Ramsdale, C. D., 1998. *Anopheles cinereus* Theobald 1901 and its synonym *hispaniola* Theobald 1903. *Europ. Mosq. Bull.* 2: 18–19.

Ramsdale, C. D., and Snow, K. R., 1999. A preliminary checklist of European mosquitoes. *Europ. Mosq. Bull.* 5: 25–35.

Ransinghe, L. E., and Georghiou, G. P., 1979. Comparative modification of insecticides resistance spectrum of *Culex p. fatigans* Wied. by selection with temephos and temephos/synergist combinations. *Pestic. Sci.* 10: 502–508.

Raymond, M., 1985. Presentation d'un programme d'analyse log-probit pour micro-ordinateur. Cah. ORSTOM, *Ser. Entomol. Med. Parasitol.* 22: 117–121.

Raymond, M., Fournier, D., Berge, J., Cuany, A., Bride, J. M., and Pasteur, N., 1985. Single mosquito test to determine genotypes with an acetylcholinesterase insensitive to inhibition by propoxur insecticide. *J. Am. Mosq. Contr. Assoc.* 1: 425–427.

Reeves, W. C., 1990. Epidemiology and control of mosqutio-borne arboviruses in California, 1943–1987. Calif. Mosq. Vector Control Assoc., Sacramento, CA, 508 pp.

Regner, S., 1969. Preliminary observation on rockpools in the region of Rovinj. *Thalassia Jugoslavica*, Vol. V: 283–292.

Reid, J. A., 1953. The *Anopheles hyrcanus* group in Southeast Asia (Diptera: Culicidae). *Bull. Ent. Res.* 44: 5–76.

Reinert, J. F., 1973. Contributions to the mosquito fauna of Southeast Asia—XVI. Genus *Aedes* Meigen, subgenus *Aedimorphus* Theobald in Southeast Asia. *Contr. Am. Ent. Inst.* 9(5): 1–218, illus.

Reinert, J. F., 1974. Medical entomology studies—I. A new interpretation of the subgenus *Verrallina* of the genus *Aedes* (Diptera: Culicidae). *Contr. Am. Ent. Inst.* 11(1): 1–249, illus.

Reinert, J. F., 1975. Mosquito generic and subgeneric abbreviations (Diptera: Culicidae). *Mosq. Syst.* 7: 105–110.

Reinert, J. F., 1999a. The subgenus *Rusticoidus* of genus *Aedes* (Diptera: Culicidae) in Europe and Asia. *Europ. Mosq. Bull.* 4: 1–7.

Reinert, J. F., 1999b. Descriptions of *Zavortinkius*, a new subgenus of *Aedes*, and the eleven included species from the afrotropical region (Diptera: Culicidae). *Contrib. Am. Entomol. Inst. (Ann Arbor)* 31(2): 1–105.

Reinert, J. F., 1999c. Restoration of *Verrallina* to generic rank in tribe Aedini (Diptera: Culicidae) and descriptions of the genus and three included subgenera. *Contrib. Am. Entomol. Inst. (Ann Arbor)* 31(3): 1–83.

Reinert, J. F., 2000a. Description of *Fredwardsius*, a new subgenus of *Aedes* (Diptera: Culicidae). *Eur. Mosq. Bull.* 4: 1–7.

Reinert, J. F., 2000b. Assignment of two North American species of *Aedes* to subgenus *Rusticoidus*. *J. Am. Mosq. Control Assoc.* 16: 42–43.

Reinert, J. F., 2000c. New classification of the composite genus *Aedes* (Diptera: Cuicidae: Aedini), elevation of subgenus *Ochlerotatus* to generic rank, reclassification of the other subgenera, and notes on certain subgenera and species. *J. Am. Mosq. Control Assoc.* 16(3): 175–188.

Reinert, J. F., 2001. Revised list of abbreviations for genera and subgenera of Culicidae (Diptera) and notes on generic and subgeneric changes. *J. Am. Mosq. Control Assoc.* 17(1): 51–55.

Reiter, P., 1978. Expanded polystyrene balls: an idea for mosquito control. *Ann. Trop. Med. Parasitol.* 72: 595.

Reiter, P., 1983. A portable, battery-powered trap for collecting gravid *Culex* mosquitoes. *Mosq. News* 43: 496–498.

Reiter, P., 1986. A standardized procedure for the quantitative surveillance of certain *Culex* mosquitoes by egg raft collection. *J. Am. Mosq. Control Assoc.* 2: 219–221.

Reiter, P., and Nathan, M. B., 2001. Guidelines for assessing the efficacy of insecticidal space sprays for control of the dengue vector *Aedes aegypti*. WHO/CDS/CPE/PVC/2001.1, 34 pp.

Reiter, P., Eliason, D. A., Francy, D. B., Moore, C. G., and Campos, E. G., 1990. Apparent influence of the stage of blood meal digestion on the efficacy of ground applied ULV aerosols for the control of urban *Culex* mosquitoes. I. Field evidence. *J. Am. Mosq. Control Assoc.* 6(3): 366–370.

Remington, C. L., 1945. The feeding habits of *Uranotaenia lowii* Theobald (Diptera: Culicidae). *Ent. News* 56: 32–37; 64–68.

Rempel, J. G., 1953. The mosquitoes of Saskatchewan. *Can. J. Zool.* 31: 433–509, illus.

Retnakaran, A., Grenett, J., and Ennis, T., 1985. Insect Growth Regulators, in: *Comprehensive insect physiology, biochemistry and pharmacology.* (G. A. Kerkut, and L. I. Gilbert, eds.), Volume 12, Pergamon Press, Oxford, pp. 529–601.

Rettich, F., 1979. A study on the mosquito fauna (Diptera: Culicidae) of the Hradec Kralove area. *Acta Univ. Carolinae—Biologica.* 377–385.

Ribeiro, H., Ramos, H. C., Capela, R. A., and Pires, C. A., 1977. Research on the mosquitoes of Portugal (Diptera: Culicidae) III—Further five new mosquito records. *Garcia de Orta, Ser. Zool.*, Lisboa 6(1–2): 51–60.

Ribeiro, H., Ramos, H. C., Pires, C. A., and Capela, R. A., 1980. Research on the mosquitoes of Portugal (Diptera: Culicidae) IV—Two new anopheline records. *Garcia de Orta Ser. Zool.*, Lisboa 9(1–2): 129–139.

Ribeiro, H., Ramos, H. C., and Pires, C. A., 1983. Contribuicao para o estudo das filariases animais em Portugal. *J. Soc. Cienc. Med.* Lisboa 147(2): 143–146.

Ribeiro, H., Ramos, H. C., Pires, C. A., and Capela, R. A., 1987. Research on the mosquitoes of Portugal (Diptera: Culicidae) IX—A new anopheline record. *Garcia de Orta, Ser. Zool.*, Lisboa 12(1–2), 1985 (1987): 105–112.

Ribeiro, H., Ramos, H. C., Pires, C. A., and Capela, R. A., 1988. An annotated checklist of the mosquitoes of continental Portugal (Diptera: Culicidae). *Actas III Congr. Iberico Entomol.*, Granada, pp. 233–253.

Ribeiro, H., Ramos, H. C., Capela, R. A., and Pires, C. A., 1989. Research on the mosquitoes of Portugal (Diptera: Culicidae) XI—The mosquitoes of the Beiras. *Garcia de Orta, Ser. Zool.*, Lisboa 16(1–2), 1989 (1992): 137–161.

Richards, C. S., 1956. *Aedes melanimon* Dyar and related species. *Can. Ent.* 88: 261–269.

Richards, G., 1981. Insect hormones in development. *Biol. Rev.* 56: 501–549.

Rioux, J. A., 1958. *Les culicides du "Midi" Mediterraneen.* Etude systematique et ecologique. *Encyc. Ent. (A)* 35: 303 pp.

Rioux, J. A., and Arnold, M., 1955. Les Culicides de Camargue (Etude systematique et ecologique). *Terre et la Vie* 4: 244–286.

Rioux, J. A., Croset, H., Pech-Perieres, J., Guilvard, E., and Belmonte, A., 1975. L'autogenese chez les dipteres culicides. *Ann. Parasitol. Hum. Comp.*, Tome I, No. 1: 130–140.

Rioux, J. A., Guilvard, E., and Pasteur, N., 1998. Description d'*Aedes (Ochlerotatus) colluzzi* n. sp. (Diptera, Culicidae), especes jumelles a du complexe detritus. *Parassitologia* 40: 353–360.

Riviere, F., Kay, B. H., Klein, J. M., and Sechan, Y., 1987a. *Mesocyclops aspericornis* (Copepoda) and *Bacillus thuringiensis* var. *israelensis* for the biological control of *Aedes* and *Culex* vectors (Diptera: Culicidae) breding in crap holes, tree holes and artificial containers. *J. Med. Entomol.* 24: 425–430.

Riviere, F. Y., Sechan, Y., and Kay, B. H., 1987b. The evaluation of predators for mosquito control in French Polynesia, in: *Proc. 4th Symposium Arbovirus Research in Australia (1986).* (T. D. St. George, B. H. Kay, and J. Blok, eds.), Q.I.M.R., Brisbane, pp. 150–154.

Roberts, G. M., 1989. The combination of *Bacillus thuringiensis* var. *israelensis* with a monomolecular film. *Israel J. Entomol.* 23: 95–97.

Robinson, G. G., 1939. The mouthparts and their function in the female mosquito, *Anopheles maculipennis. Parasitology* 31: 212–242.

Robinson, W. H., and Atkins, R., 1983. Attitudes and knowledge of urban home owners towards mosquitoes. *Mosq. News.* 43(1): 38–41.

Rodhain, F., and Hannoun, C., 1980. Present status of arboviruses in France, in: *Arboviruses in the mediterranean countries.* (J. Vesenjak-Hirjan, J. S. Porterfield, and E. Arslanagic, eds.), Zbl. Bakt. Suppl. 9, Gustav Fischer Verlag, Stuttgart, pp. 111–116.

Romeo Viamonte, J. M., 1950. Los anofelinos de Espana y de la zona espanola del Protectorado de Marruecos. Su relacion con la difusion del paludismo. *Revta Sanid. Hig. publ.*, Madr. 24: 213–295, illus.

Roush, R. T., and McKenzie, J. A., 1987. Ecological genetics of insecticides and acaricide resistance. *Ann. Rev. Entomol.* 32: 361–380.

Ruigt, G. S., 1985. Pyrethroids, in: *Comprehensive insect physiology, biochemistry and pharmacology.* (G. A. Kerkut, and L. I. Gilbert, eds.), Volume 12, Pergamon Press, Oxford, pp. 183–263.

Rutledge, L. C., and Gupta, R. K., 1996. Reanalysis of the C. G. Macnay Mosquito Repellent Data. *J. Vect. Ecol.* 21(2): 132–135.

Rutledge, L. C., Sofield, R. K., and Moussa, M. A., 1978. A bibliography of diethyl toluamide. *Bull. Entomol. Soc. Am.* 24: 431–439.

Rydell, J., 1986. Foraging and diet of the northern bat, *Eptesicus nilssoni*, in Sweden. *Holarctic Ecology* 9: 272–276.

Rydzanicz, K., Lonc, E., *et al.*, 2000. Proby integrowanej kontroli komarow na terenie miasta i okolic Wroclawia. Biuletyn PSPDDD 1/2000: 8–10.

Sabatinelli, G., Majori, G., Blanchy, S., Fayaerts, P., and Papakay, M., 1990. Testing of the larvivorous fish, *Poecilia reticulata* in the control of malaria in the Islamic Federal Republic of the Comoros. *Wld. Hlth. Org. WHO/MAL* 90.1060: 1–10 (in French).

Sabatini, A., Raineri, V., Trovato, G., and Coluzzi, M., 1990. *Aedes albopictus* in Italia e possibile diffusione della specie nell'area Mediterranea. *Parasitologia* 32: 301–304.

Sack, P., 1911. Aus dem Leben unserer Stechmücken. 42. Ber. Senckenberg. Naturf. Ges., pp. 309–322; Frankfurt/M.

Saether, O. A., 2000. Phylogeny of the Culicomorpha (Diptera). *Syst. Entomol.* 25: 223–234.

Sakellariou, M., and Lane, J., 1977. Notes on the Culicidae recorded in the Copais District of Greece, 1974–1976. *Annl. Inst. phytopath. Benaki* (N. S.) 11(4): 302–309.

Saliternik, Z., 1955. The specific biological characteristics of *Anopheles (Myzomyia) sergentii* (Theo.) and their correlations with malaria control in Israel. *Bull. Ent. Res.* 46: 445–462.

Saliternik, Z., 1957. *Anopheles sacharovi* Favre in Israel. *Riv. Parassit* 18: 249–265.

Sallum, M. A. M., Schultz, T. R., Wilkerson, R. C., 2000. Phylogeny of Anophelinae (Diptera: Culicidae) based on morphological characters. *Ann. Entomol. Soc. Am.* 93: 745–775.

Samanidou-Voyadjoglou, A., and Darsie, F. R., Jr., 1993. New country records for mosquito species in Greece. *J. Am. Mosq. Contr. Assoc.* 9(4): 465–466.

Sasa, M., and Kurihara, T., 1981. The use of poeciliid fish in the control of mosquitoes, in: *Biocontrol of medical and veterinary pests.* (M. Laird, ed.), Praeger, New York, pp. 36–53.

Sasa, M., Kano, R., and Takahashi, H., 1950. A revision of the adult Japanese mosquitoes of the genus *Aedes*, subgenus *Aedes* with description of two new species. *Jap. J. Exp. Med.* 20: 631–640, illus.

Sawicki, R. M., 1978. Unusual response of DDT—resistant housflies to carbinol analogues of DDT. *Nature* 275: 443–444.

Schaefer, C. H., and Dupras, E. F., Jr., 1970. Factors affecting the stability of Dursban in polluted water. *J. Econ. Entomol.* 63: 701–705.

Schaefer, C. H., Wilder, H., and Mulligan III., F. S., 1975. A practical evaluation of TH-6040 as a mosquito control agent in California. *J. Econ. Ent.* 68: 183–185.

Schaefer, C. H., Miura, T., Dupras, E. F., Jr., Mulligan III., F. S., and Wilder, W. H., 1988. Efficacy, nontarget effects and chemical persistence of S-31183, a promising mosquito (Diptera: Culicidae) control agent. *J. Econ. Entomol.* 81: 1648–1655.

Schäfer, M., Storch, V., Kaiser, A., Beck, M., and Becker, N., 1997. Dispersal behavior of adult snow-melt mosquitoes in the Upper Rhine Valley, Germany. *J. Vector Ecology* 22(1): 1–5.

Schaffner, F., 1998. A revised checklist of the French Culicidae. *Europ. Mosq. Bull.* 2: 1–9.

Schaffner, F., Bouletreau, B., Guillet, B., Guilloteau, J., and Karch, S., 2001. *Aedes albopictus* established in metropolitan France. *Europ. Mosq. Bull.* 9: 1–3.

Scherpner, C., 1960. Zur Ökologie und Biologie der Stechmücken des Gebietes von Frankfurt am Main (Diptera: Culicidae). *Mitt. Zool. Mus. Berlin* 36: 49–99.

Schnetter, W., and Engler, S., 1978. Oberflächenfilme zur Bekämpfung von Stechmücken, in: *Probleme der Insekten- und Zeckenbekämpfung.* (E. Döhring, and I. Iglisch, eds.), E. Schmidt Verlag, Berlin, pp. 115–121.

Schoenherr, A. A., 1981. The role of competition in the displacement of native fishes by introduced species, in: *Fishes in North American deserts.* (R. J. Naiman, and D. L. Stoltz, eds.), Wiley Interscience, New York, pp. 173–203.

Schreck, C. E., and Self, L. S., 1985. Bed nets that kill mosquitoes. *World Health Forum* 6: 342–344.

Schreck, C. E., Haile, D. G., and Kline, D. L., 1984. The effectiveness of permethrin and deet, alone or in combination for protection against *Aedes taeniorhynchus*. *Am. J. Trop. Med. Hyg.* 33: 725–730.

Schultz, G. H., Mulla, M. S., and Hwang, Y. S., 1983. Petroleum oil and nonanoinic acid as mosquitocides and oviposition repellents. *Mosq. News* 43: 315–318.

Schutz, S. J., and Eldridge, B. F., 1993. Biogeography of the *Aedes (Ochloretatus) communis* species complex (Diptera: Culicidae) in the western United States. *Mosq. Syst.* 25: 170–176.

Sebastian, A., Stein, M. M., Thu, M. M., and Corbet, P. S., 1990. Suppression of *Aedes aegypti* (Diptera: Culicidae) using augmentative release of dragonfly larvae (Odonata: Libellulidae) with community participation in Yangon, Myanmar. *Bull. Entomol. Res.* 80: 223–232.

Seguy, E., 1924. Les moustiques de l' Afrique Mineure, de l'Egypte et de la Syrie I. *Encycl. Ent (A)* 1: 1–257, illus. Paris.

Seguy, E., 1967. Dictionnaire des termes techniques d'entomologie elementaire. Encyclopedie Entomol., Serie A, Travaux generaux 41, 465 pp.

Senevet, G., and Andarelli, L., 1954. Le genre *Culex* Afrique du Nord. III. -Les adultes. *Arch. Inst. Pasteur Alger.* 32: 36–70, illus.

Senevet, G., and Andarelli, L., 1955. Races et varietes de l' *Anopheles claviger* Meigen 1804. *Arch. Inst. Pasteur Alger.* 33: 128–137.

Senevet, G., and Andarelli, L., 1956. Les *Anopheles* de l'Afrique du Nord et du Bassin Mediterraneen, Paris, Lechevalier, 280 pp.

Senevet, G., and Andarelli, L., 1959. Les moustiques de l'Afrique du Nord et du Bassin Mediterraneen. *Encyc. Ent. (A)* 37: 1–383.

Senevet, G., and Prunnelle, M., 1927. Une nouvelle espece d'anophele en Algerie, *Anopheles marteri* n. sp. *Arch. Inst. Pasteur Alger.* 5: 529–533, illus.

Service, M. W., 1965. Predators of the immature stages of *Aedes (Stegomyia) vittatus* (Bigot) (Diptera: Culicidae) in water-filled rock-pools in Northern Nigeria. WHO/EBL/33.65, 19 pp.

Service, M. W., 1968a. The taxonomy and biology of two sympatric sibling species of *Culex, C. pipiens* and *C. torrentium* (Diptera: Culicidae). *J. Zool.*, Lond. 156: 313–323.

Service, M. W., 1968b. The ecology of the immature stages of *Aedes detritus* (Diptera: Culicidae). *J. Appl. Ecol.* 5: 613–630.

Service, M. W., 1968c. Observations on feeding and oviposition in some British mosquitoes. *Entomologia Exp. Appl.* 11: 277–285.

Service, M. W., 1969. Observations on the ecology of some British mosquitoes. *Bull. Ent. Res.* 59: 161–194.

Service, M. W., 1970a. Studies on the biology and taxonomy of *Aedes (Stegomyia) vittatus* (Bigot) (Diptera: Culicidae) in northern Nigeria. *Trans. R. Ent. Soc. Lond.* 122: 101–143, illus., 2pls.

Service, M. W., 1970b. The taxonomy and distribution of two sibling mosquitoes, *Culiseta morsitans* (Theobald) and *C. litorea* (Shute). *J. Nat. Hist.* 4: 481–491, illus.

Service, M. W., 1971a. Flight periodicities and vertical distribution of *Aedes cantans* (Mg.), *Ae. geniculatus* (Ol.), *Anopheles plumbeus* Steph. and *Culex pipiens* L. (Diptera: Culicidae) in southern England. *Bull. Ent. Res.* 60: 639–651.

Service, M. W., 1971b. The daytime distribution of mosquitoes resting amongst vegetation. *J. Med. Ent.* 8: 271–278.

Service, M. W., 1973a. Study of the natural predators of *Aedes cantans* (Meigen) using the precipitin test. *J. Med. Entomol.* 10: 503–510.

Service, M. W., 1973b. The biology of *Anopheles claviger* Meigen (Diptera: Culicidae) in southern England. *Bull. Ent. Res.* 63: 347–359.

Service, M. W., 1977. Mortalities of the immature stages of species B of the *Anopheles gambiae* complex in Kenya: comparison between rice fields and temporary pools, identification of predators and effects of insecticidal spraying. *J. Med. Entomol.* 13: 535–545.

Service, M. W., 1983. Biological control of mosquitoes—has it a future? *Mosq. News* 43: 113–120.

Service, M. W., 1993. *Mosquito ecology: field sampling methods.* 2nd edition, Elsevier Science Publishers Ltd, Essex, UK, 988 pp.

Shannon, R. C., and Hadjinicolaou, J., 1937. Greek Culicidae which breed in tree-holes. *Acta Instituti et Musei Zoologici Universitatis Atheniensis.* Tom I, Fasc. 8: 173–178.

Sharma, V. P., Patterson, R. S., LaBrecque, G. C., and Singh, K. R. P., 1976. Three field release trials with chemosterilized *Culex pipiens fatigans* Wied., in a Delhi village. *J. Commun. Dis.* 8: 18–27.

Sharma, R. C., Yadav, R. S., and Sharma, V. P., 1985. Field trials on the application of expanded polysterene (EOS) beads in mosquito control. *Indian J. Malariol.* 22: 107.

Sharp, B. L., Le Sueur, D., and Bekker, P., 1990. Effect of DDT on survival and bloodfeeding success of *Anopheles arabiensis* in Norhern Kwazulu, Republic of South Africa. *J. Am. Mosq. Control Assoc.* 6(2): 197–202.

Shepard, H. H., 1951. *The chemistry and action of insecticides.* McGraw-Hill, New York, 504 pp.

Shevchenko, A. K., and Prudkina, N. S., 1973. On morphology of genitals in mosquito males from the *Aedes* genus. *Vestnik Zoologii* No. 6: 40–47, illus.

Shirako, Y., Niklasson, B. J., Dalrymple, M., Strauss, E. G., and Strauss, J. H., 1991. Structure of the Ockelbo virus genome and its relationship to other Sindbis viruses. *Virology* 182: 753–764.

Shour, M. H., and Crowder, L. A., 1989. Effects of pyrethroid insecticides on the common green lace wing. *J. Econ. Ent.* 73: 306–309.

Shtakelberg, A. A., 1937. Faune de l'URSS. Insectes, Dipteres. Fam. Culicidae (Subfam Culicinae). *Moskva AN SSSR*, Vol. 3(4), 257 pp.

Shute, P. G., 1928. A new variety of *Culicella morsitans* (Theobald) (Diptera). *Entomologist* 61: 186.

Shute, P. G., 1933. The life-history and habits of British mosquitoes in relation to their control by anti-larval operations. *J. Trop. Med. Hyg.* 36: 83–88.

Shute, P. G., 1954. Indigenous *P. vivax* malaria in London believed to have been transmitted by *A. plumbeus*. *Monthly Bull. M. of H. and P. H. Lab. Service*, 13, 48.

Sicart, M., 1940. Note sur la presence de *Culex (Culiciomyia) impudicus* Ficalbi en Tunisie. *Arch. Inst. Pasteur Tunis.* 29: 471–474, illus.

Sicart, M., 1951. Note sur la presence de *Culex mimeticus* Noe 1899, en Tunisie. *Bull. Soc. Sc. Nat. Tunisie* 4: 60–61.

Sicart, M., 1952. Description et etude de *Aedes pulchritarsis* (Rondani, 1872) recolte en Tunisie. *Bull. Soc. Sci. nat. Tunis.* 5: 95–101.

Siddall, J. B., 1976. Insect Growth Regulators and insect control: a critical apprisal. *Environ. Health Persp.* 14: 119–126.

Sinclair, B. J., 1992. A phylogenetic interpretation of the Brachycera (Diptera) based on the larval mandible and associated mouthpart structures. *Syst. Entomol.* 17: 233–252.

Sinegre, G., 1984. La résistance des Diptères Culicides en France, in: Colleque sur la réduction d'efficacoté des traitements insecticides et acaricides et problèmes de résistance, Paris, pp. 47–57.

Sinegre, G., Babinot, M., Quermel, J. M., and Gavon, B., 1994. First field occurrence of *Culex pipiens* resistance to *Bacillus sphaericus* in southern France. Abstr. VIIIth Eur. Meet. Society of Vector Ecology, Barcelona, Sept. 5–8, 17.

Singer, S., 1973. Insecticidal activity of recent bacterial isolates and their toxins against mosquito larvae. *Nature* (London) 244: 110–111.

Sirivanakarn, S., 1976. Medical entomology studies—III. A revision of the subgenus *Culex* in the Oriental region (Diptera: Culicidae). *Contr. Am. ent. Inst.* 12 (2): 1–272, illus.

Sirivanakarn, S., and White, G. B., 1978. Neotype designation of *Culex quinquefasciatus* Say (Diptera: Culicidae). *Proc. Entomol. Soc. Wash.* 80: 360–372.

Sjogren, R. D., 1971. Evaluation of the mosquitofish *Gambusia affinis* (Baird and Girard) and the common guppy *Poecilia reticulata* Peters for biological control of mosquitoes in dairy waste lagoons. Ph.D. Thesis, University of California, Riverside, 105 pp.

Sjogren, R. D., 1972. Minimum oxygen threshold of *Gambusia affinis* (Baird and Girard) and *Poecilia reticulata* Peters. *Proc. Calif. Mosq. Contr. Assoc.* 40: 124–26.

Sjogren, R. D., and Legner, E. F., 1989. Survival of the mosquito predator, *Notonecta unifasciata* (Guerin) (Hemiptera: Notonectidae) embryos at low thermal gradients. *Entomophaga* 34: 201–208.

Skierska, B., 1969. Larvae of Chaoborinae occurring in small reservoirs. I. Some observations on larvae of *Chaoborus crystallinus* (DeGeer, 1776) and on the possiblity of their predacity in relation to larvae of biting mosquitoes. *Bull. Inst. Mar. Med. Gdansk* 20: 101–108.

Sluka, F., 1969. The clinical picture of the Calovo virus infection, in: *Arboviruses of the California complex and the Bunyamwera group*. Proceedings of the symposium held at Smolenice near Bratislava October 18–21, 1966 (V. Bardos and coworkers, eds.), Publishing House of the Slovak Academy of Sciences, Bratislava, 337–339 pp.

Smith, C. N., Smith, N., Gouck, H. K., Weidhaas, D. H., Gilbert, I. H., Mayer, M. S., Smittle, B. J., and Hofbauer, A., 1970. L-lactic acid as a factor in the attraction of *Aedes aegypti* (Diptera: Culicidae) to human host. *Ann. Entomol. Soc. Am.* 63: 760–770.

Smith, K. G. V. (ed.), 1973. *Insects and other arthropods of medical significance*. British Museum Natural History, London, 561 pp.

Snodgrass, R. E., 1935. *Principles of insect morphology*. McGraw-Hill Book Company, New York, 667 pp.

Snow, K. R., 1985. A note on the spelling of the name *Orthopodomyia pulcripalpis* (Rondani, 1872). *Mosq. Syst.* 17: 361–362.

Snow, K. R., 1990. *Mosquitoes*. Naturalists' Handbooks 14. Richmond Publishing Co. Ltd., Slough, England, 66 pp.

Snow, K. R., and Ramsdale, C. D., 1999. Distribution chart for European mosquitoes. *Europ. Mosq. Bull.* 3: 14–31.

Sologor, E., and Petrusenko, A., 1973. On studying nutrition of Chiroptera order of the middle Dnieper area. *Zool. Rec. Kiew* 3: 45 (in Russian).

Spieckermann, D., and Ackermann R., 1972. Isolierung von Viren der California-Enzephalitis-Gruppe aus Stechmücken in Nord Bayern. *Zbl. Bakt. Hyg.* I. Orig. A 221: 283–295.

Spieckermann, D., and Ackermann R., 1974. Untersuchungen über Naturherde des Tahyna Virus in Süddeutschland. *Zbl. Bakt. Hyg.* I. Orig. A 228: 291–295.

Sprenger, D., and Wuithiranyagool, T., 1986. The discovery and distribution of *Aedes albopictus* in Harris County, Texas. *J. Am. Mosq. Contr. Assoc.* 2: 217–219.

Stegnii, V. N., and Kabanova, V. M., 1976. Cytological study of indigenous populations of the malarial mosquitoes in the territory of the USSR I. Identification of a new species of *Anopheles* in the *maculipennis* complex by cytodiagnostic method. Medskaya *Parasit.* 45: 192–198. [In Russian with English summary.] English translation: *Mosq. Syst* 10(1): 1–12, 1978.

Stewart, R. J., Schaefer, C. H., and Miura, T., 1983. Sampling *Culex tarsalis* immatures on rice fields treated with combinations of mosquitofish and *Bacillus thuringiensis* H-14 toxin. *J. Econ. Entomol.* 76: 91–95.

Strauss, J. H., and Strauss, E. G., 1994. The alphaviruses: Gene expression, replication, and evolution. *Microbiol. Rev.* 58: 491–562.

Strickman, D., 1980a. Stimuli affecting selection of oviposition sites by *Aedes vexans*: Moisture. *Mosq. News* 40: 236–245.

Strickman, D., 1980b. Stimuli affecting selection of oviposition sites by *Aedes vexans*: Conditioning of the soil. *Mosq. News* 40: 413–417.

Sudia, W. D., Newhouse, V. F., Calisher, C. H., and Chamberlain, R. W., 1971. California group arboviruses: isolations from mosquitoes in North America. *Mosq. News* 31(4): 576–600.

Sutcliffe, J. F., 1987. Distance orientation of biting flies to their hosts. *Insect Science and its Applications* 8: 611–616.

Suzzoni-Blatger, J., Cianchi, R., Bullini, L., and Coluzzi, M., 1990. Le complexe *maculipennis*: criteres morphologiques et enzymatiques de determination. *Ann. Parasitol. hum. comp.* 65: 37–40.

Sweeney, A. W., and Becnel, J. J., 1991. Potential of microsporidia for the biological control of mosquitoes. *Parasitol. Today* 7: 217–220.

Sweeney, A. W., Hazard, E. I., and Graham, M. F., 1985. Intermediate host for an *Amblyospora* sp. (Microspora) infecting the mosquito, *Culex annulirostris. J. Invertebr. Pathol.* 46: 98–102.

Swift, S. M., Racey, P. A., and Avery, M. I., 1985. Feeding ecology of *Pipistrellus pipistrellus* (Chiroptera: Vespertilionidae). *Myotis* 30: 7–74.

Szata, W., 1997. Zimnica w Polsce, *Przegl. Epid.* 51(1–2): 177–183.

Tabashnik, B. E., 1989. Managing resistance with multiple pesticide tactics: theory, evidence and recommendations. *J. Econ. Entomol.* 82: 1263–1269.

Tabashnik, B. E., Cushing, N. L., Finson, N., and Johnson, M. W., 1990. Development of resistance to *Bacillus thuringiensis* in field populations of *Plutella xylostella* in Hawaii. *J. Econ. Entomol.* 83: 1671–1676.

Tabibzadeh, I., Behbehani, C., and Nakhai, R., 1970. Use of *Gambusia* fish in the malaria eradication programme of Iran. *Bull. WHO.* 43: 623–626.

Takahashi, R. M., Wilder, W. H., and Miura, T., 1984. Field evaluation of ISA-20E for mosquito control and affects on aquatic non-target arthropods in experimental plots. *Mosq. News* 44: 363–367.

Takashima, I., and Rosen, L., 1989. Horizontal and vertical transmission of Japanese encephalitis virus by *Aedes japonicus* (Diptera: Culicidae). *J. Med. Entomol.* 26: 454–458.

Takken, W., 1991. The role of olfaction in host-seeking of mosquitoes: a review. *Insect Sci. Applic.* 12: 287–295.

Takken, W., and Kline, D. L., 1989. Carbon dioxide and 1-octen-3-ol as mosquito attractants. *J. Am. Mosq. Control. Assoc.* 5: 311–316.

Tanaka, K., Mizusawa, K., and Saugstad, E. S., 1975. A new species of the genus *Aedes (Aedes)* from Japan, with synonymical notes on Japanese species of the subgenus *Aedes* (Diptera: Culicidae). *Mosq. Syst.* 7(1): 41–58, illus. Addendum, 7(2): 174–177.

Tanaka, K., Mizusawa, K., and Saugstad, E. S., 1979. A revision of the adult and larval mosquitoes of Japan (including the Ryukyu Archipelago and the Ogasawara islands) and Korea (Diptera: Culicidae). *Contr. Am. Ent. Inst.* (Ann Harbor) 16: 1–987.

Taylor, M. F., Heckel, J., Brown, D. G., Kreitman, T. M., and Black, B., 1993. Linkage of pyrethroid insecticide resistance to a sodium channel locus in the tobacco budworm. *Insect Biochem. Mol. Biol.* 23: 763–775.

Taylor, P., 1980. Anticholinesterase agents, in: *The pharmacological basis of therapeutics.* (A. G. Gilman, L. S. Goodman, and A. Gilman, eds.), 6th edition, Macmillan, New York, pp. 100–119.

Telford, A. D., 1963. A consideration of diapause in *Aedes nigromaculis* and other aedine mosquitoes (Diptera: Culicidae). *Ann. Ent. Soc. Am.* 56(4): 409–418.

Tesh, R. B., 1990. Undifferentiated arboviral fevers, in: *Tropical and geographical medicine.* (K. S. Warren, and A. A. F. Mahmoud, eds.), McGraw-Hill, New York, pp. 685–691.

Thaggard, C. W., and Eliason, D. A., 1969. Field evaluation of components for an *Aedes aegypti* (L.) oviposition trap. *Mosq. News* 29: 608–612.

Thanabalu, T., Hindley, J., Jackson-Yap, J., and Berry, C., 1991. Cloning, sequencing and expression of a gene encoding a 100-kilodalton mosquitocidal toxin from *Bacillus sphaericus* SSII-1. *J. Bacteriol.* 173: 2776–2785.

Thorne, L., Garduno, F., and Thompson, T., 1986. Structural similarity between the Lepidoptera and

Diptera-specific insecticidal endotoxin genes of *Bacillus thuringiensis* subsp. *kurstaki* and *israelensis*. *J. Bacteriol.* 166, 801–811.

Tidwell, M. A., Williams, D. C., Gwinn, T. A., Pena, C. J., Tedders, S. H., Gonzalez, G.-E., and Mekuria, Y., 1994. Emergency control of *Aedes aegypti* in the Dominican Republic using the Scorpion™ 20, ULV forced-air generator. *J. Am. Mosq. Contr. Assoc.* 10(3): 403–406.

Tietze, N. S., and Mulla, M. S., 1987. Tadpole shrimp (*Triops longicaudatus*), new candidates as biological control agents for mosquitoes. *Calif. Mosq. and Vect. Contr. Assoc. Bio Briefs* 113(2): 1.

Tietze, N. S., and Mulla, M. S., 1991. Biological control of *Culex* mosquitoes (Diptera: Culicidae) by the tadpole shrimp, *Triops longicaudatus* (Notostraca, Triopsidae). *J. Med. Entomol.* 28: 24–31.

Tikasingh, E. S., 1992. Effects of *Toxorhynchites moctezuma* larval predation on *Aedes aegypti* populations: experimental evaluation. *Med. Vet. Entomol.* 6: 266–271.

Tischler, W., 1984. *Einführung in die Ökologie*. Gustav Fischer Verlag, Stuttgart.

Torres Canamares, F., 1946. Nuevos datos sobre el *Anopheles marteri* Sen. y Pru. en Espana. Se trata de una variedad? (Diptera: Culicidae). *Eos*, Madrid 22: 47–58, illus.

Torres Canamares, F., 1951. Una nueva especie de *Aedes* (Diptera: Culicidae). *Eos*, Madrid 27: 79–92.

Tovornik, D., 1974. Podatki o favni komarjev (Culicidae) v Sloveniji in ugotavljanje troficnih relacij za nekatere vrste. *Acta ent. Jugoslavica* 10(1–2): 85–89.

Tovornik, D., 1980. Podatki o prehranjevanju komarjev (Diptera: Culicidae), zbranih v kmetijskih naseljih v Ljubljanski okolici. Slovenska Akademija Znanosti i Umetnosti, Ljubljana, Razred za prirodoslovne vede. Razprave XXII/1: 1–39.

Traavik, T., Mehl, R., and Wiger, R., 1978. California encephalitis group viruses isolated from mosquitoes collected in southern and arctic Norway. *Acta Path. Microbiol. Scand.* Sect. B. 86: 335–341.

Traavik, T., Mehl, R., and Wiger, R., 1985. Mosquito-borne arboviruses in Norway: Further isolations and detection of antibodies to California encephalitis viruses in human, sheep and wildlife sera. *J. Hyg. Camb.* 94: 111–122.

Travis, B. V., 1953. Laboratory studies on the hatching of marsh-mosquito eggs. *Mosq. News* 13: 190–198.

Trpis, M., 1960. Stechmücken der Reisfelder und Möglichkeiten ihrer Bekämpfung. *Biologicke Prace*, Bratislava, 6: 117.

Trpis, M., 1962. Ökologische Analyse der Stechmückenpopulationen in der Donautiefebene in der Tschecheslowakei. *Biologicke Prace,* Bratislava 8: 1–115.

Trpis, M., 1981. Survivorship and age specific fertility of *Toxorhynchites brevipalpis* females (Diptera: Culicidae). *J. Med. Entomol.* 18: 481–486.

Tsai, T. F. F., Popovici, C., Cernescu, G., Campbell, L., and Nedelcu, N. I., 1998. West Nile encephalitis epidemic in southeastern Romania. *The Lancet.* 352(5)L: 767–771.

Twinn, C. R., 1931. Observations on some aquatic animal and plant enemies of mosquitoes. *Canad. Ent.* 63(3): 51–61.

Utrio, P., 1976. Identfication key to finnish mosquito larvae (Diptera: Culicidae). *Annal. Agr. Fenniae* 15: 128–136, illus.

Utrio, P., 1979. Geographic distribution of mosquitoes (Diptera: Culicidae) in eastern Fennoscandia. *Notul. Entomol.* 59: 105–123.

Van Thiel, P. H., 1933. Investigations on range and differentiation of *Anopheles maculipennis* races and their bearing on existence or absence of malaria in Italy. *Riv. Malariol.* 12: 281–318.

Verani, P., 1980. Arboviruses in Italy, in: *Arboviruses in the mediterranean countries.* (J. Vesenjak-Hirjan, J. S. Porterfield, and E. Arslanagic, eds.), Zbl. Bakt. Suppl. 9, Gustav Fischer Verlag, Stuttgart, pp. 123–128.

Vesenjak, H. J., Punda, P. V., and Dobe, M., 1991. Geographical distribution of arboviruses in Yugoslavia. *J. Hyg. Epidemiol. Microbiol. Immunol.* 35: 129–140.

Vesenjak-Hirjan, J., 1980. Arboviruses in Yugoslavia, in: *Arboviruses in the mediterranean countries.* (J. Vesenjak-Hirjan, J. S. Porterfield, and E. Arslanagic, eds.), Zbl. Bakt. Suppl. 9, Gustav Fischer Verlag, Stuttgart, pp. 165–177.

Vockeroth, J. R., 1954. Notes on the identities and distribution of *Aedes* species of northern Canada, with a key to the females (Diptera: Culicidae). *Can. Ent.* 86: 241–255, illus.

Vogel, R., 1933. Zur Kenntnis der Stechmücken Württembergs. II. Teil. *Jh. Ver. vaterl. Naturkd. Württ.* 89: 175–186.

Vogel, R., 1940. Zur Kenntnis der Stechmücken Württembergs. III. Teil. *Jh. Ver. vaterl. Naturkd. Württ.* 96: 97–116.

Wade, J. O., 1997. An examination of pesticide application methods and pesticide formulations. Crawling insect control and the compression sprayer. *Arch. Toxicol. Kinet. Xenobiot. Metab.* 5(2): 69–75.

Wallis, R. C., Taylor, R. M., and Henderson, J. R., 1960. Isolation of eastern equine encephalomyelitis virus from *Aedes vexans* from Connecticut. *Proc. Soc. exper. Biol. Med.* 103: 442–444.

Walters, L. L., and Legner, E. F., 1980. Impact of the desert pufish, *Cyprinodon macularius*, and

Gambusia affinis affinis on fauna in pond ecosystems. *Hilgardia* 48(3): 1–18.

Ward, E. S., and Ellar, D. J., 1988. Cloning and expression of two homologous genes of *Bacillus thuringiensis* subsp. *israelensis* which encode 130-kilodalton mosquitocidal proteins. *J. Bacteriol.* 170: 727–735.

Ward, R. A., 1984. Second supplement to "A catalog of the mosquitoes of the world" (Diptera: Culicidae). *Mosq. Syst.* 16: 227–270.

Ward, R. A., 1992. Third supplement to "A catalog of the mosquitoes of the world" (Diptera: Culicidae). *Mosq. Syst.* 24: 177–230.

Ware, W. G., 1989. *The pesticide book.* Thompson Publications, 336 pp.

Washino, R. K., 1969. Progress in biological control of mosquitoes—invertebrate and vertebrate predators. *Proc. Pap. Ann. Conf. Calif. Mosq. Contr. Assoc.* 37: 16–19.

Waterston, J., 1918. On the mosquitoes of Macedonia. *Bull. Ent. Res.* 9: 1–2.

Weathersbee III., A. A., Meisch, M. V., Sandoski, C. A., Finch, M. F., Dame, D. A., Olson, J. K., and Inman, A., 1986. Combination ground and aerial adulticide application against mosquitoes in an Arkansas rice-land community. *J. Am. Mosq. Contr. Assoc.* 2: 456–460.

Weathersbee III., A. A., Meisch, M. V., and Bassi, D. G., 1988. Activities of four insect growth regulator formulations of fenoxycarb against *Psorophora columbiae* larvae. *J. Am. Mosq. Contr. Assoc.* 4(2): 143–146.

Wegner, E., and Gliniewicz, A., 1998. Most abundant mosquito species (Diptera: Culicidae) in Poland and methods of their control. *Acta Parasitologica Portuguesa* 5(1): 30.

Weiser, J., 1984. A mosquito-virulent *Bacillus sphaericus* in adult *Simulium damnosum* from Northern Nigeria. *Zbl. Mikrobiol.* 139: 57–60.

Weiser, J., 1991. *Biological Control of Vectors.* John Wiley & Sons Ltd., West Sussex, 189 pp.

Weitzel, T., Schäfer, M., Dahl, C., and Becker, N., 1998. Studies on the species complex of *Aedes communis* s.l. by protein electrophoresis. *Acta Parasit. Port.* Lisboa 5(1): 27.

Wernsdorfer, W. H., 1980. The importance of malaria in the world, in: *Malaria*, Volume 1. (J. P. Kreier, ed.), Academic Press, New York, pp. 1–93.

Wesenberg-Lund, C., 1921. Contributions to the biology of the Danish Culicidae. *K. danske vidensk. Selsk., Nat. Math. afd.* 7(8): 1–210, illus.

Wesson, D. M., 1990. Susceptibility to Organophosphate insecticides in larval *Aedes albopictus. J. Am. Mosq. Contr. Assoc.* 6(2): 258–264.

Weyer, F., 1939. *Die Malaria Überträger.* Leipzig (Thieme), 141 pp.

Whisler, H. C., Zebold, S. L., and Shemanchuk, J. A., 1974. Alternative host for mosquito parasite *Coelomomyces. Nature* 251: 715–716.

White, G. B., 1978. Systematic reappraisal of the *Anopheles maculipennis* complex. *Mosq. Syst.* 10: 13–44.

WHO, 1975. Manual on practical Entomology in malaria, Part II: Methods and techniques, World Health Organization Offset Publication, No. 13, Geneva, 191 pp.

WHO, 1976. 22nd Report of the WHO Expert Committee on Insecticides. *Technical Report Series*, No. 585, Geneva.

WHO, 1980. Resistance of vectors of disease to pesticides. W.H.O. Tech. Rep. Ser. 655.

WHO, 1981. Report of informal consultation on standardization of *Bacillus thuringiensis* H-14. Mimeographed documents TDR/BVC/BTH-14/811, WHO/VBC/81–828, World Health Organization, Geneva, Switzerland, 15 pp.

WHO, 1982. Manual on environmental management for mosquito control with special emphasis on malaria vectors. W.H.O. Offset Publ. 66.

WHO, 1983. Integrated vector control. 7th rep. WHO Exp. Comm. VBC, Technical Report Series, No. 688.

WHO, 1988. Safe use of pesticides. *Technical Report Series*, No. 720, Geneva.

WHO, 1991. Safe use of pesticides. 14th report of the WHO Expert Committee on vector biology and control. *Technical Report Series*, No. 813, Geneva.

WHO, 1993. Tropical disease Research: Progress 1991–92, 11th programme report of the UNDP/World Bank/WHO Special Programme for Research and Training in Tropical Diseases (TDR), World Health Organization, Geneva, 134 pp.

WHO, 1995. Vector control for malaria and other mosquito-borne diseases. *Technical Report Series*, No. 857, Geneva.

WHO, 1996. Operational manual on the application of insecticides for control of the mosquito vectors of malaria and other diseases.

WHO, 1997a. World malaria situation in 1994. Part I. Population at risk. *Wkly. Epidem. Rec.* 72(36): 269–274.

WHO, 1997b. World malaria situation in 1994. Part III. Europe, South-East Asia, Western Pacific. *Wkly. Epidem. Rec.* 72(38): 285–290.

WHO, 1997c. WHO Pesticide Evaluation Scheme. Chemical methods for the control of vectors and pests of public health importance. (D. C. Chavasse, and H. H. Yap, eds.), 129 pp.

WHO, 1998. Classification of pesticides by hazard. WHO/VBC/ 88.953(10).

WHO, 1999. *Bacillus thuringiensis*, Environmental Health Criteria 217, International Programme on Chemical Safety. World Health Organization Geneva, ISBN 92 4 157217 5.

WHO, 2000. WHO expert committee on malaria. Twentieth report. World Health Organization, Geneva, pp. 71.

WHO/IDRC, 1996. Net Gain. A new method for preventing malaria deaths. IDRC, 189 pp.

WHO/IPCS, 1999. Environmental Health Criteria 217, *Bacillus thuringiensis*, 107 pp.

WHO/IRAC, 2003. Manual for Resistance Management in Public Health. IRAC/WHO (in press).

Wickramasinghe, B., and Mendis, C. L., 1980. *Bacillus sphaericus* spores from Sri Lanka demonstrating rapid larvicidal activity on *Culex quinquefasciatus*. *Mosq. News* 40: 387–389.

Wigglesworth, V. B., 1969. Chemical structure and juvenile hormone activity: comparative tests on *Rhodnius prolixus*. *J. Insect Physiol*. 15: 73–94.

Williams, C. M., 1967. Juvenile hormones, in retrospect and prospect, in: *The Juvenile Hormones*. (L. I. Gilbert, ed.), Plenum Press, New York, pp. 1–14.

Williamson, M. S., Denholm, I., Bell, C. A., and Devonshire, A., 1993. Knockdown resistance (*kdr*) to DDT and pyrethroid insecticides maps to a sodium channel gene locus in the housefly (*Musca domestica*). *Mol. Gen. Genet*. 240: 17–22.

Wilson, G. R., and Horsfall, W. R., 1970. Eggs of floodwater mosquitoes. XII. Installment hatching of *Aedes vexans*. *Ann. Ent. Soc. Am*. 63: 1644–1647.

Wojta, J., and Aspöck, H., 1982. Untersuchungen über die Möglichkeiten der Einschleppung durch Stechmücken übertragener Arboviren durch Vögel nach Mitteleuropa. *Mitt. Österr. Ges. Tropenmed. Parasitol*. 4: 85–98.

Wolz, I., 1993. Untersuchungen zur Nachweisbarkeit von Beutetierfragmenten im Kot von *Myotis bechsteini* (Kuhl 1818). *Myotis* 31: 5–25.

Wood, D. M., 1977. Notes on the identities of some common nearctic *Aedes* mosquitoes. *Mosq. News* 37: 71–81.

Wood, D. M., and Borkent, A., 1989. Phylogeny and classification of the Nematocera, in: *Manual of Nearctic Diptera*. (J. F. Macalpine, and D. M. Wood, eds.), 3:32:114: 1333–1370.

Wood, D. M., Dang, P. T., and Ellis, R. A., 1979. *The mosquitoes of Canada* (Diptera: Culicidae). Series: The insects and arachnidae of Canada. Part 6. Biosystematics Res. Inst., Canada, Dept. Agr. Publ. 1686, 390 pp.

Wood, R. J., 1981. Strategies for conserving susceptibility to insecticides. *Parsitology* 82: 69–80.

Wood, R. J., and Newton, M. E., 1991. Sex-ratio distortion caused by meiotic drive in mosquitoes. *The American Naturalist* 3: 379–391.

Woodring, J., and Davidson, E. W., 1996. Biological control of mosquitoes, in: *The biology of disease vectors*. (B. J. Beaty, and W. C. Marquardt, eds.), University Press of Colorado, USA, pp. 530–548.

Wright, R. H., and Kellogg, F. E., 1962. Response of *Aedes aegypti* to moist convection currents. *Nature* 194: 402–403.

Yamamoto, I., 1973. Mode of action of synergists in enhancing the insecticidal activity of pyrethrum and pyrethroids, in: *Pyrethrum, the natural insecticide*. (J. E. Casida, ed.), Academic Press, New York, pp. 195–209.

Yasuno, M., Kazmi, S. J., LaBrecque, G. C., and Rajagopalan, P. K., 1973. Seasonal change in larval habitats and population density of *Culex fatigans* in Delhi Villages, WHO/VBC/73, 429, 12 pp. (mimeographed).

Yen, J. H., and Barr, A. R., 1974. Incombatibility in *Culex pipiens*, in: *The use of genetics in insect control*. (R. Pal, and M. J. Whitten, eds.), Elsevier/North Holland, Amsterdam, pp. 97–118.

Yoon, S. Y., 1987. Community participation in the control and prevention of DF/DHF: is it possible? *Dengue Newsl*. 13: 7–14.

Zaim, M., 2002. Global Insecticide use for Vector-Borne Disease Control. WHO/CDS/WHOPES/GCDPP/ 2002.2, 79 pp.

Zaim, M., Aito, A., Nakashima, N. 2000. Safety of pyrethroid-treated mosquito nets. *Med. and Vet. Entomol*. 14: 1–5.

Zavortink, T. J., 1972. Mosquito studies (Diptera: Culicidae) XXVIII. The New World species formerly placed in *Aedes (Finlaya)*. *Contr. Am. Ent. Inst*. 8(3): 1–206, illus.

Zavortink, T. J., 1968. Mosquito studies (Diptera: Culicidae) VIII. A prodome of the genus *Orthopodomyia*. *Contr. Am. ent. Inst*. 3(2): 1–221, illus.

Zgomba, M., 1987. Impact of larvicides used for mosquito (Diptera: Culicidae) control on aquatic entomofauna in some biotops of Vojvodina, Ph.D. Thesis, Univ. Novi Sad, 101 pp.

Zgomba, M., Petric, D., Srdic, Z., 1983. Effects of some larvicides used in mosquito control on *Collembola*. *Mitt. dtsch. Ges. allg.angew. Ent*. 4: 92–95.

Subject Index

Taxonomic Index